Surburg's Works

Vol IV
Apologetics and
Evolution

Edited by
Herman J. Otten

LUTHERAN NEWS, INC., New Haven, Missouri

Surburg's Works

Library of Congress Card
Lutheran News, Inc.
684 Luther Lane
New Haven, MO 63068
Published 2017
Printed in the United States of America
IngramSpark, TN
ISBN #978-0-9864232-2-2

Table of Contents

Apologetics and Evolution

Seminex Operation Outreach

February 28, 1974
The Rev. Herman Otten
Trinity Lutheran Church
Box, 168
New Haven, Missouri, 63068
Dear Brother:

Dr. Klug told me that you would like to have me answer the section in "Operation Outreach" which dealt with the inerrancy of the Bible.

I have answered in detail every item dealing with both the Old and New Testament. I spent the past week in researching, writing, and typing the endorsed material.

You may use this as you see fit. I am sorry that it is not briefer, but adequately to move and refute their argument, I felt I had to go into detail. You might use it in a number of installments.

I would prefer that you do not print it with my name, but simply use the initials R.F.W.S. (The initials of my three given names and last name, Raymond Frederick Walter Surburg).

Here at Springfield, there are a good number who hate *Christian News*. Of them, if agreeable to you, I would be willing to furnish you with some other material, to any present synodical situation. Our two academic deans and others in the administration can't stomach *Christian News*. At present I believe it the part of wisdom not to reveal that I have submitted material to you.

I hope someday we can meet with each other, and I can relate more.

My God bless your effort to defend Biblical and confessional Lutheranism.

Fraternally,
Raymond Surburg
* * *

**For "Operation Outreach" January 25-31, 1974,
Concordia Seminary, St. Louis, Missouri.**

This sheet is by no means intended as an agenda for discussion but only as a guide for quick reference and documentation.

Inerrancy of the Bible

1. Authorship of the Pentateuch—Later than Moses: in Palestine already. Gen. 12:6, Dt. 1:1; Moses dead, Dt. 34:6, 10; political situation later, Gen. 14:14 ns. Jdg. 17:18. **not unified**; doublets. Gen. 28 & 35, Ex. 3 & 6; discrepancies in laws, Ex. 28:1 vs. Dt 18;7; differences in divine name (especially Hebrew), Gen. 4:26 vs. Ex 6:3; Moses' father-in-law. Ex 2:18 vs. Ex 3:1 vs. Num 10:29. references by Jesus: Mt 19:7; 10:3; John 3:14, 7:22-23. 2. The Flood—**source division**: J , Gen. 6:5-8; 7:1-5, 7-10, 12, 16b, 17b, 22-23; 8:2b-3e, 6-12, 13b, 20-22, P. the rest of Gen. 6:5-8:22. **discrepancies**: number of animals. Gen. 6:19 vs. 7:2; duration of Flood, Gen. 7:12, 17; 8:6 vs. Gen. 7:11, 24; 8:3-5, 13a, 14.

"Operation Outreach" and the Inerrancy of Scripture

Christian News, March 18, 1974

Between January 25-31 students from Concordia Seminary, St. Louis participated in "Operation Outreach," an effort to defend the theological position of Dr. John Tietjen and of the majority of Concordia Seminary, St. Louis faculty who went on strike, endeavoring to bring the Seminary's Board of Control to its knees. With one major effort it was thus purported to undue and negate the unfavorable resolution of over 500 voting delegates and of other resolutions of the New Orleans Convention to stop and reverse the liberalizing of the Synod's historical position on doctrine and practice.

Students who went on behalf of suspended president Tietjen and the striking faculty majority were give resource material for what was termed as theological response when these crusading student would meet with opposition at congregations and meeting places where they presented the theological teachings of their mentors. Students were furnished with references and quotations from Luther's voluminous writings which were designed to show that Luther actually was employing a methodology similar to that of the St. Louis majority and that if Luther had lived today he would have been in sympathy with the use of the historical-critical method. Last week we dealt with these quotations from Luther.

In addition, the same students were supplied with Biblical passages that were to show that the Synod's historic position on the inerrancy of the Bible was untenable and that the Biblical data supported the contention that the Bible was replete with errors, inaccuracies, contradictions and errors.

1. The Impossibility of the Mosaic Authorship

The employment of the historical-critical method requires the rejection of the Mosaic authorship. It is claimed that passages like Gen. 12:6, Deut 1:1 require a date after the time of Moses. Genesis 12:6 (cf also 13:7) is supposed to be post-Mosaic because of the assertion that "The Canaanite was then in the land," a statement which presupposed that the Canaanites had been driven out when this notation was made. However, the explanation of Unger that "the notice in itself merely states that they were in the country in the days of Abraham, without any implication that they were not yet there,"[1] thus shows how such a statement does not necessarily negate the Mosaic authorship of Genesis or of the Pentateuch.

Deuteronomy 1:1 is also cited against Mosaic authorship because it asserts that "these are the words that Moses spoke to all Israel beyond the Jordan in the wilderness, in the Arabah over against Suph." The phrase "beyond the Jordan. . ." an assertion which describes Moses position east of the river, has no implication that the writer was in Canaan.

That this is the case may be ascertained from the elastic meaning of the expression in Numbers 32:19, where it is used both of the east and west side of the Jordan. In other passages it is defined "beyond Jordan eastward. . ." (Deut. 4:47; 49; Josh. 1:15) and "beyond Jordan westward. . ." (Deut. 11:30; Josh. 5:1).

A matter closely related to the use of "beyond the Jordan" is the term Negebward ("southward" toward the desert of Beersheba), employed by the Lord (Yahweh) when he spoke to Abraham (Gen. 13:14, etc.) and "west" is said to be "Seaward." (Gen. 12:8; 13:14). Those opposed to Mosaic authorship have claimed that it would have been impossible for Moses to have used these terms when the Negeb (south) was to the north of him and the Mediterranean could hardly be said to be to Moses' west. Unger answers this quibbling by declaring: "The simple explanation is that the stereotyped expressions Negebward for southward and 'seaward' for westward were technical terms which had been part and parcel of the Hebrew language since patriarchal times."[2]

Chapter 34 of Deuteronomy reports the death and burial of Moses. Obviously it is asserted Moses could not write the account of his death and surely would not have written about himself: "And there has not arisen a prophet since in Israel like Moses, whom the Lord knew face to face, none like him for all the signs and the wonders which the Lord sent him to do in the land of Egypt, to Pharaoh, and to all his servants and to all his land." (34:10-11). Already in 1923, Dr. P.E. Kretzmann had suggested in his *Popular Commentary to the Bible*, Vol. I, in the introduction to the Book of Deuteronomy: "The last three sections of the book, which contain the announcement of the death of Moses, his last blessing and the narratives of his death, were probably added to Deuteronomy by the inspired author of the Book of Joshua."[3]

That the Pentateuch is post-Mosaic is deduced from Gen. 14:14 where the author wrote: "When Abram heard that his kinsmen had been taken captive, he led forth his trained men, born in his house, three hundred and eighteen of them, and went in pursuit as far as Dan." From our knowledge of historical geography it is known that Dan was known in earlier times as Laish. In other words, there was no city of Dan in Abram's time or that of Moses. However, as the Hebrew Scriptures were handed down and copied from generation to generation, the scribes modernized certain place names so that the city at one time called Laish, later on was renamed Dan. The modernization of archaic names was an attempt to make the text meaningful, but does not in any way militate against the Mosaic authorship as the critics contend.

Doublets As Evidence against Integrity of Mosaic Authorship
One of the criteria which supposedly shows the patchwork character of the five books of Moses is the occurrence of the same account in variant forms, that do not agree, showing contradictions relative to people involved, time of happening, and details of the accounts. In the material prepared for the "Operation Outreach" Gen. 28 and 35 were selected as an example of doublets, allegedly found in the writings of Moses. In Gen-

esis 28 Moses recorded Jacob's stopping at Bethel, where Yahweh granted Abraham's grandson a remarkable vision. Jacob promised Yahweh that if he returned safely that in appreciation he would erect an altar to the Lord. When Jacob was brought back safely by Yahweh's guidance, he failed to keep his vow. After experiencing much family trouble, God appeared to Jacob and told him to go to Bethel and keep his vow. He was to put away the false idols of his children and there God renewed the Abrahamic covenant. Genesis 28 and 35 are said to be two different accounts of the same event. However, a fair reading of the two narratives indicates that in the first God appeared to Jacob when he was on the way to Padan-Aram, in the second (ch. 35) after he had returned from Padan-Aram. When Jacob was at Bethel the first time he was alone, while the second time he was accompanied by his sons and daughters and their children. At least twenty years separates the two experiences of Jacob at Bethel. The manner in which critical scholarship tries to postulate discrepancies by the creation of these so-called doublet narratives show their unreasonableness when dealing with Biblical literature.[4]

Another alleged doublet given the members of "Operation Outreach" was Exodus 3 and 6, two accounts from the life of Moses that supposedly are flatly contradictory. When Moses was eighty years old, and keeping the sheep of his father Jethro, the priest of Midian, Moses relates that "the angel of the LORD (Yahweh) appeared unto him in a flame of fire out of the midst of the bush and he looked, and lo, the bush was burning, yet it was not consumed." A verse later the reader is informed that it was Yahweh who was in the burning bush, who said to Moses: "I am the God (Elohim) of your father, the God of Abraham, the God of Isaac, and the God of Jacob." In chapter 6:2,3 Moses records another theophany of Yahweh to him, one that occurred after Moses and Aaron had been to Pharaoh. Chapter 6 of Exodus begins with the information that: The LORD (Yahweh) said to Moses, "Now you shall see what I will do to Pharaoh; for with a strong hand he will send them out, yea, with a strong hand he will drive them out of his land." And God said to Moses "I am the LORD (Yahweh). I appeared to Abraham, to Isaac, and to Jacob, as God Almighty, but by my name the LORD (Yahweh) I did not make myself known to them" (Ex. 6:2,3). The last verse is said to contradict Ex. 3:6. The Exodus 6:2,3 passage has been used by Old Testament scholars in the past as justification for the Documentary Hypothesis, namely that in Genesis and Exodus at least three different documents were used by the redactor (or redactors) when composing the Pentateuch or Hexateuch. Thus H.H. Rowley employed this argument in a number of his published writings. Thus he wrote in The Doctrine of the Election: "Obviously it cannot be true that God was known to Abraham by the name of Yahweh (Footnote Ex. 6:3) and that He was known to him by that name (Footnote 2, Gen. 15:2,7). To this extent there is a flat contradiction that cannot be resolved by any shift."[5]

Dr. Raymond F. Surburg, in an article entitled "Did the Patriarchs Know Yahweh? Or Exodus 6:3 and Its Relationship to the Four Documentary Hypothesis," has given a number of different ways in which these verses of Exodus 6:2-3 can be harmonized with Exodus 3:6 and with the many assertions in Genesis where the Patriarchs, Abraham, Isaac and Jacob know Yahweh by name and where Yahweh is depicted as giving them revelations.[6] A number of scholars have pointed out the fact that the Hebrew of Exodus 6:2-6 permits rendering the verse upon which an entire theory has erroneously been built as follows: "I am Jehovah and have appeared unto Abraham, unto Isaac, and unto Jacob as God Almighty. And regarding my name Jehovah (or Yahweh) was I not known to them? Also (i.e. in addition to this) have I established my covenant with them (Namely to give to them the land, etc.)."[7] With this translation every possibility of a seeming contradiction with other Bible passages disappears entirely. The same view was expounded by W. J. Martin in his Tyndale lecture, "Stylistic Criteria and the Analysis of the Pentateuch."[8]

The materials furnished students involved in "Operation Outreach" were given as proof for the Documentary Hypothesis the difference in divine names between Gen. 4:25 and Ex. 6:3. According Genesis 4:25 after the birth of Enoch men began to call upon the name of Yahweh, thus indicating that very early in the history of the human race men worshipped Yahweh. Yahweh is mentioned over 200 times in Genesis. Yahweh spoke to Noah, Abraham, Sarah, Rebekah, and Jacob according to the Hebrew test. Now we are to believe that Moses did not know Yahweh until the supposed appearance of Yahweh for the first time as recorded in Exodus 6:2,3, although Yahweh appeared to him in chapter 3:1-6. Dr. Segal, professor of Hebrew for many years at the Hebrew University in Jerusalem, wrote concerning this matter:

> But the whole thesis, that according to E and P the name of YHWH was unknown in the world till it was revealed to Moses, has no basis in fact. It is disproved by the name Joshua in E, by the name of Jochebed in J, both names earlier than the alleged revelation of the name Yahweh to Moses, and both containing the abbreviated element of the name YHWH usual in Hebrew theophorous names. Also the patriarchal name of Joseph most probably contains this element. Moreover, it is incredible that the patriarchs, who were undoubtedly in their estimation true worshippers of God, as ignorant of the true name of deity. There could have been no true worship of God without knowledge of His true name, as it proved by the standing expression in the Bible for worship: "to call by the name of Yahweh" (Gen. iv. 26; xii. 8).[9]

In the New Testament Exodus 3:6, where it clearly states that Yahweh appeared unto Moses, is quoted by Christ to support the idea that the God of Jesus Christ is not a God of the dead but of the living. In proof of this position, Jesus quoted Exodus 3:6: "I am the God of Abraham and the God of Isaac and the God of Jacob" (Mk. 12:26; Mt. 22:32; Luke 20:37), a passage that is supposedly suspect and wrong. Strangely enough, Exodus 6:3 is never quoted in the New Testament.

In the Book of Acts Stephen also stated that God (the true God, Yahweh) appeared to Moses in the burning bush and said: "I am the God of thy father, the God of Abraham, and Isaac and Jacob," (Acts 7:32). The passage that supposedly has the wrong interpretation according to the critics is the one that Christ, Stephen and Peter (Acts 3:13) quoted as evidence that the true God from the very beginning of patriarchal history had manifested Himself to and was known by Abraham, Isaac, and Jacob.

Another contradiction given the students to show congregations that there are errors in the Old Testament is the case of Exodus 2:18 versus Numbers 10:29, 30. An examination of these two sets of passages will show that in Exodus 2:18 Reul is mentioned as the father of Zipporah, whom Moses subsequently married. In Numbers 10:29,30 Moses said to Hobab, the son of Reul the Midianite, Moses' father-in-law, "We are setting out for the place of which the LORD said, etc." Hobab obviously was Moses' brother-in-law and is not the same as Reul, so between these two passages there is no discrepancy as some critical scholars would have people believe. Attempts have been made to identify Hobab with Jethro (cf. Ex. 18:5,27) compared with Num. 10:29,30, but the text in the latter Numbers passage is plain, the son of Reul is mentioned. That Jethro and Reul (Raguel) were names of the same person is evident from Ex. 2:18,21; 3:1.

2. Contradictions in the Flood Account (Gen. 6-8).

Materials given students in "Operation Outreach" attempted to show contradictions in the story of the Biblical flood (Gen. 6-9). In these four chapters the editor or editors supposedly wove together materials from the J and P accounts. Already a hundred years ago Ewald contended that 'the composite nature of the flood story "shone as a gleaming star before all others on the horizon of the Jehovistic and Elohistic documents."[10] By the use of the names Yahweh (German scholars use the form Jahweh) and Elohim (God) critical scholars have constructed in two parallel columns the constituent elements of these two documents (J and P) and now claim that there exists discrepancies relative to the number of animals taken into the ark (Gen. 6:19 vs. 7:2); the duration of the flood (Gen. 7:12,17; 8:5 vs. Gen. 7:11,24; 8:3-5,13a,14). Eichorn, sometimes known as "the father of higher criticism," professor at the University of Jena, in the beginning of the nineteenth century, pointed out that repetitions are not only infrequent and distributed throughout Gen. 6-8, but he went so far as set down in two columns what he contended were two complete parallel narratives.

In answer to these views Dr. Unger asserts that the composite character of Gen. 6-8 can be accounted for as the product of one author, Moses. "To begin with, it must be noted that there are two marked features of Hebrew style, which, when understood, offer an easy and natural explanation of Pentateuchal literary phenomena, but, when misunderstood, lend seeming support to the idea of documentary sources, which, when separated, give accounts more or less complete."[11a] Literary phenomena that are misused by the critics according to Unger are: "The first is the exceedingly common syntactical use of the conjunction 'and' employed loosely to connect members of a compound sentence to join subordinate clauses. The second is the widespread phenomenon of elaboration and repetition. It is especially this feature of Biblical style, so apparent in the flood story, which critics abuse by wresting it out of its ancient idiom and squeezing it into the mold of modern diction to support an absurd theory."[11b]

Dr. Oswald Allis, formerly professor of Old Testament at Princeton and Westminster Theological Seminaries, has made a special study of the flood narratives and has examined in detail the critical theory relative to its alleged composite character. Allis has shown that while there is repetition and elaboration in the flood narrative, this is not due to fusion of different contradictory accounts, but is the result of a literary device to stress three main features of Genesis, being namely, the sinfulness of man as the cause of the flood, the destruction of all flesh as the purpose of the God-sent flood, and the saving of the righteous remnant by means of the flood.[12] Regarding the number of beasts that entered the ark Dr. William Arndt wrote in his book, *Does the Bible Contradict Itself?* in commenting on Gen. 7:2-3 and Gen. 7:8-9 as follows: "The superficial reader may scent a discrepancy here because the first verses of Gen. 7 say that of the unclean animals seven pairs should be brought into the ark, while the continuation of the narrative says that of the unclean beasts there went in two and two. In explaining this apparent disagreement, let me say that we must not overlook the statement, v. 5, that Noah did everything which God had commanded him. The holy writer emphasizes that Noah carried out God's command. Was it necessary, then, for him to repeat later on all the details contained in the Lord's order? Again, verses 8 and 9 do not contradict the proceeding verses in the least. They simply say that all the animals came in pairs. How many pairs of each kind entered the writer does not narrate here. Why should he? As I said before, he has covered that point by stating that Noah did just as he had ordered to do. To conclude, verses 2 and 3 are specific, verses 8 and 9 merely assert in a general way Noah's compliance with God's command. If the writer had said, vv. 8:9: Of each kind of clean animals **only** two came,' then we should have a discrepancy here; but it is clear that such is not the import of his words. "[13]

Those who divide the materials of Gen. 6-8 between two sources, J and P, claim that there are contradictions relative to the duration of the flood. Gen. 7:12,17; 8:6 are pitted against Gen. 7:11, 24; 8:3-5; 13a, 14. However, if these chapters are treated as a unity whose purpose is to give an account of the flood that is written by an intelligent author like Moses and not by bungling redactor or redactors there is no problem in setting forth a time table for the various events that occurred from the inception to the conclusion of this worldwide event. This has been done by Kevan *In The New Bible Commentary*,[14] on pages 84-85.

3. References to Adam and Eve

As further examples of the errancy of the Bible references to Adam and Eve were cited. Paul refers to both Adam and Eve in the following passages: Rom. 5:14; I Cor. 15:22.45; 1 Tim. 2:13. When these passages are studied they show that Paul believed in the historicity of Adam and Eve, and that Paul accepted the literal accounts of the formation of Adam and Eve as recorded in Genesis 1-2. Paul, also believed Gen. 3, where Eve and Adam were tempted and fell into sin (I Tim. 2:13). 1 Chron. 1:1, Luke 3:38 and Jude 14 are cited as contradictory. 1 Chron. 1:1 reads: "Adam, Seth, Enosh," Luke 3:38 reads: "the son of Enos, the son of Seth, the son of Adam, the son of God." Enosh (or Enos) is given as the third from Adam, while in Jude 14 the statement occurs; "It was of these also that Enoch in the seventh generation from Adam prophesied saying, "Behold, the Lord came with his holy myriads, etc." Concerning this verse David Payne wrote: "The phrase describing Enoch is not Jude's own, but is again drawn from the Pseudepigrapha (cf. 1 Enoch 60:8). It may be that Jude was simply arguing from grounds he knew to be acceptable to his readers or opponents. We might compare Paul's use of Gk. poetry when addressing the Athenian audience (Ac. 17:28). At any rate, the general statements drawn from the Book of Enoch, and the inferences from them, are authenticated here and elsewhere in the NT."[15] As long as the Bible reader can come up with a plausible explanation, he is not warranted to claim that there are contradictions between two sets of passages. Such an attitude is the proper and fair one in dealing with any responsible author and surely a fair minded Christian will give the Biblical writers the same consideration.

4. Discrepancies between Kings and Chronicles

The material endeavoring to show the errancy of the Holy Scriptures has pointed out discrepancies between Samuel-Kings and the Two Books of Chronicles. Thus II Sam. 24:9 is to be compared with I Chron. 21:5; II Sam. 24:24 with 1 Chron. 21:25. An examination of these passages will reveal that according to II Sam. 24:24, David paid fifty shekels for the threshing-floor, while in I Chron. 25:25 David paid Oman six hundred shekels of gold weight. These passages were discussed nearly fifty years ago by Dr. William Arndt, for many years professor of New Testament Interpretation at Concordia Seminary, St. Louis, in his book *Does the Bible Contradict Itself?* Dr. Arndt writes concerning this seeming contradiction: "The text from Second Samuel says that David bought a threshing-floor and some oxen. The text from First Chronicles declares that he bought the place. It is clear that two different transactions are spoken of. We may assume that David first bought the threshing floor and the oxen for fifty shekels of silver (about $32). Later on he may have decided to buy the whole field belonging to Oman, paying 600 shekels of gold (about $6,600). This was to be the Temple site, and, naturally more ground than merely a threshing floor was needed. It is absurd to make the charge that the two accounts are not in harmony with each other."[16]

Another example given by the proponents of the fallibility and errancy of the Old Testament is the supposed contradiction between II Sam. 24:1 and 1 Chron. 21:1. These two passages were discussed by Dr. Arndt in his *Bible Difficulties* in connection with the topic of "God's Relation to Evil-Doing", pages 34-36. The Samuel passages states that the anger of the Lord was kindled against Israel, and he incited David against them, "Go number Israel and Judah," while the Chronicle passages asserts: "And Satan stood up against Israel and provoked David to number Israel." Arndt claims that on the basis of Eph. 4:19, Rom. 1:26,28 that God withdraws his restraints that he has placed about men and permits their wicked passions and designs to run their evil course. "God moved David to number Israel" means that God allowed David to carry out the numbering which he understood for the wrong purpose. And who was it that actually moved David to this wrong census? I Chron. 21:1 informs us that it was Satan. Haley in discussing these two passages wrote: "It is consistent with Hebrew modes of thought that whatever occurs in the world under the overruling providence of God, whatever He suffers to take place should be attributed to His agency. In not preventing, as He might have done, its occurrence. He is viewed as in some sense bringing about the event."[17] To this Arndt claims should be added the Biblical teaching "that God at times removes restraining influence provided by His mercy when people in spite of His earnest pleadings eagerly fling themselves into a life of sin."[18]

5. The References to Jonah

The exegetes of the Faculty majority have rejected the historicity of the Jonah story as recorded in the writing attributed to Jonah among the 12 Minor Prophets, which are one book in the Hebrew Old Testament. The New Testament refers to different events that are reported in the four chapters of the Book of Jonah. When the Jews again asked for a miraculous sign Jesus responded: "An evil and adulterous generation seeks for a sign; but no sign shall be given it except the sign of the prophet Jonah. For as Jonah was three days and three nights in the belly of the whale, so will the Son of man be three days and three nights in the heart of the earth. The men of Nineveh will arise at the judgment with this generation and condemn it; for they repented at the preaching of Jonah, and behold, something greater than Jonah is here."

The St. Louis exegetes insist that the book of Jonah be interpreted as a parable. They argue that the story of the Good Samaritan is considered a parable and yet there is nothing in the text to indicate its parabolic nature. Since there is no indication that the story of the Good Samaritan is a parable, there would be nothing to contradict the fact that it was a true story that actually occurred. In both the Old and New Testament there are parables and there is no difficulty in recognizing them as parables. In both the Old and New Testaments parables are not very long. If the Book of Jonah were a parable, it would be one extending over four chapters, something unheard of in Biblical literature. Scholars are unfortunately not agreed as to the type of literary genre found in Jonah. Some claim it is an allegory, others a propaganda writing, a **Tendenzschrift**.[19] The real objection that critical scholars have against the Jonah book is that it is replete with miracles. But if God can perform one miracle. He can perform many.

The interpretation of the New Testament passages make impossible the non-historical explanation of the Events of the Book of Jonah. An imaginary event, which Jesus characterized as a sign, hardly can serve as a type for the events of Jesus burial and resurrection. Furthermore, imaginary people of Nineveh could hardly arise on the day of judgment to condemn the people of the cities of Galilee that had heard the proclamation of Jesus Christ, the promised Messiah of the Old Testament messianic prophecies.

However the rejection of the historicity of Jonah presents a serious moral problem. Jonah, the son of Amittai, is referred 10 in II Kings 14:24 as a prophet who prophesied during the reign of Jeroboam II. The Jonah spoken of in the Book of Jonah is the same son of Amittai. If this prophetic book is not an autobiographic account of Jonah's mission to Nineveh, his attempt to avoid going and preaching to Nineveh, Jonah's being swallowed by a great fish that God created for the occasion or saving the disobedient prophet and his belated visit to Nineveh and success, followed by great disappointment by Jonah because of Nineveh's repentance, then some unknown writer has been guilty of maligning and defaming a prophet of God about whom we know nothing unfavorable recorded in II Kings. To select a reputable servant of Yahweh and depict him as disobedient and as a person wishing for the destruction of the Ninevites, certainly is immoral and a sin against what we normally designate as the eighth commandment! Such a misrepresentation of a good prophet we are to believe was furthermore inspired by the Holy Spirit to be placed among the Minor Prophets. That is what would be involved in the rejection of the historicity of The Book of Jonah, which could only have been written by a person who really experienced what is recorded in the four chapters comprising this prophetic autobiography, written no doubt by a man who came to see the error of his way. The same lessons the parable are said to teach can also be taught on the basis of the historic episodes in the Book of Jonah. In fact, they carry more weight if based on history as opposed to fiction.

6. The Identity of the Suffering Servant

According to materials used in "Operation Outreach" the Holy Scriptures are supposed to contradict themselves about the identity of the Suffering Servant. The prophet Isaiah has Yahweh announce: "It is too light a thing that you should be my servant to raise up the tribes of Jacob and to restore the preserved of Israel: I will give you as a light to the nations, that my salvation may reach to the end of the earth" (49:6). Simeon in the Nunc Dimittis applies this verse to Jesus, whom Simeon had taken in his arms and blessed God. But in Acts 13, Paul and Barnabas in Antioch in Pisidia tell the Jews there: "For so the Lord has commanded us, saying, "I have set you to be a light for the gentiles, that you may bring salvation to the uttermost parts of the earth." When "Operation Outreach" lists Luke 2:31 as contradicted by Acts 13:47 this would mean that in the Acts passage Paul considers himself as prophet of the Gentiles predicted by Isaiah. However, that interpretation is not required as the correct one by the context. Jamieson, Fausset and Brown comment as follows on this verse: "These and other predictions must have been long before this brought vividly home to Paul's mind in connection with this special vocation to the Gentiles I have set thee — i.e., Messiah; from which Paul inferred that he was but following out his destination of his Lord, in transferring to the Gentiles those 'unsearchable riches' which were now by the Jews rejected and despised."[20]

7. The Synoptic Problem

Regarding the genealogy of Jesus "Operation Outreach" lists Matt. 1:6 and Lk. 3:31 as having a discrepancy. Matthew 1:6 reads: "And Jesse the father of David the king," while Luke 3:31 reads: "the son of Melea, the son of Menna, the son of Mattatha, the son of Nathan, the son of David." Where a problem lies here this writer cannot see.

As an example of another discrepancy is found in instruction given by Christ to His disciples. Mk. 10:9-10 are cited as compared with Mk. 6:8-9. Thus Matthew 10:9-10 depict Christ as saying: "Take no gold, nor silver, nor copper in your belts, no bag for your journey, nor two tunics, nor sandals, nor a staff; for the laborer deserves his food." Mark 6:8-9 reports that after he had called the twelve and had given them authority over unclean spirits, Jesus "charged them to take nothing for their journey except a staff; no bread, no bag, no money in their belts; but to wear sandals and not put on two tunics." While it is true that Matthew and Mark have a different version of what was not and what was to be taken along, there is no basic discrepancy. Both writers are truthful because what they report Jesus as saying was a part of the general instruction which Jesus gave his disciples before departing on their evangelistic mission. Some claim there is a discrepancy in that one account they were not to take sandals (Matthew), while the other (the Marcan) said that they should wear sandals. The point of the instruction of Christ was that they were not to take along anything but what was necessary for a short journey which they were to undertake. The instruction of Matthew is not that they were to go barefooted but that they should not take any additional sandals along. The disciples were to take nothing extra along for their journey. This was designed to train them in the practice of faith in preparation for the time when they would be on their own.

The materials used by students in "Operation Outreach" claims there are contradictions in the accounts of the Four Evangelists relative to Holy Week. Mat. 21:7 is pitted against Mk. 11:7 and Lk. 19:35; Mt. 21:10-12 against Mk. 11:11,15; Mt. 26;17; Mk. 14:12 and Lk. 22:7 against John 13:1; 19:31.

Those who have been supporting the historical critical method have been claiming that in the Passion Narratives there are discrepancies relative to the following matters: 1) Was the meal in the upper room a fellowship meal or the Passover? 2) The sequence of events related in connection with Jesus' entry into Jerusalem; the cleansing of the Temple; the anointing at Bethany: the cursing of the fig tree; the question of authority; the parable of the two sons; and the parable of the wicked tenants. All but the last of these episodes are mentioned in two or more of the Four Gospels. Four of the seven episodes adduced are found in three of the Gospels.

Critical scholars who are interested in showing that the Bible is replete with contradictions and discrepancies claim:

1. John and the Synoptics disagree concerning the days on which Jesus was anointed, when he entered the city, and when he cleansed the Temple.
2. The Gospels vary concerning the words Jesus spoke about the woman who anointed him and about words he addressed to the disciples who were sent to fetch an animal for him to ride on into Jerusalem.
3. Matthew commandeered two animals and in the other evangelists imply that there was only one.
4. Especially the Synoptics have much in common, because Matthew and Luke apparently used Mark in their own compositions. Yet the evangelists differ in what they emphasize about Jesus in these other pericopes.

The Life of Christ as reflected in the Four Evangelists has been the subject of study since Tatian composed the first harmony of Christ's Life according to the Four Gospels in A.D. 170. Down throughout the centuries differences between the Four Evangelists' have been noted long before the historical method came into use. The proponents of the historical critical method were not the first to mention them. Any student who uses the historical-grammatical method could by an inductive procedure note the fact and did that there seem to be seeming contradictions in the Passion Narratives.

The manner in which these seeming discrepancies are dealt with depends largely on the general attitude that a Biblical interpreter has when approaching God's Word. Those who believe in the reliability of the New Testament Scriptures and that the Holy Spirit would guide the apostles into all truth operate with the concept of the possibility of harmonizing the events of Christ's life where it appears there are discrepancies. Those who employ the latter principle distinguish between two different cleansings in the Temple. One (John 2:13 — 22) during the Judean ministry and a second during Holy Week. Most harmonists also distinguish two anointing's by individuals who loved Christ and therefore Luke 7:36-50 is no parallel to the anointing that took place at Bethany.

Since the seven pericopes are taken from latter parts of the Gospels a statement about the nature and relationship of the Gospels to each other ought to be made. The four records which traditionally are found in the front part of the New Testament are, properly speaking, four records of one-Gospel — "the Gospel of God. . .concerning His Son (Romans 1:1-3)." It was only in the second century that it became customary to speak of "Gospels." Bruce claims that the earlier writers use the singular, whether they are referring to a single Gospel-writing or to a set of such writings (Cf. Didache, viii, 2; Ignatius, Philadelphians viii, 2). The traditional titles of the four records imply that we have the Gospel or good news about Christ according to each of the evangelists. And the usage of the singular form to denote the fourfold record continued for long after the earliest attested instance of the plural.[21]

Jesus asserted, "I am the way and the truth." He promised His disciples that they would be able to recall the facts and sayings of His life, because the Holy Spirit would bring all things back to their remembrance. The Holy Spirit would guide them into all truth. The early Church traditions associate the Gospels with two apostles and two men who were the friends of the apostles of our Lord (Mark and Luke). In view of New Testament teaching. Christian expositors down through the centuries have worked with the presupposition that in the Gospels we actually have what the Holy Spirit wanted to have recorded and not writings that are replete with errors and contradictions.

Dr. Edward Robinson of Union Seminary, New York City, wrote in 1854:

> Matthew and Mark have in general more resemblances to each other; though Matthew, being more full, presents much that is not found in Mark or Luke, while Mark, though briefer, has somethings not contained in any of the rest. The Evangelists were led, under the guidance of the Spirit, to write each with a specific object in view, and for different communities or classes of readers. Hence while the narratives all necessarily exhibit a certain degree of likeness, they all also bear for itself the stamp of independence. [22]

The four writers vary likewise in their chronological order and character. On the one hand, it appears that Mark and John, who have little in common, follow with few exceptions the regular order of time as they have been guided by the principle of association so that, in their Gospels, events have certain relations to each other and are frequently grouped together, though they have transferred at different times and in various places.[23]

The sequence of events of Holy Week is one of the problems raised by the texts listed in "Operation Outreach." The events of Holy Week may be said to have begun six days before Good Friday when Jesus arrived at Bethany, where He usually stayed whenever He came to Jerusalem for the observance of the festivals. At the time of Jesus' arrival at Bethany many Jews also were on their way to the Holy City for the celebration of the Passover. The Pharisees and the chief priests had publicly demanded that Jesus be taken prisoner and so one of the widely discussed questions at this time was; will Jesus attend the festival? There are a number of Gospel students who believe that on Saturday evening the friends of Jesus held a feast in His honor in the home of Simon, once a leper. At this banquet Martha served. In the midst of the feast Mary anointed the feet of Jesus with a costly ointment of spikenard and dried His feet with her hair. This anointing is not the same as the one recorded in Luke 7:36-50 — which occurred during the second preaching tour in Galilee, summer and autumn of possibly A. D. 28.

The anointing of Jesus is mentioned in Matthew 26:6-13; Mark 14:3-9; and John 12:2-13. Some scholars opine that the anointing did not take place on Saturday evening but on Tuesday, which would have been a

Jewish Wednesday. Professor Shephard in his Christ of the Gospels places the anointing on Tuesday, holding that Jesus returned to Bethany after a day of intense activity, which had involved Him in meeting in various groups of enemies in the Temple, who sought to entrap Him with subtle questions, followed by His great denunciation of them and His prophetic discourse on the Mount of Olives.[24] Jesus found great consolation in that many sincere friends of His made a feast for him in the house of Simon, formerly a leper. Ylvisaker and Fahling agree with those who place the supper with the anointing on Saturday evening prior to the triumphal entry.[25] While John 12 fixes the supper on Saturday evening it is Ylvisaker's opinion that Matthew and Mark do not fix the time, even though they give the impression that the feast was held later. He claims that a more careful scrutiny indicates, however, that Matthew and Mark do not fix the time of the repast. Nor do they state that this is their intention. The two evangelists Matthew and Luke open their accounts, respectively, with the words:

Now when Jesus was in Bethany (Matthew), and being in Bethany (Mark). They mention the supper and the anointing evidently for a twofold reason. First, by way of contrast. Before this account come the report that the Sanhedrin had agreed upon Jesus' death, and after this, the story of the treason of Judas. Both are an expression of hatred, but the supper and the anointing are tokens of love. The devotion of his friends, and particularly the exuberant, well-nigh ecstatic love of Mary are the most striking contrast to the sinister picture of the leaders in Israel, and, not the least, of that disciple whose heart was even now filled with perfidious thoughts. . .Matthew and Mark would also reveal how it happened that Jesus came to his death at the feast, earlier therefore, than the council had desired at the first. The ointment and the anointing matured the plan in the mind of Judas to betray, and the offer of Judas caused the schemes of the council to miscarry. The Synoptists are thus influenced by considerations of logic when they arrange their subject-matter as they have done. There is nothing in their accounts to prevent us from following the chronology in John.[26]

Critical scholars claim that the Gospels disagree concerning the days on which Jesus was anointed, when he entered the city, and when He cleansed the Temple. The triumphal entry of Jesus into Jerusalem, recorded by all four of the Gospel writers, took place the following morning after the anointing by Mary at Bethany, on Sunday, on the First day of the week. Two of the synoptic writers tell of the anointing after recording the entrance into Jerusalem. Dr. Roney explained the reason for the anointing being recorded after the triumphal entry as follows:

Matthew and Mark give the record of this incident after beginning their accounts of the events of Passion Week because of the manner in which the true character of Judas was exposed in the complaint against the action of Mary, and because of the statement of the Savior that Mary had anointed His body in preparation for His burial. Moreover, it appears far more likely that these events of Bethany were, most appropriate during the pause between Christ's public ministry and the events of Passion Week, as given in John's Gospel.

The Triumphal entry is recorded by all four evangelists. Each of the four accounts is supplementary to each other. All four must be studied to obtain a full account of what happened. The three Synoptists state that Jesus after leaving Bethany paused at Bethpage, from where Jesus sent two disciples to fetch the colt of an ass, which He used to ride on into Jerusalem. The Synoptists practically used the same language in relating the instructions that Jesus gave. Mark and Luke mention the fact that when the two disciples found the colt and were unloosing it, they were questioned by individuals nearby why they were loosing the animal; Matthew does not mention this at all. John says nothing of sending disciples for the colt, but he merely relates that Jesus had found a young ass, without telling how it was found, and that Jesus sat upon it. Matthew and John specifically mention the fact that the riding into Jerusalem took place in fulfillment of Old Testament prophecy; Mark and Luke inform their readers that when Jesus entered into the capital multitudes spread their garments and branches before Him; Luke does not mention the branches. John tells his readers that the branches were palm branches, but does not say that garments were spread by the multitudes. Each of the four evangelists notes the royal acclaim given Jesus, however each gives the contents of the hails in different wording. The fact that divergent wording is used in no way militates against the truth that all four Gospel writers are in agreement about the significance of the royal hails. Matthew alone relates the fact of great excitement in Jerusalem among the crowds in the city and the questions that were asked about the Person riding on a colt. Luke informs us that the Pharisees complained about the demonstration and reports Jesus' answer; John mentions the aroused fear of the Pharisees because of the popular demonstration. The fact that Matthew mentions that two animals were to be brought and other evangelists mention only one animal, a colt is cited as a discrepancy or a contradiction by "Operation Outreach" and by critical scholars whose point of view is being promoted at "Seminex".

All four evangelists mention the colt and Matthew adds the additional information that the mother was to be brought with it. Matthew tells us that the ass and the colt with her were brought, because the ass was needed to quiet the previously unridden colt.

When Jesus was entering Jerusalem He wept over it because of the approaching doom that He foresaw. According to Mark 1:11 Jesus entered the Temple and at eventide returned to Bethany with the twelve. Conservative harmonies place the cleansing of the temple on Monday after Jesus on the way to Jerusalem from Bethany had cursed the barren fig tree.[28] Harmonies edited by opponents of the verbal inspiration of the Bible place one cleansing of the temple on Palm Sunday and then on Monday have another as Huck-Lietzmann do in their Synopse der Drei Ersten Evangelien, making the temple cleansing by Jesus in the days of His early ministry (John 2:13-17) a parallel with the cleansing on Monday.[29] This is surely an attempt to depict the Gospels as documents that are confused.

Those wishing to find mistakes in the Passion narratives claim that Jesus entered the city on one day, visits the Temple, and then withdrew. On the next day He returns to cleanse the Temple according to Mark. But Matthew and Luke have Jesus enter the city and cleanse the Temple on the same day. While it is true that Luke (19:45-48) and in Matthew (21:12-16) the cleansing of temple follows the triumphal entry, the day on which the cleansing occurred is not indicated. There is nothing in those lexis which forces the Biblical interpreter to pit one Gospel against the other.

The Materials for "Operation Outreach" have also raised the question of the celebration of the Passover. John 13:1 and Luke 22:1 are said to *be* contradictory. This raises the question whether the meal in the upper room was a fellowship meal, or was it truly the Passover? In two other passages John seems to indicate that the Passover had not yet come (13:29; 18:28).

From the Synoptic Gospels it appears that Jesus did eat the Passover with His disciples. E.F. Harrison suggests that "this dating in John may represent a protest on the ground following a different calendar, in line with the practice of the Qumran sect."[30] This has been suggested by Matthew Black as a possible solution to the problem raised by John in 13:1.[31]

Another possibility is that the references in John 13:29 and 18:28 to the Passover as still future are to be interpreted as references to the Feast of Unleavened Bread, which at times was called the Passover (Luke 22:1), and began immediately upon the Passover celebration and lasted for a week. Macleod states that the precise dating of the last supper and the crucifixion is the major difficulty of the Johannine chronology. While the Gospels place the supper on the 14th of Nisan, John has it one day earlier.[32] Macleod prefers the Johannine date, for he believes that the enemies of Christ resolved not to arrest Jesus on the feast day; furthermore, it would have been illegal to bear arms on a feast day or to conduct a trial on a feast day. Some scholars contend that in the year of Jesus' crucifixion the feast was celebrated a day later by the Sadducees. This may account for the appearance of disharmony.[33]

As long as there are reasonable possible solutions, it would be unwise to accuse the Gospel writers as guilty of errors and mistakes. A Bible replete with errors, contradictions and mistakes is hard to harmonize with the truth that the Biblical writers penned their writings under the guidance of the Holy Spirit. The original autographs are reliable and dependable; this, of course, does not rule out the possibility that in the copying of Biblical writings errors crept in. It is one of the functions of the science of textual criticism to remove transcriptional errors.

8. New Testament of the Old Testament

In another example of error in the Holy Scriptures students were told to show the members of the congregations where they presented their case was that the New Testament quotes a passage as found in the Old

Testament which is not there. Thus Matthew 2:23 reads: "And he went and dwelt in a city called Nazareth, that what was spoken by the prophets might be fulfilled. He shall be called a Nazarene." It is the contention of those who would ascribe inaccuracies to the Bible that Matthew erred in saying there was one specific prophecy that announced that Jesus would be called a Nazarene. However, the expression "by the prophets" prevents our seeking only one Old Testament passage, thus making doubtful any play on words based on neser, "branch," in Isaiah 11:1. "He shall be called a Nazarene" means that He would be known as an inhabitant of Nazareth. The fact that the words in question are not a particular Old Testament prophecy accounts for the indefinite way in which the fulfillment is introduced.

According to "Operation Outreach" Matthew has made a serious blunder in chapter 27:9-10, where he wrote: "Then was fulfilled that which was spoken by Jeremy, the prophet, saying: 'and they took thirty pieces of silver, the price of Him that was valued, whom they of the children of Israel did value and gave them for the potter's field, as the Lord appointed me.'" This reference by Matthew to a prophecy spoken by Zechariah 11:12-13 has evoked a number of plausible explanations. Some hold that here Jeremiah was the name of the first book in the Old Testament prophets and was taken as the name for the whole section which also contains Zechariah (just as "the Psalms" is applied to the whole section of the Writings because it is the first book) (Luke 24:44). The Babba Balhra tractate of the Talmud also makes Jeremiah the first book of the Prophets, although in most manuscripts Isaiah is the first book. Another solution to what appears to be a contradiction is the possibility that the Apostle Matthew combined Zechariah 11:12,13, with Jeremiah 18:2-12 and 19:1-15 and merely cited one source. Dr. Arndt in his book has a similar explanation.[34]

Christian News invites "Seminex" professors and students to send us any other reasons they have for rejecting the inerrancy of the Bible.

Footnotes

1. Merrill F. Unger, *Introductory Guide to the Old Testament* (Grand Rapids: Zondervan Publishing Company, 1951), pp. 262-263.

2. **Ibid.,** p. 264.

3. P. E. Kretzmann, *Popular Commentary to the Bible* (St. Louis: Concordia Publishing House, 1923), Old Testament, Vol. I, p. 303.

4. On the subject of doublets and duplicate accounts cf. Gleason L. Archer, *A Survey of Old Testament Introduction* (Chicago: Moody Press, 1964), pp. 11-124. G. C. Aalders, *A Short Introduction to the Pentateuch* (London: Tyndale Press, 1949).

5. H. H. Rowley, *The Doctrine of Election* (Philadelphia: Westminster Press, 1950), pp. 25-29. Cf. also his The Rediscovery of the Old Testament (Philadelphia: Westminster Press, 1945), p. 60.

6. Raymond F. Surburg, "Did the Patriarchs Know Yahweh? Or Exodus 6:3 and Its Relationship to the Four Documentary Hypothesis," *The Springfielder*, 31: 120-128, September , 1972.

7. **Ibid.,** pp. 126-127.

8. W. J. Martin, *Stylistic Criteria and the Analysis of the Pentateuch* (London: The Tyndale Press, 1955), p. 16-18.

9. M. H. Segal, *The Pentateuch: Its Composition and Its Authorship and Other Biblical Studies* (Jerusalem: At the Magnes Press, The Hebrew University 1967), pp. 4-5.

10. Heinrich Ewald, as quoted by John H. Raven, *Old Testament Introduction* (New York: Fleming H. Revell, 1910), p. 125.

11. As cited by Unger, **op. cit.**, p. 255.

11. As cited by Unger, **op. cit.**, p. 255.

11a. Unger, p. 246.

11b. **Ibid.**

12. Oswald Allis, the *Five Books of Moses* (Philadelphia: Christian and Reformed Publishing Company, 1943), pp. 94-99.

13. William Arndt, *Does the Bible Contradict Itself?* (St. Louis: Concordia Publishing House, 1926), p. 15.

14. E. F. Kevan, "Genesis," F. Davidson, editor, *The New Bible Commentary* (Grand Rapids: Wm. B. Eerdmans Publishing Company, 19530, pp. 84-85.

15. David F. Payne, "The Letter of Jude." in G.C.D. Howley, F.F. Bruce, H.L. Ellison, A New Testament Commentary (Grand Rapids: Zondervan Publishing House, 1969), p. 628.

16. Arndt, **op. cit.**, p. 42.

17. William Arndt, *Bible Difficulties* (St. Louis: Concordia Publishing House, 1932), p. 35.

18. **Ibid.**

19. Cf. William McKane, *Tracts for the Times.* Ruth, Esther, Lamentations, Ecclesiastes, Song of Songs (Nashville: Abingdon Press. 1965) pp. 11-26. W. Neil, "Book of Jonah, "Interpreter's Dictionary of the Bible (Nashville: Abingdon Press. 1962), E-J, p. 1967.

20. Jamieson, Fausset and Brown, *Commentary on the Whole Bible* (Grand Rapids: Zondervan Publishing House, 1962), p. 1103.

21. F. F. Bruce, "Gospels, The New Bible Dictionary", edited by J.D. Douglas (Grand Rapids: William B. Eerdmans Publishing Company, 1962), p. 484.

22. Edward Robinson, *A Harmony of the Four Gospels* (Revised edition: with additional notes by M.B. Riddle; Boston and New York: Houghton Mifflin and Company, 1885), p. 195

23. **Ibid.**

24. J.W. Shephard, 'The Christ of the Gospels (Grand Rapids: Wm. B. Eerdmans Publishing Company, 1946), pp. 53.

25. Johs. Ylvisaker, *The Gospels, Synopsis, Harmony, Explanatory Notes* (Minneapolis: Augsburg Publishing House, 1932). pp. 548-549. Adam Fahling, *The Life of Christ* (St. Louis: Concordia Publishing House, 1936), pp. 519-520.

26. Ylvisaker, **op. cit.**, p. 543.

27. Charles Patrick Roney, *Commentary on the Harmony of the Gospels* (Grand Rapids: Wm. B. Eerdmans Publishing Company, 1948), p. 355.

28. A.T. Robertson, *A Harmony of the Gospels* (New York: Harper & Brothers. 1922), p 156; Adam Fahling, A Harmony of the (dispels (Grand Rapids: Wm. B. Eerdmans Publishing House, no date), p. 157; William F. Arndt, *Bible Commentary - The Gospel According to St. Luke* (St. Louis: Concordia Publishing House), p. 395; Alfred Edersheim, *The Life and Times of Jesus the Messiah* (New York: Longmans,

Green and Co., 1899), 11, pp 374-379.

29. Albert Huck, *Synopse der Drei Evangelien, Neubearbeited von Hans Lietzmann* (Tubingen: von J.C.B. Moher (Paul Siebeck, 1950), p. 156, 30. Everell F. Harrison, "The Gospel According to John", in *The Wycliffe Bible Commentary*, edited by Charles F. Pfeiffer and Everett F. Harrison (Chicago: Moody Press, 1962), p. 1101.

31. Matthew Black. "The Arrest and Trial of Jesus and the Dale of the Last Supper," in *New Testament Studies, Essays in Memory of T.W. Manson*, edited by A.J.B. Higgins, pp. 19-23.

32. A.J. Macleod, 'The Gospel According to John", *The New Bible Commentary*, **op. cit.**, p. 889.

33. **Ibid.**

34. Arndt, *Does the Bible Contradict Itself*; **op. cit.**, pp 51-53.

Questions

1. What was "Operation Outreach?" ____
2. Students were furnished quotations and references to show that Luther ____.
3. Students were supplied Bible passages to show ____.
4. What was cited against Mosaic authorship of the Pentateuch? ____
5. The last three sections of Deuteronomy which contain the announcement of the death of Moses were probably added by ____.
6. Does the modernization of archaic names militate against Mosaic authorship? ____
7. Jethro and Reul were names of the ____ person.
8. Eichorn is sometimes known as ____.
9. Are there discrepancies between Kings and Chronicles? ____
10. The exegetes of the faculty majority rejected ____.
11. The real objection critical scholars have against Jonah is ____.
12. An imaginary event can hardly serve as ____.
13. What is involved in a rejection of the historicity of Jonah? ____
14. In the Gospels we actually have what ____.
15. The original autographs are ____.
16. *Christian News* invited the Seminex students and professors to ____.

A Response to Father Hughes' Attack on Missouri's Doctrine of Biblical Inerrancy

Christian News, September 5, 1977

Father John Jay Hughes, priest and adjunct professor of history at St. Louis University, reviewed the book of James E. Adams, entitled, *Preus of Missouri and the Great Lutheran Civil War*, in the April 3,1977 edition of the *St. Louis Post-Dispatch*. (Reprinted in April 11, 1977 *CN*, ed.) In his review Hughes continues the attack of Adams on the Lutheran Church-Missouri Synod and its President relative to the doctrine of the inerrancy and complete trustworthiness of the Bible. The tone of the review may be seen from the concluding sentence of the opening paragraph. "James E. Adams, religious editor of the *Post-Dispatch*, takes the lid off the seething cauldron in the Lutheran Church-Missouri Synod and peers inside. The result is sensational, scandalous, highly edifying-and very funny."

In order to justify the theological views of Dr. Tietjen and his Seminex faculty, which Adams has shown in his book as representing a departure from the historic position of the LCMS, Hughes is forced to attack the doctrinal position of Dr. Preus and simultaneously that of the church body he leads.

Thus Hughes writes:

The nineteenth century doctrinal ethos of the Missouri Synod centers on the conceit of an inerrant Bible. For 20 centuries believers have regarded the Bible as God's Word and hence true throughout. But from the start biblical scholars recognized that the Bible contained errors and contradictions. Very early they developed two ways of dealing with them.

The two ways adopted according to Hughes were: 1. the use of the allegorical method; 2. certain theologians adopted the view that God in the Scriptures accommodated Himself to the limited knowledge of the times of the Biblical writers.

According to Hughes the word "inerrant" is a modern coinage. "It was not used in religious context before the late nineteenth century." The St. Louis priest claims that the word is employed metaphorically and means "not wandering from the truth. This was the traditional orthodox view of the Bible: it was wholly reliable and true, despite numerous individual errors."

Hughes understanding of inerrancy, however, does not comport with the real facts. He reads the historical data through prejudiced lenses. The St. Louis professor is anything but honest when he belittles the situation relative to the "battle for the Bible" by failure to inform his St. Louis readers that the same kind of controversy has been going on in world and American Roman Catholicism. In the 20th century the Church of Rome has been embroiled in the same kind of controversy as that which has characterized the LCMS for the last two decades. The new liberal views about the Bible have been adopted by members of the Roman Catholic Biblical Association. It is their point of view which governs Hughes. *The National Catholic Register*, a Roman Catholic journal, has from time to time been attacking the theological aberrations that have appeared in Roman Catholicism. The April 10 issue of *The National Catholic Register* published the following from a letter of the late Archbishop Robert J. Dwyer of Portland, Oregon, who wrote to Pope Paul two years ago about the Catholic Biblical Association of America as follows:

This association is dominated by theologians not those deeply concerned with exegesis of a fundamental nature. The "theologians" trading as biblical scholars are using Historical-Critical method of Scriptural study and the whole seems to ignore the inerrancy of Scripture.

Scripture is being interpreted by individual theologians in a purely Protestant manner, to fit the ideas of Modernism, idealism, and relativism. They are busy "demythologizing" Scripture, thus implying or proposing that the Bible is based on myth (or else how demythologize it) and worst of all, are publishing their opinions without thought of submitting them for authoritative supervision, thus further confusing the laity, priests, religious, and unfortunately, some Bishops. . .

In the last thirty years Roman Catholic Biblical scholars have been adopting and leaning heavily upon the views and writings of Protestant scholars committed to and utilizing the historical-critical method with varying damaging results to Biblical truth. The views expressed by Father Hughes are the same arguments advanced by LCMS scholars using the historical-critical method.

Hughes is misinforming his readers by not telling them the true facts. The traditional view of the Roman Catholic Church is not the one the St. Louis professor describes in his review of Adams' book. John E. Steinmueller, in *A Companion to Scripture Studies,* published by Joseph F. Wagner, having the official Nihil Obstat and the Imprimatur of the Brooklyn Diocese, gives a different view of what the Roman Catholic Church held before the famous encyclical of Pope Pius XII, which opened the door for a change in Biblical interpretation of methodology in American Roman Catholicism. Volume I of Steinmueller has a lengthy opening section on "Biblical Inspiration," pp. 4-43.

Dr. Steinmueller, a conservative Biblical scholar, defined the inspiration of the Sacred Scriptures as involving three factors.

Thus he wrote:

First, there is the primary efficient cause, which is the Holy Spirit, who supernaturally acts upon man. The Bible is attributed to the Holy Spirit by appropriation, because every action of God ad extra is common to the three Persons of the Blessed Trinity. Secondly, there is the subject of inspiration, which is man, the instrument of the Holy Spirit. Finally, there is the end or terminus of inspiration which is the book or the word of God (p. 5).

Steinmueller does not hesitate to employ the word "inerrancy" when describing the Sacred Scriptures. Thus Steinmueller wrote in the early forties:

> From the very fact that the sacred writers were divinely inspired it necessarily follows that the Sacred Scriptures are both **de facto** and **de jure** free from error. The Catholic Church has always taught this. "For all the books which the Church receives as sacred, and canonical are written wholly and entirely, with all their parts, at the dictation of the Holy Spirit; and so far is it from being possible that any error can coexist with inspiration, that inspiration not only is essentially incompatible with error, but excludes and rejection of it as absolutely and necessarily as it is impossible that God Himself, the supreme truth, can utter that which is not true...And so emphatically were all the Fathers and Doctors agreed that the divine writings, as left by the hagiographers, are free from all error that they labored earnestly, with no less skill than reverence, to reconcile with each other those numerous passages which seem at variance ("Providentissimus Deus," Cf. E.B. 109, 112) (p. 29).

Steinmueller claimed that "this absolute inerrancy must be directly attributed to the original copies of the sacred writers." However, as Steinmueller pointed out that while all Catholic scholars admit the fact of inerrancy, differences have arisen among Roman Catholic theologians how to explain and apply this inerrancy to various parts of the Bible. This has been termed as "the Biblical Problem." The fact that in recent years many Roman Catholic scholars have adopted a position that inerrancy only applies to matter of doctrines and morals and not to science or history has resulted in the paradoxical situation that the Bible is said to be inerrant even though it contains many and numerous mistakes. In this respect the Roman Catholic Church has experienced a development which parallels that in the LCMS over the last twenty years. However, in the Roman Church this development began in the early part of the twentieth century in a limited way but became strong after Pope Pius' encyclical of 1942, called *Divino Afflante Spiritu*.

Hughes, a professor of history, does not appear to know the history of the doctrine of inspiration as it was held throughout church history, as it was taught during the periods of the ancient church, the medieval period, during the Reformation period, in the post-Trent period in Roman Catholicism or in Protestantism until the appearance of the apostasy known as rationalism, followed by the use of various forms of the higher critical method.

The History of the Doctrine of Biblical Inerrancy from Apostolic Times till Modern Times

About a hundred years ago Professor L. Gaussen of the Oratoire, Geneva, where he was professor of systematic theology, published a famous volume. *The Inspiration of the Holy Scriptures* (now available in a new, enlarged type with analysis and topical index; Chicago: Moody Press, 1949) in which he hoped with God's help and on the sole authority of His Word to set forth, establish and defend the Christian doctrine of divine inspiration. In this Christian classic Gaussen inveighed against the indifferentism and secularism of his age so frankly, firmly, and unceasingly that at last he was deposed from his office by the liberal Swiss consistoire. According to former Concordia Seminary professor, J. Theodore Mueller, "Dr. Gaussen put into his book about everything essential that can be said on biblical inspiration, both positively, and negatively in defending it against the attacks of unbelieving critics. The basic theme which he proves is that the orthodox Christian Church has always believed in and confessed verbal inspiration."

In Section II Gaussen gives a history of those who have opposed the divine inspiration and reliability of the Bible (cf. pp. 139-145); this is followed by a presentation of the arguments of the orthodox Christian Church relative to inspiration and inerrancy. On page 145 Gaussen asks the following question: Can many illustrious doctors of Church be mentioned as maintaining the plenary inspiration of the Scriptures? Answer: It is the uniform doctrine of THE WHOLE CHURCH down to the days of the Reformation. Gaussen cites professor Rudelbach, who in *Zeitschrift von Rudelbach and Guericke*, 1840, 1st vol., pp. 1-47, showed how for the first eight centuries there is "hardly a single point with regard to which there reigned. . . .a greater or more cordial unanimity." Gaussen has summarized this article of Rudelbach, in which no less than eleven different principles are cited that characterized the testimonies of the first eight Christian centuries. On page 151 Gaussen has brought together the references from primary sources, where often long passages can be found in the writings of the following ancient theologians: Irenaeus, Tertullian, Cyprian, Origen, Chrysostom, Justin Martyr, Epiphanius, Hilary, Basil the Great, Gregory the Great, Gregory of Nyssa, Theodoret, Cyril of Alexandria.

Professor R. Laird R. Harris, in his excellent volume. *Inspiration and Canonicity of the Bible* (Grand Rapids: Zondervan Publishing House, 1957), has a chapter entitled "Verbal Inspiration in Church History" (chapter 3, pp. 72-84). In it he points out that Emil Brunner, in his *Christian Doctrine of God* wrote: "The doctrine of verbal inspiration was already known to pre-Christian Judaism. . .and was probably also taken over by Paul and the rest of the apostles" (p. 11). Then Brunner goes on to say that this doctrine was not of much consequence because of the use of the allegorical method throughout the Middle Ages, a method permitting the scholastics to interpret the Bible as they desired. In commenting on what Brunner has written, Laird Harris correctly observed: "Probably Brunner does not realize the import of what he is here saying. He says that this doctrine-which he cordially castigates-was the legacy of Judaism accepted by the apostles, and held by the Early Church through the years of her strength down to the times of Origen and Augustine, who set the pattern of allegorical interpretation for a thousand years" (p. 76). Harris has pointed out that in practice, but not in theory, the doctrine was lost during the Dark Ages, but after the Reformation was Protestantism's strength until the rise of higher criticism. "It was held by the Church which spread abroad in the great missionary movement of the nineteenth century and by the laity and churchmen of the twentieth century. This terrible doctrine seems to be absent in the times of the Church's apostasy and present in the times of her power" (p. 76).

W. Sanday, not a friend of the Biblical doctrine of verbal inspiration, contributed a volume to the famous Bampton Lectures series on **Inspiration**, in which in chapter 1, Sanday gives a great number of references from the early centuries of church history, (cf. especially, pp. 28-42).

The eminent Benjamin Warfield of Princeton Seminary fame, in commenting on Sanday's book on Inspiration wrote:

> Dr. Sanday, in his recent Bampton Lectures on 'Inspiration'- in which unfortunately, he does not teach the church doctrine-is driven to admit that not only may 'testimonies to the general doctrine of inspiration' from the earliest Fathers 'be multiplied to almost any extent; but (that) there are some who go further and point to an inspiration which might be described as 'verbal'; 'nor does this idea,' he added, 'come in tentatively and by degrees, but almost from the very first.' He might have spared the adverb 'almost.' The earliest writers know no other doctrine. (B.B. Warfield, *Revelation and Inspiration* (1927), p. 54.

Alan Richardson, a notable scholar committed to the historical-critical method, stated in his *A Preface to Bible Study*(1944), that "from the second century to the eighteenth this theory (of verbal inspiration) was generally accepted as true," although Richardson finds that "few better educated Christians" hold to it any more" (p. 25).

Harold Lindsell, in *The Battle for the Bible* (Zondervan Publishing House, 1976), has devoted a chapter to the doctrine of verbal inspiration of the Bible in the history of the Church (chapter 3, pp. 40-71). The New Testament, the only reliable documents for a knowledge of the teachings of Jesus, God's only begotten Son, are the books written by His apostles whom He had promised the Holy Spirit who would guide them into all truth and bring all things to their remembrance. They clearly set forth the view that the Holy Scriptures are the inspired inerrant and infallible Word of God. This was discussed by Lindsell in chapter 2 of his book (pp. 28-40).

Kenneth Kantzer, dean of Trinity Evangelical Seminary, has written about the testimony of liberal scholars relative to biblical infallibility as follows:

> H.J. Cadbury, Harvard Professor and one of the more extreme New Testament critics of the last generation, once declared that he was far more sure as a more historical fact that Jesus held to the common Jewish view of an infallible Bible than that Jesus believed in His own Messiahship. Adolph Harnack, greatest church historian of modern times, insists that Christ was one with His apostles, the Jews, and the entire early church, in complete commitment to the infallible authority of the Bible. John Knox, author of what is perhaps the most highly regarded recent life of Christ, states that there can be no question that his view of the Bible was taught by our Lord himself.

Rudolph Bultmann, an opponent of Biblical supernaturalism, acknowledged to be one of the greatest of New Testament scholars of Germany, wrote in his well-known *Jesus and the Word* (New York: Scribners, 1934), p. 61 as follows:

Jesus agreed with the scribes of his time in accepting without question the authority of the (Old Testament) Law. When he was asked by the rich man, "What must I do to inherit eternal life," he answered, "know the commandments," and he repeated the well-known Old Testament Decalogue. . . Jesus did not attack the Law but assumed authority and interpreted it. . . .And from this time (After Jesus' day when Paul and others preached) came the well-known words, which Jesus surely cannot have said: "Do not suppose that I have come to destroy the Law and the Prophets. I have not come to destroy but to fulfill. I tell you truly, until heaven and earth vanish, no letter nor point can vanish from the Law until all be fulfilled. Whoever, erases one of the smallest commandments and so teaches others shall be called least in the Kingdom of Heaven. But whoever keeps it and teaches it shall be called great in the Kingdom of Heaven" (Matt. 5:17-19).

F.C. Grant, liberal professor and critical scholar, stated that Jesus and the writers of the New Testament, believed in an infallible Bible. In his *An Introduction to New Testament Thought* Grant wrote:

But its view of inspiration is not more advanced than that of any other part of this volume, as an examination of the passages cited in concordance advanced than that of any other part of this volume, as an examination of the passages cited in concordance (s.v. "scripture" and "written") will show. Everywhere it is taken for granted that what is written in Scripture is the work of divine inspiration, and is therefore trustworthy, infallible, and inerrant. The Scripture must be "fulfilled" (Luke 22:37). What was written there was "written for our instruction" (Rom. 15:4; 1 Cor. 10:11). What is described or related in the Old Testament is unquestionably true. No New Testament writer would dream of questioning a statement contained in the Old Testament, though the exact manner of mode of its inspiration is nowhere explicitly stated (p. 75).

As Lindsell has pointed out this statement of Grant's is important. While Grant himself did not believe in inerrancy, he acknowledged that this was a doctrine held by Christ, His apostles and the New Testament relative to the Old Testament. In our times we have evangelical scholars, who while holding a higher view on the Bible than Grant did, but deny that inerrancy is taught in the Bible. Lindsell concludes concerning this situation: "Grant was either a more perceptive scholar or a more honest one; he did not hide his unbelief by claiming the Bible does not teach the view he refused to believe" (p. 44).

"While Grant himself did not believe in inerrancy, he acknowledged that this was a doctrine held by Christ, His apostles and the New Testament relative to the Old Testament."

When Professor Hughes claims that the belief of Missouri on the nature of the Scriptures was fundamentalistic because of its insistence on the inerrancy and infallibility of a verbally inspired Scriptures, he simply shows ignorance of the views of the Christian Church during its first five centuries. The early church fathers' writings have been gathered in Ante-Nicene, Nicene, and Post-Nicene Fathers.

George Duncan Barry, in his book. *The Inspiration and Authority of Holy Scriptures, A Study in the Literature of the First Five Centuries* has surveyed the views on Scripture held by the early church fathers. Barry declared on the basis of his study:

> The fact that for fifteen centuries no attempt was made to formulate a definition of the doctrine of inspiration of the Bible, testifies to the universal belief of the Church that the Scriptures were the handiwork of the Holy Ghost. . . . It was . . . .to our modern judgment, a mechanical and erroneous view of inspiration that was accepted and taught by the Church of the first centuries, seeing that it ruled out all possibility of error in matters either of history or doctrine. Men expressed their belief in the inspiration and authority of the Bible in language which startles us by its strange want of reserve. The Scriptures were regarded as writings of the Holy Spirit, no room at all being left for the play of the human agent in the Divine Hands. The writers were used by Him as workman uses his tools; in a word, the Books, the actual words rather than the writers, were inspired.

This quotation is important because it shows that the writers of the first five centuries did believe in an inerrant Scripture, and since there "was no question about this viewpoint, it required no special statement to offset any contravening opinions," according to Lindsell. While Barry was no proponent of Biblical inerrancy and infallibility, he admits that this was the view of the Christian Church in the first five centuries of its existence. In fact, so high was their view of Scripture that they held the "words, rather than the writers inspired."

On pp. 46-54, Lindsell has given quotations from the following writers: Josephus, Clement and Polycarp, The Letter of Barnabas, the Apologists, Ireneus, Tertullian, Cyprian, Clement of Alexandria, Origen, Athanasius, Gregory of Nazianzus (c. 329-388) and Basil the Great (330-379), Chrysostom, Theodoret, Jerome and Augustine, the theologian, who marks the end of the ancient church period.

"The position of St. Augustine on the inerrancy and reliability of the canonical Scriptures continued to be the doctrine of the Roman Catholic Church for over a thousand years."

Since Augustine exercised such a great influence on the Roman Catholic Church, whom Luther and Calvin, the fathers of modern Protestant Churches, it will be well to show his stance on the question of inerrancy and infallibility. With respect to inspiration of the Bible, Augustine declares that the canonical Scriptures are "the revered pen of thy (i.e. God's) Spirit. "(**Confessiones**, Vii, 21, 27). In reading the Holy Scriptures men seek nothing more than to find out the thought and will of those by whom it was written, and through these to find out the will of God, in accordance with which they believe these men to have spoken." (**De Doctrina Christiana**, II, v. 6.). Augustine, like the author of Hebrews, asserts that a certain human writer spoke or wrote, but "the Holy

Spirit...with admirable wisdom and care for our welfare arranged the Holy Scriptures" (**De Doctrina Christiana**, II, vi. 8).

A corollary of the doctrine of verbal inspiration is inerrancy, the latter was boldly asserted by Augustine, when in a letter he wrote: "For I confess to your charity that I have learned to defer this respect and honor to those Scriptural books only which are now called canonical, that I believe most firmly that no one of those authors has erred in any respect in writing (Epistlae, 82,1,3)"

The position of St. Augustine on the inerrancy and reliability of the canonical Scriptures continued to be the doctrine of the Roman Catholic Church for over a thousand years. When the split occurred between Rome and Constantinople in A.D. 1054 it was not over the doctrine of Scripture. There were two views about the Bible that characterized Rome's theological stance on the inspiration of the Bible. One view which was repudiated was the view that the Holy Spirit "secured the writers from error only in matters of faith and morals. . ." This is the view held by many in Roman Catholic circles today. But this is a revival of a view propounded in 1685 by a man named Holden. *The Catholic Dictionary* (New York: Addis and Arnold, 1884). p. 540 reports that Holden in his *Analysis Fidel* defended limited errancy. *The Catholic Dictionary of 1910*, Vol. VIII, pp. 47-48. In the article on "inspiration" reports attempts by certain Roman Catholic theologians to advance a limited kind of inerrancy, Alfred Durand wrote: "For the last three centuries there have been authors-theologians, exegetes, and especially apologists, such as Holden, Rohling, Lenormant, de Bartolo, and others who maintained, with more or less confidence, that inspiration was limited to moral and dogmatic teaching, excluding everything in the Bible relating to history and to the natural sciences. They think that, in this way, a whole mass of difficulties against this inerrancy of the Bible would be removed. But the church has never ceased to protest against this attempt to restrain the inspiration of the sacred book." When Mgr. d' Hulst, Rector of the Institut Catholique gave a sympathetic account of opinions in *Le Correspondant*, January 25,1893, Pope Leo XIII responded in the encyclical "Provdentissimus Deus" as follows:

It will never be lawful to restrict inspiration merely to certain parts of the Holy Scriptures, or to grant that the sacred writers could have made a mistake. Nor may be the opinion of those be tolerated who, in order to get out of these difficulties, do not hesitate to support the Divine Inspiration extend only to what touches faith and morals, on the false plea that the true meaning is sought for less in what God has said than in the motive for which He has said it.

Roman Catholics rejected the concept of mechanical inspiration which they believed was espoused by Protestant writers. However, there are no reputable evangelical scholars that argue for mechanical or typewriter inspiration, although this view is erroneously ascribed to them by writers rejecting the verbal and plenary inspiration of the Holy Scriptures. Leo XIII in the "Symbol of Faith" used Augustine's classic statement in the consecration of bishops and said that God was the "one author of the Old and New Testaments."

The New Catholic Encyclopedia, which reflects a spirit of modernism, when compared with *The Catholic Encyclopedia* of 1910 (published by The Encyclopedia Press, Inc.) has a statement about inerrancy, as follows:

> The inerrancy of Scripture has been the constant teaching of the Fathers, the theologians, and recent Popes in their encyclicals on Biblical studies (Leo XIII, Ench. Bibl 124-131; Benedict XV, Ench Bibl 453-461; Pius XII, Ench Bibl 560). It is the nonetheless obvious that many biblical statements are simply not true when judged according to modern knowledge of science and history. The earth is not stationary (cf. Eccl. 1:4); Darius the Mede did not succeed Belessar (cf. Dan. 5:30-6:1). VII. II. P. 384.

While modern liberal Roman Catholic scholarship has been influenced by Protestant Liberalism, as reflected in this statement, it still makes very clear that the historical and official position of the Church of Rome has been that of total and complete inerrancy of all Scriptural words and assertions.

Adams in his book, *Preus in Missouri and the Great Lutheran Civil War* shares the new Roman liberal view about inerrancy as does Father Hughes in his favorable review of Adams' book.

In its doctrine of the complete and total trustworthiness of the Bible, the LCMS is in line with the teaching of the historic Christian Church, as held by the Roman Catholic Church, the Eastern Orthodox Churches, and Protestantism before their invasion by rationalism, scientism and various philosophies of the last two hundred years.

President Jacob A.O. Preus in 1971 contributed a valuable monograph to this highly debated topic concerning the Bible's inerrancy and reliability in Concordia's *Contemporary Theology Series* in *It Is Written*. This volume discussed the nature, authority, inspiration, truthfulness, and saving power of the Holy Scriptures. About this work. Dr. Preus said:

> It is our particular purpose in this study to consider Jesus' doctrine of Scripture, in order that as Christians we may also acknowledge His Lordship in this important area.

Questions
1. Who was Father John Hughes? ____
2. Hughes attacked ____.
3. Hughes' understanding of inerrancy does not compare with ____.
4. In the 20th Century Rome has been embroiled in ____.
5. What did John E. Steinmueller emphasize? ____
6. What began with Divino Afflante Spiritu? ____
7. What did J.T. Mueller say about L. Gaussen's *The Inspiration of Holy Scripture*? ____
8. R. Laird Harris pointed out that ____.
9. Allan Richardson stated in his *A Preface to Bible Study* ____.
10. Hughes shows ignorance of ____.
11. Adams in his book, *Preus of Missouri and the Great Lutheran Civil War* shares the new ____.
12. What Jacob A. O. Preus write in his *It is Written*? ____

Bishop John A. T. Robinson, Underminer of The Foundations of The Christian Faith

Christian News, October 10, 1977

Bishop John A. T. Robinson, now dean of Trinity College, Cambridge, England, has a flare for throwing ecclesiastical bombshells that explode with considerable noise in the ecclesiastical world. In 1963, when Dr. Robinson was Suffragan Bishop of Woolwich, southeast London, he shocked the religious world by the publication of his Honest to God, a book that was a popular presentation of liberal theology. This attack on orthodox Christian theology provoked what came to be referred to as the "honest to God controversy." The views of Honest to God came under attack from many quarters.

In 1976 Dr. Robinson published *Redating* the New Testament, a volume which spelled out theories diametrically opposed to those generally held by current critical New Testament scholarship. The so-called critical orthodoxy held as a consensus that the 27 books of the New Testament were not available before the middle of the second century. Now Robinson proposed what must have constituted a shock to his fellow New Testament scholars the view that all books, with the single exception of II Peter, were written before A.D. 70. *Redating* the New Testament contains 369 pages in which Dr. Robinson has endeavored to support his positions by painstaking argumentation. Since its publication some time last year (1976), Dr. Robinson has explained and defended his views in more popular form for the British in discussions on the British Broadcasting Corporation's cultural network, known as Radio 3.

Robinson has enunciated the view, shocking to modern Biblical criticism, that the entire New Testament was available by the time of the destruction of Jerusalem in A.D. 70. Concerning the Gospel of John, usually dated about 100, the British bishop argues, that inasmuch as its materials link up with original Christian tradition, that it is much easier to explain this fact by early authorship than by a complicated chain of transmission to an unknown writer a half century later. Thus Robinson places John's Gospel between A.D. 50-55 and 2 and 3 and 1 John in A.D. 60-65, and a final form of the Johannine prologue about A.D. 65.

Relative to the relationship of Jude and 2 Peter, the British theologian believes that Jude was writing 2 Peter in Peter's name when suddenly he broke off and penned a hurried letter of his own against a group of heretics. Jude then embodied this in the more formal encyclical letter. This constitutes Peter's last message to the Christians of Asia Minor sometime around A.D. 61-62.

By positing an early date for the writing of the Gospel of John, Robinson in a broadcast over the BBC, as reported by Jack Allen in a Religious News Service Report, claimed that "you can't consider moving one major piece in the interlocking puzzle without raising questions about the whole development of early Christianity-for the Gospel of John carried with it implications not only for the three Epistles of John and the Book of Revelation, which are regularly put in the same decade, but for the other three Gospels and indeed for the Book of Acts."

In his examination of the New Testament data, and comparing them with the current scholarly consensus, he became convinced that the conclusions of New Testament "orthodox critical scholarship" were far certain as generally supposed. There were really not many landmarks which would furnish hooks on which to hang many current views. In his reexamination of the evidence he claimed to begin with one certain date, namely, the destruction of Jerusalem in A.D. 70, an event that had traumatic effects on both Judaism and Christianity. Relative to the destruction of Jerusalem, Robinson asserted: "The remarkable thing is that this is never once mentioned in the New Testament as a past fact. There are of course predictions of it in the Gospels, particularly those of Matthew and Luke have been taken by the critics to be prophecies after the event, written up with the detail of hindsight. Yet the more one looks at them, particularly in comparison with later Jewish and Christian prophecies which undoubtedly reflect it, the more questionable this appears to be, especially since many of the details can be shown to derive not from what actually occurred but from similar description of the siege of Jerusalem in the Old Testament."

In examining the Book of Acts Robinson claims that there is no hint whatsoever of the coming persecution of A.D. 65 or of a Jewish revolt against Rome, or of the siege of Jerusalem in A.D. 68, for which Luke's first volume had prepared the reader. Obviously Acts was written before A.D. 70.

Not only does Robinson date the entire New Testament to a period prior to the fall of Jerusalem, but he would assign the Captivity Epistles as originating in A.D. 58 at Caesarea and assign the Pastoral Letters to the time of Paul's Third Missionary Journey. 1 Timothy was written from Troas, Titus while Paul was on the way to Jerusalem and 2 Timothy while he was in prison in Caesarea. These views of Robinson are very different from the generally held critical position.

In 1977, Bishop Robinson followed his *Redating* the New Testament with another book dealing with the New Testament, entitled: Can We Trust the New Testament? After expressing such relatively orthodox views about New Testament isagogics, conservative readers might have expected that here would be another book in which the English scholar would express views that would contain shocking statements for the critically-oriented readers of the 27 books of the New Testament. In his most recent book Robinson "endeavors to show that the Biblical reader needs to utilize the tools of critical scholarship in order to establish the trustworthiness of the New Testament." The publishers, A. R. Mobrary & Co., Ltd., Oxford, England, and the United States publishers and distributors, William B. Eerdmans Publishing Company, also additionally tell the readers that the English Bishop will assure everybody that the New Testament is still "the Word of Life."

However, while this volume does not compare in size with *Redating the New Testament*, yet this small volume will explode like a bombshell in the camp of those who may have believed that Robinson has experienced a change of position in the direction of traditional orthodoxy. This book propounds a rank kind of theological liberalism. At this point it might not be amiss to quote from the conclusion of Dr. Robert Hoerber's review of *Redating* the New Testament, which appeared in the *Concordia Journal*:

> That Robinson's conclusion agrees more or less with the view of students prior to the development of some higher critical methods should not be exaggerated as any proof for a traditional approach to the New Testament isagogics and theology ... Conservative chronology does not necessarily result in traditional theology, as the views of Robinson himself bear witness. (Vol. 3. Number 3, p. 144)

The Grounds of Trustworthiness according to Robinson

Bishop Robinson does not believe the New Testament to be trustworthy because its various books were written by men inspired by the Holy Spirit. The New Testament is reliable only when its trustworthiness has been humanly established by scholars having employed the tools of critical methodology. The New Testament's reliability can only be determined by the use of various types of criticism, such as have been developed in the last one-hundred years. In chapter 1 he has outlined and evaluated four different attitudes taken over against the New Testament. Those people who completely reject the historical character of the New Testament data are repudiated by Robinson as representatives of "the cynicism of the foolish" (p. 13). They are impossible. The second attitude is that of "fundamentalism of the fearful." Those adhering to this position are described as reacting to the nineteenth and early twentieth-century criticism. Robinson admits that some of this fear was justified because of some of the radical positions advocated by certain scholars. But this fundamentalist reaction was a mistake according to the English bishop, because "the defense response was however to close the hatches, and both Roman Catholic and sections of the Protestant Churches tended to seek refuge in a verbal inspiration which depended on the all or nothing mentality we have mentioned" (p. 16). The true approach to the New Testament is to have freedom to sift and to test everything.

Robinson correctly admits that no New Testament scholar approaches the Bible without presuppositions. As a proponent of the historical critical method he holds to an anti-supernaturalistic bias, which controls his interpretation of the Scriptural data relative to the miracles of the Bible. He rejects the miraculous found in the warp and woof of the life of Jesus and his bias determines his rejection of the historicity of the miracles of Christ, the disallowing of the virginal conception, the Virgin Birth, Christ's bodily resurrection, Christ's bodily ascension on the fortieth day after His resurrection and his visible Second Coming, accompanied by the holy angels.

The dean of Trinity College, Cambridge, has deified the historical-critical method, which the layman does not know how to use, and therefore he must defer to the conclusions of scholars. He believes that some scholars really have the ability impartially to investigate the evidence which leads them to trustworthy results. Thus Robinson writes:

Yet the only healthy attitude can be to trust to impartial scholarly investigation **whichever** way it comes out. For those who are genuinely committed to Christ as the truth must be prepared for the risk which God himself took when he committed himself to history, that is, to the contingency of events and to the fallibility of the records (pp. 27-28).

In addition to advocating the need for textual criticism, a type of procedure recognized by all sound Biblical scholarship, Robinson also claims the need for the employment of source criticism, redaction criticism and **Sachkritik** or content criticism. By the practice of the last three-mentioned types of criticism he ends up in this book in questioning the basic truths of the Christian faith. The Bible is an errant and untrustworthy book of revelation: And even though in chapter 4 he gives expression to a number of conservative views relative to the dating of the New Testament books, his application of the historical-critical method results in a rejection of truths set forth in the ecumenical creeds of Christendom, namely, the Apostles', the Nicene and Athanasian Creeds.

While Robinson questions the various theories put forward in the last one hundred years relative to the relationship of Matthew, Mark and Luke to each other (called the Synoptic Problem) and the relationship of the Fourth Gospel to the Synoptics, he espouses a view which asserts that it is necessary to separate and assess the various strata in the Four Gospels, so that like the archaeologist, the reader will be able "to place it and get it to 'speak' to us about the developing tradition. If some piece of material bears marks of being more distant from source, that does not mean it is useless. It may have much to tell us about the tendencies at work within the life of the church, which can then throw light on the history of other material. We can begin to see things 'in depth' and not on the surface only. It adds a new dimension to our vision and discrimination" (pp. 49, 50).

As a practitioner of form criticism Robinson contends that in the Gospels and Epistles, the New Testament reader needs to distinguish, between what Jesus said and did and what the first-century Church ascribed to Jesus as saying and doing. Thus he writes: "What the form-critics have enabled us to perceive is that the Gospels, just as much as the Epistles, are Church books and therefore sources in the first instance for its life and theology (p. 51)." What the New Testament portrays as having been spoken by Jesus is now denied and additions are often added to the parables thus rejecting the reliability of the text (cf. p. 52). In answer to the question: Is the New Testament true? Is the New Testament trustworthy? The response according to what Robinson has written, would have to be: Not all of it but only those parts acceptable to the form critic's interpretation, which differs from one form critic to another.

As a utilizer of the historical-critical method, Robinson resorts to redaction criticism, the logical successor to form criticism. It is true that he criticizes how certain scholars have used this methodology thus far, still he stated: "Yet again the perspective can yield valuable insight" (p. 54). In his demonstration of the employment of the so-called "tools of interpretation," namely, source, form, redaction and content criticisms the

English bishop does not hesitate to use the heretical Gnostic Gospel of Thomas as one of three parallel Gospel sources to show how the parable of the husbandman who sent his son to the vineyard (Matthew 21:33-46; Mark 12:1-12 and Luke 20:9-19) reflects differences that were occasioned by the passing of the original parable through the filter of Christian tradition, with the result that the application as given in the parable is made a later addition and not allowed to be a part of the original story as spoken by Jesus. While Robinson claims the New Testament is trustworthy, yet he shows by the application of the tools of the historical-critical method that just the opposite is the case.

While Robinson appears to be rather conservative on isagogical questions, still his positions on form and redaction criticisms make his interpretations of New Testament data just as vulnerable as those of the scholars whom he has criticized and with whom he disagreed. Like Bultmann and others with whom he has taken issue, uncertainty regarding the words and deeds of God's Son, Jesus Christ, is introduced. Thus the important area of Christology, one of the most vital, if not the most important in theology, is thrown into chaos by his use of form critical and reactional procedures. This result can clearly be seen from this declaration. "The other thing that the form-critics have rightly emphasized is that the Gospels are the product over a period of communities that collected and shaped material relevant to the needs of their developing life" (p. 75). The Gospels were not written at a single moment of time and not penned like the Epistles to meet some specific situation, but "they appear to have grown from combining diverse traditions and to have passed through various stages and states" (p. 75).

Robinson's Picture of Christ

The historic Christian Church has taught and confessed the truth that Jesus Christ is the God-Man, God-with-us, Immanuel. What Robinson does to Jesus is to demote Him purely to human stature, even though he calls Jesus the 'Word of Life.' The account of the Virgin Birth set forth clearly and unequivocally in the Apostles' Creed and the Nicene Creed is explained away in the following manner by Robinson: "The center of interest of the new book of Genesis' (the opening phrase of Matthew in the Greek) is not gynaecology, any more than that of the old one is geology. To search it for answers to such questions or to take the stories at that level is to misread them" (p. 96). In speaking about the infancy narratives Robinson averred that these accounts are overlaid with myth and "interpreted by stories with all the legendary and mythical beauty of folktales-whose point is not to distract from the history, nor again to be taken simply as history, but to draw out the divine significance of the history" (p. 97). The English bishop claims that by understanding the narratives of infancy in a literal manner one becomes guilty of discrediting the true meaning of these accounts. It is Robinson's contention that when "fundamentalists" explain the accounts of the visitation, the virgin birth, the coming of the wise men, the massacre of the innocents, and the flight of the holy family into Egypt as historical facts, they are to be faulted of being guilty of reading as prose what actually is poetry. But why does Robinson reject the events of Matthew 1-2 and Luke 1-2 as not containing a record of events that actually transpired? He does admit that there is some history in Luke 1-2 (p. 96). Wherever there is given an account that has in it the supernatural and miraculous, he refuses to accept it. The historical-critical method, espoused by Robinson, because of its naturalistic presuppositions rules out and cannot permit any Biblical story to be understood as it reads if there is anything miraculous in it.

In the prologue of John's Gospel the pre-existent Christ has been described. That God assumed human nature is a truth. Robinson cannot comprehend and believe, and so his interpretation of the prologue ends up denying that the LOGOS, the Word is a person, separate from the Father. This is how the true meaning of John 1:1-18 is explained:

His explanation of origin, says John, lies in the principle or 'word' of God behind it all. For he is the self-expression of that divine activity that all along has been coming into its own, first in nature, then in a people, and finally now embodied in a person. And, so perfect a reflection of it is he that the analogy that comes to John's mind is that which we used to say of a boy that he is the spit image of his father — or in the, Hebrew metaphor, the 'glory (or reflection) as of a father's only son.' (John 1:14) (p. 98).

It is the contention of Robinson that the Early Church has read back its ideas and hopes into the pre-resurrection accounts of the Gospels, thus ascribing words, titles (like the Son of Man), resurrection predictions to Jesus which he never used or uttered. If this were to be true (which it is not) then the logical and honest answer to the question of Robinson's "Can we trust the New Testament?" would have to be: we cannot!

The conservative reader will be amazed at the clever but highly erroneous manner in which the English Bishop establishes and defends his position. On the night before His crucifixion, Jesus promised His disciples the Holy Spirit who would bring back all things to their remembrance taught to them by Jesus. Now what does Robinson do with this "remembrance?" He writes: "or let us say that 'remembering' was not for the early Christians just a neutral exercise in recalling facts" (p. 103). Yet as John shows "that the Holy Spirit would bring everything to the disciples 'remembrance', a recalling of the past in such a way that did not leave it in the dead past but recreated it as present experience at a deeper level" (p. 103). "The sayings and actions of the historical Jesus 'spoke' to them as words and deeds of the living Christ in and through the Church. Yet, as John also insists, the Holy Spirit would not speak 'of himself', inventing and creating out of nothing: he would take the things of Jesus and show them in a new and living light" (p. 103). "The new and living light" are the reinterpretations and additions attributed by the Church to Jesus. Robinson made the wrong claim that Jesus did not claim anything for himself but that all He was doing was for the Father. "The one thing that is certain," asserts Robinson, "is that he did not go around 'saying he was God'" (p. 104). Relative to the teaching of John in his prologue, which clearly sets forth the deity of Christ, the reader is told that this is the language of myth and legend.'

According to the English theologian the New Testament reader must distinguish between two levels in the Gospel accounts: the historical and the theological (p. 113). Two questions with which the Gospel accounts deal are these: Who is this man? and What became of this man? Speaking on the historical level the life of Jesus of Nazareth ended with his crucifixion, death and burial. However, on the theological level something remarkable is supposed to have happened. To quote Robinson: "In one sense it is a plain story with the events of Jesus's life working themselves out to their inevitable end, and to their utterly unexpected reversal on the third day, told with an attention to detail, a restraint and lack of doctrinal elaboration which is remarkable. On the other hand, 'what became of him' was theological through and through: the Spirit, the Church, the new age, the resurrection order — that total reality — which led Paul to exclaim: "when anyone is united to Christ, there is a new world; the older has gone, a new order, has already begun" (2 Cor. 5:17). (p. 113).

The theological level according to the English divine is different from the historical one because he does not admit or permit a bodily resurrection of Jesus Christ, who was crucified under Pontius Pilate, and who on the third calendar day after his burial arose from the dead, appeared with a new glorified body, was seen and felt by reliable witnesses. On Easter afternoon the risen Jesus ate with the Emmaus disciples and also some time later with seven of His disciples on the shores of the Sea of Galilee. Robinson refuses to accept the language of the post-resurrection narratives and the resurrection accounts themselves as employing language which report true historical events. The resurrection narratives are language — so he claims — "like the language of the birth narratives, which burst the bonds of factual description" (p. 113). Robinson contends that the narratives of Jesus Christ's resurrection have been colored by interpretation, which ultimately means that the New Testament reader cannot ascertain what actually happened. According to Robinson, both in the infancy narratives and in the resurrection accounts, there is legendary material which must be separated from the factuality of the accounts. Because of this fact the scholar, according to the English Bishop, needs to use the historical-critical method and by its employment separate fact from fiction. The fictional part is then labelled "the theological interpretation".

Everything recorded in the Passion Narrative can be accepted as factual and historically true to some degree, according to Robinson. However, in these chapters reporting the events of the Holy Week, we do not possess accounts that can be assessed by "the canons of tape-recorded accuracy." There is a serious discrepancy between John and the Synoptics regarding the time when Jesus celebrated the Passover with His disciples. The rending of the veil in the Temple cannot be taken seriously as a factual happening, but according to Robinson must be understood symbolically (p. 114).

Robinson Rejects the Bodily Resurrection of Christ

The assertion that "Jesus lives" is a statement that the historian can-not make or accept as a fact, because the assertion that "Christ lives" is a theological judgment. Since Robinson claims no person saw the resur-rection happen, so he avers that it cannot be established historically. The best that the New Testament reader can do is to assert that something happened, but what that something is remains an enigma. This raises the serious question: "What are we to do with the overwhelming evidence presented by the New Testament of the Lord's resurrection?" What does one do with the statement of Peter to the congregations of Asia Minor, written in the early sixties: "Christ died once for all our sins, the Right-eous One for the guilty, to bring us to God. He was killed in His body but made alive in His spirit. In this He also went and preached to the spirits in prison" (1 Pet. 3:18, Beck translation). The angel at the grave told the women that Christ was alive and that He would meet with His disciples in Galilee. Luke, a doctor, reports that Jesus on the third day after His crucifixion appeared to Cleophas and another disciple and spoke and ate with them. On the Easter eve Jesus, the resurrected Messiah, appeared to the ten disciples and other believers and spoke to them. At least ten different times between Easter morning and the day of the Ascension Jesus gave evidence of his bodily resurrection.

The matter of the empty tomb recorded by Matthew in chapter 28 re-ceives considerable attention from Robinson. Robinson believes that the fact that the grave of Joseph of Arimathea, into which Jesus' dead body had been placed, was empty represents a reliable tradition, but why it was empty he claims, "admits of no certain explanation" (p. 123). If on Easter morn the grave was empty and this situation has been explained in various ways; either the body was stolen from the tomb (rejected by Matthew in chapter 28) or Jesus revived in the tomb, pushed the heavy rock aside, went right past the Roman guard and appeared to the women, Peter, James and the other disciples and told them a lie, namely, that He had been raised from the dead. The latter interpretation contradicts many passages in the New Testament to the effect that Jesus died. Robinson has criticized Bultmann's views of Jesus but what is the es-sential difference between Bultmann and Bishop Robinson? Robinson has a strange and irrational understanding of what the New Testament teaches by the statement: "Jesus lives!" Listen to the denial of Christ's bodily resurrection by Robinson, who writes: "The bones of Jesus could yet be lying around Palestine and the resurrection still be true. For belief in it does not depend on — let alone consist of — the fact that they do not. The empty grave cannot of itself be decisive either way" (p. 124). If the bones of Jesus could be found somewhere decomposed in Palestine then the only logical conclusion could be that Jesus never arose bodily from the dead. Then, how could the Four Gospel writers, how could Peter, Paul, John and others speak about the bodily resurrected Christ? How could Thomas place his fingers into the wounds of Jesus if Jesus was not alive. Jesus was seen by five hundred people at one time, many of whom were still living when Paul wrote his 1 Corinthian Epistle. The only reasonable explanation, for all the people who are reported as having seen Christ, is that they had hallucinations. It would be strange for so many different individuals to have the same kind of hallucinations! Yet about the appearances of Christ after His resurrection, in the opinion of Robinson, there is a great amount of subjectivity. In the end this situation makes it impossible to use the references to Christ's resurrection as reliable evidence for the fact of Christ's bodily resurrection, a historical event referred to frequently in the New Testament (some claim over 200 times), an event which Christ foretold in the days of His ministry, also foretold in the Old Testament, foreshadowed by the experience of Jonah, and depicted clearly in Isaiah 53 and Psalm 22. The very position taken by Robinson is the reason why Paul penned 1 Corinthians 15, "The Great Resurrection Chapter." In opposition to the inspired Apostle Paul, Robinson writes: "That Jesus lives, now, is a conviction that cannot finally be substantiated by any evidence from the past — only from the present."

The best antidote to this attack on the very foundation of the Christian faith is to read what Paul states in 1 Corinthians 15:12-19. Beck in his translation has rendered these all-important arguments of Paul:

If we preach that Christ rose from the dead, how can some of you say, "the dead don't rise?" If the dead don't rise, Christ didn't rise. And if Christ didn't rise, our preaching means nothing, and your faith means nothing. And we stand there as men who lied about God, because we testified against God that He raised Christ, whom He didn't raise if it is true that dead do not rise. You see, if the dead did not rise, Christ didn't rise. But if Christ didn't rise, your faith can't help you and you are still in your sins. Then also those who have gone to their rest in Christ have perished. If Christ is our hope only for this life, we should be pitied more than any other people.

On the last page of his book, Robinson makes the claim that "there is nothing fixed or final; our knowledge and our questions are constantly expanding and shifting. And who knows what new evidence may not suddenly be dug up?" (p. 134). If the New Testament believer must assume that his knowledge of religious truths is subject to change and is tentative, how can Robinson claim the New Testament can be trusted? He advocates a theology of uncertainty and ultimately of despair!

Questions
1. Bishop John T. Robinson wrote that the entire New Testament was available by ____.
2. Robinson placed the Gospel of John between ____.
3. Robinson held to an ____ bias.
4. While Robinson claims the New Testament is trustworthy, yet he shows ____.
5. Robinson does not admit or permit a ____.
6. Robinson advocates a theology of ____.

In What Language Did Christ and the Apostles Speak?
Is This Language Still in Use?

Christian News, September 4, 1978

The death of Pope Paul VI on August 6th has ushered in a period of a number of weeks during which the Roman Catholic Church and its doctrine of the supremacy of the papacy will be held before the eyes of the world. Fort Wayne's News Sentinel (evening paper) on page 1 of the Monday, August 7th edition states the view of the current Roman Catholic position of the place of the Pope in the life of the Roman Church. It gave the UPI's statement from Vatican City, which read as follows.

A Pope in the teaching of the Roman Catholic Church is Christ's representative on earth, with full powers to teach and govern the world's 700 million Roman Catholics in spiritual matters.

Catholic doctrine says St. Peter, the Galilean fisherman who was the first Pope, received his authority directly from Christ with these words: "You are Peter, and upon this rock I will build by church ... I will give you the keys of the kingdom of Heaven."

Catholics believe that God-given authority continued in all Peter's successors, the Pope and bishops of Rome.

"The Roman pontiff has full and supreme power over the universal Church in matters pertaining to the faith and morals (teaching authority) and to the discipline and government of the Church (authority of jurisdiction)," the U.S. Catholic Almanac says.

"The primacy of the Pope is real and supreme power. It is not merely a prerogative or honor - i.e. of his being considered as the first among equals. Neither does primacy imply that the Pope is just the presiding officer of the collective body of bishops. The Pope is head of the church."

Adding to the Pope's enormous spiritual powers is the doctrine of papal infallibility, which the church says has been a matter of belief since the beginning of Christianity but was explicitly defined only by the first Vatican Council.

The doctrine does not mean that anything that the Pope says on any subject is necessarily right. Papal infallibility means the Pope cannot err when speaking "ex cathedra" or "from his throne"– solemnly and intentionally on a matter of faith or morals.

In addition to his spiritual powers, the Pope also is the absolute monarch of one of the world's smallest stales — 108.7 acre Vatican City, which is surrounded by the city of Rome.

In an editorial appearing in the same paper for August 7 an editorial entitled "Pope Paul VI," among other assertions, the editor said: "He was the successor of Peter and in some respects Paul's turn as the Bishop of Rome and head of the Catholic Church was not unlike that of Peter's 2,000 years ago. The times were unsettling, to say the least" (Page 4A). When do the events of other Christian denominations constantly rate front page coverage? The Fort Wayne Journal-Gazette (a morning paper) gave a report of the Pope's burial in its Sunday edition, August 12th. As a part of the report from Vatican City, it was stated that "funeral prayers were said after the Mass by Cardinal Ugo Poletti, the Pope's vicar for Rome, who paid homage when the body was brought Wednesday from the papal summer residence at Castel Gandolfo, where the pontiff died at the age of 80 last summer after a heart attack. Other prayers were recited in Arabic and Aramaic the language spoken by Christ and his disciples by the Maronite patriarch of Antioch, Atione Pierre Khoraiche."

Since the Maronite patriarch of Antioch, Antione Khoraiche is reported to have prayed in Arabic and in Aramaic, the language of Christ and the apostles, it is important to give the origin and development of the Maronites. The Maronites have always used Syraic and not Palestinian Aramaic, the language of Christ and the apostles, and the writer of the description of the Requiem Mass for Pope Pius VI was giving a piece of misinformation and at the same time giving the false impression that the Maronites were going back to the time of our Lord and His disciples, thus supporting the alleged apostolic beginnings of the Roman papacy.

The Maronites were founded by a monk Maro who organized his followers into a separate community. Maro died in the fifth century. In 681 the Maronites separated themselves from other Syriac-speaking Melchites. When in A.D. 1182 the Maronites came into contact with the Crusaders, they accepted the supremacy of the Pope and have been in communion with Rome ever since.

The Maronites are an Eastern rite of Christians who are chiefly distributed in Lebanon, Syria, Palestine, and Egypt. This Eastern church is governed by a patriarch who is elected by the bishop of the Maronite church. The bishops are in turn nominated by the patriarch. The Maronite diocesan clergy is married, and their religious follow the Rule of St. Anthony.

The Westminster Dictionary of Church History states that "the Maronites believe that they have constantly retained perfect orthodoxy, but this is hotly disputed by other Christians of the area. The most damaging evidence comes from William of Tyre who reported that they adhered to the Church of Rome in 1182," p. 527.

However, beginning with the 16th century the Maronites' adhesions to Rome are clear. Church historians date this beginning with the Eleventh Session of the Fifth Lateran Council. A Maronite College was founded in Rome by Pope Gregory XIII, a college to which six priests could be sent by the Maronite bishops.

Al Kannobin in 1616 a general council of the Maronites passed canons against abuses and introduced mild liturgical reforms. Roman influence on the Maronite liturgy is apparent in the use of unleavened bread, Lord's Supper under one kind (Communion under bread and wine have been prohibited since 1736).

Patriarch Joseph IV in another council (1733-1742), endeavored to foster other reforms consonant with Roman theology: marriage dispensation sold for money, the remarriage of priests, the employment of Arabic in

1267

the liturgy by Maronites of Aleppo. This council also introduced the filioque in to the Creed and insisted that the altar bread was to be circular as was the case in the Roman liturgical practice.

The Uniate Churches

The Maronites are one of a number of Eastern Christian Churches who have accepted union with Rome. The beginnings of these Roman Catholic Uniate Churches have their origin in the lime of the Crusades. Other groups of Syrian Christians who were accepted in union with the Roman Church occurred in 1661, with the election of the first Catholic patriarch. In 17th century the Melkites undertook steps toward union with Rome, a union which was firmly established when Cyril IV was elected patriarch for the Melkites in 1724. "Melchites" is a term applied to the Syriac Orthodox followers of the Council of Chalcedon. The term Melchite comes from the Syriac root melek, "king." The Melchites were called "royalists" or "imperialists" because the latter were supported by the Byzantine emperors. However, since the 17th century the term Melchite has been employed primarily for that section of the Syrian Orthodox Church which has accepted papal jurisdiction.

The Jacobites, the Nestorians, the Maronites and the Melchites who shared differed views on the nature of Christ's person, all used Syriac, a major dialect of Aramaic, which is distinguished from Palestinian Aramaic spoken by Christ and the Apostles. When for example, Dr. Arthur C. Piepkorn, in *Profiles in Belief,* Vol. I, p. 5, in writing about the origin of the early Aramaic-speaking Church of the East was in the territory under Parthian rule," he was using imprecise language when calling this church which used Syriac as "an Aramaic speaking Church." That is about as accurate as if a writer would say that Pope Pius VI spoke to the Italians in a language that had a Latin base and used words that were derived from Vulgar Latin, when he wished to say that the Pope spoke in Italian.

Specialists in Semitics distinguish a number of dialects in the Aramaic subdivision of the Semitic languages. Professor Yamauchi in an article entitled, "Greek, Hebrew, Aramaic, or Syriac?" wrote: "Aramaic is a Semitic language, originally of the Arameans of Syria, a territory which extends from Damascus to the western Euphrates in northwest Mesopotamia. Aramaic, which was written in a variety of alphabetic scripts which were more efficient than the Akkadian or Persian cuneiform scripts, became widely used in the first millennium B.C." (Bibliotheca Sacra, October 1974, p. 323).

The most recent classification of available Aramaic materials is the following: 1. Old Aramaic (925-700 B.C.); 2. Official Aramaic (700-800 B.C.). During the Assyrian and Persian Empires, Aramaic was employed by the court and chancelleries from Asia Minor to Afghanistan. The Elephantine Papyri and the Aramaic of Ezra and possibly also the Aramaic of Daniel. 3. Middle Aramaic (300 B.C. -A.D. 200). Here included is the Aramaic of the New Testament, the Aramaic of Qumran and Murabba'at, the inscriptions of Nabatea, and Palmyra. 4. Late Aramaic (A.D. 200-

700). This is divided into a Western: Syro-Palestinia Christian Aramaic, Samaritan, Palestinian Jewish Aramaic; and b. Eastern: Syriac, Babylonian Talmudic Aramaic and Mandale. 5. Modern Aramaic. Aramaic is spoken in isolated villages of the Anti-Lebanon Range of Syria, in Kurdistan and Azerbaijan, and north Mosul in Iraq.

When the statement was made orally to millions of people in 49 countries that patriarch Antione Pierre Khoraiche "recited prayers both in Arabic and Aramaic, the language spoken by Christ and his disciples," this presents a problem, because the Aramaic of the New Testament is limited to a few phrases and to the Aramaic form of a number of proper nouns. While it is true that Christ and His disciples spoke Palestinian Aramaic, we have the sayings (50% of John's Gospel consists of sayings and speeches of Jesus) in the Greek New Testament. Thus according to Mark (7:34), Jesus said to the deaf man "Epphatha," He was speaking Aramaic "itpattah," open yourself ; or when he commanded the young maiden to arise and said to her: "Talitha kumi." Jesus was uttering in Aramaic "talyeta quimi" (Mark 5:41). The Hebrew word for maiden is quite different than the Aramaic. On the cross, Jesus was speaking Aramaic when he exclaimed: "Eloi, eloi lema sabaxthani," he was not uttering the verse from Psalm 22 in Hebrew but was saying in Aramaic: "Elahi, Elahi, lema shebaqtani."

Other indications of the use of Aramaic are found in the transliterations of proper names : Peter's name Cephas is from Aramaic **kepha** "rock;" Thomas is from **toma** "twin;" Matthew is from **Mattai**. The word **bar** is the Aramaic for "son," and is found in such proper names as: Bartholomew, Bar Jonas, Barsabbas, Barabbas. Golgotha is from Aramaic **golgolta** "skull. " Akeidema is from **haqel dema** "bloody ground. "The expression that Paul used at the end of I Cor. 16: 22 "Maranatha" is from "Maran," "Our Lord" and "eta" "come."

The Christian Church has no writings in Aramaic from either Christ or His apostles. There have been various ingenious attempts to reconstruct the Aramaic of the Gospels. In 1922 Charles F. Burney in his *The Aramaic Origin of the Four Gospels* (Oxford, 1922), argued that the Fourth Gospel was based on an Aramaic original. In 1933, a Semitic and Aramaic specialist, Charles Cutler Torrey, put forth the theory that all four gospels in Greek were translations from Aramaic originals in his book *The Four Gospels*. The arguments of Burney and Torrey are based on Semitisms and alleged mistranslations of the Aramaic supposedly found in the Greek text. In 1946 Matthew Black, in his *An Aramaic Approach to the Gospels and Acts*, has attempted to establish an Aramaic background for these five Biblical books. A second edition appeared in 1954 and a third edition in 1967.

The use of Syriac and Arabic accords with what is known about the history of the Maronite Uniate Church. Expert scholars have failed to establish a case for Aramaic originals for the Gospels, Acts or any other book of the New Testament.

There is a layman, G. M. Lamsa, a member of "The Church of the East," which is the Assyrian Church of Iraq, who claims his church uses the language of Jesus. In 1920 this ancient Christian Church numbered about 200,000 but with Iraq's independence, the Muslims slaughtered thousands of these Nestorians and reduced their number to 80,000. G.M.L makes the claim that the Gospels, the Epistles of Paul were written in Aramaic. The original Aramaic is alleged to be found in the Peshitta. In one of his books, *More Light on the Gospel* (Garden City, 1968), he made this claim about himself as follows: "The author through God's grace is the only one with the knowledge of Aramaic, the Bible customs and idioms, and knowledge of the English language who has ever translated the Holy Bible from the original Aramaic texts into English, and written commentaries on it, and his translation is now in pleasingly wide use," pp. xxviii-xxix . Lamsa has published an English translation of the Syriac Peshitta, entitled *The Holy Bible from Ancient Eastern Manuscripts. Containing the Old and New Testament Translated from the Peshitta. The Authorized Bible of the Church of the East.* (Philadelphia: A. J. Holman Company, 1957).

Lamsa has ignored the history of the Syriac text because scholarship has shown that the Sinaitic-Curetonian Syriac texts are older and superior to the Peshitta. New Testament critics are convinced that the Syriac Peshitta is a translation from the Greek and cannot qualify to represent the language spoken by Jesus and His Apostles.

If this statement that the prayer of the Maronite patriarch of Antioch was delivered "in Aramaic, the language of Christ and His disciples," originated with the patriarch himself, then possibly he may share a similar position as that expressed by Mar Eshai Shimun, Catholicos of the East, in a letter   published by Lamsa on page ii of the *Holy Bible - From Ancient Near Eastern Manuscripts*, which reads as follows:

> With respect to your letter concerning Lamsa's translation of the Aramaic of the Bible, and the originality of the Peshitta text, as the Patriarch and Head of the Holy Apostolic and Catholic Church of the East received scriptures from the hands of the blessed Apostles themselves in the Aramaic original, the language spoken by our Lord Jesus Christ himself and that the Peshitta is the text of the Church of the East, which has come down from Biblical times without any change or revision.

However, whether the Maronite patriarch made the assertion himself or someone else made this claim, the fact is that the language of Christ and His Apostles was not used, but what the world heard was a prayer in a related Eastern dialect of Aramaic composed many centuries later, and not in the western Aramaic of New Testament times.

The claim that Paul VI was Peter's successor and Christ's vicar is based on tradition and has no warrant in the New Testament. No passage states that Peter was the head of the church and that before his death he was instructed to appoint a successor and that this was to be an ecclesiastical practice to be continued until Christ would abrogate it. Both claims, namely that the Pope is Christ's representative on earth, and that the language of Christ and His Apostles was used at the Requiem Mass are fiction.

Questions

1. Rome teaches that the Pope is ____.
2. Papal infallibility means that the Pope ____.
3. Who are the Maronites? ____
4. "Melchites" is a term applied to ____.
5. Dr. Arthur Carl Piepkorn was using ____ language when calling a church which used Syriac as "an Aramaic speaking Church."
6. The Aramaic of the New Testament is limited to ____.
7. Christ and his apostles spoke ____.
8. What did G.M. Lamsa claim? ____
9. The Syriac Pishita is a translation from the ____.
10. The claim that Paul VI was Peter's successor and Christ's vicar is based on ____.

Apologetics and Polemics

February 24, 1979
Dear Brother Otten:
Enclosed please find a manuscript, entitled "Need We Apologize for Apologetics or Polemics." I believe this in an essay, which supports the reason for your publication of the *Christian News* and before that of the *Lutheran News*.

I believe it will also be of help to use the importance of fighting for historic Lutheranism position of our Synod. The last half of the essay deals specifically with the LCMS and would be a good preparation for delegates to the St. Louis Convention, July, 1979. You no doubt, if you decide to publish it, would reproduce it in 2 or 3 installments. Considerable research aimed effort went into the writing of this essay. I hope you use it.
Fraternally,
Raymond F. Surburg

Need We Apologize for Apologetics and Polemics?

Christian News, **March 26, 1979**

The Christian Church, its pastors, teachers, and laity may be said to have two duties: to proclaim the Word of God in its truth and purity and to defend it against the attacks of those who reject, misinterpret or redefine the Bible's truths.

Jesus Himself called attention to those who were being victimized by the Devil, who sows the seed of false doctrine and causes the weeds of unbelief and religious error to grow up. Throughout His public ministry Jesus was constantly opposed by Satan who utilized Herod, the Jews and their leaders and on occasion even Christ's own disciples to try and hinder Him from accomplishing His mission of the redemption of mankind. Thus Jesus said: "The kingdom of heaven is likened unto a man who sowed good seed in his field. But while he slept his enemy came and sowed tares among the wheat" (Matt. 13:24). There never has been a time in the history of the Church when the truth of God's Word did not need defending, because the Devil is always active and people are willing to listen and believe teachings which are novel though they may be radically opposed to Biblical doctrine and teaching.

Jesus admonished His generation against its religious leaders. The Scribes and Pharisees had added traditions to the Scriptural text as additional sources for religious teaching. Jesus warned: "In vain they do worship me, teaching for doctrine the commandments of men" (Matt. 15:9). Again he warned: "You reject the commandments of God that you may keep your tradition . . . making the Word of God of none effect"

(Mark 7:13ff). Again Jesus warned: "'Beware of false prophets that come to you in sheep's clothing but inwardly they are ravening wolves."

Between A.D. 30 and 100 false doctrines and false teachings were being advocated and promoted, as may be seen from warnings in various New Testament epistles. In his farewell address to the Ephesian elders, Paul predicted: "Take heed to yourselves and to the flock over which the Holy Ghost has made you overseers to feed the church of God which he has purchased with his own blood. For I know this, that after my departure shall grievous wolves enter, not sparing the flock. Also of your own selves shall men arise, speaking perverse things to draw away disciples after them. Therefore watch, and remember that by the space of three years I ceased not to warn every one day and night with tears" (Acts 20:28-31).

When the doctrine of justification by faith, the heart of the Good News was under attack by false teachers in the Galatian churches, Paul asserted: "I marvel that you are so soon removed from him that called you into the grace of Christ unto another Gospel... There are some that trouble you and would pervert the Gospel of Christ. But though we or an angel from heaven preach any other Gospel let him be anathema" (Gal. 1:6-8). Warnings against false teachers and doctrines are numerous in the New Testament books. Jude, who proposed to write a letter dealing with "our common salvation" was forced to pen a letter, encouraging his readers to contend for the faith once delivered to the saints" (Jude 3). Paul in his three letters sent to Timothy and Titus has at least thirteen warnings against false doctrine and gives exhortations to teach and proclaim sound and healthy doctrine.

Gnosticism, in its incipient stages, seems to be referred to in the writings of Paul and John.[1] Gnosticism was a philosophy of religion which combined elements from Judaism, Christianity, Mithraism, Zoroastrianism, and Greek philosophy. The Gnostics claimed to have a higher type of knowledge than that advocated by Christianity. The crucial issue in Gnosticism was the denigration of Jesus to the rank of one of the angelic beings (cf. Col. 1:13-17; 2:9). The chief god of unionistic and syncretistic Gnosticism was the Pieroma, the Light-God, from whom emanated angelic beings, of whom Paul said that in Jesus resided the entire fullness of (Pieroma) of the Godhead bodily. Gnosticism manifested itself in two directions in the Roman Empire: 1) in a hedonistic and 2) in a mortifying of the flesh type. Paul warns against emphasis on mortification of the body, while Jude and Peter warn against the hedonistic kind.[2] Gnosticism in Asia Minor was also characterized by the denial of the true humanity of Christ. The type of Gnosticism denounced by John in his Epistles did not accept the divine-human Christ. It did not hold to the two natures, the divine and the human in one theanthropic person, John labelled those who denied that Christ had come into the flesh as antichrists (1 John 4:3). A number of New Testament writings reflect an apologetic tone and character.[3]

Christianity has always met with opposition and Christianity must be prepared to encounter this opposition with a justification of its position to other religions as well as to the believers. Especially when we attempt to convince other people, it is always necessary to heed the apostolic injunction "to give an answer to every man that asketh a reason concerning the hope that is in us, yet with meekness and fear, having a good conscience" (I Pet. 3:15).

The renowned Reformed theologian Phillip Schaff claimed that "even if idolatry should entirely disappear, there would still be room for Apologetics to satisfy the mind of the believers. It is an integral part of theology as a science."[4] Augustine wrote: "The fact that it is *God* that giveth the increase is no reason why we need not to plant and water."[5]

There is a great need for apologetics. John D. Jess, radio preacher, wrote: "I fail to see how the element of defense can be successfully extracted from effective Gospel preaching, for we certainly live and witness in a hostile world, one which rejects, in the main, the historic Christian message. To effectively contravene through hostility, one must anticipate the question and objections of the skeptic; or to put it in another way 'exchange' minds with him. In this we identify analogously with the dexterous boxer who anticipating his opponent's style, effectively evades his blows and successfully counterattacks."[6]

Christians were accused of incestuous practices and of cannibalism. The latter charge was occasioned by the Christian doctrine of eating Christ's body and drinking His blood. When Christians refused to worship the emperor, they were accused either as being poor citizens of the state or as traitors, because they would not participate in certain state practices. The Christian's refusal to acknowledge the emperor as god, or the other gods of the Roman religion resulted in the charge of atheism being levelled against them The eschatological preaching and teaching about the destruction of the world by fire brought about the charge that Christians were sponsoring revolution and incendiarism.

After the death of the apostles there arose a group of writers in the second century (c. 120-220), who wrote in defense of the Christian faith against the attacks of Judaism, paganism and Christian heresy. The following are known to have defended the Christian faith in this period: Aristides, Aristo of Pella, Justin Martyr, Quadratus, Tatian, Miltiades, Appolinaris of Hieropolis, Athenagoras, Theophilus, Melitus, Hermias, Minucius, Felix, and Tertullian. With the exception of Tertullian, these writers were not primarily theologians.

The religious and political situation of the second century forced Christians to produce an apologetic literature. Concerning this situation Knudsen wrote: "An increased Messianic expectation on the part of the dispersed Jews prompted the Christian writers to confirm by scriptural proofs the Messiahship of Jesus Christ. Outstanding pagan writers made the charges (which became deeply imbedded in the public mind) that Christianity promoted impiety to the gods, that it engaged in immoral practices (e.g. cannibalism and sexual deviations), and that because of its rejection of emperor worship it was treasonable to the state." These charges were answered by the Apologists in this way, that they demonstrated the superiority of Christian morals as compared with the heathen and contended that this higher and superior type of morality flowed from the nature of Christian monotheism.

Nearly all of the second century Apologists fought against Gnosticism, which constituted a serious threat to Christianity.

While the term "Apologists" is frequently treated as a technical name for the second century writers who defended Christianity against attacks from paganism, the designation is also employed of later writers in defense of the Christian faith. St. Augustine's famous City of God was written by the North African church father to counteract and refute the charge that the fall of Rome was due to failure to render homage to the Roman gods, who punished the Roman Empire because of its impiety shown them.

Down through the last nineteen hundred years the varieties of the Christian religion have been subject to attack from many different religions and peoples. Schaff has listed these as follows:

> The attacks upon Christianity may proceed from rival religions, as Judaism, Heathenism, Mohammedenism; or from the various forms of infidelity within the Church, as Deism, Rationalism, Pantheism, Atheism, Materialism, Agnosticism. The former attacks have long since been overcome, or have ceased to be formidable; the latter are still going on and will continue to the end of time... Every age must produce its own apologies adapted to prevailing tendencies and wants.[7]

Some theological writers have distinguished between popular apology and scientific apologetics. It is probably better to employ the word apology as the defense against a particular set of objections and apologetics as the science of the principles on the basis of which apologies are to be made.

Apologetics may be defined as a philosophical and historical science, because it derives its proofs from that which is internal, reason and conscience, as well as from the external history. By contrast, polemics, is the science of theological warfare, which is directed against error and misapprehension of the Biblical truth within the church. Irenics has as its objective to present points of agreement among Christians with a view to ultimate union.

The nineteenth century scholar Ebrard in his classic *Apologetics, the Scientific Vindication of Christianity* distinguished as follows between apology and apologetics:

> Christian apologetics is distinguished from the mere apology in this, that it is not determined in course and method by attacks appearing casually at any point of time but from the very nature of Christianity itself and deduces the method of the defense of the same, and consequently the defense itself. Every apologetics is an apology; but every apology is not an apologetics.[8]

Apologetics is a theological discipline that has a special value for members of the church because the uncommitted and unbelievers do not as a rule read theological literature. The writings of Christian apologetics aids in strengthening and firming up the faith of the believer, provides him with a clearer grasp of the foundations of this faith, and also supplies him with weapons wherewith to confute the adversaries.

The History of Apologetics since the Apostolic Age

Between A.D. 30 and 313 Christians had to defend themselves against the false accusations made by heathen and Jews and this fact explains why the first type of Apologetics was of a defensive character. Among the early apologists were Justin Martyr, Athenagorus, and Clement of Alexandria, all who wrote in Greek; and among those writing in Latin were Tertullian, Cyprian, and Lactantius. Beginning with Constantine the Great to about A.D. 450 Christian apologetics experienced further development. In this period Apologetics became polemical. Athanasius, Augustine and Cyril of Alexandria were outstanding apologists for the Christian faith during this time span.

In the Reformation days Luther and the Reformers of other Protestant lands found it necessary to engage in polemical and apologetical teaching, preaching and writings. Eventually Luther came to defend the great *solas, sola Scriptura, sola gratia, sola fide, and Solus Christus*, against the Church of Rome which excommunicated him on December 10, 1521. Over against Zwingli he upheld the doctrine of the Real Presence of Christ in the Lord's Supper. Luther opposed the Anabaptists and enthusiastic groups because of their failure to do justice to the Means of Grace.[9] The *Formula of Concord* was written to defend the correct Biblical position on a number of doctrines, which had been attacked and misrepresented by Lutherans between 1546 and 1577.[10] The Lutheran Confessions do not only set forth what Lutherans believe to be the correct Biblical doctrine, but they also denounced wrong and heretical doctrinal positions. Apologetics and polemics characterize the writings and confessions found in *The Book of Concord* of 1580.

In the 400 years that have elapsed since the adoption of the *Book of Concord*, Lutherans have found it necessary to defend the Biblical teachings set forth in the Lutheran Confessions. Dr. Spitz writes about the period following the signing of the *Formula of Concord*: "After a long series of theological controversies, the Lutherans had finally achieved doctrinal unity in a collection of symbolical writings. A spiritual blessing which had been acquired with such great efforts had to be properly safeguarded against threats from within and without. This determination accounts for the development of the large theological literature during that period."[11]

In the seventeenth century the peace and calm that had been effected by the signing of the *Formula of Concord* was broken by two Christological Controversies, the Cryptist and the Kenoticist Controversies between the Giessen and Tuebungen theologians (1619-1626) and the so-called Luetkemann Controversy (Luetkenunn, died 1655).[12]

A far greater threat to sound Lutheranism was the theology propounded by Georg Calixt (1586-1656). He became the sponsor of Melanchthonian syncretistic theology in German Lutheranism. According to *The Lutheran Cyclopedia* (1954). Calixt's "main idea was that the prime object of theology was not so much purity of doctrine as a Christian life; hence his unionistic tendency towards the Catholic and Reformed Churches. At the Convention of Thorn he sided with the Reformed delegates where also as before he advocated, as a basis for union, the teachings of the Church of the first five centuries."[13] Calixt took the position that only the doctrinal content of the Bible was inspired, while in other matters the writers had merely been governed and kept from error by the Holy Spirit. Strict Lutherans opposed this as syncretism and crypto-Calvinism.

Johann Andreas Quenstedt (1617-1685), a nephew of Johann Gerhard, was educated under Calixt but afterwards, at Wittenberg refuted the syncretistic tendencies of his former teacher. Quenstedt has been called the "Bookkeeper of Lutheran orthodoxy." His most famous writing was the *Theologica Didactico-Polemica sive Systema Theologicum*, a work which came to be recognized as a standard of Lutheran orthodoxy. The definitions and those of Quenstedt were based on J. F. Koening. Quenstedt became noted especially for his quiet, mild and irenic disposition."[14]

Apologetics as well-organized discipline came into its own in the seventeenth century with the work of Hugo Grotius, *Truth of the Christian Religion* published in 1627. Before that time the Reformers had been mostly preoccupied with the great questions that arose within the Church because of the break with Rome. Grotius' book was designed for seamen who came into contact with Mohammedans and heathen. Pascal in his *Thoughts on Religion (1669)* has given defenders of the faith hints on the true nature of apologetics.[15]

In the various and voluminous writings of the Lutheran dogmaticians of the 16th and 17th centuries there were to be found apologetical and polemical assertions in defense of the Biblical faith as well as attacks upon erroneous teachings. This holds true of the published works of the following theologians: Luther, Melanchthon, Chemnitz, Hunnius, Hutter, Gerhard, Koenig, Calov, Quenstedt, Hollaz and Baier.

The Age of Orthodoxy was followed by emphasis on correct living (pietism) to the detriment of correct formulation.[16] Lutherans of whom the latter judgment were true were: Spener, Francke, and Freylinghausen.[17] The over emphasis on man led to rationalism, which first endeavored to explain doctrines in a rational number. During the period between 1650 and 1825 Deism and Naturalism had a great influence in England, France, and Germany, and a large number of apologetical works were published, especially in England. In England among the most prominent defenders of the Christian religion were: Cudworth (d. 1688), Stillingfleett (d. 1699), Lock (d. 1704), Waterland (d. 1740), Butler (d. 1752), the author of the famous *Analogy*, Lardner (d. 1768), Warburton (d. 1779), the founder of the "Warbutonian Lectures," George Campbell (d. 1796) and Paley (d. 1805)."[18]

Rationalism became the great foe of orthodox Christianity. In the 17th century attempts were made to portray Christianity as a reasonable faith, as Joseph Glanvill did in his Anti-Fanatical Religion and Free Philosophy; John Tillotson; John Locke, *Reasonableness of Christianity*;

Samuel Clarke, in *The Truth and Certainty of the Christian Revelation*. While revelation was not rejected, the stance was taken that revelation was in harmony with reason. (This came to be known as rational supernaturalism).[19]

With the emphasis then upon place of reason in Christian theology, the next step was making reason supreme and thus led to pure rationalism. A step in the transition from rational supernaturalism to deism was the work of John Toland's *Christianity Not Mysterious*, a book that argued that revelation could offer nothing above reason. Thus there came the rejection by rationalistic deism of Biblical revelation, although the deistic system was somewhat tempered by the five doctrines of "universal religion."[20]

The next step in the downward trend of Biblical theology on the European continent was the rejection of all dogmatic assertions, which meant that rationalism progressed to a skeptical position on matters of religion and theology.[21] The following were men who fostered and supported this phase of rationalism: Thomas Hobbes, David Hume, Berhard de Fontenelle, Pierre Bayle, Francois Marie Arouet de Voltaire, Denis Diderot, Michel Eyquem Montaigne, Claude Adrien Helvetius, Frederick the Great, Ethan Allen, Elihu Palmer, Gouverneur Morris and others.[22]

Germany, the home of the Reformation, saw the attacks of Reimarus (d. 1768) on traditional Christianity by the former's Wolfenbuettel Fragments, which were published by Lessing in 1774-1778.[23] These called forth a number of apologetical writers, the most prominent authors of whom were men like A.F.W. Sack, Lilienthal (16 vols., 1750-1782), Noesselt, Haller, Less, Roos, and Kleuker.[24]

The chief theologians of the 18th century in the development of rationalism were J. A. Ernesti (d. 1781), who specialized in New Testament interpretation; J. D. Michaelis (d. 1791) in Old Testament interpretation;[25] and Johann Semler, who was prominent in Biblical and historical criticism; and Wilhelm Gesenius (d. 1842), who trade some outstanding contributions to Old Testament and Semitic philology.[26] The latter's *Hebrew Grammar* and *Hebrew Dictionary* was very influential and became the basis for numerous subsequent grammars and dictionaries.[27]

With Rationalism there came to be allied a type of literary Biblical criticism of both the Old and New Testament which had devastating effects on Biblical studies and on Biblical doctrines. Dr. Kramer of Springfield, gave the following evaluation of Semler's work:

> Johann Salomo Semler (1725-1791) is acknowledged as the father of the modern historical and literary criticism of the Bible. He was raised in pietistic surroundings, but drifted more and more into rationalism although he despised the vulgar rationalists, and according to his own testimony at least desired to hold to the fundamental Christian doctrines.[28]

Dr. Kramer pointed out that while Semler had forerunners in the matter of the historical and literary criticisms of the Scripture, he is generally acknowledged as the man who helped these theories to triumph in the Protestantism of his age. Professor Kramer in his essay. *The Introduction of the Historical-Critical Method and Its Relationship to Lutheran Hermeneutics* showed how Semler's rationalism affected his hermeneutics and his understanding of the canon, and how it influenced his grasp of Biblical doctrine. In his summarization Dr. Kramer evaluated the Lutheran Semler as follows:

> From all this it is evident that while Semler thought of himself the foe of rationalism, who would defend the Christian truth against the attacks of the vulgar rationalists, he himself gave up one article of the faith after the other, and played himself into the hands of the rationalists. Above all things Semler denied the divine inspiration of the Scripture as a whole, brought in a conception of inspiration that differs not only from the exaggerated concept of orthodoxy, but also from the claim of Holy Scripture itself.[29]

The result of the employment of the historical critical method developed by the English deists, the pantheistic Spinoza and the rationalists Lessing and Reimarus have been correctly stated by Kuhl:

> They changed the old conception of the authority of Scripture and especially the idea of the inspiration: henceforth not the thing, but the man, the biblical writers, were held to be inspired. The Bible remains it is true, the word of God, but now in the different sense that it was bound up with man, experienced and proclaimed by man, at the most diverse times and under the most diverse conditions. The authors as persons are considered to be inspired, intimately linked with God by their piety and speaking and writing on His behalf: the books of the Bible are no longer regarded as self-revelation by God, but as historical records of revelation experienced by man or better, as testimonies of revelation.[30]

Rationalism suffered a serious set-back at the hands of two of its most famous disciples. Immanuel Kant, who while elevating human reason, also showed its limitations in spiritual matters.[31] In *Kritik der Reinen Vernunft* which is a critical, destructive argumentation, Kant proceeded to demonstrate that the transcendent world, the existence of God and immortality of the soul were unknowable to pure reason. In *Kritik der praktischen Vernunft* he endeavored to undo the damage brought about by the argumentation of the *Kritik der Reinen Vernunft*, by showing that freedom of man, immortality of the soul, the existence of God (the three great principles of the Enlightenment) are postulates of the practical reason, i.e. of conscience. *In Religion innerhalb der Grenzen der blossen Vernunft Kant* claims that morality is the essence of religion. Saving faith he made equal with a God-pleasing life.[32]

K. A. von Hase (d. 1890), a brilliant church historian, brought about the death of vulgar rationalism with his *Hutterus Redivivus*.[33] The nineteenth century witnessed other attacks upon Biblical faith, emanating especially from the side of idealistic philosophies. F. D. E. Schleiermacher (d. 1834) contributed to the death of rationalism by his emphasis on feelings as a source and criterion for religion. Dr. Spitz says about Schleiermacher's effect on rationalism as follows: "F.D.E. Schleiermacher, sometimes styled the father of modern religious liberalism, made his contribution to the decline of the Old Rationalism by identifying religious belief with religious feeling, thereby getting away from Rationalism's— one-sided emphasis on reason."[34]

However, while new winds of philosophy and teaching were blowing in Europe and in Germany, it must not be forgotten that the old Lutheran orthodoxy was not dead. This was forcefully shown by Claus Harms (d. 1855) who protested against religious indifference with his "Ninety-Five Theses."[35] Harms had grown up under rationalistic influences at Kiel and had passed from rationalism to positive Lutheranism. Influenced by Schleiermacher's *Reden ueber Religion*, he was led to a sound Lutheranism as a result of Scripture study which brought about his complete conversion. Since Harms was convinced that the Lutheran Church had forsaken the heritage of the Reformation, he published for the tercentenary jubilee of 1817 Luther's Ninety-Five Theses together with 95 of his own, directed against rationalism.[36] However, Rationalism did not die so that two distinct streams of theological thought are found side by side in Lutheran lands of Europe in the 19th and 20th centuries.

Influential theological thought in Germany and other West European countries derived its pattern of development from the influences of Immanuel Kant (d. 1804), G.F.W. Hegel (d. 1831) and Schleiermacher (d. 1834). Kant emphasized the importance not of creed but of moral precepts.[37] Although not considered a member of the idealistic coterie, Kant by his emphasis on the idea of God, freedom, and immortality gave impetus to the movement that came to play an important role in the 19th and 20th centuries, namely, idealism. Schleiermacher gave theologians and religionists a system of dogmatics which many found useful, despite its anthropocentric emphasis in theology.[38]

The great leading thinker in philosophical circles was Georg Wilhelm Friedrich Hegel (1770-1831), the main exponent of Absolute Idealism in philosophy.[39] *The Concordia Cyclopedia* (1926) characterized Hegel's philosophy as follows: "Everything that exists is the result, ultimately of the development of one absolute thought or idea, or, expressed in terms of religion, the world, including nature and humanity, is only the self-manifestation of God."[40] Although the claim was made by Hegel and his followers that his interpretation of religion and the Bible was in harmony with the Christian doctrines of the historic Christian Church, yet many considered his writings on religion and theology as the most rational explanation of Christianity and that Hegel had perfectly united religion and philosophy. Hegel's idealism and the variant positions developed from it, were pure pantheism, and therefore a complete negation of Christianity. Hegel did not believe in a concrete historical Jesus and in the Neo-Hegelian school his philosophy led to a destruction of the historical foundations of the Christian faith. Spitz correctly claims "that Hegel converted historic religion into philosophical and rational ideas and stimulated the tendency to pantheism in theology."[41]

To the Hegelian camp belonged D. F. Straus (d. 1874), whose Leben Jesus clearly showed what kind of clash was involved when trying to explain the Gospels along Hegelian philosophical lines.[42] Wellhausen responsible for the popularization of the critical approach to the Old Testament clearly showed the influence of Hegel's development evolutionism, which the former applied to the history of Israel. Another offspring of Hegel was to be seen in the views of Ferdinand Bauer and the Tuebingen school of Biblical criticism. According to F. Bauer Christianity is not a perfect religion, but a religion in the process of development according to Hegel's triadic law of thesis, antithesis and synthesis. Other scholars under the influence of Hegel were O. Pfleiderer (d. 1909), L. Feurerbach (d. 1872) and Bruno Bauer (d. 1882).[43]

The influence of Kant and Hegel were not the whole story in theology in Germany. The first half of the 19th century also saw a revival or religion throughout Germany. Lutheranism became active and manifested itself in two forms: In the more conservative "Repristination Theology" and in the more liberal Erlangen School.[44] There were a number of scholars who endeavored to restore Lutheranism to its 16th and 17th century form. To this group must be assigned such scholars as C. P. Caspari (d. 1892), A. Vilmer (d. 1868), E.W. Hengstenberg (d. 1869). Th. Kliefoth (d. 1895) and W. Loehe (d. 1892).[45] A polemical controversy broke out between these two representative schools of Lutheran theology. The Erlangen School endeavored to combine Lutheran theology with the new learning, and differentiated between the Reformation and post-Reformation theologies in favor of the former.[46] The University of Erlangen has exerted a strong influence on the Lutheran Church through a number of strong personalities that became associated with the Erlangen University. The leaders of this school have been von Hofmann, and later Frank. Other scholars belonging to this school were: Harless, Hoefling, Thomasius, Th. Harnack, Plitt, v. Zezschwitz, Th. Zahn, Ihmels and others. The literary organ of the Erlangen School was **Zeitschrift fur Protestantismus und Kirche**.[47] Concerning this school's influence the *Lutheran Cyclopedia* asserts: "This school has manfully combatted rationalism in its old form, as well as its modern guise of liberalism, and has made some valuable contributions to Lutheran theology. But, though claiming to represent conservative, confessional Lutheranism, it has forsaken the Lutheran base."[48] According to the proponents of this school, confessional theology was not to remain static but was to be dynamic in the Church, which was understood to be a dynamic fellowship. Members of the Erlangen school proceeding to develop doctrines according to the lines of "scientific theology." They repudiated the principle that Scripture alone is the source of theology (principium cognoscendi), and substituted therefore the believing ego, the Christian consciousness, the theologian himself, thus really being influenced by Schleiermacher instead of by Luther.

One of the stalwart defenders of historical Biblical and Lutheran Christianity was Ernst Wilhelm Hengstenberg (1802-1869), son of a Reformed clergyman, who by private study had found in Christ his Savior.[49] In the Confessions of the Lutheran Church he was convinced he saw the clearest expression of true Biblical theology. Hengstenberg was mainly a Biblical scholar, whose specialty was the Old Testament, which had come under serious attack by the new Biblical criticism. This criticism rejected the Mosaic authorship, the Isaianic unity, the sixth century date for Daniel and was responsible for denying the historical character of many Old Testament books and also repudiated Biblical prophecy and the miracles of the Bible. The orthodox Lutheran Church owes Hengstenberg a great debt, because of his warfare against liberalism, negative higher criticism, rationalism and unionism as well as the mediating theology of his day. In the Old Testament field his chief writings were: **Christologie des Alten Testaments, Beitrage zur Einleitung in das Alte Testament, Evangelium Johannes, Offenbarung Johannis.**[50]

In 1827 Hengstenberg founded *Die Evangelische Kirchenzeitung*, a magazine that became a powerful organ for the defense of Biblical Christianity and confessional Lutheranism. For 41 years Hengstenberg edited and contributed to its pages; without fear or favor he attacked the errors of his time. The authorities in Berlin disliked his orthodox stand and tried to have him transferred from the University of Berlin to other universities and positions under the pretext of advancing him as a reward for his labors. However, till the end of his life he remained in Berlin, but was the object of much slander and abuse because of his orthodox stand and because of his uncompromising defense of the Holy Scriptures. Although he was opposed to unionism, he still remained within "the Union, defending his action by claiming: What God hath joined together, let no man put asunder." While Hengstenberg was opposed to rationalization in theology, yet he defended a certain measure of freedom in theology. He waged a great battle with members of the famous Erlangen school, especially with von Hofmann, whose view on the Word of God and hermeneutics were not in agreement with those of the Lutheran Confession and of Luther.[51]

From the very beginning the conclusions of higher criticism were challenged and vigorously contested by conservative, Bible-believing scholars. The supernatural origin of the Holy Scriptures had always been accepted by the Christian Church and it had been an axiom of the Christian faith, as found both in the Old and the New Testaments. The arguments of the liberal and doubt-creating critics have been challenged on their own grounds. European and American scholars have shown that merely from the perspective of literary and historical studies the critics' stance was inadequate and erroneous.

Dr. Roehrs wrote concerning this matter:

> Investigations as detailed as those of the higher critics have been made to disprove the validity of the linguistic canons upon which rest the negative theories. On the basis of secular literature it was demonstrated that the application of the principles of the higher critic lead to absurd results. Why should not a Biblical writer be permitted to use synonyms, especially since synonyms do have their own connotation and are used deliberately by discriminating writers?[52]

The 19th century may be said to have been the century when Apologetics became a separate science, and the literature that developed in the

19th and 20th centuries was immense. The apologetical works produced in these two centuries might be said to have concerned themselves with three main areas:

I. Fundamental Apologetics. Under these Weidner placed:
1. The Being and Nature of God;
2. The Cosmological Question;
3. The Anthropological Question;
4. The Ethical Question;
5. The Question of Man's Immortality.

II. Historical Apologetics. Under this division the following topics can be placed:
1. The Supernatural in History in General;
2. Special Forms of the Supernatural in History;
3. The Bible in History;
4. Christ in History, and
5. The Church in History.

III. Philosophical Apologetics. Under this caption the following subjects are placed by Weidner:
1. Philosophy of Religion;
2. Philosophy of History; and
3. Philosophy of Christianity.

In the 19th century the following produced significant works in the area of apologetics: C. A. Auberlen, *The Divine Revelation.* Edinburgh, 1867; Franz Delitzsch, System der *Christlichen Apologetik.* Leipzig, 1869. Of this work Weidner wrote: "A valuable work, which deserves to be better known than it is."[54] In it Delitzsch argued that: 1. Christianity satisfies the religious needs and cravings of man. 2. The historical actuality of Christianity and of the Bible. 3. The historical actuality of the agreement and correspondence of Christianity and the Bible.

During the last quarter of the 19th century Christlieb published his *Modern Doubt and Christian Belief*, New York, 1875. *The Lutheran*, J. H. A. Ebrard issued a three volume Apologetics in Edinburgh, 1886. *The Lutheran* Luthardt published the following volumes, dealing with apologetics: *Fundamental Truths, 1869, Moral Truths, 1873* and *Saving Truths, 1868*, all three printed in Edinburgh. Among other German apologetic writers we find men like: Kurtz, Reusch, Steinmeyer, Tholuck, Ullmann, Zeschwitz, and Zoeckler.

Scotland produced the following scholars who wrote on apologetical subjects: Chalmers, Calderwood, and Flint. A very famous Lecture series was known as the Bampton Lectures, which was given in London between 1780-1888, of which altogether 108 volumes saw the light of day. Two other series were the Boyle and the Hulsean Lectures, established for the defense and propagation of historic Christian faith.[55] Rev. John Bampton, Canon of Salisbury, in his last will and testament, provided for eight lectures to be given yearly on the First Tuesday after Easter at Oxford University. A part of his will stipulated as follows: "Also I direct and appoint, that eight Divinity Lecture Sermons shall be preached upon either of the following subjects — to confirm and establish the Christian Faith, and confute all heretics and schismatics-upon the Divine authority of Holy Scriptures-upon the authority of the writings of the primitive Fathers, as to the faith and practice of the primitive Church — upon the Divinity of our Lord and Savior Jesus Christ — upon the Divinity of the Holy Ghost — upon the Articles of the Christian Faith, as comprehended in the Apostles' and Nicene Creeds. [56]

Other English writers contributing the apologetic literature in England were: Liddon, Row, Lightfoot and Ellicott. Among French Protestant authors the following made worthwhile contributions: Gaussen, Godet, Guizot, Janet, and Pressense, and among American writers Joseph Cook, Harris, Fischer, McCosh, Schaff and Storrs.

The Evangelical Lutheran Church-
The Missouri Synod and Apologetics and Polemics

The literature emanating from the theologians and pastors of what came to be known as The Lutheran Church-Missouri Synod always has had in it works and sections in doctrinal books that were either polemical or apologetical in character. The founders of the main contingent of what first was known as the Evangelical Synod of Missouri, Ohio, and other states came from Saxony in Germany in 1839 at a time when rationalism was rampant in the state church of Saxony. Church authorities in Saxony had made it impossible for true Lutherans to worship God according to the Lutheranism set forth in the Lutheran Confessions. Philosophy and negative criticism had been adopted by many Lutherans in Germany, and the religion once held by Luther and the theology as explicated in the Lutheran Confessions was abandoned or completely reinterpreted.

Stephan, the Walthers, and others leaving on five different ships, embarked for America where they hoped to be able to worship God according to what they believed was the true Biblical and Lutheran faith. Before embarking they adopted the following *Emigration Regulations*:

All the undersigned acknowledge with sincerity of heart the pure Lutheran faith as contained in the Word of God, the Old and the New Testaments, and set forth and confessed in the Symbolical Books of the Old and New Testaments, and set forth and confessed in the Symbolical Books of the Lutheran Church. After deliberate and mature counsel they can, humanly speaking, see no possibility of retaining in their present home this faith pure and undefiled, of confessing it and transmitting it to their posterity. Hence they feel in duty bound to emigrate and to look for a country where this Lutheran faith is not endangered and where they can serve God undisturbed in the way of grace revealed and ordained by Him, and where they can enjoy, without being interfered with fully, without adulteration, the means of grace ordained by God for all men unto salvation, and can preserve them in their integrity and purity for themselves and their children.

Such a country as they are looking for is the United States of North America; for there, as nowhere else in the world, perfect religious and civil liberty prevails.[57]

Under the leadership of Pastor Stephan about 750 people left their homes and friends in November, 1838 and arrived in St. Louis early in 1839. Those who arrived in Missouri had given up much in order to preserve the true Lutheran faith. Theological and religious conditions in Germany made many of the Saxon immigrants feel it impossible to worship God according to the Lutheran faith and to raise their children so that they could embrace and hold fast to the teachings of the Lutheran Confessions.[58]

The founding fathers were determined to practice a type of Lutheranism in harmony with that expressed in the *Book of Concord* of 1580. The various symbols, the *Augsburg Confession*, the *Apology of the Augsburg Confession*, the *Smalcald Articles*, the two *Catechisms* of Luther and the *Formula of Concord* clearly gave expression to what the Lutheran Church believed as of 1580. The century following recorded many theologians and pastors who taught and proclaimed these truths faithfully.

When the Evangelical Synod of Missouri, Ohio and other states was organized in Chicago in 1847, the founders were determined to follow the theology of Luther, Chemnitz, and Gerhardt. A reading of the objectives of this new Lutheran federation of congregations shows that they were seriously concerned that the doctrine taught be Biblical and Lutheran.[59]

Upon arrival in America, the Saxon fathers found a type of Lutheranism, known as "American Lutheranism." It was essentially Calvinistic, Methodistic, Puritanic, indifferentistic, and hence unionistic, hence not Lutheran. Proponents of "American Lutheranism" attacked what was most prominent in the Lutheran Confessions.[60]

While the founding fathers, among them, C.F.W. Walther, claimed that the justification of the sinner by faith was the central doctrine of the Christian faith, that Law and Gospel must be correctly distinguished as to origin, nature and effect and that the *Augsburg Confession* is the chief of the Lutheran confessions,[61] and although these views constitute important major emphases, they do not represent Walther's nor the Missouri Synod's complete doctrinal position.[62]

From its very inception the Lutheran Church-Missouri Synod considered the entire Word of God as inspired, not merely some passages, and because they were the Word of God they were to be accepted, and it was immaterial whether the doctrines were classified as fundamental or nonfundamental. C. F. W. Walther wrote in Thesis XVII of *The Evangelical Lutheran Church the True Visible Church on Earth*: "The Evangelical Lutheran Church accepts the whole written Word of God (as God's Word), deems nothing in it superfluous or of little worth but everything needful and important, and also accepts all teachings deduced of necessity from the word of Scripture (Matt. 3:18,19; Rev. 22:18,19; Matt. 22:29-32)."[63]

In Thesis XIII of the same book, Walther asserted: "The Evangelical Lutheran Church recognizes the written word of the apostles and prophets as the only and perfect source, rule, norm, and judge of all teachings—a) not reason, b) not tradition, c) not new revelations; (Deut. 4:2; Josh 23:6; Is. 8:20; Luke 16:29; 2 Tim. 3:15-17; I Cor. 1:21; 2:4,5; Col. 2:8, Matt. 15:9).[64] This is the teaching of the Lutheran Church.

Until his death in 1887 C. F. W. Walther was the great theological leader of the Evangelical Luther Synod of Missouri, Ohio, and other states. In 1844 Walther began with the backing of Trinity Congregation, St. Louis, to publish *Der Lutheraner*, which was published uninterruptedly until December 1974. This periodical served to bring Lutherans in various sections of the country together. *Der Lutheraner* in clear tones set forth the true character of confessional Lutheranism. It was instructive and doctrinally educational but also warned against erroneous theological opinions, as held by other Protestant groups and by errant Lutherans.[65]

When the Altenburg college was moved from Perry County to St. Louis, Walther was elected as its theological professor and president and labored zealously to educate men for the ministry who were resolved to be faithful to the Word of God and the Lutheran Confession. This, by God's grace he succeeded in doing, grounding students in a sound Lutheran theology. In 1851-5? Walther accompanied Wyneken to Germany in order to have conferences with Wilhelm Loehe, who had done so much for Missouri. Loehe began to deviate from the Lutheran doctrine of the Church and Ministry. The mission failed of its purpose.[66]

The Lutheran Church—Missouri Synod from the very beginning, yes even before its organization, was forced to engage in a number of doctrinal controversies with other Lutherans, so that during the first seventy years of its synodical history, the Missouri Synod was involved in activities that must be characterized as polemical and apologetical. Other individuals attacked the doctrinal stance of Missouri and so the latter was obligated to defend its doctrinal position and also endeavor to show the unscriptural character of the position of its opponents.[67]

At Walther's suggestion the Missouri Synod in 1855 founded *Lehre and Wehre*, a theological monthly, edited at first by Walther, later by the faculty of Concordia Seminary. *Lehre und Wehre* was a Lutheran journal that proclaimed the doctrines of the Lutheran Confessions clearly and defended the truth and attacked error as it appeared upon the European and American theological scene. *Lehre und Wehre*, which published its articles in German continued until 1910, when it was supplanted by the *Theological Quarterly*. The latter was replaced by *The Theological Monthly*, which in 1930 had its place taken by *The Concordia Theological Monthly*. All these journals pursued the same course, promulgating Biblical doctrines and having many articles that dealt with the Lutheran teachings as set forth in the various Lutheran Confessions. Apologetics and polemics are very much in evidence in the many issues that appeared between 1855 and 1965, when the character of the *Concordia Theological Monthly*, under the editorship of Herbert Mayer changed its character radically, taking a position that was the very opposite of that represented for over one hundred years of publication by the faculty of Concordia

Seminary, St. Louis.

The Lutheran Witness which came to occupy an important place next to *Der Lutheraner* was for many decades an advocate of Lutheran orthodoxy and of sound Biblical interpretation. But during the 1960's its caliber changed and it became a different kind of church magazine as compared to what it had been under the editorship of theologians like Martin Somner, Theodore Graebner, Blankenbueler and Sohn. Both *The Lutheran Witness and Der Lutheraner* served the Synod well in keeping its leadership and membership loyal to God's Word and simultaneously warning them against theological errors and aberrations. Calvinism, Romanism, pseudo-Lutheranism, Neo-orthodoxy, the errors of the cults as well as anti-Christian philosophy and anti-Scriptural views were delineated and examined in the light of God's Word.

Biblical polemics and apologetics presuppose the truth that the Holy Scriptures are clear, that a church body can have and know the truth. This is the presupposition of Article II of the LCMS's constitution. Biblical polemics and apologetics are disciplines that are eschewed and deliberately avoided in the majority of today's theological seminaries. The twentieth century has been called "The Ecumenical Century." The ecumenical movement is not at all concerned with purity of doctrine. The ecumenical movement is based on the presupposition that as long as one believes in the Lordship of Jesus Christ, and that acceptance of this one theological belief is sufficient reason for cooperation among the Christians of the world. The World Council of Churches and its affiliates, the Lutheran World Federation and the National Council of Churches do insist, in fact, they are opposed to the necessity of having doctrinal consensus on the teachings of the Scriptures. At the Synodical convention of the LCMS at Detroit, Dr. Martin Kretzmann submitted a document which proposed that the Missouri Synod join the Lutheran World Federation, the National Council of Churches and the World Council of Churches. During the presidency of Dr. Harms the forces favoring ecumenicity were beginning to put their new philosophy of church fellowship into practice.

Recent Developments in the Lutheran Church-Missouri Synod (1959-79)

The last twenty years have seen doctrinal developments which would have been impossible had all members of the Lutheran Church-Missouri Synod chosen to abide by the Synodical constitution and its by-laws. When the LCMS celebrated its centennial in 1947, the essays published in the first two volumes of The Abiding Word showed that the doctrine and practice of the Synod had not changed in a hundred years,[68] although there had been rumblings of discontent in *The Statement of the Forty-Four and Speaking the Truth in Love.*[69]

However, the late fifties and early sixties saw changes occurring in the LCMS that were planned by a group of people who were convinced that the historic doctrinal position of the Synod was not correct and who decided to change the course of the history of the LCMS. To bring one of the most orthodox Lutheran churches in the world into the mainstream

1287

of American religious life would take some doing. The plan called for securing control of the thought influencing agencies of the Synod, such as the following: the two theological seminaries, the two teachers colleges, the senior college, the junior colleges, the Board of Higher Education, The Board for Parish Education, the Synodical Board of Directors, the Board of Home and Foreign Missions, *The Lutheran Witness*, The Lutheran Woman's Missionary League, The Lutheran Layman's League, *Advance Magazine*, and Concordia Publishing House. In fact, all administrative positions were to be taken over by individuals sympathetic to the new program. The idea was to control the direction of the Synod in every way possible. A number of members of the influential Council of Presidents was won over to the concept of "a new Missouri."

Between 1955 and 1969 many of the revisionistic goals were being realized and gradually the character of Missouri was being changed. Between 1962 and 1969 Synodical Conventions were passing resolutions implementing the new program.[70] In 1965 the LCMS joined the Lutheran Council in the USA (LCUSA),[71] and in 1969 altar-and-pulpit fellowship were established with the American Lutheran Church.[72] Such memberships were impossible for those who understood the true nature of Lutheranism and who supported the objectives of the Synodical constitution.

The group of men responsible for the changes that were introduced into the LCMS was committed to a world-wide forum of so-called Christian ecumenism. It was the aim of this group to have the LCMS not only to join the Lutheran World Federation, but also to affiliate itself with the National Council of Churches and the World Council of Churches. Such memberships were impossible for those who understood the true nature of Lutheranism at the St. Louis Seminary, prior to the walk out of winter 1974, the historical-critical method had been introduced with resulting attacks on the inerrancy and historical reliability of the Bible and with the repudiation or questioning of many Biblical miracles.

The LCMS from its very beginnings believed in sending out missionaries to foreign countries. Synod's foreign mission program, undertaken because of the Great Commission of Matthew 28, always operated on the belief that it was the purpose of foreign missions, as well as home missions, to win unbelievers for Jesus Christ. The LCMS always taught that Christ was the only Savior of the world. To bring people to Christ was the main purpose of mission work. This was not the view of The World Council of Churches. The erroneous views about the nature and purpose of missionary worked out by Dr. Martin Kretzmann, was adopted and his theological position exerted great influence on the Mission Board of the Synod, with the result that the mission policy was changed without the rank and file suspecting that the entire direction of the mission work was being changed. Doctrinal differences among Lutherans were being sidetracked and the Synod was being pushed into the ultimate formation of the inter-Lutheran Church, of which The Lutheran Council of the United States was to be a forerunner.

Adams, a working journalist for the *St. Louis Post-Dispatch* in his book dealing with Dr. J.A.O. Preus admits that "the moderates," as he called those who had embarked on a new program in the LCMS, were promoting it surreptitiously, and did attempt to bring the LCMS out of its self-chosen isolationism but did not set forth their program honestly.[73] The picture they endeavored to paint was that what they were doing was in accord with the tradition the Synod had known for at least a century. In order to hide what their goals were, they tried to soothe grassroots concerns by claiming that no changes were involved, that what was being done was to present the old truths unchanged in language more meaningful to the American public. In the judgment of Hughes "this seek-no-change, hear-no-change, speak-no-change was ecclesiastical disaster."[74] Father John Jay Hughes in his April 3, 1977 review of Adams' book made this evaluation about the moderates' policy: "For an intellectual elite to work for change while denying that any change was taking place seemed to many to come close to confidence artists fleecing a crowd of yokels in a shell game."[75]

Doctrinal Problems Recognized by Synodical Leadership

The leadership in the LCMS knew that there were theological problems and new practices in vogue not in harmony with the LCMS' traditional understanding of church practice which prompted two Presidents of Synod to deal with them by having joint meetings with the two seminary faculties, the district presidents and the Presidium of the Synod. Beginning with November 27, 1961 such three-day gatherings were held annually on the campus of Concordia Seminary, St. Louis.[76] At these meetings essays were delivered which tried to deal with theological issues that were disturbing the Synod. Evolution, revelation and inspiration of Scripture, the historical-critical method (discussed at a number of conferences), the new hermeneutic, the obligation of Synodical teachers to teach Synod's doctrinal stance and other topics were the subjects of discussion and debate. Despite these annual meetings which were held at the end of November or beginning of December between the years 1961 and 1970, the proponents of theological change continued on their planned course of controlling the Synod through a clergy brainwashed by men who practiced and made the student use the historical-critical method, were encouraged to be ecumenical-minded and ignore the historic doctrinal position of the LCMS.

On December 2, 1963 Dr. Roland Wiederaenders delivered an essay before the Council of Presidents, in which he said in part:

Despite repeated efforts we have not dealt honestly with our pastors and people. We have refused to state our changing theological position in open, honest, forthright, simple and clear woods. Over and over again we said that nothing was changing when all the while we were aware of changes taking place. Either we should have informed our pastors and people that changes were taking place and, if possible convinced them from Scripture that these changes were in full harmony with "Thus saith the Lord!" or we should have stopped playing games as we gave assurance that no changes were taking place. With increasing measure the synodical trumpet has given an uncertain sound.[77]

The results of the program carried on by "the moderates" after a number of years was described by Dr. Wiederaenders as follows:

Quite generally our pastors and almost entirely our laity became more and more confused. Confusion led to uncertainty. Uncertainty led to polarization. Polarization destroyed credibility. Loss of credibility destroyed the possibility for meaningful discussion. The loss of meaningful discussion set the stage for a head-on collision.[78]

Before 1969 the polemics were mainly waged by those who saw the historic synodical doctrinal stance and its position on anti-Scriptural unionism go down the drain. Polemical memorials were in evidence in the *Workbooks* for the various Synodical conventions of 1962, (Cleveland), 1965 (Detroit), 1967 (New York), and 1969 (Denver). Although resolutions were passed dealing with substantive theological issues that were polarizing the Synod when resolutions were unfavorable, they were ignored by those securing control of the thought-influencing and policy-determining agencies of the LCMS. Finally, the conservative people in the Synod began to organize themselves with the purpose of halting and changing the theological direction the LCMS was being forced to take.

With the Denver convention (1969) the beginning was made by the conservatives to reverse the trend that had been in evidence since the San Francisco convention of 1959. The period between 1969-1979 has witnessed a theological civil war within the Synod, because now the so-called moderates began to go on the attack and finally came out into the open and in the decade between 1969 and 1979 set forth their theological positions on such topics as the errancy of Scripture, the defense of the use of the historical-critical method, the denial of the Third Use of the Law, advocating a different doctrine on the nature and purpose of the church, ecumenism, unionism, and a different concept of missionary work.

Before 1969, an important year in the history of the LCMS, numerous voices were raised by various individuals to alert the constituency of the Synod as to what was happening to its doctrinal heritage. Over twenty years ago Pastor Burgdorf began the publication of a magazine, *The Confessional Lutheran*, in which he cited evidence that in the LCMS false doctrine was creeping in and was being promoted. Many considered him and the organization which published his magazine as stirring up trouble where actually there was none. Time, however, appears to have vindicated many of the predictions he made and the consequences, Burgdorf claimed, would follow, unless checked.

Conservative professors at both the St. Louis and Springfield seminaries wrote essays and gave lectures on issues that eventually came to a head at conventions in 1969. Dr. John Warwick Montgomery delivered a number of incisive essays pointing up problems in the LCMS as reflected at a number of Synodical institutions. These Essays were published by him in *Crisis in Lutheran Theology*, volume 1, which contains four excellent essays relative to the doctrine of inspiration, the interpretation of

Scripture, theological issues in the LCMS, and current trends in the LCMS.[80] Volume 2 of *Crisis in Lutheran Theology* contains a reprint of essays on revelation and inspiration (six essays) and six dealing with Biblical interpretation and ecumenicity in the light of *Luther and the Confessions*. Contributors to this volume were: Hermann Sasse, Robert Preus (4 essays), Raymond Surburg (2 essays), L. W. Spitz, Sr., Douglas Carter, Ralph Bohlmann, H. Daniel Friberg, Donald R. Neiswender.[81]

In 1962 Pastor Herman Otten began publishing a magazine called *Lutheran News* and later changed to *Christian News*. Both its foes and its friends admit that this apologetical and polemical magazine contributed to the waking up of laymen and clergy alike as to what really was going on in the LCMS. *A Christian Handbook of Vital Issues* has republished materials the editor of *Christian News* considered worthy of retaining in a more permanent form.[82]

In 1964 the Commission on Theology and Church Relations (CTCR) was voted into existence, a new prestigious committee which was to deal with doctrinal and ethical problems which were there and those which should develop. The composition of this august group consisted of six professors from the two seminaries, pastors, teachers and laymen. Some members were elected by the Synod in convention, others appointed by The Council of Presidents and still others by the Synodical President. The history of the CTCR has two periods, 1964-1969 and 1969-1979. Materials produced during the period of 1964-69 seemed to aid the program of the socalled "moderates." Studies produced by the CTCR sounded an uncertain and confusing sound were: "Revelation," "Theology of Fellowship," "A Lutheran Stance toward Contemporary Biblical Studies," and "The Witness of Jesus and Old Testament Authorship."[83]

News and Views of Wheaton, Illinois in Vol. 24, Numbers 1, 2, 7 of 1961 and May, 1962 published a series of studies and analyses of World and American Lutheranism and showed how there had been serious departures from historical Lutheranism as set forth in the ecumenical creeds and Lutheran Confessions gathered in *The Book of Concord of 1580*. Part IV also pointed out the theological problems and trends toward a more unionistic-oriented approach to church practice. It is interesting to see an evaluation by non-Lutherans who recognized that significant changes were occurring also in the LCMS.

The Failure of the Core Hermeneutics Committee

The Detroit convention of The Lutheran Church-Missouri Synod authorized its Commission on Theology and Church Relations (CTCR) "to conduct a comprehensive study of Biblical hermeneutics" and was instructed to engage sufficient full-time personnel and "to make provision for leaves of absence for men appointed to participate in the study, in order to expedite the study." Since it did not seem feasible to engage full-time personnel as had been envisaged, the commission appointed the following persons to serve as a core group for carrying out the Synod's assignment: Walter Bartling, Robert Bertram, Martin Franzmann, Heino Kadai, H. Armin Moellering, Robert Preus, Walter Roehrs, and Raymond

Surburg. Richard Jungkuntz, executive secretary of the CTCR, was appointed covener and ex officio member of the core committee. Between January 1966 the committee met at 12 different times.

In *A Project in Biblical Hermeneutics* Dr. Jungkuntz has given a history of the hermeneutics project.[84] The CTCR's executive secretary wrote: "The committee is fully convinced that the basic hermeneutical issues seem to be the source of tension and confusion in our church today are not uniquely a problem of the Lutheran Church-Missouri Synod and that their solution or removal from the area of Christian concern will not be achieved by our unique contribution alone."[85] After three years of meetings it became apparent that the assignment of the Synod to settle hermeneutical problems that were bedeviling the Synod would not be solved because the committee was affected by the very problems it was to resolve. Some members were adhering to the traditional hermeneutics which the Synod had followed since 1947 and that was consonant with the hermeneutical principles employed by the formulators of doctrine in the Lutheran Confessions, while another group leaned toward or were proponents of the new hermeneutic.

A Project of Biblical Hermeneutics contains four hermeneutical studies together with responses by members of the core committee. For a radical approach to the interpretation of Pauline literature the interested reader should read Walter J. Bartling's "Hermeneutics and Pauline Paranesis."[86]

Since the core committee could not agree on many basic issues as they applied to the writing of sound hermeneutical documents, the project was discontinued by recommendation of the CTCR.

Professor Kurt E. Marquart, who spent many years as a pastor in Australia, watched from the Southern Hemisphere the theological scene as it was unfolding in Europe and America. Even though his incisive book. *Anatomy of an Explosion* (1977), published in the great Lutheran civil war years (1969-1979), nevertheless contains a discussion of the roots of the theological problems and shows how step by step the LCMS was being taken down the primrose path of neo-orthodoxy and theological liberalism. Not only does the book tell about the internal tension of the seventies, but shows how the fifties and sixties prepared for the catastrophic events that were to shake the Synod, beginning especially with the New Orleans Convention (1973).[87]

The Lutheran Congress of 1970

When Dr. Preus assumed the Synodical presidency the "moderates" were firmly entrenched and at first nothing could be done to stop the progress that the "new Missourians" had been making. In the early 1970's it was not clear what shattering developments the religious world would witness as a great doctrinal battle was to be fought in the LCMS. In 1970 leaders from various Lutheran denominations issued a call for a Lutheran Congress. From August 31 to September 2, 1970 a number of hundreds of Lutherans from all over the United States gathered at Chicago. Calm and positive expressions were given to eternal truths of the Scriptures, the truths that are reflected in the Lutheran Confessions.

The editors of the volume containing the essays delivered at the Lutheran Congress stated: "We must remember that most Lutherans have never walked this way before. Most of us have never experienced a situation where God's Word is openly questioned, where eternal truths are relativized, traditional theological terms are emptied of their Biblical meanings, and the process of normal communication between brothers in faith is made difficult with endless ambiguity."[88]

It was the conviction of the participants that the Lutheran Church was facing a new humanism, a new theology, a new hermeneutic, and a new approach to missions which could not be allowed to be options for Lutherans. All participants were Lutherans from the three major Lutheran Churches in America. The only non-Lutheran was Rev. Francis A. Schaeffer, who came to warn his Lutheran friends about his experiences in connection with the great Presbyterian doctrinal struggle. Prof. Warth from Brazil and the Rev. Gunnar Stalsett of Oslo, Norway, Elmer Reimnitz from Porto Alegre, Dr. Manfred Roensch from Oberursel, Germany, Dr. Hermann Sasse from Adelaide, South Australia read essays or sent greetings. Drs. Klann, Robert Preus, Ralph Bohlmann from the St. Louis seminary delivered essays as did Vice-President Edwin C. Weber, who also issued the Congress' Call to Order. Dr. Erich Kiehl and Paul Zimmerman of Ann Arbor, Michigan read pertinent essays. Dr. L. Greene, non-Missourian, a member of another Lutheran Synod, spoke on behalf of sound Lutheranism.

Other clergy who delivered important essays were: Alvin E. Wagner, George Wollenburg (now 4th Vice-president of Synod), Waldo J. Werning, Wilbert Sohns. Dr. Eugene Bertermann spoke on World Missions. Laymen who testified were Wm. Gast, Alfred Tessman, George Mohr.

The essays of the Lutheran Congress dealt with 1. The nature of Scriptural Truth; 2. Faithful Confessional Life in the Church; 3. Evangelical Communication of the Word; and 4. General Topics. All the participants recognized the seriousness of the theological situation which faced world Lutheranism, and especially also that of the LCMS.[89]

Between the Milwaukee Convention (1961) and the New Orleans Convention (1973) the "moderates" began to engage in polemics and apologetics on a scale heretofore not done by them as when they were in the saddle of the synodical horse. In the year before the New Orleans convention there was much apologetical and polemical activity by both parties in their respective efforts to either hold the gains they had made ("moderates") or to restore the Synod to its former stance (conservatives). What transpired between 1972-1979 needs a separate treatment, which if adequately dealt with, will easily fill a good-sized volume.

Footnotes

1. John Rutherford, "Gnosticism," in James Orr, John Nuelsen and Edgar Y. Mullens, editors. *The International Standard Bible Encyclopedia* (Grand Rapids: Wm. B. Eerdmans Publishing Company, 1939), II, pp. 1241-1243.
2. **Ibid.**, pp. 1242-1243.
3. Cf. Francis Pieper, Index *Christian Dogmatics* (St. Louis: Concordia Publishing

House, 1957), vol. IV. Prepared by W.M.F. Albrecht.

4. Schaff as quoted by Revere Franklin Weidner. *Theological Encyclopedia and Methodology* (Chicago: Wartburg Publishing House, 1910), II, p. 132.

5. Quoted by John J. Jess, "Need We Apologize for Apologetics?" *The Chapel of the Air*, 7:1, July 1967.

6. **Ibid.**, p. 1.

7. Schaff as quoted by Weidner, **op. cit.**, p 132.

8. J.H.A. Ebrard, *Apologetics: or the Scientific Vindications of Christianty*. Translated by John Macpherson (Edinburgh: T. & T. Clark, 1886-1887), 3 volumes.

9. Francis Pieper, *Christian Dogmatics* (St. Louis: Concordia Publishing House, 1951), I, p. 208.

10. Edwin Luecker, "Lutheran Confessions," *Lutheran Cyclopedia* (St. Louis: Concordia Publishing House, 1954), p 633-634.

11. L. W. Spitz, "Lutheran Theology after 1580," *Lutheran Cyclopedia*, **op. cit.**, p. 638.

12. **Ibid.**

13. "Calixt, George," *Lutheran Cyclopedia*, **op. cit.**, p. 154.

14. "Quenstedt, Johann Andreas," *Lutheran Cyclopedia*, **op. cit.**, p. 876.

15. Weidner, **op. cit.**, pp. 135-136.

16. Kenneth Scott Latourette, *A History of Christianity* (New York: Harper It Brothers, 1953), pp. 731, 894-895.

17. Otto Zoeckler, "Pietism," Henry Eyster Jacobs and John W. Haas, *The Lutheran Cyclopedia* (New York: Charles Scribners Sons, 1899), p. 381.

18. Weidner, **op. cit.**, p. 136.

19. "Rationalism," *Lutheran Cyclopedia*, **op. cit.**, p. 882.

20. **Ibid.**, p. 882.

21. **Ibid.**

22. **Ibid.**

23. Weidner, **op. cit.**, p. 136.

24. **Ibid.**

25. Spitz, "Lutheran Theology after 1580," *Lutheran Cyclopedia*, p. 639.

26. **Ibid.**, p. 639.

27. Edward Frederick Miller, *The Influence of Gesenius on Hebrew Lexicography*, (New York: Columbia University Press, 1927), p. 105.

28. Fred Kramer, "The Introduction of the Historical-Critical Method and Its Relationship to Lutheran Hermeneutics," An Essay Delivered Before the Council of President of The Lutheran Church-Missouri Synod and Joint Faculties of the Two Seminaries, November 30, 1965, p. 19.

29. **Ibid.**, p. 49.

30. Eric Kuhl, *The Old Testament, Its Origin and Composition* (Richmond: The John Knox Press, 1961), p. 11.

31. Spitz, **op. cit.**, p. 639.

32. "Kant, Immanuel, (1724-1804)," *Lutheran Cyclopedia*, **op. cit.**, p. 545.

33. Cf. Karl Hase, Evangelische Dogmatik. Vierte verbesserte Auflage (Leipzig: Druck and Verlag von Breitkopf und Hartel, 1850), 535 pp.

34. Spitz, "Lutheran Theology after 1580," *Lutheran Cyclopedia*, p. 639.

35. Andrew O. Voigt, "Harms, Claus," Jacobs and Haas, *The Lutheran Cyclopedia* (1899), p. 214.

36. **Ibid.**

37. Spitz, **op. cit.**, p. 640.

38. "Schleiermacher, Friederick Daniel Ernst," *Lutheran Cyclopedia* (1954), **op. cit.**, p. 953; cf. also J. T. Mueller, "Schleiermacher, His Theology and Influence," *Concordia Theological Monthly*, 15:73-93.

39. Edgar Sheffield Brightman, "Hegel, Georg Wilhelm Friedrich," Vergilius Ferm, editor. *Encyclopedia of Religion* (New York: The Philosophical Library 1945), pp. 327-328.

40. L. Fuerbringer, Th. Engelder and P. E. Kretzmann, editor-in-chief. *The Concordia Cyclopedia* (St. Louis: Concordia Publishing House, 1927), p. 314.

41. Spitz, **op. cit.**, p. 640.

42. **Ibid.**

43. **Ibid.**

44. **Ibid.**

45. **Ibid.**

46. "Erlangen School," *The Lutheran Cyclopedia* (Concordia, 1964), p. 344.

47. **Ibid.**

48. **Ibid.**

49. Otto W. Heick, *A History of Christian Thought* (Philadelphia: Fortress Press, 1966), pp. 200-201.

50. Cf. Otto Zoeckler, "Ernst Wilhelm Hengstenberg," Die Christliche Apologetlick im 19. Jahhundert (Guetersloh: Bertelsmann, 1904.; "Hengstenberg, Ernst Wilhelm," *The Lutheran Cyclopedia* (1954), pp 454-455.

51. "Hengstenberg, Ernst Wilhelm," *The Lutheran Cyclopedia*, **op. cit.**, p. 454.

52. W. Roehrs, "Higher Criticism," *Lutheran Cyclopedia* (1954), p. 648.

53. Cf. Weidner, **op. cit.**, pp. 143-145.

54. **Ibid.**, p. 157.

55. **Ibid.**, p. 157.

56. This statement is found in the front of each volume in the Bampton Lecture Series of Books.

57. As given in the lengthy article on "The Lutheran Church-Missouri Synod," *Lutheran Cyclopedia* (Concordia, 1954), pp. 606-607.

58. **Ibid.**, p. 607.

59. **Ibid.**, p. 609.

60. Cf. "American Lutheranism," *Lutheran Cyclopedia* (1954), p. 28.

61. Wm. Dallmann, W.H.T. Dau, and Th. Engelder (editor), *Walther and the Church* (St. Louis: Concordia Publishing House, 1938), pp. 125-126.

62. **Ibid.**, p 125.

63. **Ibid.**, pp. 122-123.

64. **Ibid.**

65. *Der Lutheraner* (St. Louis: Gedruckt bei Weber u. Olshausen 1844-1845), IIII, especially pp. 1-17 of Vol. 1.

66. "Loehe, Wilhelm," Erwin L. Lueker, editor *Lutheran Cyclopedia*, Revised edition; (St. Louis: Concordia Publishing House, 1975), p. 477.

67. Cf. "XVIII. Doctrinal Discussion and Controversies," in the article on Lutheran Church-Missouri Synod," Erwin L. Lueker, *Lutheran Cyclopedia* (St. Louis: Concordia Publishing House, 1954), pp. 623-628.

68. Theodore Laetsch, *The Abiding Word. An Anthology of Doctrinal Essays for the Year 1945* (St. Louis: Concordia Publishing House, 1945), 393 pages; Theodore

Laetsch, *The Abiding Word. An Anthology of Doctrinal Essays for the Year 1946* (St Louis: Concordia Publishing House, 1947). Vol 2, 783 pages.

69. *Speaking the Truth in Love.* Essays Related to A Statement (Chicago: The Willow Press, 1945), 80 pages.

70. Cf. Proceedings of the 45th Regular Convention of The Lutheran Church-Missouri Synod. Cleveland, Ohio, June 20-29, 1962. Proceedings of the 46th Regular Convention of The Lutheran Church-Missouri Synod, Detroit, Michigan June 16-26, 1965, pp 79-82. Proceedings of the 48th Regular Convention, Proceedings, The Lutheran Church-Missouri Synod, Denver, Colorado July 11-18, 1969. pp. 96-99.

71. Proceedings of the 46th Regular Convention of The Lutheran Church-Missouri Synod, Detroit, Michigan June 16-26, 1965, pp. 109-110.

72. Proceedings of the 48th Regular Convention of The Lutheran Church-Missouri Synod, Denver, Colorado July 11-18, 1969, pp. 96-99.

73. James E. Adams, *Preus of Missouri and the Great Lutheran Civil War* (New York: Harper & Row Publishers, 1977), p. ix.

74. John Jay Hughes, Book Review of Adams', *Preus In Missouri, St. Louis Globe Democrat*, April 3, 1977.

75. **Ibid.**

76. The Programs and Reports from these Conferences may be found in some Synodical College or Seminary Libraries.

77. Quoted from a Synodical Press Release, January 24, 1974, p. 2.

78. **Ibid.**, p. 3.

79. *Missouri In Perspective*, which began publication in 1973, in the years following (1973-79) has in editorials and in articles attacked the historic LCMS' stance on these Theological issues; Cf. also Report of the Advisory Committee on Doctrine and Conciliation (St. Louis: Concordia Publishing House, 1976).

80. Dr. John Warwick Montgomery, *Crisis In Lutheran Theology* (Grand Rapids: Baker Book House, 1967), Vol. I. 133 pages.

81. Dr. John Warwick Montgomery, *Crisis In Lutheran Theology* (Grand Rapids: Baker Book House, 1967), Vol. II. 194 pages.

82. *A Christian Handbook on Vital Issues Christian News,* 1961-1973 (New Haven, Missouri: Leader Publishing Co., 1973), 854 pages.

83. All CTCR Study Documents are available from Concordia Publishing House, 3558 S. Jefferson Avenue, St. Louis. MO 63105.

84. Richard Junkuntz, editor, *A Project in Biblical Hermeneutics* (St. Louis, Published by The Commission on Theology and Church Relations of The Lutheran Church-Missouri Synod, 1969), pp. 1-11.

85. **Ibid.**, p. 7.

86. **Ibid.**, pp. 57-83.

87. Kurt Marquart, *Anatomy of An Explosion* (Grand Rapids: Baker Book House, 1978).

88. Erich Kiehl and Waldo J. Werning, Evangelical Directives for the Lutheran Church (Chicago: Evangelical Directions for the Lutheran Church, 2751 Karlov Avenue, 1970), from the Foreword.

89. **Ibid.**, the volume contains 166 pages.

Questions

1. What are two duties of the Christian Church? ____

2. Why were Christians accused of cannibalism? ____
3. Apologetics may be defined as ____.
4. Polemics is the science of ____.
5. Who is acknowledged as the father of modern historical and literary criticism of the Bible? ____
6. Claus Harms protested against ____.
7. Hegel did not believe in ____.
8. Why does the orthodox Lutheran Church owe Hengstenberg a great debt? ____
9. Stephans, Walther's and others embarked for America to ____.
10. What was "American Lutheranism?" ____
11. Thesis XIII of Walther's "The Evangelical Lutheran Church the True Visible Church on Earth" says: ____
12. *Der Lutheraner* set forth ____.
13. Loehe deviated from the Lutheran doctrine of ____.
14. *Lehre und Wehre* was ____.
15. *Lehre und Wehre* was supplanted by ____, replaced by ____, which became ____.
16. When did the *Concordia Theological Monthly* change its character? ____
17. When did the *Lutheran Witness* change? ____
18. The ecumenical movement is not concerned about ____.
19. What happened at the LCMS convention in Detroit in 1965? ____
20. The two volumes of *The Abiding Word* showed that ____.
21. Between 1955 and 1969 influential members of ____ were won over the concept of "a new Missouri."
22. When was historical criticism of the Bible introduced in the LCMS? ____
23. What happened to LCMS missions under Martin Kretzmann? ____
24. What did Adams of the *St. Louis Post-Dispatch* in his book dealing with J.A.O. Preus recognize ____.
25. Father John J. Hughes in his review of the Adams book wrote: ____
26. What did Roland Wiederaenders say on December 2, 1963? ____
27. What did Paul Burgdorf show in *The Confessional Lutheran*? ____
28. Dr. John Warwick Montgomery delivered some essays pointing up ____.
29. What did friends and foes admit about the Lutheran and then *Christian News*? ____
30. *A Christian Handbook on Vital Issues* contained ____.
31. What is the CTCR? ____
32. What did *News and Views* of Wheaton, Illinois publish? ____
33. What happened after three years of meetings in the CTCR's "A Project in Biblical hermeneutics?" ____
34. What did Kurt Marquart's *Anatomy of an Explosion* show? ____
35. It was the conviction of The Lutheran Congress of 1970 that the Lutheran Church was facing a ____.

The Defective Theology of C.S. Lewis

September 29, 1979
The Rev. Herman Otten
Christian News
New Haven, Missouri

Dear Brother Otten:
In view of the great influence C.S. Lewis is asserting through his many published books, you may wish to publish the enclosed evaluation of Lewis' defective theology.

The seminary is again sponsoring a tour to Greece and Asia Minor, beginning Feb. 21, 1980 lasting two weeks. I believe you were so kind as to advertise it in *Christian News* last year. Would you kindly do so again?

Thanking you in advance and wishing you God's blessing in your variegated ministry.

Fraternally,
Raymond Surburg

Ed. *Baal or God* by Herman Otten, published in 1965, noted that a study on "Justification by Faith in Modern Theology" by Henry Hammann published in 1957 by the Graduate School of Concordia Seminary in 1957 showed that C.S. Lewis denied the scriptural doctrine of justification by faith alone.

Women and C.S. Lewis (Lion Books, Oxford, England, 2015), reviewed in the November 8, 2015 *Christian News,* lists a long line of those who have commended C.S. Lewis as a great faithful Christian theologian, even though the book reveals Lewis was long involved in a sexual relationship in England with Mrs. Jamie Moore to whom he was not married. After her death he married Helen Joy Davidman, an unscripturally divorced mother of two. She was born of well-educated Jewish parents from the Bronx, New York.

When the 2008 Higher Things Conference led by William Cwirla held up C.S. Lewis as a great defender of the Christian faith, Higher Things declined to publish what Surburg wrote about C.S. Lewis. ("Hamann, C.S. Lewis, Cwirla, and Higher Things," *CN*, November 11, 2015). "A Bully Bewitched," a review of *C.S. Lewis: A Biography* by A.N. Wilson published in the *Sunday Times*, London, England, February 11, 1990 and reprinted in the November 9, 2015 *CN* exposed the unscriptural lifestyle of C.S. Lewis. This English writer has had many supporters in Lutheranism. Almost all of the "conservative" Lutheran speakers at a Schwan Foundation financed seminar held at the elaborate Schwan resort in Treago, Wisconsin, had nothing but praise for C.S. Lewis and his theology. The February 20, 2017 *Christian News* reviewed *From Atheism to Christianity* published by The Lutheran Church-Missouri Synod's Concordia Publishing House in 2017. Author, Joel D. Heck, presents C.S. Lewis as a faithful Christian highly praised by many Christian scholars.

An Evaluation of the
Theological Stance of C.S. Lewis

Christian News, October 8, 1979

It has been said by some that the late C. S. Lewis (1898-1963) may
well be the greatest Christian writer of the 20th century. Interest in the
writings of Lewis has been on the increase lately. According to sales reports, business is booming. In 1978 over 850,000 copies of such books as
*Mere Christianity, Chronicles of Narnia, The Space Trilogy, Screwtape
Letters* and other of Lewis's books have been sold. Books about Lewis are
also appearing on the book market. It is estimated and predicted that
sales will in 1979 even exceed those of 1978. One of the most recent books
to appear has been that of James T. Como, one of the founders of the New
York C. S. Lewis Society, which contains a series of 24 essays titled, C.
S. *Lewis at the Breakfast Table ... and Other Reminiscences*, whose purpose is to give insight about Lewis never published before.[1]

Although Lewis has been dead now for sixteen years he is exercising
a great influence in that his books and essays are being read by millions
of people around the world. Clyde S. Kilby claims that this is the case because of the profound truths Lewis has uttered and spoken them with
strength and grace. Thus recently Kilby wrote:

> Paradoxically, Lewis is an original not after the modern fashion of
> seeking originality and a name but by a profound desire simply to re-
> capture in all their original meaning, the old platitudes. Many theolo-
> gians and philosophers are finding in him evidences of true greatness
> of mind and spirit. He is a thinker with a warm heart.[2]

1. The Life of C. S. Lewis[3]

C. S. Lewis was born in 1898 in Belfast, Northern Ireland, the son of
parents of culture. His early education was partly private. When he was
ten years old he was sent to school in England for a period of two years.
Then he returned home for a short time and this was followed by a return
to England to Cherbourgh House, a preparatory school for Malvern College. Here he underwent a conflict between his belief in atheism and his
anger that there was no God. Already beginning with his sixth year Lewis
experienced a constant yearning for joy. To fulfill his yearning or **Sehnsucht**, he began reading Norse mythology and he assimilated the beauty
of nature. When he was sixteen years old, Lewis was ready for university
study under a tutor.

Next Clive Staples Lewis came under the influence of the Scotsman
W. T. Kirkpatrick, who was an atheist and a logical positivist. Lewis's
exposure to Kirkpatrick resulted in Lewis having respect for reason and
also strengthened him in his atheism. Nevertheless Lewis's romantic
spirit was not drowned and he reacted imaginatively to fantasies and ro-
mances which helped his romantic spirit to flourish. His reading of ro-
mances satisfied his romantic longing.

An important change occurred in Lewis's life as a result of reading George MacDonald's *Phantasies*, a work which made a great impression on his imagination. Through the reading of MacDonald Lewis claimed that his imagination was "baptized," enabling him eventually to find God with renewed romantic vision.

At nineteen Lewis is found at Oxford, for which he had won a classical scholarship. There he met Owen Barfield, whom he called "the wisest and best of my unofficial teachers." At Oxford he also met J.R.R. Tolkien and a number of others who helped to break down his prejudices against Christianity. At twenty-six Lewis was elected a fellow of Magdalen College. While lecturing to his students, he was challenged by them to reevaluate his philosophical presuppositions and at the same time his quest for Joy was intensified. Lewis was fellow and tutor of Magdalen College (1925-54). From 1954 to 1963 he was professor of medieval and Renaissance English at Cambridge University.

In the Trinity terms of 1929 Lewis claimed that he gave in to the Hound of Heaven and admitted that God was God and that he knelt and prayed during the night, "the most dejected and reluctant convert in all England." Lewis's *Pilgrim's Regress* (1930) had as its subtitle an "Allegorical Apology for Christianity, Reason and Romanticism" and it was intended to set forth his quest for what he always called "Joy." By the term romanticism Lewis meant **Sehnsucht.** *Pilgrim's Regress* is the story of pilgrim Lewis and in it he portrayed allegorically his intellectual journey "from popular realism to philosophical Idealism; from Idealism to Pantheism and from Pantheism to Theism and from Theism to Christianity."[4]

Lewis believed that the highest religious truth could only be communicated symbolically and be received imaginatively. It was Lewis's contention that reason alone could not lead to truth, but on the other hand, truth could not be comprehended apart from reason. For, Lewis reason and imagination were necessary for the meaningful apprehension of truth. But he admitted that it was not easy to reconcile these two to each other.

At Cambridge Lewis's reputation grew and his academic reputation was enhanced by numerous scholarly writings.[5] Lewis's published writings fall into two main categories. His literary studies include *The Allegory of Love* (1936), *A Preface to Paradise Lost* (1942), and *English Literature in the Sixteenth Century* (1954). His majority of over 50 published books deals with religion, the area in which fame has come to Lewis. His books treating the problems of Christian theology and conduct are sometimes in the form of allegorical fiction, e.g. *Out of the Silent Planet* (1943), *Perelandra*(1943), *That Hideous Strength* (1945). Sometimes his writings were more direct: *The Problem of Pain* (1940), *Christian Behavior* (1943), *Mere Christianity* (1952). One of his most effective publications was *The Screwtape Letters* (1942). In this latter book using an older devil and a younger one Lewis analyzed with insight and shrewdness the many weaknesses and temptations to which the Christian is exposed. Other important works of Lewis are: *The Chronicles of*

Narnia, The Great Divorce, Miracles, Reflections on the Psalms, Surprised by Joy, Letters to Malcolm, The Literary Impact of the Authorized Version, The Abolition of Man, God in the Dock and others.

An unexpected facet of Lewis's talent was revealed in books for children. *The Lion, the Witch and the Wardrobe* (1950), *Prince Caspian* (1951) and *The Voyage of the Dawn Trader.* In these children's volumes Lewis demonstrated his inventiveness.

2. Clive Staples Lewis: Conservative, Liberal or What?

In his *Reflections on the Psalms Lewis* informed his readers that some have suspected him of being a fundamentalist.[6] Liberals accuse Lewis of being a fundamentalist, while conservatives believe that he is not a fundamentalist at all. Liberals cannot appreciate the fact that Lewis rejected source and redaction criticisms, that he repudiated skeptical historicism, that a priori he refused to accept many positions espoused by modern liberal theology, and that he was ignorant of the conclusions of negative Biblical criticism.

In view of the fact that Lewis's books are being so widely read and undoubtedly his well-written volumes will be influencing the theological and religious thinking of hundreds of thousands of people, it might not be amiss to evaluate his religious views in terms of sound Biblical theology and see exactly where Lewis stands.

Edgar W. Boss in a Th.D. thesis submitted and accepted by the Northern Baptist Theological Seminary, Chicago, 1948 scrutinized Lewis's theological and philosophical writings to see whether or not they were in agreement with orthodox Christian teachings.[7] Boss found that in a number of his statements as to how he understand various Biblical teachings that Lewis did not adhere to sound Biblical theology but his views would have to be classified as deviation from orthodox teaching.

3. The Doctrine of the Holy Scriptures

Traditionally Christian theology has gone to the Scriptures of the Old and New Testaments for the source of its doctrinal formulations and for the stating of its ethical directives. For historic Protestantism the Bible and the Bible alone was the source for dogmatics and ethics. The Sola Scriptura principle distinguished Lutheranism and some forms of Protestantism from Roman Catholicism, which had two sources: Scripture and Tradition. Later on Protestantism, when it became adulterated and even apostatized used human reason (as found expressed in a number of philosophical systems) and human feelings and experience as additional sources for theological formulations. Lewis also claimed that reason and feeling were to be employed.[8] He thus did not follow the traditional *Sola Scripture* principle of early Lutheranism and of Protestantism. The conversion of Lewis in 1929 did not result in his shedding his previous theological baggage which he had obtained from Kant with his insistence on the magisterial use of reason in theology or the Schleiermachian emphasis on feeling for determining theological beliefs. The fact that Lewis would not abide by the nude text of Scripture accounts for some of the theological aberrations that will presently be enumerated.

The doctrine of inspiration is directly connected with the topic of the Bible's place in religious teaching. Lewis's view about the nature of inspiration is not in agreement with that found in the Scriptures. In a personal letter, dated May 7, 1959, Lewis suggested to Clyde Kilby that an adequate theory of divine inspiration ought to account for among six considerations given for the following, which are listed as numbers 5 and 6. They read: 5. "If every good and perfect gift comes down from the Father of Lights then all true and edifying writings, whether in scripture or not, must be in some sense inspired." and 6. "The paradoxical phenomenon of divine inspiration operating 'in a wicked man without his knowing it,' so that he utters the untruth he intends ... as well as the truth he does not intend." (See John 11:49-52).[9]

Lewis's concept of writings outside of Scripture as being inspired is not in harmony with Biblical teaching. Since Christ is essential for salvation and since the Old Testament in over 300 prophecies foretold many facts about the Messiah and in the New Testament both Christ and His apostles identified Jesus of Nazareth with the promised Messiah, the Bible is the only book to contain the way to salvation and the only book which sets forth the manner of how to get to heaven. The Jews of Christ's time believed in Yahweh and yet Jesus insisted that people needed to believe in Him to have eternal life. Jesus taught both the exclusiveness as well as the inclusiveness of His religion. John 3:15, the Gospel in a nut shell, proclaims both aspects of the Christian faith. The exclusiveness of Christ and His religious teachings are not acceptable to Lewis, because he wants tolerance for other religious beliefs. He claimed that not all religions are totally in error in what they teach. Lewis contended that God has not revealed to us what His final attitude will be toward those who do not know about Christ.[10]

In his fairy tale for children, *The Last Battle*, Lewis has offered his ideas as to what God's arrangements might be for people of other religions. In this tale Emeth, the young prince of Calormene, after the last battle has been fought toward the end of Narnia's existence, is depicted as entering the afterlife, expecting to meet Tash and not Asian, who was a Lion, the true ruler of Narnia. But instead Emeth meets in the Kingdom of Asian the Lion himself. Emith relates how this happened. Toward the end of the dialogue Lewis has Emeth say: "I have been seeking Tash all my days. Beloved said the Glorious One, unless thy desire had been for me thou wouldest not have sought so long and so truly. **For all find what they truly seek.**"[11] (Emphases supplied)

4. Lewis's View of Revelation

According to the Bible God revealed Himself through nature and through the revelation found in the Old and New Testaments. In *Mere Christianity* Lewis listed four methods of revelation-conscience (or "The Universal Ought"), the chosen people of God, Israel (or "Election"), pagan mythology (or "Great Dreams") and the Christ event (or "Incarnation").[12] In the *Problem of Pain* Lewis added two others: Immortal Longing

("**Sehnsucht**") and the Idea of the Holy (or "The Experience of the Numinous").[13]

Numinous as revelation was a concept that Lewis took from Rudolf Otto's *The Idea of the Holy*, a book in which Otto endeavored to show that man has an awareness of the Divine or Holy, described by Otto as "the experience of the Numinous."[14] Literature attests to man's belief in the existence of supernatural beings before whom men manifest a dread. In the presence of the Holy God man wants to hide his face in shame and also echo the words of Isaiah in 6:5: "Woe is me! for I am undone; because I am a man of unclean lips, and I dwell in the midst of a people of unclean lips." When men realize that they are in the presence of a holy God fear fills them. Thus Lewis held that the Numinous was a medium of revelation.

Another medium of divine revelation for Lewis was the Universal Ought.[15] He believed that moral responsibility was universally recognized; the sense of right and wrong is found in every culture testifying to the fact that this was a method using by God to speak to people. Moral awareness Lewis argued led to religious consciousness. Thus he said: "When the Numinous Power of which (men) feel awe is made the guardian of the morality to which they feel obligation," then religion comes forth.

In his discussion of heaven in The Problem of Pain Lewis listed as a medium of revelation the central motif of romanticism — the longing for something immortal, which he called by the German word **Sehnsucht**.[16] This concept of Sehnsucht is found in many of Lewis' writings. Thus he wrote: "If I find in myself a desire which no experience in this world can satisfy, the most probable explanation is that I was made for another world. If none of my earthly pleasures satisfy it, that does not prove that the universe is a fraud. Probably earthly pleasures were never meant to satisfy it, but only to arouse it, to suggest the real thing."[17] The fact that people are not satisfied with mere earthly pleasures is an indication that this is God's way of giving men a foretaste of heaven. Corbin Scott Carnell compared it to "the hound of heaven, relentlessly pursuing man that he may find his true identity and home."[18]

Lewis believed that God chose Israel as a medium of revelation and pounded into their heads the kind of God he was. In Mere Christianity he wrote: "Those people were the Jews, and the Old Testament gives an account of the hammering process."[19] In The Pilgrim's Regress he put forth the view that God's revelation to Israel "the Shepherd People," came by means of the Law.[20] To the Pagans God made himself known through myths, or pictures, so that through them pagan people were granted glimpses of God. Concerning Israel and the Pagans Lewis had this to say: "The truth is that a Shepherd is only a half man, and Pagan is only a half man, so that neither people was well without the other, nor could either be healed until the landlord's Son came into the country."[21]

Lewis argued that in both Israelite culture and in pagan culture God's revelation was gradual. God gave each class of people the amount of revelation they could stand. Among pagan peoples revelation took the form of mythology; among the Chosen People it took the form of Law and Prophets. In the New Testament, in Christianity the world will find the culmination of progressive divine revelation.

Lewis contends that in pagan literature there are premonitions of the coming of Christ. These "pagan dreams," as Lewis called them, were for him another medium of revelation. In Mere Christianity he claimed that in ancient literature are archetypal patterns, stories, rituals, and religious motifs "about a god who dies and comes to life again, and by his death, has somehow given new life to men."[22] Lewis expressed the view that the Roman poet Virgil (70-19 B.C.) in one of his ecologues may have given expression to a dim foreknowledge of Christ.[23] He also found a premonition of the Passion of Christ in Plato's Republic.[24] Lewis' book contains a number of passages in which he endeavored to create parallels between pagan mythologies and Christianity. He claimed that either pagan mythology is essentially demonic and therefore counterfeit revelation, or otherwise they contain premonitions of God's final revelation in Jesus Christ.[25]

Those acquainted with the New Testament will see at once that Lewis has not correctly comprehended the true relationship between the Old and New Testaments. Both Law and Gospel are found in the Old Testament. There is only one way of being saved since the Fall and that is by faith in Jesus Christ and through justification by faith. Paul's Letter to the Romans clearly teaches that the Gospel Paul was to preach and for which Paul was set apart, was the Gospel, "which through his prophets he promised before time, in holy writings, this Gospel is concerning his Son Jesus Christ our Lord" (1:2-3). In Acts 17:29-30 Paul in his Areopagus address called the time before the coming of Christ in the pagan world "the times of ignorance"; there is no hint that the myths of Greek and Romany mythology were premonitions for Christ's coming. Paul told the Greek philosophers that God commanded people to repent; inasmuch as he has fixed a day in which he will judge the world justly, by the Man whom he has ordained, and he has given proof of all this by raising him from the dead. The extra-Biblical revelations which Lewis assumed may be in harmony with the speculations of human reason but they have no basis whatever in the nude text of Holy Writ.

The stance held about the nature of the Bible will determine a theologian's religious beliefs and ethical practices.[26] Historical Christianity has held that the Bible is the inspired, infallible and the inerrant Word of God. Only a book which in all of its parts and all of its words is errorless and reliable in its pronouncements can carry full authority.[27] Modern evangelicalism is a divided house on this important issue today. The Bible is the inscripturated revelation of God. The Bible as the Word of God Written carries authoritative truth about God and man.

Various statements appearing in Lewis' writings would indicate that his views about the Bible are not in agreement with those of historic Christianity or with declarations found in the Anglican 39 Articles, the Westminster Confession and other Calvinistic and Arminian symbols.

It is difficult to classify Lewis' views about the nature of the Bible.[28]

Neo-orthodoxy holds that the Bible contains the Word of God and under certain circumstances may become the Word of God. The Bible is said to contain a record of God's revelation but in itself it is not a Written Revelation of God.

On the one hand Lewis believes that the Bible is not merely a witness to God's Word and that real spiritual truths are conveyed by the words of Scripture, yet on the other hand, he comes close to the neo-orthodox stance when he wrote that the Bible "carries" the Word of God.[29] For Lewis the Bible was a book of human literature which "has been raised by God above itself, qualified by Him to serve purposes which of itself it would not have served."[30]

Paul wrote Timothy that the Old Testament Scripture was God-breathed or "Spirit-spirited," which means that the entire Old Testament is the product of God's activity, in which the Divine predominates over the human and not vice versa as Lewis has envisioned the relationship between the human and divine in the Bible. The promise of Christ to His disciples that the Holy Spirit would guide them into all truth assumes that the books and epistles that the Holy Spirit would cause them to write would be errorless and not contain mistakes. Lewis does not believe in plenary inspiration, he does not advocate the errorlessness of God's Word. Lewis, for example, would not have agreed and his writings would not support Lindsell's assertion: "God the Holy Spirit by nature cannot lie or be the author of untruth. If the Scripture is inspired at all it must be infallible. If any part of it is not infallible, then that part cannot be inspired. If inspiration allows for the possibility of error then inspiration ceases to be inspiration."[31]

A reading of Lewis' letter to Clyde Kilby, will reveal that Lewis does not believe in an errorless and inerrant Bible.[32] In this letter he asserted that not all statements made in Holy Writ must be historical truth. He claimed that the kind of truth certain people demanded of the Bible was never "envisaged by the ancients."

4. Lewis' Views about the Nature of Man

Lewis did not accept Genesis 1:1-2:3 and 2:4-25 as historical accounts which give reliable   information about the creation of the universe, the planet earth, the creation of life itself, the creation of sun, moon and stars, the creation of all forms of life, animal and human. He does not recognize the account of Adam and Eve's creation as the definitive record of their human origin. Just as Genesis 1-2 were not regarded as factual and historical accounts so he misinterpreted Genesis 3 as myth. He espoused theistic evolution, although he argued against emergent evolution. In *The Problem of Pain* he wrote as follows: "For long centuries God perfected the animal form which was to become the vehicle of humanity and the image of Himself. He gave it hands whose thumbs could be applied to each of its fingers, and jaws and teeth and the throat capable of articulation, and a brain sufficiently complex to execute all material motions whereby rational thought is incarnated. The creature may have existed for ages in this state before it became man."[33]

How did man become an intelligent being according to Lewis? He speculated as follows: "In the fullness of time God generated in this organism 'a new kind of consciousness which could say 'I' and 'me,' which knew God . . . which could make judgments of truth, beauty, and goodness."[34] But this is not what the Bible says. Man was uniquely created, by a special act of God and not merely by that command, the method used to bring into existence all sea, land and air animals. The Bible especially states that man was made in the likeness and image of God, never said about the animals.

Lewis portrayed early man as a savage: "I do not doubt that if Paradise man could now appear among us, we should regard him as utter savage, a creature to be exploited, or, at best, patronized. Only one or two, and those holiest among us, would glance a second time at the naked, shaggy-bearded, slow-spoken creature; but they, after a few minutes, would fall at his feet."[35]

5. Lewis' Conception of Man's Fall

From a Biblical perspective the third chapter of Genesis is the saddest chapter of the Bible, because in it Moses records by inspiration how man and woman became subject to all forms of death: spiritual, physical and eternal. From the state of innocence Adam and Eve passed into the state where now man is a sinner. What is Lewis' teaching about man's fall? Lewis, following the evolutionary thinking, postulated many male and female animals becoming human at the same time. How many, he claims, he does not know. Primitive man did not sin at first. But in the course of time Lewis would have us believe that man was tempted to become as gods — "to call their souls their own." In his book, *Pain* the Cambridge professor describes man's fall as follows: "This act of self-will on the part of the creature, which constituted an utter falseness to its true creaturely position, is the only sin that can be conceived as the Fall."[36]

According to Lewis' views about the nature of man, his development is still going on — either toward, or away from God, the Creator. Possibly Lewis' acceptance of the views of Darwin expressed in *The Ascent of Man* caused him to look forward to immortality as the goal of human development. Lewis even speculated that animals who are related to man may have immortal souls. He argued that just as Christians are said to be "in Christ, and Christ is in God, so animals are in man." The immortality of animals does not lie in animals themselves but the fact that man once was an animal.[37] When Lewis was asked where he would put all the mosquitoes he replied: "A heaven for mosquitos and a hell for men could very conveniently be combined."[38]

When Lewis calls into question the historicity of the fall narrative this has serious implications for the doctrine of original sin. The New Testament gives indications that it understood Genesis 1-3 as chapters recording actual historical events. In the Scriptures following Genesis there are at least 22 references which clearly show that the Biblical writers after Moses considered these chapters as giving actual and factual events. Romans 5:12-19 clearly contrasts Adam and Christ. "As by one man sin entered into the world, and death by sin; and so death has passed upon all men" (Romans 5:19). To question the historic Genesis account relative to the fall narrative nullifies Paul's comparison between the first Adam and Christ, the second Adam.

6. Lewis' Christological Views

As an Anglo-Catholic Lewis accepted the Gospels as giving valuable information about Jesus Christ, the Son of God. He opposed the liberal interpretation of the Gospels and essentially rejected source criticism which resulted in "the quest for the historical Jesus." He had no sympathy for those who denied the resurrection of Christ from the dead. He accepted the miracles of Christ reported in the Gospels.

However, Lewis' view of the atonement was not that of the Bible. The vicarious and substitutionary death of Christ is a vital and important doctrine of the Christian faith. St. Paul informed the Corinthians that God made Christ to be sin for us that we might receive the righteousness of God through Jesus, who on the cross was man's substitute and was punished for the world's sins. Isaiah 53:3-5 is a classic passage describing what the Messiah, whom Jesus claimed to be, did for mankind. "He is pierced for our transgressions and crushed for our sins; He is punished to make us happy and wounded to heal us" (Beck). Christ as the Lamb of God took away the sins of the whole world. In the history of Christian thought there have been those in the church who balked at the doctrine of the substitutionary and vicarious atonement. This refusal to accept the clear Scriptural teaching has led to what has been termed "theories of the atonement." Lewis, unfortunately went along with the idea that there was no one correct meaning of what Christ accomplished by His sufferings and death. He claimed that theories of the atonement were not as important as the fact of the atonement itself. He argued that such theories as Penal substitution, limited atonement, vicarious atonement, ransom paid to the Devil, satisfaction made to the Father, were not as important as the reality of what Christ did on the cross. What happened when Christ suffered and died on Calvary? Paul gives the answer: "God was in Christ reconciling the world unto himself not imputing their trespasses unto them and hath committed unto us the word of reconciliation." (2 Cor. 5:19). "Christ was delivered for our offenses and raised again for our justification" (Romans 4:25).

Lewis chose from among the different views that developed in the course of church history that one which is usually known as the "moral influence theory." He favored the view that Christ died to give an example of divine repentance. Christ died for us to show us how we are to die to our fallen selves. Edgar Boss in his doctoral study of C. S. Lewis's theological pronouncements on the atonement claimed that the Britisher espoused "the example theory of the Atonement." In *Mere Christianity* Lewis wrote that the important fact about the death of Christ is to accept what Christ has done on the cross. Here is one of Lewis's explanations of the meaning of Christ's death:

Of course, you can express this in all sorts of different ways. You can say that Christ died for our sins. You may say that the Father has forgiven us because Christ has done for us what we ought to have done. You may say that we are washed in the blood of the Lamb. You may say that Christ has defeated death. They are all true. If any of them do not appeal to you, leave it alone and get on with the formula that does. And, whatever you do, do not start quarrelling with other people because they use a different formula from yours.

The doctrine of justification by faith is not understood by Lewis or its relationship to Christ's vicarious atonement. The Bible teaches that all men are dead in trespasses and sins. All have fallen short of the glory of God. According to 1 John 2:2 God sent His son to be the Savior of the world. Because of the redemption effected on Calvary God no longer imputes sins to men (2 Cor. 5:19). God does not charge their transgressions against them, but credits the merits of Christ to men's account. For the sake of Christ's complete satisfaction God "justifies the ungodly" (Romans 4:5). Those, who by nature and by their own deeds are ungodly, were because of the death of Christ pronounced just and righteous. Therefore, "by the righteousness of one the free gift came upon all men unto justification of life" (Romans 5:18).

The fruit of Christ's redemption is not that He merely opened for man the way to reconciliation with God, and that God is now ready and willing to forgive sins, pending certain conditions man must first fulfill. The fruit of Christ's redemption is that Christ actually did effect reconciliation, that God no longer imputes sins, but has in His heart forgiven all sins to men. This reconciliation and forgiveness of sins is an accomplished fact, it is an objective reality. There is not a single sinner whose sins have not been forgiven. This is called universal or objective justification. Dr. C. H. Little expressed it this way: "Objective justification may be defined as God's declaration of amnesty to the world of sinners on the basis of the vicarious obedience of Christ, by which He secured a perfect righteousness for all mankind, which God accepted as a reconciliation of the world to Himself, imputing to mankind the merits of the Redeemer."[40]

The fact of God's reconciliation to mankind and that all men's sins have been atoned are offered through and in the Gospel, "for therein is the righteousness of God revealed from faith to faith" (Romans 1:17). This righteousness is not God's righteousness but that of Christ. When the Holy Spirit through the Law causes men to see their sins and need for Christ's redeeming work, and when the same Spirit creates faith in Christ, then they are declared righteous and their sins are forgiven. The latter is called subjective justification.[41] Faith in Christ's saving and atoning work is wrought in the heart of man by Word and Sacrament, for faith cometh by hearing, and hearing by the Word of God.

The true place of the vicarious atonement and its interrelationship with other central doctrines was not grasped nor set forth by Lewis.

Christ and Other Religions According to Lewis

The exclusiveness of Christ and His religious teachings was not accepted by Lewis, because of the latter's love for other literatures and myths led him to not exclude pagans from salvation. Not all religions are totally in error in what they teach. Lewis contends that God has not revealed to us what his final attitude will be to those who do not know Christ. While Lewis believed that salvation comes only through Christ, he held that not only those who accept Christ but others also could be saved. It was his contention that a person's religious experience transcends superficial appearances. Lewis did not care for the distinction between Christian and non-Christian. To appreciate Christianity properly from Lewis's perspective he felt that it would be better to look upon Christianity as a developmental process. According to the *Britisher* there are people who are on the way to becoming Christians, even though they have not fully accepted the Christian doctrine about the person of Christ. In Mere Christianity he wrote: "There are people in other religions who are being led by God's secret influence to concentrate on those parts of their religion which are in agreement with Christianity, and who thus belong to Christ without knowing it."[42]

Lewis does not accept the assertions of Christ in John's Gospel and in Acts that there are not many roads to heaven, but there is the narrow road, which is only found in and through Christ. To justify his position he argued that there are Christians who are so confused that it was unprofitable for them, so Lewis claimed, and it was helpful to distinguish between Christian and non-Christian people.

Lewis's theologizing in *Mere Christianity* in other of his theological writings failed to understand the true nature of conversion and what is truly characterized Christian conversion, namely, that this work of the Spirit involves an instantaneous and radical change. His approach is synergistic and not monergistic when he describes what actually happens when a person becomes a new man in Christ. The real difference between justification and sanctification is also not understand by Lewis.

7. Lewis's Theological Views about Man's Ultimate Destiny
a. His Views about Heaven and Hell

According to clear Biblical teaching man's ultimate destiny will be determined by whether or not he has accepted Christ and believed in Him. Men will either spend their existence beyond the death in heaven or hell. Before the day of judgment the souls of the believers will be with the Lord, while the souls of the unbelievers will be with the devil in Hell. After the resurrection the same will hold true according to body and soul. The historic account of the life and death of Dives and Lazarus teaches this in Luke 16:19-31.

Lewis speaks of heaven and hell both as conditions and as places. According to the late Cambridge professor men are progressing toward either a state of heaven or hell. In *Mere Christianity* Lewis made it clear that heaven and hell are states. Thus he wrote: "To be one kind of creature is heaven; that is, it is joy and peace and knowledge and power. To be the other means madness, horror, idiocy, rape, impotence, and external loneliness. Each of us at each moment is **progressing to one state or the other**."[43] (Emphases supplied) In *The Great Divorce* Lewis wrote:

"And every state of mind, left to itself, every shutting up of the creature within the dungeon of its own mind — is, in the end Hell."[44] Again in *The Problem of Pain* Lewis theologized about hell as "the darkness outside" — "the outer rim where being fades away into nonentity."[45]

Hell was not eternal for Lewis, because he believed that men were in hell because of their own choice. However, those in hell can get out because the "doors of hell are locked on the inside."[46] Concerning the wicked men's soul we are "to hope that he may, in this world or another be cured." Since the Bible taught that God desires all men to be saved, Lewis believed that there would be a second chance offered the denizens of hell. While all will not receive eternal life because of their deliberate resistance, in *The Great Divorce* he envisages man in hell being offered a second chance. "You have been in Hell," a bright Spirit says to a ghost from hell, "though if you don't go back you may call it Purgatory."[47]

The Gospel of the Second Chance is a figment of Lewis's imagination, because it has **no** basis in Holy Writ, which declares: "It is appointed unto man once to die, and then the judgment" (Hebrews 9:27). In Matthew 18:8,9 the Savior speaks of the everlasting hell-fire. Paul speaks of "everlasting destruction from the presence of the Lord" (2 Thess. 1:9). When, therefore, on the Last Day the sentence of condemnation: "Depart from Me, ye cursed, into everlasting Fire, prepared for the devil and his angels," is passed, it will be carried out immediately, for it reads: "And these shall go into everlasting punishment" (Matt. 25:41,48).

b. Lewis's Belief in Purgatory

Since according to Lewis Christianity was an ongoing process, that men and women are either potential gods or devils who someday will reign in heaven or hell, it is not surprising that he advocated a belief in a purgatory, a theological position consonant with Anglo-Catholic tradition.

According to Lewis's understanding of Christ meant the necessity of purgation. That a person's sins were totally forgiven and that by repentance and faith a Christian's sins are daily forgiven and that a Christian who dies in a state of grace needs no further forgiveness or purgation beyond the grave, was not apprehended, and if it was understood, rejected in favor of the Roman Catholic doctrine of purgatory.

Thus in *Mere Christianity* Lewis wrote: "Whatever suffering it may cost you in your earthly life, whatever inconceivable purification it may cost you after death, whatever it costs Me, I will never rest, nor let you rest, until you are literally perfect."[49] In another of his writings Lewis wrote: "that purgatory is a place for purification of the saints, where at the very gates of heaven the saved soul 'begs to be taken away and cleansed.'"[50] In *The Great Divorce* Lewis claimed "that if we insist on keeping Hell (or even earth) we shall not see Heaven; if we accept Heaven we shall not be able even to retain even the smallest and the most intimate souvenirs of Hell."[51]

c. Prayers for The Dead

It is not surprising that a person who believed in a purgatory should believe in prayers for the dead, a doctrine held by Roman Catholicism and one rejected by historic Protestantism. Lewis was aware that Protestantism believed in an afterlife, in which people are either saved or lost, and thus there is no point in offering up prayers for the dead. But Lewis's retort was "our souls demand Purgatory, don't they?" [52]

8. Lewis's Views on the Eucharist

As an Anglo-Catholic he had views about the Lord's Supper that are not in harmony with Calvinism, Arminianism and Lutheranism. In Mere Christianity He expressed the view that the life of Christ comes to people by baptism, by belief and by Holy Communion.[53] Lewis warned his readers that he held views about the Eucharist that some would consider "magical."[54] Lewis taught that the physical bread and wine were transposed into spiritual vehicles which carried the life and grace of God. Like the Anglo-Catholic theological position Lewis believed in the "real" presence of Christ in the Eucharist.[55] He stated that Christ was present when the communicant worthily receives the Eucharist. On the other hand, Lutheran theology teaches that all who attend the Lord's Table receive the Body and Blood of Christ "in, with and under" the earthly elements, not only the worthy, but the unworthy, but the benefit, however, of the personal forgiveness of sins is bestowed on only those who are repentant sinners. Lewis is willing to accept the "magical" nature of the Eucharist because Jesus said: "Take eat" and not: "Take, understand."[56]

On the basis of the doctrinal positions that he has espoused Lewis cannot be said to be orthodox in his theological pronouncements. While he defended many truths of the Apostles' Creed, other basic doctrines were not accepted. In the true sense of the word Lewis was not a fundamentalist. Christensen claimed that despite his religious speculations, occasional pronouncements of unorthodoxy, and departures from theological conservatism, C.S. Lewis in no way identifies himself with what he calls Liberal Christianity. In his Preface to *Mere Christianity*, as well as in other writings Lewis clearly states his adherence to and defense of common or "mere" Christianity. To identify with the traditional certitudes of Christian faith is to be at odds with much of modern theology.[57]

The stance which Lewis took on Biblical criticism satisfied neither fundamentalists nor liberals. Lewis was inconsistent, allowing for a very liberal use of the historical-critical method when dealing with Old Testament data, as may be seen from reading his *Reflections on the Psalms*, and his writings in which he castigated modern Biblical criticism for its methodology when dealing with the Gospels, the main source for our knowledge about Christ's life and teachings. Richard Cunningham wrote: "Lewis' approach to biblical criticism offers a middle ground between radical criticism and literalism. Consequently, he can please neither the liberal nor conservative."[58] Lewis's autobiography, *Surprised by Joy* (1955) frankly described his spiritual development. He never became a Catholic, although his sympathies lay with the Catholic position, and he had many Catholic friends. Corbishley believes that "his early conditioning in ultra-Protestant Belfast may have been responsible for his reluctance to take the final step but his autobiography seems to imply that a distaste for many of the externals of organized religion was a stronger influence."[59]

Professor Paul Holmer of Vale made this observation about Lewis's theological stance: "Lewis' discovery of Christianity was plainly a rather momentous event for him. But he seems not to have been converted to a theological scheme at all, and he refused all his life to think than an understanding of Christianity would necessitate that he adopt an elaborate theology."[60]

Footnotes

1. As reported in *Christian News*, Vol. 17, No. 24, August 20, 1979. p.2
2. Introduction to C. S. Lewis in Michael J. Christensen, *C. S. Lewis on Scripture* (Waco, Texas: Word Books, 1979), p. ii.
3. Based on Roger Lancelyn Gree and Walter Hooper, *C. S. Lewis, A Biography* (New York and London: Harcourt Brace Jovanovich, 1974), 320 pages.
4. C. S. Lewis, *The Pilgrim's Regress* (2nd ed., Grand Rapids: Wm. B. Eerdmans Publishing Company, 1958), p. 5.
5. For a bibliography of Lewis's writings cf. Joselyn Gibb, editor, *Light on C. S. Lewis* (New York: Harcourt, Brace and World. 1965). pp. 117-160. This complete bibliography was worked up by Walter Hooper.
6. C. S. Lewis, *Reflections on the Psalms* (London: Geoffrey Bles, 1958), p. 109.
7. Edgar W. Boss, "The Theology of C. S. Lewis" (Th.D, dissertation submitted to Northern Baptist Theological Seminary, Chicago, 1948). This dissertation was reviewed by Clyde S. Kilby, *The Christian World of C. S. Lewis* (Grand Rapids; Wm. B. Eerdmans Publishing Company. 1964), 191-192.
8. Lewis, *Pilgrim's Regress*, **op. cit.**, p. 13; C. S. Lewis, Poems (New York: Harcourt, Brace & World. 1964), p. 81.
9. This letter is published in Christensen, **op. cit.**, pp. 97-99.
10. C. S. Lewis, *Mere Christianity* (New York: The Macmillan Company. 1960), p. 43.
11. C. S. Lewis, *The Last Battle* (New York: Macmillan Company, a Collier Book. 1960). pp. 164-65.
12 Lewis, *The Problem of Pain* (New York: The Macmillan Company. 1948), pp. 11-12
13. **Ibid.**, p. 4.
14. Rudolph Otto, *The Idea of the Holy*. Translated by W. Harvey (2d ed., London: Oxford University Press, 1950), 232 pp.; Lewis, The Problem of Pain. p. 4.
15. Lewis, *The Problem of Pain*, **op. cit.**, p. 10.
16. **Ibid.**
17. Lewis, *Mere Christianity*, **op. cit.**, p. 120.
18. Corbin Scott Carnell, *Bright Shadow of Reality* (Grand Rapids: Wm. B. Eerdmans Publishing Company, 1974), p. 144.
19. Lewis, *Mere Christianity*, **op. cit.**, p. 54.
20. Lewis, *Pilgrim's Regress*, **op. cit.**, p. 154.
21. **Ibid.**, pp. 152-55.
22. Lewis, *Mere Christianity*, **op. cit.**, p. 54.
23. C. S. Lewis, *Reflection on Psalms* (London: Geoffrey Bles), p. 104.
24. **Ibid.**, p. 108.
25. Douglas Gilbert and Clyde S. Kilby, *C. S. Lewis: Images of His World* (Grand Rapids: William B. Eerdmans Publishing Company, 1973), pp. 20-21.

26. Cf. Raymond F. Surburg, *How Dependable Is the Bible?* (New York: Lippencott-Homan, 1972), pp. 48-65.

27. Cf. Harold Lindsell, *The Battle for the Bible* (Grand Rapids: Zondervan Publishing House, 1976). pp. 25-26; Harold Lindsell, *The Bible In the Balance.* 1979), p. 14.

28. Christensen, **op. cit.**, p. 88.

29. **Ibid.**

30. Lewis, *Reflection on the Psalms*, **op. cit.**, p. 111.

31. Lindsell, *The Battle for the Bible*, **op. cit.**, pp. 30-31

32. Cf. the Letter of Lewis as given in Michelsen, pp. 97-99.

33. Lewis, *The Problem of Pain*, **op. cit.**, pp. 65.

34. **Ibid.**

35. **Ibid.**, p. 67

36. *The Problem of Pain*, **op. cit.** p. 68.

37. **Ibid.**, pp. 127-128.

38. **Ibid.**, p. 125.

39. Lewis, *Mere Christianity*, **op. cit.**, p. 157.

40. C. H. Little, *Disputed Doctrines* (Davenport: Lutheran Literary Board, 1933), p. 60.

41. Edward W. A. Koehler, *A Summary of Christian Doctrine* (Distributors: The Reverends Louis H. Koehler and Alfred W. Koehler, 1939). pp. 146-47.

42. C.F. Lewis, *Mere Christianity*, **op. cit.**, p. 176.

43. **Ibid.**, pp. 86-87.

44. C. S. Lewis, *The Great Divorce* (New York: The Macmillan Company. 1946). p. 65.

45. Lewis, *The Problem of Pain*, **op. cit.**, p 115.

46. **Ibid.**

47. Lewis, The Great Divorce, **op. cit.**, p. 32.

48. C. S. Lewis, *Letters to Malcolm: Chiefly on Prayer* (New York: Harcourt, Brace & World, 1964), p. 108.

49. Lewis, *Mere Christianity*, **op. cit.**. p. 172.

50. Lewis, *Letters to Malcolm*, **op. cit.**. p. 108

51. Lewis, *The Great Divorce*, **op. cit.**, p. vi

52. Lewis, *Letters to Malcolm*, **op. cit.**, p. 108.

53. Lewis, *Mere Christianity*, **op. cit.**, p. 62.

54. Lewis, *Letters to Malcolm*, **op. cit.**, p. 9.

55. **Ibid.**, p. 103.

56. **Ibid.**, p. 104.

57. Christensen, **op. cit.**, p. 38.

58. Richard B. Cunningham, *C. S. Lewis: Defender of the Faith* (Philadelphia: The Westminster Press. 1957), p. 101.

59. T. Corbishley, "Lewis, Clive Staples," *The New Catholic Encyclopedia*, VIII, 686.

60. Paul L. Holmer, *C. S. Lewis: The Shape of His Faith and Thought* (New York: Harper & Row. Publishers, 1976). pp. 100-101.

Questions

1. What is said about C. S. Lewis ____?
2. The writings of C. S. Lewis are being read by ____.
3. Liberals cannot appreciate that Lewis rejected ____.
4. Lewis did not follow the traditional ____ principle of early Lutheranism.
5. Lewis's concept of writings of Scripture as being inspired is not in har-

mony with ____.

6. Lewis contended that God has not revealed to us what his final attitude is toward ____.
7. Lewis held that ____ was a medium of revelation.
8. The views of Lewis about the Bible are not in agreement with ____.
9. Lewis does not believe in a ____ Bible.
10. Lewis did not accept Genesis as ____.
11. Lewis espoused ____ evolution.
12. The doctrine of ____ by faith is not understood by Lewis or its relationship to Christ's ____.
13. Lewis did not exclude ____ from salvation.
14. The Gospel of the Second Chance is a figment of Lewis's ____.
15. What was Lewis's view of purgatory? ____

Do the Coptic Gnostic Texts Preserve Authentic Traditions About Jesus and First Century Apostolic Christianity?

Christian News, November 11, 1980

Phillip Nobile, a syndicated writer for *Roto Magazine*, whose columns or articles are usually found in metropolitan newspapers, usually discourses on some secular or religious subject on what he terms the "Uncommon." When writing on religious subjects he endeavors to acquaint his readers with views that challenge traditional orthodox teachings. Such was the situation regarding the column which appeared in *Roto, The Fort Wayne News-Sentinal* on Saturday, January 12, 1980. His column reports the answers given by Dr. Elaine Pagels, professor of religion at Barnard College, New York City and head of the college's religious department. The interview was prompted by the publication of her book, *The Gnostic Gospels* by Random House (201 East Fiftieth Street New York, N. Y. 10022) on November 26, 1979, The book has 182 pages and sells for ten dollars.

Mr. Nobile states in his "Uncommon" column that "*The Gnostic Gospels* should be required reading for all Christians." He further claims that the views found in these Coptic Gnostic writings challenge the orthodoxy of primitive Christianity. After a brief historical introduction, Nobile addressed seven significant questions to Dr. Pagels, the answers of which are found in his column, and may be said to center upon some of the issues and conclusions she discussed in her book. Her interpretations challenge the reliability of the New Testament as well as the position taken by second century Christian theologians over against Gnosticism.

The Sources for Pagels The Gnostic Gospels

In The Gnostic Gospels there is to be found an interpretation by one of the female scholars, who for a decade or so has been involved in the translation of the Coptic books which were discovered at Chenoboskion, the Greek name of a village, now known as Qasr-es-Sayad. It lies near Nag Hammadi, some 30 miles northwest of Luxor. At Chenoboskion in 1945 natives accidently discovered 13 well-preserved papyrus codices written in Coptic. One of these documents found its way into the Jung Institute at Zurich, Switzerland, while the others eventually were deposited in the Coptic Museum in Cairo. There are altogether 52 works, a few of them extent in two or more versions.

The Jung Codex is in Subachmimic, one of the five dialects of Coptic, used in Egypt by Christians and pagans until the Muslim conquest of Egypt. The other documents are in the Sahidic dialect, which was the main dialect of Coptic, the other important dialect being Bohairic. The Chenoboskion documents of the Gnostics are translations made in the 3rd and 4th centuries A.D. The 13 codices comprise treaties, letters, dialogues, discourses, apocalypses, apocryphal "gospels" and acts, and the like which come from different branches of Gnosticism. The Gnostic Coptic writings found in the cemetery at the foot of Jewel-el-Tarif, near Nag Hammadi, were probably a part of the library of the Sethian sect, a group which ceased to exist when monasticism became prominent in this area. The names of many of the Nag Hammadi codices were known from earlier references in the apologetical writings of Ireneus, Hipaulutus and other Church Fathers. From early writings it was also known that many of the Coptic Gnostic Treatises were written by Valentinus (middle of the second century A.D.) and other of the Gnostic codices were attributed to the Gnostic sects of the Sethians, Archontes and Barbelognostics.

It was only last year that those interested in the history of Gnosticism finally had placed at their disposal an English translation of the Nag Hammadi Gnostic materials. At first scholars had at their command only a few of them. The first to be published (both Coptic text and translation) was *The Gospel of Thomas*, which contained a collection of 114 sayings of Jesus, some of which were known to students from the Greek papyri found at Oxyrhynchus in Egypt. Some critical scholars went so far as to consider these sayings of Jesus (called agrapha) as authentic as those found in the Four Gospels. The second was *The Gospel of Philip*, which contains a collection of sayings strongly dualistic in character and which emphasized the four elements of water, earth, wind and air. The third work was *The Gospel of Truth*, a book which is composed of a conglomeration of different phases of Gnostic philosophy.

In 1966 The Institute for Antiquity and Christianity of Claremont Graduate School undertook "The Coptic Gnostic Library Project." Scholars conversant with Coptic and Gnosticism were assigned the task to make translations of the codices found at Nag Hammadi. The program director was Dr. James M. Robinson. In 1977 the documents from Chenoboskion were published by Harper & Row as *The Nag Hammadi Library*. It is a translation of Gnostic works to which no less than thirty-one scholars contributed. This translation volume therefore makes available for the first time in English the secret Gnostic writings as they were current in one community in Egypt, where Later Pachomius established Christian monasteries.

Dr. Elaine Pagels was associated with this project and participated in the translation work. Her interest with Gnosticism and Nag Hammadi began in 1965 when she was made aware of the existence of the Chenoboskion codices. According to the information given by Dr. Pagels to Robert Dahlin editor of "Trade News" in *Publishers Weekly*, October 15, 1979, Mrs. Pagels was warned as a child by her father, a doctor of medicine, that religion was something to be avoided so that she had not received a Christian upbringing. Her father's prohibition aroused in her a curiosity to investigate religion. During her youth for a year she was associated with an evangelical group because she liked the emotional power of the music.

Later she attended college at Stanford where she studied Greek and this was followed by graduate study at Harvard, where she earned a

Ph.D. in the history of religions. Her doctoral dissertation was entitled: *The Johannine Gospels in Gnostic Exegesis: Heracleon's Commentary on John* and two years later Fortress Press printed her book, *The Gnostic Paul, Gnostic Exegesis of the Pauline Letters*. In both of these works she shows her preference for Gnostic thought as compared with the interpretation which John and Paul set forth in their writings in the New Testament.

Pagels' Presuppositions

The views expressed especially in and the answers given to Nobile's question show that Dr. Pagels must be classified as belonging to the School of Comparative Religion, a school of interpretation that believes in the evolution of religion and does not hold that the religion of the Old and New Testaments is in a class by itself. For her The Judaeo-Christian faith represents the religious beliefs of Jews and Christians which were developed by human beings who either borrowed their religious ideas from other sources or created their own religious views. That the Bible is unique among the "bibles" of other Near Eastern religions is not recognized nor accepted. In her interpretation of the New Testament writings she follows the radical conclusions of radical literary criticism and the speculations of form criticism. New Testament higher criticism does operate with the view that in the New Testament we do not have a reliable record of Jesus sayings and deeds. What we actually have according to form critical scholars are the interpretations of what the early apostolic church wanted Jesus to say and do. Till recently critical scholars were engaged in the "third quest for the historical Jesus." That Pagels shares this view may be seen from the answer given to Nobile's question: "Is there any reason to speculate whether the Gnostics were closer to the Lord's actual words and thoughts than orthodox Christianity?" Thus the Barnard-Columbia professor responded: "In fact, it's very hard to know what Jesus really said on earth. Many scholars theorize that lists of Jesus sayings predated the Gospels themselves. There are many sayings in the Gnostic gospels that parallel the New Testament, and others don't."

Dr. Pagels shares the views of liberal unorthodox savants who place the sayings of Jesus on a par with the sayings in *The Gospel of Thomas*. Thus she quoted from the Gospel of Thomas the following saying: "If you do not bring forth what is within you, what is within you will destroy you." Nobile asked Mrs. Pagels why this saying is not found in the canonical Gospels, the latter responded: "If you were trying to build the institutional church or turn the Jesus movement into an organizational structure by the year A.D. 200 there are certain things you wish Jesus hadn't said. Specifically, this saying does not serve the purpose of this institution."

According to Pagels the orthodox church is based on the premise that God and Christianity are very different. However, it is her conviction that Gnosticism is closer to Buddhism in that the human and the divine are identical at some point, and if this is true, then she claims a person does not need to have a church between him or her and God.

Pagels and New Testament Christology

Nobile asked Professor Pagels two questions about Jesus. The first query was: "Did the Gnostics trace their beliefs back to Jesus?" Her answer was: "They claimed access to a secret oral tradition, that is, teachings of Jesus outside Matthew, Mark, Luke and John." The other question was: "Did the Gnostics believe that Jesus was divine." She replied: "Gnostics felt they were divine. Gnostics felt they were identical with Christ. According to the Gospel of Philip, you became not a Christian but a Christ." To this we reply: "These Gnostic views are erroneous." Today church historians cannot pronounce favorably for Gnosticism, because how can judgments be made concerning secret oral traditions as to their validity? The Gnostics ascribed heretical views to Jesus which he never uttered. Making the Gnostics initiates to be like Christ would mean ascribing deity to them, for the New Testament teaches clearly that Christ was God.

According to the information vouched to the writer of the "Story Behind the Book," in Publishers Weekly some of the views propounded in *The Gnostic Gospels* were occasioned by a woman's group who asked Mr. Pagels to speak on the relationship of her work to other members of her sex. Pagels as she thought of the symbolism of the Gnostic teachings, she had studied, said she became aware of the possibilities for her work. She averred: "I began to see that these religious ideas and symbols had political and social consequences." With this new insight her attitude toward her work changed. She claims that the Gnostic writings conceived God as Father and Mother. This discovery prompted Pagels to deliver a lecture at Barnard College, entitled: "Whatever Happened to God the Mother?" It was materials from this lecture which were developed into chapter of her current book, a chapter which bears the title: "God the Father/God the Mother." (Cf. *The Gnostics Gospels,* pp. 48-69). The fact that women are kept out of the priesthood by the Roman Catholic Church is attributed by Pagels to the fact that God was conceived as masculine. In the review of her book in *Publishers Weekly*, she is reported as declaring: "It was Pope Paul who said there can be no woman priests because our Lord was a man. But when you look at Nag Hammadi and its sexual references, it leads you to question this." The identical evaluation of Pope Paul's prohibition on women in the priesthood was also given to Nobile as reported in his interview with Pagels.

Dr. Pagels shows her liberal and anti-New Testament feelings when she approves of the view of the Gnostics which they held about the resurrection. The Coptic Gnostic writings interpret the resurrection of Christ symbolically and not literally. This is in the best tradition of historical-critical liberalism which denies the deity of Christ and so the life of Christ ends with Him in the grave. Pagels supports the symbolic interpretation of the resurrection rather than the literal. If there is one truth taught by the New Testament it is the fact of Christ's bodily, corporal resurrection. Over 200 times the various New Testament writers refer to and attest Christ's physical resurrection from the dead. Paul in 1 Corinthians 15:12-19 shows the clear implications that flow from the denial of Christ's bodily resurrection. What Pagels is advocating is paganism and heresy.

In dealing with the New Testament Professor Pagels claims that she is dealing with the New Testament data not as a theologian but as an historian. But the truth of the matter is, that she does not deal responsibly with the data of the Gospels. The Biblical text asserts: "Christ died and arose again." "Christ was delivered for our sins, but raised again for our justification." As a member of the School of Comparative Religions she simply treats the historical data as traditions which may not be true. The early church in her view attributed sayings and deeds to Jesus which He may not have spoken or done respectively. According to Pagels the church created a tradition to support its preconceived theological notions. The Gnostics also developed their own tradition and allegedly ascribed to Jesus sayings they believed he might have spoken. We ask: Why does a modern student of early religion not prefer the New Testament in preference to the Gnostic understanding of religion? For the twentieth century the implication of favoring Gnosticism would be that the church has the right to create its own religious tradition and may select ideas from past traditions that suit the cultural demands of our day. Since Coptic Gnosticism favored women, ideas of the Gnostics are more useable today in view of the Equal Rights Amendment. Woman should and ought to serve as priests, ministers or rabbis just as well as the men. Pagels believes that Gnosticism has more to offer that is timely than orthodox Christianity.

Pagels' Interpretation of Early Church History

In her *The Gnostic Gospels* Pagels has advanced interpretations about early Christianity that are highly speculative. The doctrine sponsored by the orthodox church, according to the Barnard professor, was monotheism, a view she claimed promoted not only one God but one bishop. The Gnostic writings envisioned more than one God, and in many cases proposed that the source of divine power was to be found in "the depth" of being. In harmony with the latter view, the Gnostics eliminated an organized hierarchy and. established a social structure which was based on the equality of all participants. The description of God as Father was repudiated by the Gnostics who conceived God as embracing both male and female elements, and in harmony with this theological conception they argued for the equality of men and women both spiritually and politically. With the ERA's demands for total equality, it is not difficult to comprehend that Pagels, who does not recognize Paul as an inspired apostle of Jesus Christ, an apostle who writes under the inspiration of the Holy Spirit, that she would favor Gnostic teachings and argue for their superiority as compared with New Testament teachings which do not support the ERA's demands.

Many scholars who have discoursed on Gnosticism have called this philosophical and theosophical speculation "a heresy." When Nobile asked Dr. Pagels about the heretical character of Gnosticism, she replied: "Of course that's the view of orthodoxy-their belief survived because that the Holy Spirit guides the church. But as an historian, not a theologian, I could suggest other reasons for the successes of orthodoxy. I am interested in why certain beliefs became orthodox and others heretical." Nobile had suggested to Pagels that since the Gnostic gospels were lost to history and the New Testament gloriously triumphed that therefore, the orthodox church has a larger claim on truth. With this reasoning Pagels would not go along and seems to reject it.

What Pagels and Nobile do not appreciate is the fact that the teachings of the New Testament are God-given and because of this, they are true. The action of Christians did not make the teachings authentic or true. The New Testament teachings are correct, are orthodox, because they are true to begin with. Pagels really shows what her position is over against the New Testament as compared with Gnosticism when the following query of Nobile's which said: "Although Gnosticism is heretical and strikes at the heart of Christian dogma, can believers profit from reading these secret gospels?" she answered as follows: "Christianity is not monolithic. After reading the Gnostic gospels, many Christians may wonder whether the belief in the resurrection is necessary. John Paul II denied women the priesthood, insisting there is no tradition of women in orders. The Gnostic gospels show rituals in which women played the roles of priest and bishop. The Gnostics also spoke of God in feminine terms as well."

Pagels has attempted to find important implications in the Coptic documents for orthodox Christianity. Among a number of invalid deductions is the theory that the early Christians placed the crucifixion at the center of their creed because by means of it they affirmed and justified their persecution and martyrdom at the hands of the Romans. This identification of their persecution with Christ's crucifixion, she thinks, helped in the consolidation of various branches of orthodoxy and those who were sympathetic to their suffering were allegedly attracted to the Christian movement. According to Pagels this emphasis on the crucifixion led to the adoption of a literal interpretation in place of a symbolical interpretation as the Gnostics understood the meaning of Christ's death. In response to this theory of Pagels what are the facts? The crucifixion and the resurrection were two doctrines which were proclaimed together. Of what good was a dead Christ? Because of the great hope for a happy eternity after death the Christian faith spread and grew. The Biblical message of Christ crucified and Christ risen was powerful, it was a dynamite which changed people's lives and ultimately conquered the Roman Empire.

According to her interpretation of Biblical data and the facts of church history, Pagels contrasted the concept of "the true church" and the Gnostic attitude toward it. As the orthodox saw it, the church was the actual community of worshippers. Any person who was willing to subscribe to the teachings of the Scriptures and the theological assertions of the Apostles' Creed was welcome to join the fellowship. However, according to the Coptic Gnostic conception of the "true church", the latter consisted of the elite and of those who demonstrated a "spiritual maturity," which meant that not all people were welcome. Gnostic religion tended to become esoteric and exclusivistic.

Pagels also claimed that the orthodox church held that one could only find salvation through Christ, that the way to God the Father was through Jesus Christ and "that outside the church there was no salvation." By contrast, the Gnostic treaties suggest that humanity had created God and thus man had written him the potential to discover truth and God. Thus religion was self-discovery and the church was unnecessary for salvation.

Gnosticism and Heresy

Full-blown Gnosticism is a development of the 2nd and 3rd centuries A.D. That varieties of Gnosticism were beginning early is evident by statements in 1 Corinthians, Colossians, the Pastorals, the Johannine Epistles, Jude and Revelation. This type has been called "incipient Gnosticism." Thus the Apostles John, Paul and Jude had need to warn the members of various congregations against the heresies propounded by Gnostic teachers or by teachers which appear to have advocated views that were part and parcel of later Gnosticism. Pagels speaks of the Gnostics as among the first Christians, but from the perspective of New Testament religion the Gnostics were false teachers, who under all circumstances were to be avoided.

The Gnostics endeavored to use many Christians ideas and concepts which they misused. G.W. MacRae has labelled Gnosticism as a heresy. He wrote that Gnosticism essentially was a form of paganism.

The Westminster Dictionary of Church History, edited by Jerald C. Brauer (Westminster Press, 1971) said about Gnosticism: "The most significant feature that strikes anyone reading the Gnostic texts for the first time is the vast and bizarre mythology that characterizes the formulation of the Gnostic systems, but these strange systems all have at their center a doctrine of redemption, a doctrine that differs from the Christian doctrine of redemption in its view of man and the world. The world, according to the Gnostic view, is intrinsically evil, having been created by an inferior deity or Demiurge who fell from the supreme Deity. Man, as a part of the created world, is thus radically alienated from God, but there is in man a divine spark or spiritual element constituting his true inner self, which longs to free itself from alienation and ascend to the ultimate God of light and goodness" (pp. 361-362).

Between the Light God and the material evil world Gnostics placed a vast systems of emanations or aeons, which became less pure as they were removed from the Light God and approached the Demiurge, the creator of the evil world. Jesus was identified with one of these angelic aeons. This hierarchy of aeons, or cosmic powers, angels and semidivine beings, differed in the various Gnostic systems, highly mythological in character. Since man is in alienation from the Light God (called pleroma) he cannot redeem himself. Redemption according to Gnosticism is effected by gnosis, or knowledge. By contrast in Christianity man is saved by grace and faith in the atoning work of Christ, in whom dwells all of fullness (pleroma) of the Godhead bodily (Col. 2:9). Gnostic systems taught that if a man really knows in his inner self what his true condition is, and what his destiny is, namely to become part of the Light God, he is redeemed.

In some Gnostic systems a redeemer came to this earth to reveal to man the knowledge by means of which he can save himself. Christ was seen as one of the aeons who descended upon the human Jesus in order to reveal to man the human knowledge for self-salvation (autosoterism). Jesus really did not become man and He did not die upon the cross. The Westminster Dictionary of Church History correctly asserted: "It was this teaching about Christ, set against the background of the dualistic teaching of the intrinsically evil nature of the world and of man's full human nature, which made Gnosticism the greatest single threat to the Christian church in the 2nd century" (p. 362).

Gnosticism's doctrine of God, the Trinity, creation, hamartiology, anthropology, soteriology, cosmology, the sacraments, Christology, the church, and eschatology were totally different from the these loci of theology as given in the Old and New Testaments.

Questions

1. The interpretations of Elaine Pagels challenge ____.
2. Pagel must be classified as ____.
3. The Gnostic views are ____.
4. The Coptic Gnostic writing interpret the resurrection of Christ ____.
5. Pagel is advocating ____ and ____.
6. The teachings of the New Testament are ____ given.
7. Gnostic Systems taught that if a man really ____.

The Untenability of Ecummenisms Attempt to promote the Apocrypha as Word of God

Christian News, November 17, 1980

The nineteenth and twentieth centuries have been characterized by a great amount of Bible translation activity by various Christian missionaries and linguists.[1] The different European and American Bible Societies were primarily involved in this type of activity of making the Bible, or at least portions of Holy Writ, available to many different peoples of the world, who according to linguists are using somewhat between 2,500 and 4,000 different languages and dialects.[2] The Old Testament was probably first translated into one of the Aramaic languages, but the oldest translation known of the Old Testament was the Septuagint (LXX).

With the spread of Greek as a result of the conquests of Alexander the Great there developed a need for translation among the Jews who were dispersed throughout Egypt, the Mesopotamian Valley, Persia, Elam. The result of the deportations by the Assyrians and Babylonians was that the exiled Hebrews needed a Bible translation in the language of the lands where they lived. More Jews lived outside of Palestine than in it. Communities of Jews were to be found in Alexandria (Egypt), Antioch (Syria), Asia Minor, Greece and Italy. Throughout Mesopotamia heavy concentrations of the descendants of Abraham could be found living. During the time of Alexander the Great, as well as after his death, Greek came to be the International language (lingua franca) of the conquered peoples. Van Bruggen claims that "no one is precisely sure of the history of the Septuagint, but in the synagogues of Greek-speaking Jews, it attained a wide acceptance long before the birth of Christ."[3] It is interesting to note that the Jews did not offer opposition to the idea of rendering their sacred Scripture into another language. Once the Septuagint had been made, an activity which was probably took over a century (c. 250-130 B.C.), they regarded it with the same reverence as they did the Hebrew text.

The Letter of Aristeas, now considered as belonging to the paeudepigraphical writings of Judaism, gives a popular account about the origin of the Septuagint. At one time the facts reported in this letter were accepted as factual. In the last few hundred years real questions have been raised about the factuality of the data and events reported in the Letter of Aristeas, which was written for Philocrates. The latter writing claims that the Greek Old Testament translation was made at Alexandria in Egypt by seventy-two translators, who had been brought from Jerusalem by King Ptolemy II of Egypt. The Egyptian king accepted the translation and threatened to punish those who would dare lo make a change in it. Thus the Septuagint was regarded as a fixed translation and was thus to be immune from changes and textual fluctuations. The Letter of Aristeas gives an account and report of the alleged table conversations between the Jewish translators who in their work were led by the Spirit and wisdom of God. The impression given the reader of The Letter of Aristeas was that the Palestinian translators exhibited the wisdom of Solomon, which amounted to a wisdom of revelation and faith.

In the synagogues where the Septuagint was employed, it was regarded as the Word of God. Philo, the Jewish philosopher and exegete, used the Septuagint went so far as to consider the Jewish translators u prophets of God.

Thus at the beginning of the Christian era the LXX was the version of Holy Writ employed by the Jews throughout the diaspora. The first Christians were Jews who saw no reason not to continue to use the LXX in their worship services and in their missionary propaganda. The LXX differed from the Hebrew Bible in that the former contained both canonical as well as apocryphal writings. Some scholars claim that at Christ's time there existed two different canons: the Palestinian and the Alexandrinian, which contained more than the 22 of 24 books of the Hebrew canon. The number 22 is arrived at by adding Ruth to Judges and Lamentations to the prophetic book of Jeremiah. Both Jewish and conservative Christian scholars refuse to accept this theory of the existence of two different Jewish Bibles. Concerning those books not found as a part of the Hebrew Old Testament, but found in the Septuagint, F.F. Bruce has written as follows:

> There is no evidence that these books were ever regarded as canonical by any Jews, whether inside or outside Palestine, whether they read the Bible in Hebrew or in Greek. The books of the Apocrypha were first given canonical status by Greek-speaking Christians, quite possibly through the mistaken belief that they already formed part of the Alexandrian canon. The Alexandrian Jews may have added the books to their versions of the Scriptures, but that was a different matter from **canonizing** them.[6]

Bruce further suggested that possibly the Alexandrian Jews may have added what are known as the apocryphal books because of her bibliographical conditions, He argued that "when each book was a papyrus or parchment roll, and a number of such rolls were kept together in a box along with canonical documents, without acquiring canonical status. Obviously the connection between various rolls in a box is much looser than between documents which are bound together in a volume."[7]

The New Testament and the Apocrypha

Some of the New Testament writers give evidence of a knowledge of various apocryphal writings. Some believe that the author of Hebrews (1:1-3) may have had in mind the Wisdom of Solomon when he penned the opening words and verses of his letter addressed to the Hebrew Christians in the dispersion.[8] However, that supposition is doubtful. In Hebrews 11:35-36 the author of Hebrews appears to have knowledge of the two books of the Maccabees. Relative to the issue as to whether or not the New Testament writers regarded the Apocrypha as canonical Unger wrote: "Despite the fact that New Testament writers quote largely from the Septuagint rather than from the Hebrew Old Testament, there is not a single clear-cut case of citation from any of the fourteen apocryphal books, eleven of which Rome receives as canonical. The most that can be said is that the New Testament writers show acquaintance with these fourteen books and perhaps allude to them indirectly, but in no case do they quote them as inspired Scripture or cite them as authority."[9] C.C. Torrey, deceased Semitic scholar of Yale, in his *The Apocryphal Literature* has attempted with great effort to sort out apocryphal quotations and allusions in the New Testament, but he has to concede "that in general, the apocryphal scriptures were left unnoticed."[10]

The Stance of the Early Church
Over Against The Apocrypha as Canonical

Among the Apostolic Fathers Clement of Rome (A.D. 95) incorporated quotations from the Wisdom of Solomon in his writings. He is witness to many Old Testament books, including Judith. Polycarp of Smyrna (died A.D. 156) quoted from Tobit. Some of the early Church Fathers quoted and used the Apocrypha as Scripture in public worship. Certain of the Church Fathers accepted various of the Apocrypha as canonical, such is the case with Ireneus, Tertullian, and Clement of Alexandria.

Those who support a broader Alexandrian canon also cite as evidence the fact that the catacombs depict episodes from the Apocryphia.[11] Furthermore, the great Greek manuscripts (Aleph, A. and B) interpose the Apocrypha among the Old Testament books. The Syriac Church of the fourth century accepted the Apocrypha as Scripture. Some of the apocryphal books have been found among the MSS written in Hebrew in the Dead Sea Community at Qumran.[12] Jerome of Hieronymous, translator of the Vulgate, even though he grudgingly includes a translation of the Apocrypha in his work, declared all books outside the Hebrew canon as apocryphal.[13]

The Position of the Eastern Church toward the Apocrypha

The Christians of Palestine and other lands were concerned about the extent of the Old Testament canon. Melito and Origen investigated this important issue.[14] Melito has left us the oldest catalogue of the canonical books of the Old Testament, originating from the second Christian century. His listing of his research on the Old Testament canon (ca. A.D. 171) indicates that the Jews did not accept the Apocrypha as canonical and he listed the same books as those found in the Hebrew Old Testament manuscripts.

The same canon that Melito found was that employed by Justin Martyr, who studied widely and travelled extensively. He wrote voluminously but never cited the Apocrypha as Word of God. Justin Martyr, one of the Apologists of the second century A.D., became involved in a controversy with Trypho, an Ephesian Jew, but in their use of Scripture there was no difference, because both used the Palestinian canon.

The first translation of the Hebrew Old Testament into Syriac, a related Semitic language to Hebrew, was the Peshitta which did not accept the Apocrypha as a part of the Old Testament canon. However, later Syriac-speaking Christians incorporated the Apocrypha in the fifth century in their Old Testament Bible.

Another voice from the early Eastern Church testifying to the Palestinian of Hebrew Old Testament canon is Origen (died A.D. 254), one of the most learned of the Church Fathers. According to Eusebius' testimony as given in his *Ecclesiastical History* (VI:25) Origen found only the Hebrew canon as the one recognized by the Jews. At the close of his report Origen stated: "And apart from these (i.e., not comprising a part of the O.T. canon) are the Books of the Maccabees."

Basil of Cappadocia considered the Old Testament canon as comprised of twenty-two books, thus agreeing with Melito and Origen. About A.D. 363 the Council of Laodicea passed the following canon: "Books not admitted into the canon, but only the canonical books of the New and Old Testaments" are to be "read in the Church." John Crysostom taught that only the books written in Hebrew should be adhered to, thus providing evidence for the Hebrew Palestinian covenant in the East.

The Position of the Western Church toward the Apocrypha

There appears to be a difference of opinion on whether Tertullian accepted the Apocrypha as canonical or not. Both Unger[15] and Archer[16] claim the North African theologian did not while Fritsch[17] claims he recognized some of the Apocrypha. The situation varied among the Church Fathers. However it was Augustine who at two North African Synods prevailed upon Latin-speaking Christianity to recognize the Apocrypha. Geisler and Nix note that even in several instances in his writings there is not that firm commitment to the Apocrypha that one would expect.[18] Thus Augustine omits Baruch and included I Esdras, and thereby took a position opposite to that taken by the Council of Trent. From the City of God it is apparent that Augustine distinguished between "secondary canonicity" and "primary canonicity." The former term he ascribed to the Apocrypha; the latter term was reserved for the Hebrew canon. The councils of Hippo and Carthage were local councils and do not qualify as ecumenical gatherings where qualified persons could examine the arguments and evidence for the claim of the canonicity of the Apocrypha. Augustine was not a trained Hebrew scholar as was Jerome of Hieronymus.[19]

Jerome (340-42) was an accomplished scholar, who translated the Bible from Hebrew into good flowing idiomatic Latin. He opted for "the recognition of only the Hebrew canon, excluding the Apocrypha." In the *Prologus Galeatus* he rejected the Apocrypha as being of the same quality and authority as the 22 of 24 Hebrew Old Testament canon (39 in the English Bible).[20] Other contemporaries of Jerome and Augustine were Hilary of Poitiers, France (305-366), and Rufinus of Aquila, Italy (died 410), both of whom left lists of a 22-book Old Testament canon.

The Historical Witness from Augustine to the Council of Treat

The sponsorship of the Apocrypha by Augustine had the effect that in the succeeding centuries many subscribed to the "larger canon" of forty-four books, while others continued to adhere to the Hebrew canon. There was a tendency, due to Augustine's theological influence, to break down the distinction between "inspired" and "uninspired." In the writings of Cassiodorus (*The Institutes*) (556) and Isidore of Seville (636) are to be found two lists side by side without a suggestion which is to be followed.[21]

In the seventh century Gregory the Great came out for the Palestinian canon and was not in favor of the Apocrypha. In the fifteenth century Cardinal Ximenes, publisher of a polyglot Bible, the *Complutention Polyglot*, in his introduction claimed that all books written in Greek were not in the canon. These were Tobit, Judith, Wisdom of Solomon, Ecclesiasticus, Baruch, Maccabees with the additions to Esther and Daniel. The latter were not to be used for doctrine but could be read for the people's edification. The dedication of the Complutensian Polyglot was made to Pope Leo X and accepted by him.

Cardinal Cajetan, who at one time examined and dealt with Luther, was reputed as a theologian; he praised Jerome for his distinction between Canonical and uncanonical books. In his dedication to Pope Clement VII Cajetan wrote like this about the Apocrypha: "These are not canonical books, that is, they do not belong to the rule for confirming those things which are of faith; yet they can be called, that is, belonging to the rule for the edification of the believers. With this distinction what is said by Augustine and written by the Council of Carthage can be rightly apprehended."[22]

The Stance of the Holy Eastern Orthodox Church

This church has the official title: "The Holy Orthodox Catholic Apostolic Eastern Church." Its Bible has been the Septuagint since time immemorial. Like the Roman Catholic Church this church, whose theological development closed with John of Damascus, has also granted a degree of religious authority to the Apocrypha.

At the Councils of Synods of Constantinople (1638), Jaffa (1642) and Jerusalem (1672) the Apocrypha were declared to be canonical. Still according to H.S. Miller theologians in the Eastern Orthodox Church have refrained from citing them as authority in doctrinal matters. A writer like Patriarch Metrophanes (died 1640) omits the Apocrypha as does the Large *Catechism* (1839), one of the most authoritative doctrinal standards, because "they do not exist in the Hebrew."[23]

The Attitude of Lutheranism and
Protestantism to the Apocrypha

The Protestant Reformation represented a return to the, true source of religious authority . When Luther and other reformers gave the people of Germany, England and other European countries the Bible based upon the Hebrew Old Testament and the Greek New Testament, the Apocrypha were rejected as Word of God and relegated to a position between the two testaments.

In the Counter Reformation the Roman Church was practically forced to declare the Apocrypha canonical. In 1546, twenty-nine years after Luther posted his famous Ninety-Five Theses, the Council of Trent rendered the dogmatic decision, known as "Sacrosancta." The Decree reads as follows:

The Synod ... receives and venerates ... all the books (including the Apocrypha) both of the Old and of the New Testament-seeing that one God is the Author of both ... as having been dictated, either by Christ's own word of mouth or by the Holy Ghost... if anyone receive not as sacred and canonical the said books entire with all their parts, as they have been used to be read in the Catholic Church ... let him be anathema.[24]

The session at which "Sacrosancta" was passed was one the prolonged ones at which only fifty-three prelates were in attendance, not one of whom was a scholar reputed for historical knowledge. The Vatican Council of 1969-70 reaffirmed the dogmatic decision of the Council of Trent.

The Protestant Churches from their inception accepted as canonical the canon used by Christ and the apostles and the canon of the Early Christian Church. The Church of England (1562) held to the view of Jerome, who had said that the Apocrypha were useful "for example of life and instruction of manners; but yet doth it not apply them to establish any doctrine." The Presbyterian Westminster Confession declared: "The books commonly called Apocrypha, not being the same divine inspiration, are no part of the canon of Scripture, and therefore of no authority in the Church of God, not to be otherwise approved and made use of than other human writings."

The Difference between the Protestant and Roman Catholic Old Testament Canons

Luther, Calvin, Beza, John Knox and other Protestant reformers returned to a 22 of 24 (39 in English) - book Old Testament canon, while Rome sponsored a forty-four-book canon. The Protestants did not accept as authoritative Scripture the following: I Esdras, II Esdra, Tobit, Judith, the Additions to Esther, the Wisdom of Solomon, Ecclesiasticus, Baruch, the Prayer of Azariah, the Song of the Three Men in the Fiery Furnace, Susannah, Bel and the Dragon, I and II Maccabees. The Roman Catholic Church at Trent affirmed these books as deuter-canonical with the exception of the following books: III and IV Esdra, (I and II Esdras of the Protestant list) and the Prayer of Manasses (Manasseh).

This means that the Roman Catholic Church and the Greek Orthodox Eastern Church recognize a number of books as canonical which orthodox Protestants do not. This fact has serious implications for the establishment of doctrine and ethical practice, as will be shown somewhat later in this article. Since Protestant churches are members of the World Council of Churches, in which also the Greek Orthodox Eastern Church bolds membership one needs to ask how can there be doctrinal unity when such a foundational issue as the extent of the canon is not settled for agreement exists on this important matter, as to the source for doctrine and life. Furthermore, how can there be honest dialogue and agreement reached between the theologians of Lutheranism and Roman Catholicism who have been involved in the so-called Roman Catholic-Lutheran dialogues, when there is this basic difference on the extent of what actually constitutes the Word of God?

The Position of the Apocrypha in Protestant Bibles

The first Protestant Bibles included the Apocrypha and considered them as appendices to the Books of the Old Testament. They were not considered as an integral part of the Bible, which they are in the Vulgate, The Jerusalem Bible and The New American Bible of the Confraternity of Christian Doctrine of the American Hierarchy.

The Luther Bible had them as an appendix between the Old Testament and the New Testament, but over the Book of Judith was to be found the statement: "Das sind Buecher, so der heiligen Schrift nicht gleich gehalten, und doch nuetzlich und gut zu lesen sind."[26] (These are books which are not equal to those of Holy Scripture, nevertheless are useful and good to read).

In 1950 when Bruce issued the first edition of his *The Books and the Parchments* he wrote: "It is not easy now to buy a copy of the Bible-A.V. that includes the Apocrypha. Very few are produced, probably because very few are demanded. But in the early days of English Bible-printing the Apocrypha was included as a matter of course."[27]

It is interesting to trace the stages by means of which the Apocrypha were eventually excluded from English Bibles, published both in England and America. The King James had a translation of these non-canonical books, although the rendering of the apocryphal books was not of the same accurate quality as were the 39 books of the Old Testament and the 27 of the New. According to Bruce, the Zurich Bible of 1524-1529 and the Coverdale Bible of 1535 were the first Bibles to place the Apocrypha after Malachi with a special statement that the Apocrypha were less authoritative. However, there was an interesting exception, because Baruch was placed after Jeremiah. The 1937 edition of Coverdale, removed Baruch and placed it after Todit. The Coverdale Bible was followed by Matthew's Bible of 1537, but the Prayer of Manasseh was added, which was turned into English on the basis of the French of Olivetan. The next Bible, the Taverner's of 1539 did not include the introduction to the Apocrypha found in the Coverdale and Matthew Bibles. The Great Bible of 1539, a revision of the Matthew Bible, retained the introduction to the Apocrypha, but called them Hagiographa and not Apocrypha.

The revision of the Taverner Bible, known as Beck's Bible, in 1549-1541, had a translation of I Esdras, Tobit and Judith and also added 3 Maccabees to the other Apocrypha. This was the first time that 3 Maccabees appeared in English. Beck's Bible contained a new introduction to the Apocrypha, and commended its books "for example of life."

The Geneva Bible of 1560 printed a preface to the Apocrypha which was more severe than those of previous Bibles. It gave the prayer of Manasseh as an appendix to 2 Chronicles, but specified that it was apocryphal in character. By contrast the Bishop's Bible of 1568 dealt with the Apocrypha as if they were true canonical books. The Puritan Party rejected the attitude of the Bishop's Bible toward the Apocrypha.

The Geneva Bible was the first English Bible to omit the Apocrypha. In a number of copies of this 1599 Bible the printer seems to have omitted the Apocrypha. The Authorized 1611 Bible had them, and Archbishop Abbott forbade any stationer from issuing a Bible that did not print in it the Apocrypha. But in 1640 a Geneva Bible published in Amsterdam deliberately omitted the Apocrypha. A defense for its omission was placed by the issuers between the two testaments. In 1644 Parliament ordered the apocryphal books not be read in public worship services. Among Nonconformists this position became popular. The first English Bible printed in America in 1782-83 did not have the Apocrypha. In 1826 the British and Foreign Bible Society began the practice of only printing the 39 Old Testament books and the 27 New Testament books and no Apocrypha. Since 1827 the Apocrypha have been omitted from practically all editions. In the Revised Version of 1885 and the American Standard of 1901 they were not included. In 1895 the Apocrypha were revised and issued separately.

The Comeback of the Apocrypha

However, in the second half of the twentieth century the apocrypha is making a comeback. In 1965, the Second Vatican Council stated that in certain cases Roman Catholics would be allowed to undertake translations "in cooperation with the separated brethren."[29] As a result of these assertions negotiations were undertaken between The United Bible Societies and the Vatican's Secretariat for Promoting Christian Unity of the Vatican.[30] In 1968 there was issued a document entitled "Guiding Principles for Interconfessional Cooperation in Translating the Bible."[31] Before the publication of these "principles" scholars were busy in France on a joint Catholic-Protestant translation project. According to a report appearing in the Bulletin of United Bible Societies (98, 1975, p. 7) there were 134 of 630 translation projects in process reported that where being done in cooperation with Roman Catholics.

The insertion of the Apocrypha between the Old and New Testaments was one of "the guiding principles" adopted jointly by the United Bible Societies and the Roman Catholic hierarchy. According to van Brueggen this represents a "drastic principle."[32] On the face of it the adoption of this translation guideline, does not appear to be different from the practice adopted by the revisers of the Revised Standard, or of the New English Bible all of which made translations of the Apocrypha and included them in these translations. But van Bruggen contends that there is a difference. We wrote: "In the preparation of the ecumenical Bible translations, the Apocrypha is not merely an appendix translated along with the Bible. Rather it is treated as part of the Bible, its translation is put on the same level as that of the canonical books, and its position between the Old and New Testaments implies its right to be there."[33]

This "Guiding Principle" means that the ecumenical Bible will appear with the Roman Catholic imprimatur in a Bible supplied with the Apocrypha, books Protestants have rejected as canonical. However, we are informed that the imprimatur will only appear in Bible that Roman Catholics want for other persons desiring it. Those Protestants who do not wish to have the Apocrypha may have a Bible minus the Apocrypha. It is the judgment of van Bruggen that for a Protestant to ask for a Bible without Apocrypha is tantamount to requesting a Bible which is not complete and thus would constitute only a larger portion of a more complete Bible. Van Bruggen claims: "The question must continue to be asked: What is the extent of the Bible? The answer to that question is crucial for the future of the Bible. How can a Bible be truly ecumenical while people are divided on its extent? Sooner or later a new conflict over the Apocrypha will arise because of modern ecumenical translations."[34]

Those liberal scholars in Protestantism and Lutheranism who are proponents of the historical-critical method believe that it was the Jewish Synagogue, specifically the Synod of Jamnia which determined the extent of the Old Testament canon. It would be the contention of these scholars that the church in the twentieth century can decide which books it would wish to use as source for theological ideas, which are also determined by the religious consciousness of the individual Christian and by the canons of logic. The church under the influence of the Holy Spirit can add to and detract from the Biblical canon. It is not surprising therefore, to find liberal Protestant scholars willing to recognize the Apocrypha as part of the Old Testament canon, especially when such an attitude aids and abets the union of Christendom.

Footnotes

1. Eric M. North, *The Book of A Thousand Tongues* (New York: Harper & Brothers, 1938), pp. 20-35.
2. Victoria Fromkin and Robert Rodman, *An Introduction to Language* (New York: Holt, Rinehart and Winston, 1978), p. 329; Louis H. Gray, *Foundations of Language* (New York: The Macmillan Company, 1939), p. 417.
3. Jacob van Bruggen, *The Future of the Bible* (New York and Nashville: Thomas Nelson Inc., 1978), pp. 37-38.
4. H.T. Andrews, "The Letter of Aristeas," in R. H. Charles, *The Apocrypha and Pseudepigrapha* (New York: Oxford University Press, 1913), I, 83-122.
5. Artur Weiser, *The Old Testament: Its Formation and Development* (New York: Association Press, 1961), p. 341; Gleason Archer, *A Survey of Old Testament Introduction* (Chicago: Moody Press, 1974), p. 73.
6. F. F. Bruce, *The Book and the Parchments* (London: Pickering & Inglis, 1950), p. 157.
7. **Ibid.**, p. 157.
8. R.K. Harrison, "Apocrypha," Merrill C. Tenney, *The Zondervan Pictorial Bible Encyclopedia* (Grand Rapids: Zondervan Publishing House, 1975), I, 208.
9. Merrill F. Unger, *Introductory Guide to the Old Testament* (Grand Rapids: Zondervan Publishing House, 1951), p. 101.
10. C. C. Torrey, *The Apocryphal Literature* (New Haven: Yale University Press, 1945), p. 101.
11. Norman L. Geissler and William E. Nix, *A General Introduction to the Bible*

(Chicago: Moody Press, 1968), p. 170.

12. Archer, **op. cit.**, p. 74.
13. Unger, **op. cit.**, p. 173.
14. **Ibid.**, pp. 101-103.
15. Unger, **op. cit.**, p. 103.
16. Archer, **op. cit.**, p. 72.
17. C.T. Fritsch, "Apocrypha," G.A. Buttrick (ed.) *The Interpreter's Dictionary of the Bible* (New York and Nashville: Abingdon Press, 1962), I, 164.
18. Geissler and Nix, **op. cit.**, p. 171.
19. William H. Gree, *General Introduction to the Old Testament* (New York: Scribner's and Sons, 1899), The Canon, p. 56.
20. **Ibid.**, p. 66.
21. Unger, **op. cit.**, p. 105.
22. Green, **op. cit.**, p. 177.
23. Geisler and Nix, **op. cit.**, p. 172.
24. Philip Schaff (ed.), *The Creeds of Christendom*, 6th edition, II, p. 81.
25. Article VI of *The Westminster Confession*, 1647.
26. Cf. Die Bibel oder die ganze Heilige Schrift des Alten und Neuen Testaments nach der deutschen Uebersetzung Dr. Martin Luthers. 501. Auflage (Halle: a.d.s.: Druck und Verlag der v. Cansteinischen Bible-Anstalt, 1922), S. 917.
27. Bruce, **op. cit.**, p. 165.
28. In the following the writer has utilized the history of the Apocrypha as found in the English Bibles from 1535-1611, as given by Bruce, **op. cit.**, pp. 165-167.
29. *Dogmatic Constitution of Divine Revelation* ,VI, 22.
30. Cf. "The Bible Societies and the Roman Catholic Church," *Bulletin of the United Bible Societies*, 71 (1967); also cf. W. M. Abbott, "Easy Access to Sacred Scriptures for All," *Catholic Biblical Quarterly*, 30 (1968), 60-75.
31. For the text of "Guiding Principles etc.," cf. Bible Translator, 19 (1968), 101-110.
32. Van Bruggen, **op. cit.**, p. 33.
33. **Ibid.**, p. 34.
34. **Ibid.**, p. 35.

Questions

1. The oldest translation of the Old Testament was ____.
2. When did Greek become the international language of the conquered peoples? ____
3. In the synagogues where the Septuagint was employed it was regarded as ____.
4. What was the Peshitta ?____
5. What did Jerome do? ____
6. What was the attitude of Luther toward the apocrpha? ____
7. The Council of Trent decreed that the apocrypha ____.
8. What does the Greek Orthodox Church teach about the canon? ____
9. Liberal Protestant scholars are willing to recognize ____.

Recent Events in Egypt Focus Attention on the Ancient Coptic Church

Christian News, October 28, 1981

Recent historical happenings in early September and early October have focused attention on an ancient Christian Church in Egypt, namely the Coptic.[1] Pope Shenouda III, leader of the Coptic Christians, was placed under house arrest in a desert monastery by now deceased President Anwar Sadat, charged by the latter as having "political ambitions" and contributing to the violence between Christians and Moslem fundamentalists. Sadat also placed other Christian and Moslem leaders under arrest. At the celebration on Monday October 6, when Sadat and other high officers reviewing a parade in commemoration of a victory in a war against Israel, one of the high officials on the review stand was Coptic Bishop Samuel, who had earlier visited the United States.

Many Christians probably have never heard of the Coptic Church of Egypt, which happens to be one of the oldest Christian communities in the world! The Coptic Church has followers in the United States and should not be confused with a small cult which calls its self the Coptic Brotherhood International. The latter was the name "Coptic" and also has alleged to have had its origin in the secret Coptic temples of Egypt. One of the purposes of this essay will be to compare the two.

The Beginnings of the Coptic Church

The Egyptian Coptic Church traces its spiritual heritage to the preaching of Saint Mark. The word "Coptic" is ultimately derived from the Greek *Aigyptos*.[2] The name is applied to the body of Christians of Egypt, who were subject to the patriarch of Alexandria, but who were divided at the Council of Chalcedon into two opposing factions: the Jacobites, or dissident Copts who adhered to the Monophysite view of Christ's person, and the Melchites or Catholic Copts.[3] The latter remained faithful to the Holy See of Rome.[4]

At the Council of Laodicea, Dioscorus was the patriarch of Alexandria (c. 454), and he defended the Monophysite teaching (i.e. that Christ had only one nature) being convinced that he was defending the orthodox teaching of St. Cyril.[5] When Dioscorus refused to recant, he was deposed from his ecclesiastical throne and sent into exile, where he died. The majority of Egyptians looked upon their disgraced patriarch as a martyr to his faith. The stance of the Council of Laodicea on the person of Christ was rejected by the Egyptians and the result was that the Egyptians embraced the Monophysite view on the person of Christ.

Church historians contend that it was ecclesiastical and national jealousy that kept the Egyptians from embracing the Christological beliefs held by the Western Church, namely that there were two different natures, the human and the divine, found in the one person of Christ. Since the emperors of Byzantium (modern Constantinople) favored the decrees of Chalcedon, the Egyptians named those "the Melchites," or "King's men." At times the emperors of Byzantium kept a Melchite patriarch on the Alexandrian ecclesiastical throne by power, however, most of the time the patriarch was a Monophysite. For the next century and a half the history of the Church in Egypt was characterized by a struggle between those who adhered to the decrees of Chalcedon of 451 and those who rejected them. Finally, for the sake of peace Emperor Justin in 567 recognized two patriarchs of Alexandria, one a Melchite and the other a Monophysite. From that date forward the Egyptian Church embraced the Monophysite understanding of the person of Christ, rejected by the Western Church as understanding of the person of Christ, rejected by the Western Church as a heresy. The ritual of the church was a modified form of the Alexandrian rite and its liturgical language was Coptic.

The Coptic Language

The Coptic language is the liturgical language of the Coptic Church. Coptic is the last stage of the ancient Egyptian language which goes back to the Pyramid Texts.[6] In contrast to earlier stages of Egyptian, which were written in monumental hieroglyphics, hieratic script, or demotic script, Coptic was written in the Greet alphabet with the aid of seven letters borrowed from the demotic writing.[7] Many words from Greek were employed by the Christian writers of Coptic books.

Coptic has come down to modern times in five distinct dialects without counting the multiple interplay of one dialect on the other. The five dialects with their symbols are: Sahidic, (S), Bohairic (B), Fayumic (F), Akhmimic (A), and ub-Akhmimic (A2). Sahidic was spoken and written in Upper Egypt, whose capital was Thebes. Bohairic was as the dialect of Lower Egypt. Bohairic became the language of the Coptic liturgy, while Sahidic became the literary vehicle in which many ecclesiastical and pagan works were composed.[8] The Gnostic heretics also published their writings in Coptic; the most sensational find of these writings was made in 1945, when at Chenoboskion about 13 Gnostic codices were found. From a grammatical and literary viewpoint, Sahidic was the most important of the five dialects. Translations of the New Testament exist both in Sahidic and Bohairic. Various books of the Old Testament were translated from the Greek Septuagint into some of the Coptic dialects, and so from the viewpoint of Old Testament textual criticism, the Old Testament translations in Coptic do not have the same value as the New Testament versions in Coptic have for the New Testament textual critic.

In 640 Egypt was conquered by the followers of Mohammed who used Arabic, the sacred language of the Koran. Arabic became the spoken language of Egypt and Islam the state religion. The Islamic conquest had a devastating effect on the fortunes of Christianity in Egypt. Thousands, yes, hundreds of thousands apostatized to Mohammed's religion and within a few hundred years Coptic ceased to be a spoken and written language, only being kept alive as the liturgical language of the Coptic Church's liturgy. Probably between the 13th and 14th centuries Coptic became a dead language, although it appears to have lingered on in certain villages till the seventeenth century.

A considerable literature exists now in Coptic: most of it was popular in nature-moralizing, homiletical and writings of Coptic monastic writers. The surviving corpus of Coptic texts contains quite a proliferation of apocryphal writings and lives of the saints written in legendary and epic style.[9] Coptic literature also reflects the influence of the Greek Church Fathers, for many of their writings were translated into Coptic.[10] Not all Coptic literature consists of Christian writings; the language was also employed to perpetuate the theological beliefs of the Gnostics and of the Manicheans, both considered as heretical forms of Christianity, if not outright heresies.[11]

Throughout the centuries since 640 the Coptic Christians have suffered discrimination and at times persecution, the result of which were, that frequently under this type of pressure Christians in Egypt embraced the faith of Allah as reached originally by Mohammed. Estimates vary about the numerical strength of the Coptic Church in Egypt, some give a figure of two million, some even claim there are as many as six million.

The Monophysites Christians formed churches not only in Egypt, but also in Ethiopia and Syria. These national churches were separated from those churches who accepted the Chalcedonian Christology. When therefore Egypt was invaded in the seventh century, A.D., neither Constantinople or Rome came to the defense of the Monophysite Christians as they were attacked by Moslem forces.

There are about 100,000 Coptic Christians in the United States and Canada, who are to be found organized into 26 congregations, served by 28 priests. These churches worship, in English, Arabic, and Coptic.[12] Charles Austin in an article appearing in *The New York Times*, September 9, 1981 claimed that "Egyptian Christians in this country have been concerned about anti-Christian violence in their homeland, but have disagreed on how they should respond."[13] When President Sadat visited the United States late in summer, the Christian Coptic Church look an advertisement in *The New York Times* welcoming him and declaring that they would "pray almighty God for the success of the peace mission for the Middle East and prosperity of Egypt." However, the American Coptic Association bought an advertisement in ***The Washington Post*** **in order to draw attention to the harassment of Christians in Egypt. Dr. Shawky Karas, president of the Association, charged that Egyptian authorities ordered the church to place pro-Sadat advertisements and to have bishops and priests on hand when Mr. Sadat visited New York, and Washington. Karas also claimed that when Moslem extremists attacked a Christian suburb of Cairo and killed more than a hundred Christians in three days of rioting, that the police did nothing to protect the Christian Copts.[14]**

The Coptic Church Not Related to the
Cult Called "Coptic Brotherhood International"

The Coptic Church of Egypt and Its counterpart in the United States are Christian Churches. Apart from the Monophysite question, the Coptic and Eastern Orthodox Churches agree in doctrinal matters.[15] The doctrine of the Trinity is central to the Christian theological system, whether it be Roman Catholic, Eastern Orthodox or Protestant. According to an interview of Jens Sorenson,[16] staff writer for the *Journal Gazette*, Fort Wayne, Indiana, who interviewed John Davis, the spiritual master of the Coptics, a group with about a membership of a few thousands in the United States, there are radical differences between the Coptics and the teaching of the Coptic Church.

The Coptic faith of this American Cult group did not take hold in the United States until 1927, when Hamid Bey, a master teacher of the secretive Coptic Temples of Egypt, came to the United States in response to a challenge of Houdini. Davis claimed that Bey had complete control of his body to a point where he could reduce his body to two or three beats a minute and go into suspended animation at will.

The beliefs of this twentieth-century cultic Coptic Brotherhood International are opposed to what Davis has termed the negative aspects of Christianity. The cultic Coptic Brotherhood claims that every person is destined for heaven on earth. The existence of the devil ls rejected. The dire predictions of Revelation need not go into fulfillment. The Second Coming of Christ has already occurred and will not still be a future event of world history.

According to Davis, as stated to Sorenson, the world is now entering upon a new age. Davis claimed that he now is engaged in preparing the way for the new messiah's ministry. The age in which mankind now lives is a transitional period between the Piscean Age and the Aquarian Age. Each new age is introduced by an avatar, or god coming to earth in bodily form. The last avatar was Jesus. The Second Coming of Christ occurred when Jesus appeared on November 11, 1949 near Mount Shasta, California and thus the United States will become the new Holy Land.

The new Christ is now in the company of protectors and travelling the world to judge the world's consciousness level and when enough people have reached this Christ-consciousness, then Christ's ministry will begin. What did Davis mean by Christ-consciousness? According to him, it is a person's awareness that he is in essence one with God. It is man's destiny to attain to Christ-consciousness. This is, so Davis claimed, what Jesus meant when he said to Nicodemus: "Ye must be born again."

The Coptic Brotherhood International's View of Christ

This cult does not believe in the divinity or deity of Jesus Christ. The cult distinguishes between Jesus and Christ. "Jesus was not the Christ when he was born in Bethlehem," said Davis to Sorenson, "He was only the Son of Man." "However, when Jesus was baptized in the Jordan River, He became the Son of God." Then Jesus had attained Christ-consciousness. Sorensen wrote in the *Ft. Wayne Journal Gazette*, August 25, 1979: "Coptics also believe in reincarnation and the law of karma-that every action has an equal and same reaction." These are basic teachings that were thrown out of the Bible in the third century A.D. by orthodox Christians.

Davis claimed that the first Christian Church was established in A.D. 70 by St. Mark and it was Coptic. All other denominations have branched from the Egyptian Church. Church historians, of course know, that this is not true. The Coptic Brotherhood International claims to have historic roots that go back even farther in time than A.D. 70, averring that their roots really extend back to 22,000 B.C., when philosophy flourished in Atlantis, the mythical island of which Plato wrote. Davis claims that Atlantis was a place with an advanced civilization, both technologically and spiritually advanced, where a tremendous power existed. However, Atlantis was destroyed when this magic power was overextended. Some of the inhabitants are alleged to have escaped to Egypt and Mexico. These transplanted people from Atlantis built pyramids, which to this day are the continual source of amazement by men today as to how the people of Egypt and Mexico built such massive structures.

According to Davis an important part of the Coptic philosophy is to promote a balance between the physical, mental, emotional and spiritual life. A Fort Wayne priestess, named Loma, of this cult was also interviewed by Sorensen and she told the former: "We teach people to live in harmony with themselves." Again she stated: "We try to teach people to live in harmony with themselves." Again she stated: "We try to teach people how to live in the world." The Coptic Church of Egypt it should be noted, has no women priests functioning in the church. To help individuals to control their lives the cultic Coptics emphasize certain practices to help strengthen techniques for the physical body to meditate on the spiritual.

Davis also claimed that an important concept of the Coptic Brotherhood International is to promote the universal brotherhood of man and goodwill among peoples and religions. Every religion is basically good and has certain elements of truth in it. According to Davis all religions constitute a beautiful picture puzzle. They all need to be fit together and it would appear that Bey and Davis have been able to accomplish this much needed task.

An Evaluation of the Cultic Coptic International Brotherhood

There is absolutely no relationship whatsoever between the Christian Coptic Church and the Coptic Brotherhood International. The one is Christian and the other is a form of paganism. The Christian Coptic Church derived its theology from the Bible, the cultic Coptic Brotherhood from various sources. The latter is an amalgam of heterogeneous ideas woven together by a fertile mind. The Coptic Brotherhood International borrowed its basic ideas from India and promotes a world view which is not that of the Bible. James Sire has written a book, *Scripture Twisting - 20 Ways the Cults Misread the Bible*.[17] An analysis of the religious philosophy of this small American cult will show that they are guilty of a number of the Bible ways cited by Sire in his useful and informative book.[18] The use of the cultic Coptics reveals that they misquote the Bible, tearing verses out of their context to support totally unbiblical views. Using the name "Coptic" is also deceptive because there is no proved connection between the it and the true Coptic Church of Egypt. Even though they reject Christ's claims, they still find it useful to use the name of Jesus to give religious respectability to their philosophical speculations and thereby lead astray gullible and uninformed Christian people, who seem to have a penchant to become the victims of all kinds of cult leaders.

It is truly to be hoped that no Christians will permit themselves to be misled by this pagan and soul-destroying movement active ln certain places in the United States.

Bibliographical Footnotes and Reference

1. For a history for the Coptic people, fc. W. H. Worrell, *A Short History of the Copts* (Ann Arbor: University of Michigan Press, 1945).
2. For a history of the Coptic Christian Church, cf. David Attwater, *Eastern Christian Churches* (Milwaukee: Bruce Publishing Company, 1961), Vol. 1; Edward R. Hardy, *Christian Egypt* (New York and London, 1952), Ibrahim Nosby, *The Coptic Church, Christianity in Egypt* (Washington: Ruth Sloan Associates, 1955); Th.E. Dowling, *The Egyptian Church* (London: Cope & Fenwick, 1909); Betold, Spuler, Die morgenlaendlschen Kirchen (Leiden: E. J. Brill, 1964).
3. George A. Maloney, "Coptic Church," *The New Catholic Encyclopedia*, cf. IIV, 769.
4. **Ibid.**, p. 769.
5. Kenneth Scott Latourette, *A History of Christianity* (New York: Harper & Brothers, 1953), pp. 171-173.
6. George Steindorff, *Lehrbuch der Koptischen Sprache* (Chicago: The University of Chicago Press, 1951), pp. 1-5.
7. William F. Edgerton, "Coptic Language and Literature," *The New Americana Encyclopedia*, III, 770.
8. Anton Maumstark, *Die Christliichen Literatures des Orients*, I. Das christlicharamaeische und das koptische Schriftum (Leipzig: G. J. Goeschensche Verlagshandlung, 1911), pp. 107-121.
9. Edgerton, **op. cit.**, p. 771.
10. P. Beilet, "Coptic Language and Literature," *New Catholic Encyclopedia*, IV, p. 312.
11. **Ibid.**
12. Charles Austin, "Coptic Church Is Ancient and Isolated," *The New York Times*, Wednesday, September 9, 1981.
13. **Ibid.**
14. **Ibid.**
15. "The Coptic Church," *The Encyclopedia Britannica*, Micropedia Section, III, 138.
16. Jens Sorenson, "Coptics await ministry's fulfillment," *The Fort Wayne Journal Gazette*, August 25, 1979.
17. James W. Sire, *Scripture Twisting, 20 Ways the Cults Misread the Bible*. Downers Grove Illinois: Intervarsity Press, 1980), 180 pp.
18. Cf. **Ibid.**, pp. 55-160.

Questions

1. Where is the Coptic Church? ____
2. Why did the Coptic Church begin? ____
3. What is the monophysite teaching? ____
4. Byzantium is modern ____.
5. Who conquered Egypt in 640?____
6. What happened to the Coptic language? ____
7. What is the numerical strength of the Coptic Church? ____
8. There are about ____ Coptic Christians in the United States.
9. What is the Coptic Brotherhood International View of Christ? ____

The Forgotten Sesquicentennial of Charles Darwin's Beagle Voyage

Christian News, April 19, 1982

December 27, 1981 marked the 150th anniversary of the beginning of the celebrated voyage of *H.M.S. Beagle*, a voyage famous for the influence it had on Charles Darwin (1809-1882). David Crosson, former director of the Allen County Fort Wayne Historical Society, lamented in a letter printed in *The Fort Wayne Journal Gazette*, January 2, 1982, that this centennial was being ignored. Mr. Crosson believes that our intellectual heritage reflects Darwin's influence; just think how often the word "evolution" appears in American political thought and also in our scientific literature.[1]

The Importance of the Beagle Voyage

There is no doubt about it that this Beagle voyage was very important to Darwin's beliefs and determined the direction of his life between 1836 and 1882.[2] Darwin was engaged as a naturalist and his duties were to study rocks and life of the places visited and also to collect specimens. Darwin began the voyage as a divinity student who hoped to collect evidence to support the Biblical explanation of creation. In fact, the captain of the Beagle, a vessel of the Royal Navy, had taken Darwin along for that specific purpose and he never agreed with the logic of the conclusions that Darwin later drew from his observations and the theories that were based on his collected data as stated in his major published writings.[3]

The Purpose of the Beagles's Long Journey

The purpose of the sea journal was to survey the coasts of Patagonia, Tierra del Fuego, Chile, Peru and to visit some of the Pacific Islands as well as to establish a chain of chronological stations around the world. The H.M.S. Beagle landed at Falmouth England on October 2, 1836 after having been away from England nearly for five years and had been involved in a voyage that entailed a great deal of danger. On one occasion the ship foundered when attempting unsuccessfully to go around Cape Horn and at times hostile Indians presented dangers and problems when the crew went ashore. While on this voyage, Darwin was bitten by a large bug (Triatoma infestans) of the Pampas, which carries the trypanosome that causes Chagas disease. As a result, after the return from the ship, Darwin fell into ill health and became a semi-invalid. No doctor was able to cure Darwin's illness.

When Darwin with a B.A. to his name began his scientific observations, he was by no means a finished naturalist. Darwin made both geological and biological observations. In writing about his geological observations Sir Gavin de Beer, author of *Charles Darwin, A Scientific Biography*, wrote:

1339

Darwin's first geological observations were on the structure of the Caper Verde islands. He was able to unravel the past history of Sao Tiago Island and to show evidence indicating the past elevations and subsequent subsistence of the island. In Argentina, Darwin's observations of rock strata enabled him to interpret correctly certain previously misunderstood features of the way sedimentary rocks crystallize when they are metamorphized by the pressure of overlying rocks.[4]

In Chile Darwin had occasion to notice the effects of earthquakes and volcanic eruptions in causing the ground to rise. His study of the Andes led him to conclude that strata in these mountains had been uplifted a several thousand feet. His observations of geological elevations and subsistence led him to formulate a theory as to the manner the coral reefs were formed.

It was Darwin's biological observations and interpretations that eventually led him to postulate his form of the theory of evolution. Gradually he rejected the current view of his time that held that species of plants and animals were immutable. When he returned to England he recorded his observations in *The Journal of Researches*, in which he described the Beagle voyage, a book said by some to be one of the best travels books ever written. The concept of evolution, the idea that species change with the passage of time, had been proposed already by Darwin's grandfather Erasmus Darwin (1731-1802).Already in 1809 Lamarck set forth the view that bodily characteristics acquired as a result of use or disuse of bodily organs could be transmitted by hereditary. The accumulation of acquired characteristics could eventually change a species entirely was the stance Lamarck fostered. A number of naturalists were dissatisfied with the belief in the fixity of species, a notion propounded by Linnaeus. The Linnean view erroneously was imputed to Genesis 1. In philosophical circles the idea of evolution was in the air and scholars were leaning toward this concept. That was the atmosphere when Darwin began to think about what he had seen and written down. It was on May 14, 1850 that Darwin began to formulate his theory of evolution, which was to stress "modification by descent." On May 14,1858 Darwin received a letter from Alfred Russel Wallace from the Malay Archipelago in which the latter set forth a theory identical with the one Darwin held, one which Darwin had kept in abeyance and not published. Darwin was greatly upset to think that he would not receive credit for its conception and delineation. On July 1, 1858 both Wallace and Darwin delivered a joint paper at the Linnean Society, which was published August 20, 1858. However, it created no stir or ripples; it went unnoticed.

Thereupon Darwin sat down and wrote an abstract of his theory, *On the Origin of Species*, published on November 24, 1859. It was an instant success and the edition sold out immediately. Five more editions appeared in the author's lifetime. Darwin's book was hailed as a writing that showed the untenability of the Biblical view of special creation and deemed to demonstrate that evolution could explain the diversity of species found in nature. In the Origin evolution was hailed as the mechanism by which diversity of species has been effected. The mechanism of evolution was held to be natural selection, which worked automatically so that there was no need for God.

De Beer has given the following as a summary of the argumentation proposed by Darwin in his epoch-making book:

(1) All species produce more germ cells, pollen, or spores than ever reach maturity.

(2) The number of individuals in a species remains fairly constant.

(3) There must be a high rate of mortality, since more young are produced than ever reach maturity.

(4) The individuals of a species are not all identical but show variation.

(5) Some variations are better adapted than others to fit into the ecological niche in nature.

(6) The better-adapted variations will have a better chance of surviving and producing offspring than will the less well-adapted variants.

(7) Offspring resemble parents by heredity.

(8) Therefore each successive generation will tend to have an increased proportion of better-adapted variations until finally the population consists only of better adapted, and the less well adapted will have been entirely "eliminated."[5]

As gradual accumulations of adaptions occurs, a new species may be formed when a population splits into two portions isolated from each other.

Darwin wrote in *The Origin of Species:*

If it could be demonstrated that any complex organism existed which could not have possibly been formed by numerous successive slight modifications, my theory would absolutely break down.[6]

Darwin contended that the evidence for his theory was not direct but only indirect. De Beer quoted Darwin as writing:

I am actually weary of telling people that I do not pretend to adduce (direct) evidence of one species changing into another, but I believe that this view is in the main correct, because so many phenomena can thus be grouped and explained.[7]

According to De Beer, Darwin's argument was that if evolution had occurred, it is understandable that vertebrates, descended from a common ancestor, are built on the same plan, start their development in the same way, have similarities in behavior, and fossils become more developed at higher levels of the geological record. Darwin's method was that of hypothesis.

Darwin's subsequent books expanded the ideas incorporated in *The Origin of Species*. His *Descent of Man*, and *Selections in Relations to Sex* (1871) endeavor to give evidence from comparative anatomy, embryology, and behavior for deriving man's body and mind from that of animals. These animals if they were alive today would be apes. Darwin's views of natural selection were in contradiction to the Bible, which taught that plants and animals had been created according to their min, "kind," a concept which involved much more than species as once erroneously held.[8] Darwin's views did replace the work of the Creator by a mechanistic law that also denied to man his uniqueness in creation. De Beer, an advocate of evolution, does assert about Darwin's position as follows:

There was a gap in Darwin's system. Although he proposed natural selection working on naturally occurring variations within a population as the mechanism of evolution, Darwin could not explain the source of the variations. Biologists, who quickly noticed the flaw in Darwin's system, soon became divided into two schools: Darwinian and Lamarkian. The Darwinists stressed only natural selection as the mechanism of evolution, while the Lamarkians thought that characters acquired during a lifetime could be passed on to offspring and bring about variation within a population.[9]

An Examination of Some of Darwin's Assumptions and Their Untenability

One of Darwin's postulates was as stated by Fringe and Fringe: "There was variability between individuals and that variability is unlimited." These variations were said to produce beneficial changes in the structure of the DNA molecule. But such a change would involve a chemical change which follows certain laws of physics. Concerning this issue Harold Bluhm has written:

Whatever the nature of mutation, it will have to follow certain lines that are determined by molecular pattern and energetic relationships. Mutation, then is not random, but may occur only within certain restricting limits and according to certain pathways determined by thermodynamic properties of the system. Thus, to state the case in a somewhat animistic fashion, the organism cannot fit itself to the environment by varying unrestrictedly in any direction.[10]

This view of Bluhm restricts the direction in which an organism can evolve. If the laws of chemistry were to allow the production a new gene, then the odds are that it will be detrimental to the organism. Winchester has said that 99 per cent of all mutations that have been studied in various forms of life have been found to be harmful.[11] Evidence has shown that mutation and variability are limited, and that no beneficial mutations are known in normal environments. Scientific evidence thus disagrees with one of Darwin's basic postulates.

Another postulate of Darwin's theory is that there is competition among animals for life, food and mates. This is considered as the driving force of evolution. In the section, entitled: "Struggle for Life Most Severe Between Individuals and Varieties of the Same Species," Darwin did not give any examples to demonstrate this contention. Since Darwin's time the phenomenon of territoriality has been studied. Robert Ardry, *The Territorial Imperative* has shown that territorially, which has been defined as the tendency of an animal to hold and defend territory against all undesireable intruders, has two different implications regarding evolution. Territoriality spreads out the population, so that there is generally enough food in relationship to the population, and second, when battles take place among animals occupying the same area there are certain rules that operate.[12] Killing each other is a rare phenomenon, but instead displacement activity occurs. That competition results in the "survival of the fittest" has not been borne out by the observations of zoologists. Darwin assumed that the world was overcrowded with animals which as a result would necessitate competition among them for the same space, food and mates. This concept is crucial to Darwin's theory. He stressed the fact that their' would be competition between animals of the same species.

Thus Darwin wrote:

> But the struggle will almost invariably be most severe between individuals of the same species, for they frequent the same districts, require the same food, and are exposed to the same dangers.[13]

The Russian prince Peter Kropotkin in the early 1900s decided to investigate this theory for himself in Siberian Russia and in Manchuria. In his book, *Mutual Aid* he stated that he came to find out that just the opposite was the case. He concluded from his studies:

> I was persuaded that to admit a pitiless inner war for life within each species, and to see in that war a condition of progress, was to admit something which not only had not yet been proved, but also lacked confirmation from direct observation."

Strangely enough, Darwin contradicted his own theory when in his book on The Origins he cited examples of animal warning each other of danger, which conflicts with the competitive survival view. Many examples are known to naturalists where animals aid each other and do not try to eliminate each other. Competition is essential to Darwin's theory and observed facts do not necessarily support this contention.

Do Animals Evolve through Small Modifications?

It was essential to Darwin's theory that animals evolve through successive small modifications. Thus he wrote:

> If it could be demonstrated that any complex organism existed which could not possibly have been formed by numerous successive, slight modifications, my theory would absolutely break down.[15]

Darwin as he formulated his hypothesis argued that if evolution had occurred, it is understandable how vertebrates, descended from a common ancestor, are built on the same plan, start their development in the same way, have similarities in behavior, and that fossils become more highly developed at higher levels of the geological record. This has become proof for many evolutionists to this day. The idea that the embryo on its way to adulthood recapitulated the evolutionary heritage of the species has been rejected by other scientists. It was Ernst Haeckel in 1869 who especially developed this concept as strong proof for evolution.[16] This has become known as "the recapitulation theory."

Leo Berg in his book, Nomogenesis has disputed this theory and claimed that an animal embryo does not tell where the animal came from, but in what direction it was heading.[17] Shumway and Adamstone have asserted:

> It has been found very difficult, if not impossible, to draw up a ge-

neological tree of the vertebrates based solely on embryological data. Hence, the recapitulation theory is not accepted and applied so unreservedly as formerly.[18]

Two other scientists, Dott and Batten have admitted: Much research has been done in embryology since Haeckel's day, and we know that there are all too many exceptions to this simple analogy, that ontogeny does not accurately reflect the course of evolution. For example, we know that teeth developed before the tongue in the vertebrates, yet in the embryo the tongue appears first.[19]

Does the Paleontological Record Favor Evolution?

After a hundred years of the examination of the geological record the proof for evolution is still missing, since there are many gaps in the fossil record which are said by specialists still to exist just as they did in Darwin's day. A hundred years ago the view was expressed that further research in the paleontological record would find the transitional animals needed to definitely prove the evolutionary theory of the evolvement of man ultimately from a one-celled amoeba. After the 1959 Darwin centennial celebration A.C. Olson wrote:

A third fundamental aspect of the record is somewhat different. Many new groups of animals suddenly appear, apparently without any close ancestors . . . This aspect of the record is real, not merely the result of faulty or biased collecting. A satisfactory theory of evolution must take into consideration and provide an explanation.[20]

Olson, does not of course, admit that because the paleontological evidence is insufficient, that evolution, did not occur. Darwin's view that all organs and organisms arose as a result of slow, small modifications, if it were true, should require that organisms that show these small changes should exist in the geological record, since they occurred over long periods of time. But these transitional forms are not found in the geological record. Strange creatures are postulated as having existed, but no scientific proof of their real existence, apart from imaginary ones, has been forthcoming. Evolutionists have a strong faith or belief that this is the way it must have happened. In fact de Beers has admitted that for at least fifty years Darwin's theory had no real proof. But Faith and belief are supposed to be the characteristics of religionists not of scientists!

Are There Symbiotic Relationships?

According to *Webster's New World Dictionary* symbiosis is defined on this wise:

Symbiosis-in biology, the living together of two dissimilar organisms in close association or union, especially where this is advantageous to both, as distinguished from parasitism.[21]

The fact that two animals or plants have lived in an advantageous relationship to each other has constituted a problem for the evolutionary theory. Thus Darwin conceded:

If it should be proved that any part of the structure of any one species has been formed for the exclusive good of another species, it would annihilate my theory, for such could not have been produced through natural selection.[22]

Bolton Davidheiser in his excellent book. Evolution and Christianity, has given some examples of symbiotic relationships.[23] Josh McDowell and Don Stewart give two interesting examples in their book: *Reasons, Skeptics Should Consider Christianity*, of which one is given here:

A plant and animal in New Zealand also developed a mutually beneficial relationship. The dodo bird ate the leaves and seeds of the plant Calvaria Major. The animal got food while the seeds of the plant, passing through the dodo's gizzard, were scattered and became able to germinate. Only the seeds which had been scratched in this manner were able to germinate. When the dodo became extinct, the plant nearly did also. It now can be grown only after the seeds are artificially scratched.[24]

Various examples given of interdependence of forms of life upon each other are detrimental to Darwin's theory, who admitted that proof of the interrelationship of life upon each other would annihilate his theory.

Evaluation and Criticisms of Evolution

Josh McDowell and Don Stewart have devoted the third part of their book. *Reasons Skeptics Should Consider Christianity* to the evaluation and rejection of the evolutionary hypothesis (pp. 104-221). Of this presentation Dr. Henry Morris said: "One who would choose to remain committed to the evolutionary philosophy after reading this concise and cogent critique must do so on the basis of pure faith, not science."[25]

Rusch shows in his chapter in *Darwin, Evolution and Creation* (Ch. 1) that Darwin himself had doubts about the universal efficacy of natural selection.[26] Because of a serious criticism made against his theory by a Scottish engineer Darwin was forced to fall back upon Lamarckinism, around which he built his theory of pangenesis and gemmules as carriers of acquired characteristics. Many of his followers ignored the changes which Darwin made in theory of evolution and continued to promote his abandoned views. Thus Dodson wrote some years ago about the period immediately following Darwin that it was:

Characterized by extreme enthusiasms, together with an uncritical acceptance of whatever data were claimed to support Darwinism. Negative evidence was given little weight, while absurd extremes of interpretations, in order to make observed facts fit Darwinian theory were quite common."[27]

Rusch claims that "Darwinism became more a religion than a science, with a goodly amount of dogma to be accepted on faith is the opinion of such men as Von Vexhall, W. R. Thompson, and Sir Arthur Keith."[28]

The theory of evolution has come to tyrannize scientific and religious thought. The articles on evolution in the various standard encyclopedias reveal that there is still disagreement on a precise explanation as to how evolution may have occurred, evolution itself is regarded as a fact and a law of life, and is advocated with almost religious devotion and for any person to challenge evolution is to be guilty of serious heresy. Today we have the anomaly that religious people manifest this same devotion to the evolutionary hypothesis and are forced thereby to attack the Bible's teaching on creation (the how of it) in Genesis 1:1-2:3.

While Darwin, so it is claimed, did not deny the existence of God, his theory of evolution by means of natural selection from accidental variations, promoted the way to eliminate God the Creator, which means God's Son, the Redeemer as well. His followers led on to the logical conclusion and end materialism.

It is true that Darwin did exercise a great influence on his generation as well as subsequent generations. There were important results of Darwin's Beagle voyage. But, sad to say, they were detrimental! Dr. W.R. Thompson, in his introduction to The Origin of Species, asserted: "I am not satisfied that Darwin proved his point or that his influence in scientific and public thinking has been beneficial."[29] Again in the same work Thompson declared: "The success of Darwinism was accompanied by a decline in scientific integrity."[30]

The triumph of mechanistic materialism, of pragmatism, the prominence of humanism, the ethical and sex revolution characterizing our time, are partly the result of the evolutionary philosophy of life. The life of the Christian Church has been damaged because many religionists, who accommodated their religious thinking to the evolutionary hypothesis, have changed their interpretation of the Bible to harmonize with the vagaries of evolution. Genesis 1-3 were declared to be myth and not to be understood as a true and literal reading of these chapters does yield. With this demythologizing of the opening chapters of the Bible, the New Testament teachings were also reinterpreted, resulting in destroying the fiber of Christian faith.

In chapter 6 of Darwin's *Evolution and Creation,* the writer of this essay has shown the negative and hurtful effects on a number of disciplines for which Darwin's evolutionary ideas were responsible.[31]

Footnotes

1. Wilbert H. Rusch, "Darwinism, Science and Bible," in Paul A. Zimmerman (editor), *Darwin, Evolution and Creation* (St. Louis: Concordia Publishing House, 1959), p. 19.

2. Sir De Beer, "Darwin, Charles," *The Encyclopedia Americana*, 8:508, 1979 edition.

3. **Ibid.**

4. **Ibid.**

5. Gaven De Beer, **op. cit.,** p. 509. Compare also the sixfold postulates as stated by Fringe and Fringe, *Concepts of Evolution* (New York: McMillan Company, 1970), pp. 53, 54; reproduced by Josh McDowell *Reasons Skeptics Should Consider Christianity* (Here's Life Publishers, San Bardino, Cal., 1981), p. 149.

6. Charles Darwin, *The Origin of Species* (New York: New American Library, 1959), p. 171.

7. De Beer (p. 510) does not cite the reference in the *Origin of Species.*

8. cf. Rusch, **op. cit.,** pp. 14,15,19,20,25,28; Also Raymond F. Surburg, "In the Beginning God Created," Zimmermann, **op. cit.,** pp. 64-67.

9. De Beer, **op. cit.,** p. 511.

10. Harold F. Bluhm, Time's *Arrow and Evolution* (Princeton: Princeton University Press, 1968), p. 150.

11. A. M. Winchester, *Genetics* (Dallas; Houghton and Mifflin, 1966), p. 405.

12. Robert Ardry, *The Territorial Imperative* (New York: Dell Publishing Company, 1966), pp. 51, 87.

13. Darwin, **op. cit.**, p. 83.

14. Peter Kropotkin, *Mutual Aid* (Doubleday Page and Co., 1909), p. x.

15. Cf. footnote 6

16. Rusch, **op. cit.**, p. 18.

17. Leo Berg, Nomogenesis, translated by J. N. Rostovtsow (Cambridge: MIT Press, 1969), pp. 108-109.

18. Waldo Shumway and F. B. Adamstone, *Introduction to Vertebrate Embryology* (New York: John Wiley and Sons, 1954), p. 5.

19. Robert H. Doll and Roger L. Batten, *Evolution of the Earth* (St. Louis: McGraw Hill Book Co., Inc. 1971), p. 86.

20. E.C. Olson. *The Origin of Life* (New York: Mentor Books, 1965), p. 94.

21. *Webster's New World Dictionary* (New York: World Publishing Company, 1964), p. 1477.

22. Darwin, *The Origin*, **op. cit.**, pp. 186-187.

23. Bolton Davidheider, *Evolution and Christian Faith* (Grand Rapids: Baker Book house, 1969), p. 200.

24. McDowell and Stewart, **op. cit.**, pp. 165-166.

25. As quoted on the back side of the book referred to in footnote 24.

26. Rusch, **op. cit.**, p. 22

27. Edward O. Dodson, *A Textbook of Evolution* (Philadelphia: Saunders, 1952), p. 91.

28. Rusch, **op. cit.**, pp. 23-24.

29. In Charles Darwin, *Origin of Species* (Everyman's Library: New York: E. P. Dutton & Co., 1956), Introduction, p. xii, N. 32.

30. **Ibid.**, p. vii.

31. Zimmerman, Klotz, Rusch and Surburg, *Darwin, Evolution, and Creation* (St. Louis: Concordia Publishing House, 19S9), pp. 168-204.

Questions

1. Darwin began the Beagle voyage as a divinity student to collect evidence to support ____.
2. Darwin's biological observations and interpretations led him to ____.
3. The concept of evolution has already been proposed by ____.
4. Linnaeus propounded ____.
5. Darwin's *On the Origin of Species* was published on ____ .
6. Darwin's view replaced the work of ____.
7. Winchester has said that 99% of all mutations have been found ____.
8. What concept is crucial to Darwin's theory? ____
9. What is the recapitulations theory? ____
10. What are symbiotic relationships? ____
11. The theory of Evolution has come to ____ W.R. Thompson in his introduction of *The Origin of Species* said: ____
12. What are some of the results of evolutionary philosophy of life? ____

The Biblical Answers Versus the Evolutionary Speculations About the Origin of Language and Languages

Christian News, October 4, 1982

Linguistics is the science of language.[1] Today a distinction is made between philology and linguistics. Philology, besides treating of language, also concerns itself with the culture and the history of languages, the traditions behind languages, and the literary output of languages.[2] Linguistics, by contrast, "concentrates on language itself, with only occasional reference to cultural and literary values."[3] Robbins divided the whole field of linguistics into the study of scientific language and the study of individual languages.[4] Robbins claims that "general linguistics is concerned with human language as a universal and recognizable part of human behavior and of the human faculties, perhaps one of the most essential to human life as we know it, and one of the most far reaching of human capabilities in relation to the whole span of mankind's achievements."[5] Linguistics which includes the study of phonology, morphology, syntax, semantics, and lexicography has from time immemorial been interested especially in two basic questions: when did **homo sapiens** (man) become **homo loquens** (speaking man) and is **monogenesis** or **polygenesis** the basis of origination for the three to four thousand languages and dialects found in the world today. Two diametrically opposite answers have been given to these questions. One set of answers is based upon the speculations and theories propounded by human reasons and human experience and the other set of answers is derived from the Bible, which claims to give true answers because they come from God Himself.

Those who are humanists, materialists and naturalists insist that only human evidence and human experience can furnish adequate and scientific answers to these queries that have intrigued men for thousands of years. Agnostics, atheists and users of the historical-critical method reject the answers Genesis has given to these queries and categorically label Genesis 2 and 11 as at best primitive speculation. Genesis 1-11 is considered as containing saga, myth, legend of aetiologies and therefore is not to be taken seriously as giving any kind of reliable information on language origin of any kind.

The history of linguistics reveals that the ancient Greek philosophers gave some attention to the problem of the origin of language and how man received the "gift of language."[7] Mario Pei, Columbia University language expert, stated that "practically all primitive accounts claim a divine origin, a gift of God to man, and to man alone."[8] The first recorded written attempt at dealing with the origin of language is Western civilization was proposed by Cratylus (Kratylos).

Plato raised the question whether language arose from physis (by nature) or from **nomos** (from convention). Feeling no urge to consider the word for "horse" was something other than Greek, Plato opted for physis.

Since the Greek paid no attention to other languages, regarding them as barbaric attempts at Greek, it is not difficult to understand why Plato was led to his conclusion. Aristotle (384-322 B.C.), more of a taxonomist than Plato, and at heart a biologist, divided the Greek language into three classes or parts of speech-names (onomata) or nouns, sayings (rhemata) or verbs, and connectives (syndesmain). Hermogenes passed over the problem of language origin by proposing his conventional thesis according to which "if a man agrees to use certain sequences of sound to symbolize certain objects, the sound sequences must have their inception somewhere."[9] Pythagorus, Plato and the Stoics held that language had come out of "inherent necessity," out of "nature," which was really begging the question Democritus, Aristotle and the Epicureans claiming that language had arisen by "convention" or "agreement." The manner by which this agreement had been reached they failed to answer.[10]

Leibnitz in the beginning of the eighteenth century proposed the view that all languages came not from an historically recorded source but from a proto-speech. In the nineteenth century Wundt advocated the view that language was an outgrowth of emotional expression. A twentieth century linguist, E.H. Stutevant, presented the novel theory that language was invented for the purpose of lying and deception. He argued that "since all real, intentions and emotions get themselves involuntarily expressed by gesture, look or sound, voluntary communication such as language must have been invented for the purpose of lying and deceiving."[11]

I. The Origin of Language Itself According to Evolutionists

Linguists are all agreed that it is impossible for modern linguistics to give a scientific answer to the problem of the origin of language in man. Sturtevant, a one time professor at Yale University in the area of linguistics, wrote: "After such futile discussions linguists have reached the conclusion that the data with which they were concerned yield little evidence or no evidence about the origin of human speech."[12] La Societe de Linguistic de Paris has long had the standing rule that no papers on the subject may be presented at its sessions. The recorded history of language, even when supplemented by the prehistoric reconstructions by means of the comparative method, covers only a small fractions of the development to which language has been subjected since its origin. Since according to evolutionists language goes back only to about 3500 B.C. and man supposedly originated millions of years ago, it is not difficult to see how impossible it is for evolutionistic linguists to establish what language originally was like and how it originated. Although Sturtevant admitted that there is no reliable evidence to answer the question: "How did human speech, human language originate?," still he did not hesitate in a book of his, in chapter 5, 'The Origin of Language' to make suggestions and set forth views which definitely place him in the camp of evolutionists.[13]

Another well-known former linguist, Bloomfield asserted:

When scholars ceased to view language as a direct gift from God, they put forth various theories as to its origins ... It began in man's attempt to imitate noises (The bow-wow theory), or in his natural sound-producing responses (the ding-dong theory), or in violent outcries and exclamations (the pooh-pooh theory).[14]

The evolutionary explanation for the Origin of Speech

When Charles Darwin and his followers pointed out the alleged biological relationship of man with the lower animal world students of language re-examined their attitude about the Origin of language. It became fashionable to look for "evolution" in all areas that dealt with living beings. In view of the popularity of the evolutionary theory in all other disciplines modern linguists jumped on the bandwagon of that world view which many people considered to be "scientific," and thus the Biblical account was rejected, yes even ridiculed. Thus Margaret Schlauch wrote:

It was formerly assumed (with typical human conceit) that man, as a special and separately created being, had received the gift of language ready-made from the Creator. Just as woman was to have appeared suddenly, by a swift, if unaccountable exit from Adam's side, so speech was supposed to have begun abruptly on the day when Adam named the animals and other creatures under God's tutelage. Many religions contain a myth about the origin of speech at a given moment under divine instruction . . . Some writers, particularly those of theological bent, deny the kingship today.[15]

In the last one hundred years, since the appearance of Darwin's *Origin of Species* (1959) and *The Descent of Man* (1871), many different explanations have been advanced for the origin of speech of the time when evolving man supposedly became a **homo loquens**, a speaking being. In the quotation cited from Bloomfield, three of the former theories were already mentioned.

At least seven different theories, all attempting to link the origin of human speech with animal sounds, have been advanced. Of the first, the ding dong theory, it has been said that it came dangerously close to the physics idea of the ancients, as it was predicated upon some kind of link between sense and sound.

The "bow-bow" theory, proposed by evolutionists was popular for a time, and held that language arose.

The second "bow-wow" theory, proposed by evolutionists was popular for a time, and held that language arose in limitation of sounds heard in nature, as when a child says baa-baa to describe a sheep after hearing it bleak.

A third theory, the "ding-dong", was based on the description of a train with its different noises. This theory some linguists believed explained the origin of words in the dictionaries as echoic (crash, clang, buzz etc.), but whether the echoic process can be adduced as proof for the origin of all words is highly questionable. Mario Pei lists as objections to this once popular theory the following:

There is a further obstacle to its acceptance in the known fact that speakers of different languages seem to hear natural sounds in different ways, and imitate them in such a fashion that altogether different words come into being. English hears a bow-wow or woof-woof what French hears as oua-oua, and Italian as bu-bu; the cocks crowing is variously cock-a-doodle-do, chichirichi, cocorico; only the cat's meow seems to be international.[16]

A fourth theory was set forth by evolutionists and has been dubbed the "poh-pooh" theory, which is nothing but an extension of the bow-bow theory advanced to account for the origin of language in man. The essence of this view was that words began to be formed out of original ejaculations of pleasure, fear, surprise, such as produced by apes; this would make the interjection the original language unit and part of speech.

Yet a fifth theory for language origin was advanced and it commonly was known as the "song-song" theory, which claimed that the vocalization of primitive rhythmic chants as the source for words that ultimately turned into language. Since this did not satisfy students of language yet another, a sixth has been proposed, denominated the "yo-ho-ho" theory, which postulated a kinesthetic origin, namely, that language arose in a series of reflex exertions. A seventh theory has been called the "ta-ta" theory, which advanced the suggestion that language arises as the vocal chords endeavor to imitate movements performed by other parts of the body.

These seven theories cannot be proven and are unprovable. While they contain a kernel of truth as explaining certain aspects of language, they certainly cannot be applied either to a group or to the individuals comprising the group, as valid explanations for the entire language process. Mario Pei claimed that there is an imposing fact that militates against the notion that language originated in a series of isolated monosyllabic grunts or wheezes which men later refined and combined to form words. Thus Pei wrote:

> We might then expect to find such a language in use among primitive and backward groups with a low standard of civilization. Such is emphatically not the case. The opposite is rather true. The tongues of primitive groups are, as a rule complex in structure, while the languages of the more civilized groups appear to be more complex and involved the farther we go back into their history, and tend to simplicity as we approach their modern stage.[18]

It is difficult to justify the view that all languages developed from the grunt-and-groan stage to the complicated languages as seen by diachronic linguistic history, especially when anthropologists have found people living in the Stone Age with structurally complex languages.

Cr. Hockett and Robert Asch, in an article, entitled "The Human Revolution" have proposed a more sophisticated explanation for the origin of human speech,[19] a theory which Harry Moijer has summarized in the symposium *Linguistics Today*. The article starts with the assumption that evolution is a fact and on that basis they use their imagination to speculate on how in the past man acquired or developed the ability to speak. The author of the article on "Language" in the 1962 edition of *The Encyclopedia Britannica* claims that there must have been a time in the past when the biological ancestors of man had not yet developed language. He claimed that speculation regarding the origin of language has existed for ages but modern linguists did not concern themselves with it.[21] Joshua Whatmough devoted a number of pages in his *Language - A Modern Synthesis* to the subject of the origin of languages. Like others before him, Whatmough resorted to pure speculation when he wrote:

> The factors, whatever they were, that led to the evolution of man, led at the same time step by step to the evolution of language. Upright posture, the freeing of the hands and arms, getting the tongue away from the throat, the development of stereoscopic vision and the emergence of the entire frontal area of the brain, led to new powers of perception and these in their turns to an appreciation and designation (i.e. linguistic symbolization) of objects, events, and qualities in the external world-mass and velocity and space and time, form, color, texture — which is a specifically human attainment. Many of these objective and the subjective abstractions of them, not to mention such higher level abstractions as freedom, justice, liberty (and their opposites). Cannot be symbolized, given a meaning that is, without the help of language. But it still **requires an effort of imagination** to see how the element of convention arose (emphasis supplied).[22]

Some authors of textbooks on liguistics avoid both the question of the origin of language itself and also the question of monogenesis versus polygenesis. Although the question about the origin of language and languages has interested linguists, psychologists, students of animal behavior, paleontologists, Hoenigswald admitted: "By necessity the discussion has remained somewhat speculative, perhaps with a touch of phantasy, in the constructive, creative sense of the word."[23] Concerning these questions former Columbia University linguist Gray asserted: "None of these guesses here summarized can find a place in scientific linguistics."[24]

II. The Biblical Teaching Concerning the Origin In Language In Man

The Biblical teaching concerning man's ability to speak is that Adam the first human being was created by the LORD God (Yahweh Elohim) with the ability to speak. Chapter 2:19-22 of Genesis states: "When the Lord God had formed out of the ground every animal in the field and every bird in the air He brought them to the man to see what he would call them, and whatever the man called a certain living being that became its name. And so the man gave a name to all cattle, the birds in the air, and every wild animal, but no helper was found for the man."[25]

This passage and the one in chapter 11 according to Waterman are among "the most ancient expressions of linguistic interest known to the western World."[26]

The ability to speak of the gift of language was a part of the image of God with which man was originally endowed. The great difference between men and animals is to be found in this fact that man was made in "the likeness and image of God " a fact not ever asserted about the animal world. The ability to speak and communicate with God on man's part can only be understood in the light of what is meant by "the image of God."

Just what is meant by the Scriptural teaching that man was made in the image of God? Since God is a spirit (John 4:24), the likeness thus referred to cannot be of a bodily nature but must be spiritual in character. In this discussion of "the image of God" Joseph Stump distinguished between the metaphysical and the moral image in man, with both of which Adam was originally created. Stump distinguished the two and related them as follows in relationship to the fall of man:

> The term is used in two senses, which should be carefully distinguished. There are passages which speak of the image of God as something which man still possesses (Gen. 9:6; James 3:9), and there are other passages which speak of it as something which must be restored, and hence as something which has been lost (Col 3:10, Eph. 4:24). The image of God which has not been lost but still retained by man refers to this very constitution as man, and is called the metaphysical or substantial image.[27]

Stump distinguished between the metaphysical and the moral as follows:

> The Moral image refers to the concreated perfection which belonged to man in his original state. He was not only created a person, but he was created a morally perfect and finite personality. He was morally the image of God because he was without fault or sin. The particular factors of which this moral image consisted may be learned from the two classic passages in the New Testament which refer to the need of restoration of that image (Col. 3:10; Eph. 4:24).[28]

When the Fall occurred man lost the Moral image but not his metaphysical. Even after the Fall man continued to be a person with whom God continued to communicate. Thus language was an endowment bestowed on man when God created him and which he did not lose, even though he had lost the state of innocency. From the very beginning of man's existence language was man's unique possession, a gift God did not bestow on any animals including the primates, from whom man is alleged to have developed. In Genesis 3 the divine writer tells us that Eve and Adam were tempted by the Devil, who employed the serpent as the instrument of temptation. The Devil is depicted as speaking to Eve, and she responds by speaking. Later God called them to account and announced judgment as well as uttering the first Gospel promise (Gen. 3:15).

The second chapter of Genesis, when it records the fact that Adam named the animals which were brought to him, evidently pictures man as endowed with wisdom and speech at his creation. In opposition to this record, however, evolutionary scientists have declared that man's speech has been derived from the cries of animals and that it was only through a long process that the human race acquired the gift of developed speech.

This latter assertion has met with emphatic denial by some competent philologists. Prof. Max Mueller, for example, declared many years ago:

> There is one barrier which no one has yet ventured to touch, the barrier of language. Language is our Rubicon and no brute will dare

cross it. . . No process or natural selection will ever distill significant words out of the notes of birds and animals.[29]

In contrast to Simlar's theories H. W. Morris, in his book *Work Days of God* wrote:

To enhance and elevate the social happiness of His offspring, the Creator bestowed on them one other gift of immeasurable worth and importance — the gift of language. Of all the living tenants of the new — made world, speech was given to a man alone, as he alone had reason to employ it; and it is impossible to estimate the advantage and the pleasure that flowed to the happy pair through its faculty. It is in itself a most striking display of the Divine wisdom that contrived it. In the wonderful system of man there have been inserted two or three little organs, so exquisitely contrived, which, by a few scarcely perceptible motions, can shape and form the air into sounds which express the kinds, properties actions and relations of things, under thousands of aspects, in forms infinitely more recondite than those in which they present themselves to his senses.[30]

Whether evolutionists do admit it or not, the fact is that language is the great barrier between mankind and the animal kingdom. The capacity for abstract, personalized thought and the ability to articulate and communicate that thought in symbolic sounds to other individuals also, is, no doubt the most obvious difference between men and animals. The difference between animal instincts and human reason, and the grunts and barks of animals and the intelligent speech of man, are practically infinite.

Even the evolutionist Simpson had to admit:

Human language is absolutely distinct from any system of communication in other animals. That is made most clear by a comparison with other animal utterances, which most nearly resemble human speech and are most often called 'speech.' Non-human vocables are, in effect interjections. They reflect the individual's physical or, more frequently, emotional state. They do not, as true language does, name, discuss, abstract or symbolize.[31]

The creationist asks the evolutionist to explain how animal noises ever evolved into human language. If there ever was a mystery that evolutionists have been unable to explain, it is the supposed development of language from animal chatter. J. B. Lancaster, a leading authority in the field of man's origins, has concluded: "The more that is known about it, the less these systems seem to help in the understanding of human language."[32]

The evolutionary model is unable to account for the origin of language. Of all the unbridgeable gaps which characterize man's constructed evolution from amoeba to man and there are many — the language gulf between man and animals is unexplained. The anthropologist Ralph Linton has said this about language:

The use of language is very closely associated within the superior thinking ability of humans in his ability to communicate man differs even more from other animals than he does in his learning or thinking. . . . however, man is the only species which has developed communication to the point where he can transmit abstract idea . . . It is a curious fact that there is no mammalian species other than man which imitates sounds . . . In this respect, humans are truly unique.[33] Again Linton declared: "We know nothing about the early stages in the development of language."[34]

The *Revised Standard Version* appears to support a simple and uncomplicated origin of language by a questionable translation of Genesis 11:1 Thus it rendered 11:1 this way: "The whole earth was of one language and of few words."[35] The RSV is the only translation or version that this writer has found which has this strange and inaccurate translation. Other translations interpret the Hebrew text like this: "The whole earth was of one language and of one speech."[34] The RSV rendering would be in line with the evolutionary view that language at first was characterized by the greatest simplicity. If the theory of the evolutionists is correct such people having simple language should be found in remaining primitive cultures. Anthropologists inform us that there are still people living in the Stone Age. Do these people have very simple languages with a limited vocabulary and simple syntax? Linton claimed that there is no evidence among primitive tribes to the effect that they possessed simple languages. Thus he wrote: "The so-called primitive languages are actually more complicated in grammar than the tongues spoken by civilized people."[37] Robert Hall, Jr. asserted about the nature of "primitive" languages: "It is also widely believed that the language of 'primitive' people must somehow be inferior to those of more advanced, 'civilized' communities ... Here again, careful investigation has shown these notions to be unfounded. Every study that has been made to date on a 'primitive' language has shown it to have the same type of structure and as rich a vocabulary as to other languages."[38]

There is no clue available to support the evolutionary theory of the origin of man's language. Hall in his chapter on "Child Language" engaged in trying to draw a parallel between the manner in which a child learns to speak, and the way primitive or the first speakers of the human race learned how to speak. His presentation is highly speculative, offering no experimental evidence for his views.[39] The people of Terra del Fuego, whom Darwin classified as the lowest human scale and whole language was considered crude, had more than 32,000 words, or more than twice as many as Shakespeare used. The language of some tribes in the Congo district of Africa are more complex than Greek. Similarly, elder languages are often more complex than more recent languages. This is true, for example, of modern Greek and Latin, when compared with their ancient and classical forms.[39a] Objective historical linguistic data show that as far as history is concerned highly complex languages have been in existence from the very beginning of time. Leslie and Kathee Bruce, missionary linguists in Papua-New Guinea report that the Alembaks in northern Papua-Guinea live in organized villages and have an exceedingly complex as well as intricate social structure.

III. The Origin of the World's Numerous Languages: Monogenesis or Polygenesis?

According to Genesis 11:1: "The whole earth was one language and of one speech." "The Immediate descendants of Noah all spoke the same language. What this language was no one knows. Possibly it was a Semitic language, since the proper names of men and places names in the pre-Babel period all have meanings in Hebrew and its cognate languages. Others have suggested it was Fumsrian, the oldest language known from about 3500 B.C.[40]

The Source of Today's World Languages

In the history of linguistics the views have ranged around either monogenesis or polygenesis as to the answer: Did all languages today come and develop from one? or did many different languages spring up simultaneously? The problem of a single versus a plural origin for today's world languages was explained by the ancients by their belief that the language spoken by them was the original language from which all others were derived. The people of India and Greece did not discuss language diversity because to them their respective languages were the proto-types for all other existing languages. The Greeks took the position that Latin was descended from Greek. During the Middle Ages teachers of language accepted the Biblical account of Babel and taught that Hebrew was the original language and therefore all languages were descended from it. The controversy between monogenesis or polygenesis was a question students did not discuss, because it would have involved them in the theological question of rejecting Scripture if they argued for polygenesis.

Pei stated that:

> The religious tradition on the one hand, the belief in a universal grammar on the other, combined to postpone any scientific discussion of the problem until after the discovery of the unity of Indo-European languages at the beginning of the nineteenth century. At that time, the question of monogenesis was temporarily settled, not so much by argument as by default.[41]

It was held by nineteenth century linguistics that just as the Indo-European languages constitute a unity and so other language groups like the semitic, which also had a common ancestry and common relationship, were different from the Indo-European and thus the question of monogenesis was answered, it was believed, in favor of polygenesis. By the year 1900 linguists had classified and pigeonholed most of the world's languages into the following major language families: Indo-European, Semitic-Hamitic, Ural-Altaic, Sino-Tibetan, Dravidian, Japanese-Korean, Malayo-Polynesi, Bantu and so forth.[42] Somewhere between 3,000 and 4,000 languages and dialects are known to have existed, of which some now are extinct.[43] Is it possible that they go back to one language? Or do the various language families indicate that numerous languages originated independently in various parts of the world? Professor Pei claimed that if a linguist were interviewed and asked: "Do you believe that all the world's languages stem from a single prototype, or do you believe that the various languages arose independently of one another in various parts of the world," he would either refuse to answer these queries or he would refer you to his fellow-scientists, the anthropologists to answer them.[44]

What are the implications of the different schools of thought relative to origin of the world's languages? The question whether all languages come from one original language is definitely linked up with the issue of the monogenesis or polygenesis of man. If the human race lived in one place and as descended from one individual, and later left this central area, then logically monogenesis would be the answer to the problem issues from where the present-day languages originated. This is the view of the Apostle Paul, when in an address to a group of philosophers in Athens he said. "He hath made of one blood every nation of men to dwell on all the face of the earth etc. (Acts 17:26)." Genesis clearly sets forth the view that there was one language spoken after the Flood by people in the Mesopotamian Valley, the area from which men began to spread over the face of the earth.

The linguist Hall made this statement about the issue of monogenesis versus polygenesis:

> There is no possibility of ever recapturing the details of the first prehistoric development of human language, because it lies too many hundreds of millennia in the past. We shall never know, even, whether language developed only once in the history of the race, at one specific time and place (according to the monogenetic theory), or at several times and places, and among different groups of incipient humans (as the polygenetic theory holds).[45]

Many linguists, however, who are committed to the evolutionary theory would logically argue for polygenesis.

One outspoken opponent among linguists to the polygenetic theory was the Italian A. Trombetti, who in his **D'origine del linguaggio** (Bologna, 1905) and later in his **Elementi di glottologia** (Bologna, 1922-1923) upheld the view that all languages originally came from one. Both Gray and Schuhart have rejected Trombetti's position.[46] However, Pei is not willing to jettison the thesis of Trombetti, who in his books has accumulated an immense amount of material. Pea asserted:

> But perhaps what would not be accomplished by a single stroke of genius may be attained by painstaking, point-by-point comparison among various supposedly independent language families. There is some evidence of a possible link between Indo-European and Semitico-Hamitic, between Indo-European and Uralic, between Munda, Mon-Xmer, Malayo-Polynesian, Sino-Tibetan, Uralic and Australian, between Semito-Hamitic and Bantu.[47]

If these links are definitely established, Pei is convinced (and much more work needs to be done in these areas) the case for monogenesis will be at least half proven in the meantime, Pei contended, students of the history of languages will have to wait patiently until all the evidence is in. "But watchful reserve does not mean," said Pei, "outright skepticism or a refusal to look at least at the available evidence, as, if and when it unfolds."[48]

Genesis 11:1-9 and the Spread of Languages

Old Testament scholars, committed to the use of the historical critical method, do not regard Genesis 11:1-9 as an account recording true history but they claim that it contains a legend of aetiology, a story invented to account for the existence of so many languages in the world. All critical scholars reject the historicity and factuality of chapters 1-11 because the supernatural is embedded in them throughout. The New Testament does not share the view of the historical-critical rejecters of the truthfulness and dependability of Genesis 1-11 for Luke in his genealogy of Jesus accepts the persons mentioned in chapters 5 and 11 of Genesis. Hebrews 11 also considers Enoch, Noah, Cain and Abel as historical persons, some of these individuals lived before and some after the Deluge.

The Great Flood occurred as far as Jesus was concerned. The manner in which Moses employed the phrase "these are the generations" (elleh toledoth) also clearly shows that Moses regarded the events recorded in Genesis 1-11 as factual and as having taken place.

For those people who do not accept Genesis 11 as a record of true historical happenings the confusion of tongues at Babel wall be fable. But those people who accept Genesis as God's inspired Word, the happenings recorded by Moses. In Genesis 11 are true. This means that the controversy between monogenesis and polygenesis regarding the origin of the world's languages is settled.

The events of chapter 11 are probably earlier in their occurrence than the resulting division of mankind which is outlined in chapter 10. The ark had alighted in Mt. Ararat and from there the sons of Noah, Ham, Shem and Japhet headed eastward, finally arriving at the land of Shinar and the fertile Mesopotamian plain, where they decided to settle and build a city. Morris suggests that "perhaps the region reminded them of their antediluvian home, and they thought they might even be able to restore the conditions of Eden itself, for they named the rivers Tigris and Euphrates after two of the streams that had once flowed from the Garden."[49]

Here in the fertile Tigris-Euphrates plain the population began to grow and reach a point where Noah's descendants did not need to spend their time with food production, but it became possible for them to think of urban development. According to chapter 10:8-11, Nimrod, a man of great ability and energy, was evidently the leader of a group that built Babel (11:4,8,9), which then was the capital city over this region over which Nimrod became king. The latter embarked upon the building of an empire comprising a complex of cities which had Babylon as its center. The cities included in Nimrod's empire were Erech, Accad, Calneh in Shinar; Erech was also known as Uruk, a city one hundred miles southwest of Babylon, also the legendary home of Gilgamesh, the hero of the Babylonian flood epic. In the ruins of Uruk archaeologists have found samples of ancient writing, written long before Abraham's time. The name Akkad and Agade are found in Mesopotamian monuments as Accad and Agade, and were situated north of Babylon. Accad gave its name to the Accadian Empire, which is essentially the same as the Sumerian empire.

From Babel Nimrod "went forth" into Assyria, where the "mighty hunter" built Nineveh, Rehoboth, Resen and Calah. Nineveh was located on the upper Tigris River just as Babylon was situated on the Euphrates. Nineveh was nearly two hundred miles north of Babylon. Later on both Babylon and Nineveh were conquered by Semitic people, but Nimrod, a Hamite, was the founder and emperor of both cities, later to become the capitals of Babylonia and Assyria respectively.

Morris believes that the episode in Genesis 11:1-9 may be linked up with Genesis 10:8-10. As Nimrod speculated on possible future developments for his empire, there were two possibilities open to him: (1) systematic colonization and development of all parts of the earth, each with its own local government, in accordance with God's command (Gen. 1:28; 9:11, or (2) establishment of a strongly centralized society, which, with controls over resources and occupations, would soon be able to produce a self-sufficient civilization capable of similarly controlling the entire world.[50]

Of these two proposals, it would seem, that the second would serve the plans of Nimrod and his fellow rebels and of the invisible satanic inspiration behind such a program of action. Thus to quote Morris: "A self-sufficient society, integrated under a powerful and brilliant leader would be a society no longer dependent on God and this was Nimrod's aim."[51]

After the establishment of this community at Babel, Nimrod called a council of family leaders of the Babylonian community. A formal decision was made to first develop a brick-making industry. Building stone was not conveniently to be had, and wood was not durable enough for the structure that was planned. Clay, which was very plentiful in the Mesopotamian valley, could be made into durable brick after proper treatment in a kiln. Archaeological excavations have revealed that the Babylonians used asphalt (slime in Genesis), abundant in the Tigris-Euphrates Valley asphalt pits, as mortar and with it they produced good strong bricks, different from the bricks made in Egypt and Assyria.

Verse 4 of Genesis 11 records that a decision was reached to build a permanent community in Babel, especially after the brick-industry had been established. The reason given for this action was "lest we be scattered abroad upon the face of the earth."

A carefully planned urban center, each part made for the maximum permanence was envisaged. A great tower was to dominate the city both in terms of architecture and culture. This new building complex would be the center for both the political and religious life of the city. The tower had a religious and astrological significance probably. The tower was to signify their unity and strength the Hebrew text says "Whose top in the heavens," and this may mean that on it were depicted the signs of the zodiac and other drawings of the heaven. Deuteronomy 1:28 and 9:1 speak of cities "fortified up to heaven." Some have suggested that accounts in Babylonian literature relative to the building of Babylon and its lower may be a prototype of the ziggurat or temple-mound, first found in classical form early in the third millennium B.C.

In the land of Shinar archaeologists have discovered many ziggurats or "holy lowers." These ziggurats may have been copies of the tower (Hebrew word: midgal) of Babel.

Three matters emphasize that sin had become serious in the post-Flood period. First, there was the attitude of self-centeredness. The Genesis text used the word "us" a number of times in Genesis 11:1-9. The people planned to make a name for themselves. The second reason why they wanted to build this tower was that they wanted to unify themselves so that they would not be scattered. They thus placed themselves in opposition against God's command given in Genesis 9:1,7: "Be fruitful and multiply, and fill the earth." They were thus in clear rebellion against God's plan. Third, from God's perspective the situation was serious enough to put a hall to man's haughtiness and defiance.

The Judgment of God, Genesis 11:5-9

The reaction of Yahweh to this situation and planned action of the people of the Shinar plain is stated in Genesis 11 by the author: "Look one people!" the LORD said. And this is only the beginning of what they'll do. Now nothing they plan to do will be too hard for them. Come, let us go down and confuse their language, so that nobody can understand what the other one is saying."[52]

As little as sinful man deserved a habitation like Eden, so the possession of one language they no longer deserved. Though mankind had one language, no good could come from it. Leupold wrote: "As long as the medium of one language is theirs, just as long they will be able to carry through reasonable though ungodly projects that they may have taken in hand, God discerns that similar undertakings will follow this one."[53]

The Confusion of Languages

To thwart man's evil plans Yahweh scattered them abroad from there all over the earth and they left off building the city. Wherefore it is called Babel because there God made a babble of the languages of all the earth, and from thence Yahweh scattered them abroad over the earth. Even though the people of Babel did not want to fill the earth, they are now forced to carry out God's command. The word "city", a use of legitimate synecdoche, is employed for the entire building project.

Opinions differ as to which came first, the scattering of men or the confusion of languages. Jerry A. Grieve suggested that "to bring about the diversity of language the Lord first caused the dispersion. The real miracle at the Tower of Babel, then, is not the confusion of tongues but the scattering of the people." Grieve argued that linguistic science has shown that different languages develop when a single community and when groups migrate to new terrorists.[54] However, a more likely view is that the scattering from Babel occurred as a result of the confusion of tongues, a miracle which Yahweh caused to happen.

The manner in which God confused the speech of the people who built the city and tower is not known. Some have explained it as being effected by a change in the speech organs, others have suggested that "it had a foundation in the human mind."

Aalders in his commentary on Genesis stated that a famous Assyriol-ogist, whose name he does not mention, made the amazing discovery that there is a clear relationship between the languages of the native people in Central and South America and some of the Islands, on the one hand, and the ancient Sumerian and Egyptian languages, on the other.[55] Aalders claimed: "This scholar, who formerly considered the account of Genesis 11:1-9 to be no more than a myth, came to the conclusion that the biblical narrative is more credible than has been supposed."[56] The oldest language known from the Near East is Sumerian, an agglutinative language, whose style of writing was borrowed by the Old Babylonians and Old Assyrians. The Documents from Ebla, Tell Mardikh, are written 80% in Sumerian and used by a Semitic people.[57] Aalders claimed: "As far as our present understanding goes, Sumerian would be the most apt to reveal certain root words from which the words in other languages could have developed. The fact that it was there that a relationship was discovered with languages spoken by native peoples spoken on the other side of the world is certainly of more than passing interest."[58]

Max Mueller of Oxford in his book *Science of Language* wrote:

We have examined all possible forms which languages can assume, and we now ask, can we reconcile with these three distinct forms of the radical, the terminational, and inflectional, the admission of one common origin of human speech? I answer decidedly, Yes.[59]

We conclude this presentation with the observation that many have, do and will continue to deny the historicity of Genesis 11:1-9, just as they reject the historical events recorded in the previous ten chapters. It is considered scholarly to deny that the two Genesis accounts that give the reader information about the gift of language as well as the manner in which the confusion of tongues occurred as fictional. However, what Aalders noted a number of years ago, all in Genesis will be determined from what point of view Scripture is approached. Wrote the Old Testa-ment professor of the Free University of Amsterdam: "When one acknowl-edges the Scriptures to be the very word of God in its entirety then everything in the Scriptures is accepted without doubt. Then the Tower of Babel and the confusion of speech can also be readily accepted as de-scribing actual events. When, on the other hand, one does not accept the divine authority of Scripture, the way is opened for all kinds of critical questions regarding the truth of that which is recorded."[60]

The New Testament records another miracles relative to the confusion of languages, namely, the outpouring of the Holy Spirit, when the Spirit enabled the apostles to speak in many different languages on Pentecost. Luke the historian reported in his church history, Acts: "And how does every one of us hear his own language he was born in-Parthians Medes Elamites and people living in Mesopotamia, Judea and Cappadocia, Pon-tus, and the provinces of Asia, Phrygia, and Pamphylia, Egypt and the country near Crene in Libya, the visitors from Rome, Jews and those who have accepted the Jewish religion, people from Crete and Arabia? In our own language we hear them tell about God's wonderful things" (Acts 2:7-11).[61]

Footnotes

1. John B. Carroll, *The Study of Language* (Cambridge: Harvard University Press, 1953), p. 3.
2. Mario Pei, *Invitation to Linguistics* (Garden City, New York: Dobleday & Company, 1965), p. 1.
3. **Ibid.**
4. R .H. Robins, *General Linguistics: An Introductory Survey* (Bloomington, Indiana: Indiana University Press, 1968), p. 2.
5. **Ibid.**
6. Alan Richardson, *Genesis I-XI - The Torch Bible Commentaries* (London: SCM Press, 1953), pp. 27, 129.
7. John P. Hughes, *The Science of Language* (New York: Random House, 1965), pp. 39-40.
8. Mario Pei, *Language for Everybody* (New York: The Devin-Adair Company, 1957), p. 122.
9. Mario Pei, *The Story of Language*(Philadelphia and New York: J. P. Linnencott Company, 1969), p. 19.
10. **Ibid.**
11. Edgar H. Sturtevant, *An Introduction to Linguistic Science* (New Haven: Yale University Press, 1947), p. 40.
12. **Ibid.**
13. **Ibid.**, pp. 40-50.
14. Leonard Bloomfield, *Language* (New York: Henry Holt and Company, 1932), p. 6.
15. Margaret Schlauch, *The Gift of Languages* (New York: Dover Publications, 1955), p. 4.
16. Mario Pei, *Voices of Man – The Meaning and Function of Language* (New York: AMS Pess, 1972), pp. 19-20.
17. **Ibid.**, p. 20.
18. **Ibid.**, p. 21.
19. C.F. Hockett and Robert Aacha, "The Human Revolution," *Current Anthropology*, V, No. 3. (June, 1961).
20. Harry Hoitjer, "The Origin of Language," *Linguistics Today*, edited by Archibald A. Hill (New York: Basic Books, 1969), pp. 55-58.
21. "Linguistics," *Encyclopedia Britannica,* 1962 edition, XIII, 696.
22. Joshua Whatmough, *Language – A Modern Synthesis* (New York: New American Library, 1937), p. 157.
23. Henry M. Hoenigswald, "Our Own Family of Language," in *Linguistics Today*, **op. cit.**, p. 59.
24. Louis H. Gray, *The Foundations of Language* (New York: The Macmillan Company, 1938), p. 40.
25. William F. Beck, *The Holy Bible – An American Translation* (New Haven: The Leader Publishing Company, 1978), 3rd edition, p. 2.
26. John T. Waterman, *Perspectives in Linguistics* (Chicago: The University of Chicago Press, 1963), p. 1.
27. Joseph Stump, *The Christian Faith* (New York: The Macmillan Company, 1932), p. 103.
28. **Ibid.**, p. 103.
29. Max Mueller, *Lessons on the Science of Language*, pp. 23,340,370 as quoted by

Walter A. Maier, Sr., Notes on Genesis (St. Louis: Concordia Mimeo Co., 1929), p. 58.

30. H.W. Morris, *Work Days of God*, p. 58 as quoted by Maier, Sr., **op. cit.**, p. 58.

31. George Gaylord Simpson, "The Biological Nature of Man," *Science*, 15, 152 (April 22, 1966), p. 476.

32. J. B. Lancaster, "The Origin of Man," Symposium, edited by P. L. Devons, (New York: Wenner-Gren Foundation), 1965.

33. Ralph Linton, *The Tree of Culture* (New York: Alfred A. Knopf, 1955), pp. 8-9.

34. **Ibid.**, p. 9.

35. *The Holy Bible – The Revised Standard Version* (New York: Department of Christian Education of the National Council of Churches), a number of publishers printed the RSV, Cf. the pertinent passages.

36. Cf. *The New International Version* (Grand Rapids: Zondervan Bible Publisher, 1979), p. 11.

37. Linton, **op. cit.**

38. Robert A. Hall, *Introductory Linguistics* (Philadelphia: Chilton Company, 1964), p. 13.

39. **Ibid.**, pp. 277-282.

39a. A cited by W. A. Maier, Sr., *Genesis Notes*, p. 57.

40. "Sumer," D.S. Parlett, *A Short Dictionary of Languages* (London: The English University Presses, 1967), p. 118.

41. Pei, *The Voices of Man*, **op. cit.**, p. 51.

42. Guliano Bonfante, "The Languages of the World," in *Colliers Encyclopedia* (New York: Crowell-Collier Publishing Company, 1961), XI, 440-441.

43. Siegfried H. Muller, *The World's Languages – Basic Facts of Their Structure ,Kinship, Location and Number of Speakers* (New York: Frederick Ungar Publishing Company, 1964), pp. 1-123.

44. Pei, *The Voices of Man*, **op. cit.**, pp. 19-26.

45. Hall, **op. cit.**, p. 281.

46. Gray, **op. cit.**, p. 454.

47. Pei, *The Voices of Man*, **op. cit.**, pp. 53-54.

48. **Ibid.**, p. 54.

49. Henry Morris, *The Genesis Record* (Grand Rapids: Baker Book House, 1976), p. 266.

50. **Ibid.**, p. 268.

51. **Ibid.**

52. Beck, **op. cit.**, p. 11.

53. H. C. Leupold, *Exposition of Genesis* (Columbus: Lutheran Book Concern, 1942), p. 389.

54. Jerry A. Grieve, "The Origin of Languages," *Ashland Theological Bulletin*, No. 3, (1970), p. 17.

55. G. Ch. Aalder, *Bible Student's Commentary – Genesis*. Translated by William Heynen (Grand Rapids: Zondervan Publishing House, 1981), I, p. 254. This writer knows that this was the belief of Cyrus Gordon of Dropsie University and Brandeis University.

56. **Ibid.**

57. Giovanni Pettinato, "The Royal Archives of Tell Mardikha-Ebla," *The Biblical Archaeologist*, May 1976, p. 50.

58. Aalders, **op. cit.**, pp. 254-255.

59. Max Mueller, *Science of Language*, p. 329.

60. Aalders, **op. cit.**, p. 251.

61. Beck, An American Translation, **op. cit.**, New Testament, p. 150.

Questions

1. Linguistic is ____.
2. Philology concerns itself with ____.
3. How many languages and dialects are found in the world today? ____
4. How do users of the historical-critical method regard Genesis 2 and 11? ____
5. According to evolution languages goes back to the year ____.
6. What seven theories of language cannot be proven? ____
7. What does Genesis 2 teach about the origin of language? ____
8. Evolutionary scientists are declared that man's speech began with ____.
9. Language is the great barrier between ___ and ____.
10. The RSV rendering of Genesis 11:1 is in line with ____.
11. Anthropologists inform us there are still people living in the ____ Age.
12. The language of some tribes in the Congo district of Africa are more complex than ____.
13. What was the language of Noah and his descendants? ____
14. Genesis sets forth the view that after the flood there was ____ language.
15. Moses regarded the events recorded in Genesis 1-11 as ____.
16. What was to be built in Babel? ____
17. Whose "top in the heavens" may mean ____.
18. It is called Babel because God made a ____ of languages.
19. The oldest language from the Near East is ____.
20. What happened at Pentecost? ____.

A Reaction to UPI Reporter Anderson's "Easter Controversy"

Christian News, April 29, 1985

The second stage in the state of exaltation of Christ is the resurrection. Correctly Stump has stated about the resurrection: "The resurrection is the act by which Christ according to His human nature came forth from the tomb in which He had lain dead and buried. It is the crowning event of redemptive history and a fact with which Christianity stands or falls."[2] Josh McDowell has noted:

> All but four major world religions are based on mere philosophical propositions. Of the four that are based on personalities rather than philosophical systems, only Christianity claims an empty tomb for its founder. Abraham, the founder of Judaism, died about 1900 B.C., but no resurrection was ever claimed for him.[3]

The resurrection was the fulfillment of Old Testament prophecies and of Christ's own. According to the Jewish reckoning, which counted a part of a day as a day, Jesus rose on the third day. The first day ended and the second began on Friday evening at six o'clock. Thus Jesus rose in the third day "according to the Scriptures."[4]

David E. Anderson, a writer for the United Press International, published an article entitled "Easter Controversy. Resurrection stirs profound theological debate." This article appeared in major metropolitan newspapers just before Easter. It purported to be "a Commentary" on discussions and debates going on in the world of New Testament scholarship relative to the nature of the Easter event. Anderson contrasts the traditional resurrection of Christ view with current views quite dissimilar from that traditionally held. Anderson admits that the traditional Biblical view has been believed for the last 1900 years and that the belief that Christ bodily rose from the dead, as a belief has been "inspiring the worship of millions of plain folk and has unleashed the genius of musicians as Johann Sebastian Bach, Ludwig von Beethoven, T.S. Eliot and Jonne Donne."[5]

Because of the central place which the doctrine of the resurrection holds in Christianity,[5a] it has been the favorite point of attack by the enemies of the Christian faith.[6] Thus the chief priests and elders disseminated the slander that the disciples of Jesus had stolen the body of Christ from the tomb. In modern times the effort has been made to explain away the resurrection account by various theories, chief of which were: the swoon, the hallucination and fraud theories.[7] **The Swoon Theory** taught that Christ did not really die on the cross but merely swooned, and in the tomb revived. After having somehow gotten out of the tomb, rejoined his disciples. But the theory is utterly discredited in the face of the clear facts of Christ's crucifixion and death.[8] Would such a debilitated and pathetic looking Master have produced the great change that occurred in the disciples after the Day of Pentecost?

The Vision Theory claims that the Disciples of Christ, by brooding over Christ's death and his promised resurrection, became the victims of hallucination, in which they imagined that they had seen Him arisen from the dead.[9] Koehler has refuted this theory by calling attention to the following textual and historical facts:

The disciples themselves were slow to believe that Christ had risen (Luke 24:25; John 20:25; Luke 24:11). However, at different times under different circumstances, He appeared to different groups of them. They saw Him, ate with Him, handled Him, and were thus in ever y conceivable way convinced of the fact that Christ had arisen (Acts 1:3; 2:32; I Cor. 15:4-8; Luke 24:36-43). In view of these facts, the hallucination theory, according to which the disciples are said to have imagined they had seen the Lord, is itself a hallucination.[10]

The fraud theory claims that the disciples deliberately fabricated the whole account of his resurrection and the alleged post-Resurrection appearances successfully convinced people of their deception. In answer, it might be said that the Jews easily could have disproved this theory.[11] In addition, would the disciples have been willing to have endured persecution and even death for a dead Jesus (Acts 5:41; 7:56; 2 Cor. 11:23-27), unless they had actually seen the living Christ?

Down through the centuries, at least till the time of rationalism, only the enemies of Christianity have attacked the resurrection of Christ. But with the coming of the **Aufklaerung** the deniers of the Lord's resurrection have also been found in the church, especially in the theological departments of the European universities and American theological seminaries.[11] Hand in hand with the rejection of the doctrine of the Trinity went the rejection of Christ's resurrection. The last two hundred years have produced a number of unbelievers, who called themselves Christians, who proposed novel theories for explaining away the clear statements of the New Testament.[12] The twentieth century has produced its own brand of resurrection deniers. Anderson, in his "Easter controversy," has supplied us with a few more. Anderson's article is comprised of twenty paragraphs, of which seventeen are devoted to the unscriptural views of Hans Kung, Ian Wilson, and Pinchas Lapide. Anderson's article seems to be more concerned with questioning Christ's bodily resurrection, than with citing scholars in support of it.[13]

The Real Reason for the Rejection of Christ's Resurrection

Anderson wrote: "It is also, for many reared in a rationalist and secular world, an obstacle, a holdover from a Greco-Roman mythological worldview in which gods regularly visited Earth." Biblical scholars since the beginning of the age of rationalism have been utilizing the historical-critical method, which operated with a number of wrong presuppositions. One is that the Bible is not the inscripturated revelation of God, which is inerrant and infallible.[14] Further, it is a given of the historical-critical method that miracles do not happen and the supernatural element in the Bible is to be ruled out as inadmissible, and wherever it is found in the Old and New Testaments, it must be eliminated or reinterpreted.[15] There are over a hundred miracles in the Bible, which are unacceptable to human reason. The Life of Christ is interwoven with many miracles which cannot be accepted by critical scholars as having occurred. The resurrection of Christ is the miracle of miracles and cannot be recognized, so when Christ died He was placed in the tomb of Joseph of Arimathea, where it remained or possibly His body was removed to a new tomb. But from the dead he could not have been raised and did not arise. Historical-critical scholars have rejected the literal account of the historical resurrection narratives.[16]

Anderson depicted the manner in which the story of the resurrection allegedly was promulgated and accepted as fact and not fiction as follows: Mary Magdalene, the mother of James and John "encountered the empty tomb, but had their story dismissed by the apostles as 'an idle tale.'" In other words the disciples when they first heard it did not believe. But, Anderson does not report that these disciples, who at first refused to believe, later, on Easter evening did believe, and eight days later also Thomas became convinced, who at first when told by his fellow apostles: "We have seen the Lord" refused to believe.

Anderson writes: Nevertheless, the story spread and testimonies of encounters with the risen Lord gained currency and credibility and later it was confessed in the Apostles' Creed, the earliest formulation of Christian belief.[17]

Kung's Views about the Resurrection

Kung, a very controversial Roman Catholic theologian, has rejected the literal historical account of Christ's resurrection.[18] He does not believe that a dead body in the case of Christ was reanimated; because that would break the continuum of nature.[19] Anderson quoted Kung as teaching: "To be exact (the resurrection) is not a historical but nevertheless a real event." Is Kung not guilty of writing nonsense, for if Christ never arose, how could He then appear to people? Kung does not believe the grave was empty on Easter morning, and the only logical conclusion that would follow from such a statement is that never thereafter was Christ seen alive again. In setting forth Kung's views, Anderson quotes him to the effect: "For Kung, the meaning of Easter is not in the empty tomb but rather in the post-resurrection encounters Jesus' disciples had with their risen Lord."[20] The logic of Kung is strange, for on the one hand Christ never left the tomb, but on the other hand He makes appearances to His disciples.

Furthermore, Kung claims that the Apostle Paul never mentions the empty tomb. However, Paul frequently speaks of the resurrection as a real event that occurred in calendar time and in his famous 15th chapter of I Corinthians gives evidence of individuals who saw Christ alive on Easter. How could Jesus have made these appearances if he had not been raised from the dead? Kung's exegetical views contradict the historic belief of the Roman Catholic Church.[20a] All Christian churches that subscribe to the Three Ecumenical Creeds of Christendom accept Christ's resurrection. We cite but one official statement of the Roman Catholic Church. The Theologians Creed, in the *New Catholic Encyclopedia*, volume 12, asserted this about those who have rejected the corporeal resurrection of Christ:

Many objections against the reality of the resurrection have been formulated by rationalistic critics on grounds that are purely subjective and that involve such arbitrary rejection of the witness of the early Christians that it amounts to a rewriting of the Gospel accounts by men coming 2,000 years later. The objections hardly need to be taken seriously, for they arise, not from an unprejudiced investigation of the Gospels, but from a rejection of the possibility of miracles. (p. 404).

Neoorthodox Theologians on Christ's Resurrection

Kung has simply adopted views, which sad to say, were propounded by Lutheran Bultmann, and Reformed Emil Brunner and Karl Barth. Thus Bultmann, whose demythologization hermeneutic would not permit any interference with the laws of nature, wrote this about the Lord's resurrection: "Christ's death and resurrection are cosmic occurrences, not incidents that took place once upon a time."[21] In yet another book he declared: "And although it presents the Cross and Resurrection of Jesus in mythological terms, the preaching of the Cross is nevertheless a decisive summons to repentance."[22]

After claiming that the resurrection "is not an event of past history with a self-evident meaning," this Lutheran wrote:

Yet it cannot be denied that the resurrection of Jesus is often used in the New Testament as a miraculous proof. Take for instance Acts 17:31. Here we are actually told that God substantiated the claims of Christ by raising him from the dead. Then again the resurrection narratives: both the legend of the empty tomb and the appearances insist on the physical reality of the risen body of Christ (see especially Luke 24:39-43). But these are most certainly later embellishments of the primitive tradition. St. Paul knows nothing about them. There is however one passage where St. Paul tried to prove the miracle of the resurrection by adducing a list of eye-witnesses (1 Cor. 15:3-8). But this is a dangerous procedure, as Karl Barth has involuntarily shown. Barth seeks to explain away the real meaning of I Cor. 15 by contending that the list of eyewitnesses was put in not to prove the fact of the resurrection but to prove that the preaching of the apostle was like the preaching of the first Christians, the preaching of Jesus as the risen Lord. The eye witnesses therefore guarantee St. Paul's preaching, not the fact of the resurrection. An historical fact which involves a resurrection from the dead is utterly inconceivable.[23]

Bultmann even went so far as to assert:

I do not indeed think that we can now know almost nothing concerning the life and personality of Jesus since the early Christians sources show no interest in either are moreover, fragmentary and often legendary and other sources about Jesus do not exist.[24]

Dr. Gerhard Gloege, a "post-Bultmannian" wrote in a volume published a number of years ago by the Lutheran Church in America: "The ground of faith in the biblical view is not isolated and objectified matters of fact, not facts like what was falsely presented and passed-off as the objective fact of the resurrection of Jesus."[25]

Again the same scholar wrote:

Living by Easter faith we take up the stone anew each morning 'with joy', as the hymn says. This Easter faith does not mean believing in the correctness of certain narratives and assertions or in the reality of certain events. In fact, strictly speaking, it does not even mean believing in the resurrection, but in the Risen One personally.[26]

Emil Brunner in his book *The Mediator* rejects a real physical resurrection of Christ. He claims that the empty tomb plays no part whatever in the New Testament the foundation for faith in the Resurrection.[27] Edwin Lewis claimed that "The Resurrection does not mean that a dead man came to life again."[28]

David Anderson further adduces the British writer Ian Wilson, who recently published a book, "Jesus: the Evidence." In this volume Wilson is reported as writing that the various (biblical) accounts of the scene of the empty tomb on the first Easter morning are so full of inconsistencies that it might be easy to deride them.[24] Wilson is simply following past critics of the Gospels who have claimed that the variant details in the four Gospels argue against the reliability of actually what transpired on Easter mom. However, Bible believing scholars contend that the Holy Spirit would not cause the writers whom he selected and inspired to write unreliable and confusing accounts of what actually transpired. Thus William Arndt noted:

" ... every well-informed Bible reader will admit without hesitation that no one of the four accounts of the resurrection is complete, reporting all the facts. Neither is there one among them which makes the claim of being exhaustive. Each one reports actual occurrences. It will be allowed by all fair-minded persons that reports may be fragmentary, incomplete, and yet true. If this simple principle is borne in mind, most of the difficulties contained in the resurrection will vanish."[30]

Arndt has sketched the events on that great First Easter day as occurring as follows.[31] At dawn, on the first day of the week, a number of faithful women hurried to the grave of Jesus with their ointments. But before arriving at the tomb of Joseph of Arimathea, an earthquake occurred, at which time there may have taken place the resurrection of Christ. An angel rolled away the stone before the sepulcher, and the Roman guards placed there to prevent any person from removing Jesus' body, in terror fell to the ground. After having seated himself on the stone, the angel withdrew into the tomb. Upon coming to the tomb, Mary Magdalene, seeing the stone rolled away, without making further investigation, at once ran back to tell Peter and John what she had seen. She assumed that the Lord's body had been stolen. In the meantime the other women entered the tomb, but did not find the body of Jesus, but were accosted by two angels, one of whom told them that Jesus was risen. They left the grave in great joyous excitement. Mary Magdalene in the meantime returned, with Peter and John preceding her. The two apostles found the grave empty but noticed that the grave clothes were lying in order, thereby showing his body had not been stolen. After they left the tomb, Mary Magdalene saw the angels in the grave, and a minute or two later she beheld Jesus Himself. According to Mark's account in 16:9, Mary Magdalene was the first to see the risen Jesus. Some, or possibly all the women, met Jesus as they were returning to their dwelling places. When the apostles were told by the women that they had seen Jesus, they refused to believe. In the afternoon Jesus appeared to the two Emmaus disciples. It would seem that Cleophas and his friend had not spoken to the women but heard a report that Jesus was alive. That same afternoon, Jesus probably appeared to Peter, Luke 24:34. On Easter evening Jesus was seen alive in Jerusalem by the ten, with Thomas absent On Easter Sunday the glorified Jesus was seen five times.

One week after Easter Jesus appeared to the Ten plus Thomas. Altogether there were ten different appearances before Easter morning and the day of Christ's ascension. Paul in I Corinthians 15 mentions six different appearances of the resurrected Jesus, including Paul meeting with Jesus on the Damascene road. The Evangelist Luke writes: "To whom he shewed himself alive after his passion by many infallible proofs being seen of them forty days and speaking of the things pertaining to the kingdom of God" (Acts 1:3).

Despite the confusion and contradictions supposedly found in the resurrection narratives, Ian Wilson would have us believe that somehow, "whatever its origin, caught on very soon after the crucifixion and spread like wildfire. And it was embraced by an extraordinary diversity of people."[32]

The Views of Pinchas Lapide, Jewish New Testament Scholar
Anderson's "Easter Controversy" also refers to Pinchas Lapide, the Jewish New Testament scholar, who, says the UPI reporter, "finds it possible for Jews to embrace the resurrection of Jesus within a Jewish context and without leaving the Jewish faith."[33] Lapide rejects the hallucination theory. Although the Jewish scholar claims there are legendary embellishments in the oldest records there is a recognizable historical kernel that cannot be demythologized. Something must have happened to turn frightened and discouraged disciples into men who became obsessed with winning the world for their Master. Still, while accepting the resurrection fact, Lapide says that he cannot "accept neither the messiahship of Jesus for the people of Israel nor the Pauline interpretation" of the Easter event. Still, Lapide is convinced of the fact that the disciples took their views about Jesus to the Gentile world "must surely be interpreted as a God-willed encouragement in a world that so often is hopeless."[34]

Lapide suffers from the same kind of spiritual illness as did the contemporaries of Jesus, as depicted in the Johannine Gospel and by the Jews of Paul's day who rejected Christ as the Messiah, Son of God and Son of Man.

Anderson, in setting forth the traditional stance of the Christian Church claimed that Easter is "an affirmation that in the death and resurrection of Jesus, God says 'yes' to life, to the world and 'no' to death, to annihilation and the alienation from God."[35] However, the Bible, especially the New Testament, which has at least two hundred references to Christ's resurrection, has more wonderful things to say about the accomplishments of Christ's resurrection. Stump has expressed it well, when he wrote: The resurrection guarantees that Jesus is the Son of God, as He claimed to be for God would not have raised an impostor from the dead. It guarantees the actuality and sufficiency of the redemption, because He who paid the ransom for us is proved by the resurrection to be God as well as man and to have paid the price of our redemption in full. It guarantees the certainty of our own resurrection from the dead, because as Christ rose, so we shall also rise with Him. It guaranteed the coming of the Holy Spirit who was poured out on the disciples on the day of Pentecost, and who equipped them for the preaching of the Gospel in all the world.[36]

What gave power to the preaching of the apostles in the days after Pentecost? The Book of Acts, which contains a number of apostolic sermons, reports consistently that the apostles preached "Christ and Him crucified," risen from the dead, who could and did change human lives. For the New Testament writers the resurrection was the crowning and irrefutable proof that Jesus is the Son of God and the Savior of the world.

The Resurrection the Activity of the Triune God

The resurrection is ascribed in the Bible to the activity of the Father and the Holy Spirit, as well as that of the Son Himself. Paul described the Father as raising His Son from the dead (Col. 2:12; Acts 2:24) and by the glory of the Father (Rom. 6:4), also by the Holy Spirit (Rom. 8:11), and by Christ's own act (John 10: 18; 2:19).[37] The explanation lies, of course, in the fact that the Father, Son and the Holy Spirit are one Triune God. Neither Kung and Lapide believe these inspired and clearly-stated truths. G.B. Hardy has said, "Here is the complete record:

Confucius' tomb .... occupied
Buddha's tomb .... occupied
Mohammad's tomb .... occupied
Jesus' tomb .... EMPTY."[38]

Footnotes

1. Joseph Stump, *The Christian Faith, A System of Christian Dogmatics* (New York: The Macmillan Company, 1932), p. 173.
2. **Ibid.**, pp. 173-174.
3. Josh McDowell, *Evidences that Demand A Verdict* (San Bernadinio: Campus Crusade International, 1972), p. 816.
4. Stump, **op. cit.**, p. 174.
5. David E. Anderson, "Easter Controversy," (Pathways/Religious News, Fort Wayne News-Sentinel, Saturday, April 6, 1985, p. 5C.
5a. C. F. Henry, *Christian Faith and Modern Theology* (New York: Channel Press, 1964), p. 265.

6. Stump, **op. cit.**, p. 177.

7. Cf. McDowell, **op. cit.**, pp. 241-260.

8. **Ibid.**, pp. 241-247.

9. **Ibid.**, p. 258.

9. Edward W.A. Koehler, *A Summary of Christian Doctrine* (Detroit and Oakland: Louis H. and A. W. Koehler, Distributors, 19390, p. 102.

10. **Ibid.**, pp. 102-103.

11. Koehler, **op. cit.**, p. 102.

12. Cf. C. Brown, "The Resurrection in Contemporary Thought," *The New International Dictionary of New Testament Theology* (Grand Rapids: Zondervan Publishing House, 1978), vol. 3, pp. 281-305.

13. Anderson, **op. cit.**, p. 5 C.

14. Edgar Krentz, *The Historical-Critical Method* (Philadelphia; Fortress Press, 1975), pp. 6-32.

15. Merrill C. Tenney, "The Historicity of the Resurrection," Carl F. Henry, ed.) *Jesus of Nazareth: Savior and Lord* (Grand Rapids: William B. Eerdmans Publishing Company, 1966), p. 136.

16. Brown, **op. cit.**, 3, pp. 281ff.

17. Anderson, **op. cit.**, 5C.

18. **Ibid.**

19. **Ibid.**

20. **Ibid.**

20a. Xavier Leon-Dufour, *Dictionary of Biblical Theology* (New York – Seabury Press, 1970), p. 496.

21. Rudolf Bultmann, *The Theology of the New Testament*, Translated by Kendrick Grobel (New York: Scribner & Sons, 1955), II, p. 299.

22. Rudolf Bultmann, *Primitive Christianity* (New York: Meridian Books, 1956), pp. 201-202.

23. Rudolf Bultmann, *Kerygma and Myth* (New York: Harper & Brothers, 1961), pp. 38-39.

24. Rudolf Bultmann, *Jesus and the Word* (New York: Scriber's & Sons, 1958).

25. As quoted by Herman Otten, *Baal or God* (New Haven, Missouri: Leader Publishing Co., 1965), p. 107.

26. Gerhard Gloege, *The Day of His Coming* (Philadelphia: Fortress Press, 1963), p. 281.

27. Emil Brunner, *The Mediator* (Philadelphia: The Westminster Press, 1947), p. 574.

28. Edwin Lewis, *The Philosophy of the Christian Revelation* (New York: Harper & Sons, 1940), p. 65.

29. Anderson, **op. cit.**, 5C.

30. William Arndt, *Does the Bible Contradict Itself?* (St. Louis: Concordia Publishing House, 1930), p. 70.

31. **Ibid.**, pp. 69-76; William Arndt, *New Testament History* (St. Louis: Concordia Publishing House, 1940), pp. 69-70.

32. As quoted by Anderson, **op. cit.**, 5C.

33. **Ibid.**

34. **Ibid.**

35. **Ibid.**

36. Stump. **op. cit.**, p. 178. Cf., also Koehler, **op. cit.**, pp. 103-105.

37. E. Hove, *Christian Doctrine*, (Minneapolis: Augsburg Publishing House, 1939), p. 202.

38. G. B. Hardy, **Countdown** (Chicago: Moody Press, 1970), as given by McDowell, **op. cit.**, p. 270.

Questions

1. What is the crowning event of all history? ____
2. Only Christianity claims an ____ tomb?
3. The resurrection was a fulfillment of ____.
4. Jews counted a part of a day as a ____.
5. What has been a favorite point of attack of the enemies of Christianity? ____
6. What is the Swoon Theory? ____
7. How do Christians respond to this theory? ____
8. What is the Vision Theory? ____
9. How do Christians respond to this theory? ____
10. Would the disciples have been willing to die for a dead Jesus? ____
11. Where have the deniers of the resurrection been found? ____
12. The historical critical method has ___ the resurrection.
13. There are over a hundred miracles in the Bible which are unacceptable to ____.
14. Reporter Anderson does not report ____.
15. The Roman Catholic theologian Hans Kung has rejected ____.
16. Does Paul speak of the Resurrection as a real event? ____
17. Kung has adopted the view which were propounded by the Lutheran ___ and Reformed ____ and ____.
18. What did Bultmann say about Jesus? ____
19. What did Gerhard Gloege say about the resurrection? ____
20. Edwin Lewis claimed that the Resurrection does not mean ____.
21. The Jewish scholar Pinchas Lapide suffers from ____.
22. The New Testament has at least ___ references to the Resurrection.

God and Gender

The Loyal Lutheran, **Fall, 1985**
Christian News, **January 6, 1986**

A few weeks ago we asked Professor Raymond Surburg. Ft. Wayne, IN. to comment on grammatical usage of the feminine gender in certain references to God in the original languages of the Bible.

Dear Pastor Steenbock: First of all the Father and the Holy Spirit are spirits while the Second Person of the Trinity, according to His human nature, is a man, who is referred to by the masculine nouns of prophet, priest and king. The two most frequently used words for God in the Old Testament are Elohim (employed 2550 times) and Yahweh or Jehovah (over 6000), and both are used with masculine adjectives and masculine verb forms, which does not support the feminist idea of God. The Hebrew word **ruah** (for spirit) is feminine usually. When spirit is used of the Holy Spirit, the fact that it is feminine conforms to the grammatical rule that **inanimate** objects like **fire** are feminine. In Proverbs 8:1 and 9:1, the Messiah is called Wisdom **(chochmah)** which is feminine.

In the Septuagint (Greek translation of the Hebrew Old Testament), the Greek word for Holy Spirit is To Pneuma To amon which is neuter, not feminine. In the Lord's Prayer, Jesus calls God, Father, a masculine gender noun.

The "Woman's Creed" (particular example of the feminine usage of "she" for the Holy Spirit, published in a Lutheran campus bulletin) is really espousing a Gnostic heresy. For in Gnosticism, God is referred to as "God the Father/God the Mother." (Cf. Elaine Pagel's *The Gnostic Gospels*, New York Vintage Books, 1979, pp. 57-83.)

Yours in Christ,

Raymond Surburg.

Professor Surburg has a doctorate in Semitic languages. His comments show the implied substantive difference between a grammatical form and the conferring of the personal pronouns "she" or "her" which are never used for the Holy Spirit in Scripture: "He" and "Him" are frequently used.

Questions
1. What does not support the feminist idea of God? ____
2. In the Lord's Prayer Jesus calls God ____.

An Unscriptural Image of Christ

A Book Review of *and Commentary* on *Pelikan's* Jesus
Through the Centuries *(His Place in the History* of *Culture)*

Christian News, February 17, 1986

In the late fall of 1985 Yale University Press of New Haven, Connecticut released a volume authored by the renowned Yale church history professor, Jaroslav Pelikan, entitled: *Jesus Through the Centuries.* A number of reviews appeared in newspapers in November, December, and January because undoubtedly of the fame of the author and also because of the proximity of the Christmas season, when Christians think about Christ's incarnation.

Pelikan, Sterling Professor of History and William Clyde DeVane, Lecturer of Yale University, author of numerous scholarly volumes, dedicated this volume to the Benedictine monks at St. John's Abbey, Collegeville, Minnesota with the words: "Nihil amori Christi praeponere."

This scholarly volume is supposed to be not so much about Jesus as it is about the differing ways in which He has been perceived and been interpreted down through the centuries. "Regardless of what anyone may personally think or believe about him," writes Pelikan, "Jesus of Nazareth has been the dominating figure in the history of western culture for almost 20 centuries" (p. 1). Again he avers: "If it were possible, with some sort of super magnate, to pull up out of that history every scrap of metal bearing at least a trace on his name, how much would be left? It is from his birth that most of the human race dates its calendars, it is by his name that millions curse and in his name that millions pray" (p. 1).

The volume has an introductory chapter; after which there are 18 chapters, totaling 233 pages, followed by scholarly references and notes (pp. 234-258); and the book concludes with an index of Proper Names and an Index of Biblical References (pp. 259-269).

The author of the Epistle to the Hebrews asserts about Jesus: "He is the same yesterday, and today and forever" (13:8). Pelikan claims that that assertion came to have a metaphysical and theological significance. Wrote the Yale historian: "The same yesterday and today and forever" eventually came to have a metaphysical and theological significance, as "the same" was taken to mean that Jesus Christ was, in his eternal being, "the image of the unchangeable God, and therefore likewise unchangeable" (p. 1). But in this book Pelikan claims to concern himself with the historical happenings and not with the metaphysical or the theological. The ways Jesus has been viewed in art, music and literature, in theology and politics have differed sharply over the centuries. In fact, it would be possible to construct a history of western civilization in terms of the varying answers to Christ's question: "What think you of Christ?" The readers of Pelikan's book will see that this is precisely what the Yale professor has attempted to do. He has accomplished this by giving attention to the three major periods of church history, namely, that of the ancient church, and of the medieval and the modern church ages. However, not all periods receive the same attention. A good deal of attention, considerable space was devoted to the early and later centuries of the Christian era than to those intervening. He also dealt with the era during which Christ lived, the time of the New Testament. In dealing with the art, literature, politics of nearly two thousand years of history since the ascension of Christ and the outpouring of the Holy Spirit on Pentecost, the author has found different "images" of Christ, which are not characterized by sameness, but are said to exhibit a "kaleidoscopic" variety, which the author believes constitutes their most conspicuous features (p. 2). Pelikan claims that his book is "neither a life of Jesus nor a history of Christianity as a movement or an institution" (p. 6).

In his presentation of the 18 different images of Jesus across the centuries, Pelikan believes that "the most conceptual framework for this range of images is provided by the classical triad in the Beautiful, the True and the Good, which itself played a significant role in the history of Christian thought" (p. 7). To this classical trio one might place, although not identical within the Biblical triad of Jesus Christ as the Way, the Truth and the Life (John 14:6). It is interesting to note that in the Ravenna Mosiac (reproduced in the volume, p. 30) Christ as the Way, the Truth, and the Life is summarized as "the Beautiful, the True, and the Good."

Since the first-century interpreters of Jesus' teachings and life have stressed one or another aspects of Jesus as either a teacher or a poet, as a philosopher and even as in the 20th century as a liberator. The four images from which all others arise, Pelikan claims, are rabbi, prophet, Christ, and Lord with the latter two eventually triumphing.

Pelikan alludes to the fact that in the history of Western thought there appeared great men like Pinto and Socrates, who have been studied by people since their deaths. Yet no cathedrals have been erected to Socrates, nor are there great composers like Handel or Bach who have written music honoring and praising pagan historical personages, but about Jesus Christ they have written immortal music in praise and adoration.

The following are the titles of the 18 chapters of Pelikan's book, which cover nearly two thousand years of church history. Chapter 1, The Rabbi; chapter 2, The Turning Point in History; chapter 3, Light to the Gentiles; chapter 4. The King of Kings; chapter 5, The Cosmic Christ; chapter 6, The Son of Man; chapter 7, The true Image; chapter 8, Christ Crucified; chapter 9, The Monk Who Ruled the World; chapter 10, The Bridegroom of the Soul; chapter 11, The Divine Human Model; chapter 12, The Universal Man; chapter 13, The Mirror of the Eternal; chapter 14, The Prince of Peace; chapter 15, The Teacher of Common Sense, chapter 16, The Poet of the Spirit; chapter 17, The Liberator: chapter 18, The Man Who Belongs to the World.

The Cross - Symbol of Christianity

Inasmuch as the cross has become the symbol of Christianity and during the medieval days no other symbol was so revered, the author has begun each of his chapters, including also the introduction, with one of the many different shape crosses created by Christians down the centuries. Here is a listing of the initial crosses: Introduction Russian Orthodox; 1. Passover; 2. Alpha and Omega; 3. Light and Life; 4. Labarum of Constantine; 5. Universe; 6, Golgotha; 7. Byzantine; 8. Monogram or Charlemagne; 9. Right Hand of Our Lord; 10. Christus Noster; 11. Cross of Peter; 12. Branch; 12; 13. The Evangelists; 14. Crusaders; 15. Latin Cross; 16. Fleur-de-lis; 17. Jerusalem; 18, Rainbow.

Interspersed through the volume are 18 illustrations, being various kinds of art objects, paintings, mosaics busts sculpture pieces, engravings, title pages of a book and woodcuts, some found in public collections; others in private collections, all selected to illustrate and support points Pelikan was making about a certain "image of Christ."

According to Carlin Romano, a writer for Knight-Rider Newspapers (November 30, 1985), "Pelikan's study does not focus on specific cultural genres. Nor does he pay much attention to the cultural exploitation of Jesus through TV evangelism and the commercialization of Christmas. He expertly captures archetypal intellectual images. The details mount until one understands why Pelikan categories an era's reigning notion of Jesus in a particular way.

"For each age, the life and teachings of Jesus represented an answer to the most fundamental questions of human existence and of human destiny, and it was to the figure of Jesus ... that these questions were addressed." (Fort Wayne News Sentinel, Sal Nov. 30, 1985) p. 4C.

In Jesus through the Centuries Pelikan seems to be distinguishing between "the religion of Jesus" and "religion about Jesus." The Yale historian observed that "Jesus is far too important a figure to be left only to the theologians and the church-traditional specialists in promoting rigid ecclesiastical concepts."

David Anderson of United Press International in his review of Pelikan's book wrote: "Pelikan, nearly always a graceful writer is also to use judiciously his wide ranging scholarship from the very early manuscripts of the church through Feodor Dostoevsky and Martin Luther King — has written a book that will challenge academics even as it pleases and informs the lay reader." (Fort Wayne News Sentinel, Sat., Jan. 11, 1986, p. 4C).

The Source for the Different Images

One has to go to the Gospels to the first four books of the New Testament for the different images to which individuals and cultures responded during the centuries. The Four Gospels, which strictly speaking are not lives of Christ are the chief sources for those who have been enamored of Christ to follow Him or adopt his teachings. Other books of the New Testament add to, supplement data found in the Four Gospels. Pelikan appears to follow the view of critical New Testament scholars who do not believe that the actions and words of Jesus recorded in the Four Evangels can be accepted exactly as set forth in the Gospels. Thus He wrote: "For the presentation of Jesus in the New Testament is in fact itself a **representation**: it resembles a set of paintings more closely than it does a photograph" (p. 9). (Emphasis is mine.) Prior to the publication of the Four Gospels in written form there developed an oral tradition, which according to Pelikan was written down after Paul's Epistles appeared. However, it should be noted in opposition to Pelikan, there are critical scholars who believe that all Four Gospels could have been in existence by A.D. 60.

In true higher critical fashion Pelikan writes: "Everyone must acknowledge, therefore, that Christian tradition bad precedence, chronologically and even logically, over Christian Scriptures; for there was tradition of the church before there ever was a New Testament, or any individual book of the New Testament." In the statement of Pelikan which now follows we find advanced the type of criticism, known as form criticism, when he asserted: "By the time the materials of the oral tradition found their way into written form, they had passed through the life and experience of the church which had laid claim to the presence of the Holy Spirit of God... it was to the action of the Holy Spirit that Christians attributed the compositions of the books of the 'new testament,' as they began to call it, and before the 'old testament', as they referred to the Hebrew Bible" (p. 10).

While Pelikan is not sympathetic toward those scholars who question whether or not Jesus ever lived, and so far there have been no less than three quests for "the historical Jesus," yet he states it is a hermeneutical principle: "Because the narrative or the sayings of Jesus and the events of his life and ministry had come down to the evangelists and compilers in this context, anyone who seeks to interpret one or another saying or story from the narrative must always ask not only about its place in the life and teaching of Jesus, but also its function within the remembering community. It is necessary to begin with the caution that every later picture of Jesus is in fact not a picture based on an unretouched Gospel original, but a picture of what in the New Testament is already a picture" (p. 10). In writing about whether Jesus ever travelled far from home, Pelikan said: "Whatever may be the historical status of the story of the night to Egypt, etc." he seems to be noncommittal about the historicity of the flight of the Holy Family into Egypt. The account of the Flight into Egypt cannot be challenged on textual grounds, then on what grounds should it legitimately be questioned?

When one opts for the hermeneutic of the form and redaction critics of Gospel interpretation, it would imply that the normal layperson and even ordinary pastor is not able to really know what the four Gospels correctly teach. It might parenthetically be noted that there are about 250 scripture quotations from both testaments scattered throughout Pelikan's book. One wonders whether Pelikan has checked all passages cited with the interpretation given by both Old and New Testament scholars committed to the historical critical Method. Each chapter at its head is supplied with a Bible verse, either taken from the Old Testament or from the New. The verses are chosen to illustrate the image of Christ depicted in that chapter, as persons discussed in the particular chapter seem to reflect that image. Pelikan has cited about 120 verses from the Gospels, and one wonders whether or not he is quoting the specific verse in harmony with higher critical scholarship's understanding of the particular verse. Sometimes Pelikan quotes a Biblical verse as spoken by Christ, and yet at other times he portrays the disciples of Jesus as attributing certain sayings and beliefs to Jesus (p. 95).

Relative to Pelikan's allegation that tradition existed before the New Testament Scriptures, and thus tradition preceded the Scripture of the New Testament, it should be noted that Christ and His apostles operated with the Old Testament Scriptures, and found many happenings in the life, passion, death and resurrection of Christ foretold in the Old Testament, so that the church Christ was founding was in harmony with the Old Testament and had a Scripture. Furthermore, it should not be forgotten that Christ was God, who had come to reveal the Father's will and therefore was a LIVING authority on a par with the Old Testament. Furthermore, Jesus had promised His disciples the Holy Spirit who would guide them into all truth both when they preached, taught and wrote, the same spirit who had inspired the Old Testament authors. The church after Pentecost faithfully recorded what Jesus had said and done and whatever tradition was utilized was factual and what each evangelist recorded under the guidance of the Holy Spirit was true. If Pelikan and the critics he follows and agrees with are correct, it would mean that the normal reader must be a scholar well versed in the different methods of accepted critical procedure to know what is true, because geographical statements may not be true, words ascribed to Jesus may not have been spoken by the Master but were placed by the early Christian community in His mouth. But when one reads the theologians of the centuries after the apostolic era, they seem to know nothing of the presuppositions and conclusions of current Gospel criticism.

The Interpretation of the Old Testament in Pelikan

The Yale professor uses about 30 verses from the Old Testament where he engages in interpretation of exegesis, again he follows Old Testament critical scholarship. Here is a case in point Deuteronomy 18:15 was understood in the New Testament as being a prophecy of the coming of Christ the great Prophet, of whom Moses was a type. The Deuteronomy statement in 18:15f., of it Pelikan averred: "In the context, this is the authorization of Joshua as the legitimate successor of Moses, but already within the New Testament itself, and then at greater length in later Christian writers such as Clement of Alexandria around the year 200, the promise of the prophet to come is taken as reference to Jesus who had the same name as Joshua" (p. 16). Thus the New Testament and Clement gave Deuteronomy 18:15 a new meaning, one that was not the original meaning of that passage, but Peter claims that Deuteronomy 18:15 was a prophecy about Christ!

Psalms quoted in the New Testament as Messianic (like 2, 8,16,40,45) are interpreted by critical Old Testament scholarship as "enthronement psalms," psalms which predicted nothing about Christ. Thus Pelikan claims that the Christians who saw the goal of Old Testament history as being on the life, death and resurrection of Christ, were in his view manipulating the Jewish prophetic tradition to serve their own ends. The Christian evangelist completely misinterpreted the Hosea statement: "Out of Egypt have I called my son" as a prediction of the flight into Egypt by the Holy Family and then the later instruction to this family to leave Egypt and return to Nazareth (p. 26). Thus the early Christians and subsequent church fathers are said to have followed this misinterpretation of the New Testament.

The Images Found by Pelikan in the New Testament

The four images from which all others arise, Pelikan asserts, are rabbi, prophet, Christ and Lord — with the latter two eventually triumphing. So beginning with the earliest — but often ignored in the name of anti-Semitism the portrayal of Jesus as rabbi, Pelikan believes the neglect of this image in subsequent centuries led to anti-Semitism. Thus he asked: "Would there have been such anti-Semitism, would there have been such programs, would there have been an Auschwitz, if every Christian church had focused its devotion on icons of Mary not only as Mother of God but as the maiden and Queen of Heaven but as the Jewish maiden and the new Miriam, and on icons of Christ not only as Pantocrator but as Rabbi Jeshua bar-Joseph, Rabbi Jesus of Nazareth, the son of David, in the context of a suffering Israel and a suffering humanity?" (p. 20).

In the interpretation of the Gospels, Pelikan accepted the view that there are similarities between rabbinic materials and Christ's teachings and that "Jewish materials have been taken over by Christian tradition and ascribed to Jesus" which seems to assert that Jesus is reputed to have uttered certain sayings that he did not speak. One might ask what does that do for the truthfulness of the New Testament and its dependability? How can this be harmonized with a Biblical doctrine of verbal inspiration?

In describing the teaching methods of Jesus Pelikan speaks about Jesus' parables. Thus Matthew states: Jesus said to the crowds in parables, "Indeed without a parable did he not speak to them (Matt. 13:34)." However, avers Pelikan that the interpretation of Jesus' parables makes no sense unless the parables are interpreted in the light of their Jewish background. The newer methodology employed by critical Bible scholarship is to be preferred to traditional ways of parable interpretation (p. 13). Declared the Yale professor: "Recent interpretations of his parables on the basis of that setting have fundamentally altered the conventional explanations of the point being made in many of these comparisons between the kingdom of God and some incident from human life often homely in its outward appearance" (p. 19). The new and better interpretation of the Parable of the Prodigal Son (Luke 15:11-12) would be the following: the younger prodigal son is the Christian church and the older son is Israel. The father says to the older son: "Son, you are always with me and all that is mine is yours. It is fitting for me to make merry and be glad, for this hour brother was dead, and is alive; he was lost and is found." Thus it follows that the covenant God made with Abraham and later with Israel was a permanent one and was into this covenant that other people were introduced. "This parable of Jesus affirmed both the tradition of God's relationship with Israel and the innovation of God's new relation with the church — a twofold covenant" (p. 14).

Thus there would be two ways of salvation: the way of salvation by good works and character apart from Christ and the Christian, which according to Paul, requires justification by faith apart from the deeds of the law. The problem with this view is that it makes the mission of Christ to his own people unnecessary and their rejection of His messiahship deity and claims was not sinful or reprehensible. John 14:6 and Acts 4:12, as well as other assertions of Christ and the apostles, would be wrong!

Another new interpretation accepted by Pelikan is the view that in the Sermon on the Mount there is oscillation between tradition and innovation.

The oscillation is between Jesus as rabbi and Jesus as prophet. According to Pelikan Jesus did not set himself in opposition to first-century rabbinical teaching, because early in the Sermon on the Mount, Jesus asserted "Think not that I have come to abolish the law and the prophets; I have come to fulfill them. For truly I say unto you, till heaven and earth pass away, not an iota, not a dot will pass from the law until all is accomplished" (Matt. 5:17-18). This is interpreted by Pelikan as giving permanent validity to the law of Moses as given by the people of Israel on Mt. Sinai. In rebuttal to this position, it must be stated that at Mount Sinai, God gave the Hebrews three different kinds of law: the moral, ceremonial and political. The New Testament has repeated the moral laws, with the exception of the Sabbath commandment (belonging to the ceremonial), but after Pentecost the ceremonial law no longer was binding, as may be seen from the decree of the Apostolic Council in A. D. 48. The Judaizers at Colosse were rebuked by Paul, when the latter wrote: "Therefore let no one judge you in food or in drink, or regarding a new moon of sabbaths, which are a shadow of things to come, but the substance is of Christ" (2:6-17).

The Reception of Christ as Son of God and LORD

"Pelikan discusses the images of rabbi and prophet as they were understood by Jesus' contemporaries, but he does not take up what response Christ's Jewish countrymen made to His claim to be the Son of God and LORD, Kyrios, used to translate Yahweh in the Septuagint. The documents of the New Testament are of first century A.D. origin and reflect what Christ's followers believed and taught about his deity or divinity. When Jesus asked His disciples at Caesarea Philipp: "What think ye of Christ," Peter responded "You are Christ the Son of the living God." Jesus told Peter that that answer had been given him from above. Various images were floating around among the people of Christ's time. But

1381

only one was correct, namely, that Christ was the Son of God.

What was the general response of Christ's fellow countrymen? Jesus said that he had been primarily sent to the lost sheep of the house of Israel. The Jews of Christ's time had this choice: they could either accept His claims or reject them. The New Testament writings, especially the Gospels deal with this all important image, of Christ. There are a number of passages that record what happened to Jesus' claims. St. John in his Gospel wrote that Jesus was the Logos, through whom the world was created and sustained, and that when He came into the flesh and lived among his countrymen they rejected Him. "He came into His own, but His own received Him not" (1:11). According to the Prologue, Christ is not only the Creator of all life, He is the source of the life of all people, and so Christ is the Light that lights all creatures. The fact that Christ is the source of life and light for all humanity has been interpreted by the Roman Church and Pelikan that the false religions also share and reflect this light from God.

What does the New Testament teach about the reception of Jesus' unique claims? His whole public ministry was characterized from the very beginning by opposition and persecution. It began early during the Galilean ministry (Mark 2:6-7). In Mark 3:6 it is recorded "And the Pharisees went forth, and straightway took counsel with the Herodians against him, how they might destroy him" (Mark 3:6). In John 6, where there is a record of "The Bread of Life Sermon," the people refused to accept His claim that He was the Living Bread which came down from heaven, and that they needed spiritually to receive Him to have eternal life; sadly it is recorded that the people walked away (John 6:60). In Luke 4:28 it is stated: "And they in the synagogue, when they heard these things, were filled with wrath, and rose up and thrust him out of the city, and led Him to the brow of the hill where the city was built and they might cast him down headlong." In Luke 22:2 it is reported "The chief priests and scribes sought how they might kill him; for they feared the people." In John 5:16 it is asserted "Therefore did the Jews persecute Jesus, and sought to slay him." Matthew 26:9 says: "Now the chief priests and elders, and all the council sought false witness against Jesus, to put him to death." When Pontius Pilate gave the Jesus a choice between Christ and Barabbas, they chose Barabbas, and said all the people, His blood be on us and on our children" (Matt, 27:25). The Jewish Sanhedrin condemned Jesus to death, because he was guilty, in their estimation of blasphemy, having made himself God. John 19:6 reports: "When the chief priests therefore and the officers saw him, they cried out, saying "Crucify him, crucify him. Pilate said unto them, Take ye him away and crucify him for I find no fault in him." Both Peter and Paul in addresses reported in Acts, claim that it was Jewish persons who demanded the death of Jesus and that they were responsible for Christ's crucifixion (Acts 3:13; 13:27) Christ's claim to be the Son of God was repugnant to the Jews in Christ's time as well to Jews in the centuries that followed His crucifixion and death.

The passages just cited are considered biased and any person who cites them is considered as being guilty of anti-Semitism. There are those who want to rewrite history and claim the Gospel contain interpolations of hostile statements by biased Jewish writers against their own people. Nearly every New Testament writer was a descendant of Abraham and thus the New Testament is Jewish. How can it be anti-Semitic?

The many passages in the New Testament that speak of Christ's rejection are ignored and it appears that Pelikan is adopting the current position of the National Conference of Christians and Jews which exonerates Jesus's own countrymen as innocent of the death of rabbi-prophet of Nazareth. Paranthetically it should be noted that there were Jewish people who accepted the claims of Christ. In the Christian congregations established in Asia Minor, Greece, and Italy there were Jewish believers of Christ. The first Christian congregation in Jerusalem was Jewish. The Book of Acts also reports that many priests embraced the Christian faith (6:7). A minority of Jews in the last 2000 years have accepted Christ as their Savior.

Pelikan follows those Christian scholars who deplore the fact that Christians have been guilty of "the de-Judaizing of the Gospel and even for the transmutation of the person of Jesus from rabbi in the Jewish sense to a divine being in the Greek sense" (p. 18).

An example of the transmutation is cited by Krister Stendahl as seen in his misinterpretation of Paul's Letter to the Romans. According to Stendahl the purpose of Romans has been completely misunderstood by conservative scholars (and this includes Luther). Heretofore, Romans was used as "the charter for their supposed declaration of independence from Judaism." Pelikan follows Stendahl (now primate of Sweden) who claimed that the main focus in Romans is not the need to be justified by faith, "because all have sinned and fallen short of the glory of God," but the relationship between Jew and Gentile. According to the former head of the Harvard Divinity School the main point in Romans is not the attitude of the Gospel versus the law, but with the struggle of the synagogue with the church, and the outcome of it. Paul announced in a promise and a prediction: "and so all Israel will be saved" (11:25-26). The Gentiles were grafted as a wild olive branch into the Jewish tree. Thus it would follow that Judaism needs not Christ nor His Gospel, but Christians are greatly indebted to Judaism.

Evangelism or Accommodation to Non-Christian Faiths

In the chapter 18, his last chapter, "The Man Who Belongs to the World," Pelikan emphasizes a view that has been set forth by the Roman Catholic Church and which is promoted by the World Council of Churches. The old conception of Christianity and of Roman Catholicism that Jews need to accept Christ has been rejected with the Roman Catholic Decree, issued October 28, 1965, the Declaration of the Church to Non-Christian Religions, Nostra aetate of the Second Vatican Council. Relative to Buddhism, Hinduism, Islam, the primitive religions, Vatican II declared:

The Catholic Church rejects nothing which is true and holy in those religions. She looks with sincere respect upon those ways of conduct and of life, those rules and teachings which, though differing in many particulars from what she holds and sets forth, nevertheless often reflect a ray of that Truth which enlightens all men (John 1:9). Indeed, she proclaims and must ever proclaim Christ "the way, the truth and the life (John 14:6), in whom men find the fullness of religious life, and in whom God has reconciled all things to Himself" (*Living Documents of Vatican*, 11, pp. 660-668).

It is argued that since Judaism has the Old Testament, it has a knowledge of the true God, Yahweh. Paul and the apostles were Jews and owed their religious beliefs to their Jewish ancestors. The fact that the contemporaries of Christ, Paul Peter, Stephen, John, Jude, and James rejected the distinctive teachings of Christ, does not mean that they were wrong; they simply are following a different road to heaven. The repudiation of Christ's vicarious death, His resurrection and ascension and the outpouring of the Holy Spirit are of no important consequence for a Jew or Moslem or Hindu.

Pelikan appears to endorse the argumentation and theological stance of the writers of the book Rethinking Missions, which was a frontal attack on the evangelization program of those Christian churches who believe the object of Christian missions was to bring about the conversion of people to Christ, to His Gospel and to His way of life; therefore of those who had not been brought to faith in the atoning work of Jesus Christ, the world's only Savior needed to be saved. This book was the work of seven American Protestant denominations. *Rethinking Missions: A Layman's Inquiry After One Hundred Years* agreed with Nathan Soederblom's stance that even though Christ is unique and the Revealer of God, yet the latter reveals Himself in history, outside of the Church as well as in it. Pelikan stated that in the seven volumes of this massive *Laymen's Report* the whole philosophy and practice of evangelization was challenged as well as its strategy. Averred the Yale professor "They concluded that the stress upon the particularly of Jesus and the absoluteness of his message had been, though perhaps necessary a temporary element in the program of missions." A major conclusion of this new view about missions was that it was the duty of Christian denominations to see the best in all religions, to help these non-Christian religions discover or rediscover the best in their own tradition. Conversion no longer should be to bring about the emergence of a world fellowship in which each religion will find its appropriate place.

Pelikan in his last chapter appears to subscribe to this ecumenical position, the view of the school of comparative religions that there is no real difference among religions in kind but only one in degree. Thus Christianity has no right to claim that Jesus is the only Savior for mankind and that without belief in Him and commitment to Him, eternal life is impossible. Thus Pelikan opposes the clear teachings of John's Gospel: "he that believeth in him shall be saved, but he that believeth not shall be damned" (3:16). Immediately after John 3:16 are the words in verses 17-18. In verse 18 God caused John to write: "He who believes in Him is not condemned, but he who does not believe is condemned already, because he has not believed in the name of the only begotten Son of God."

In writing about 20th century religious development the Yale University history professor wrote: "A growing feature of the debate has been the stress on cooperation rather than opposition between the disciples of Jesus and those who follow other ancient Teachers of the Way. These followers of Jesus who advocate such cooperation insist that they are no less committed to the universality of this person and message than are the advocates of the traditional methods of conquest through evangelization" (p. 228). Other religions beside Christianity house elements of light and truth. The other world religions are said to have their ultimate source in God, in the same God whom Jesus called Father. To hold otherwise, the Yale professor avers, would make the oneness of God an empty phrase or concept. So while Christians can teach other religions much, Christianity also in its encounter with other faiths can learn much from them. The Jewish opposition to the Trinity, the Muslims who reject the deity and resurrection of Jesus Christ, the teachings of Zoroaster with his two opposing deities from the beginning of time, the various gods of Hinduism are supposedly the reflections of the same God and the different names for deity in the non-Christian religions are simply different designations for the same God Christ worshipped.

This universalism is flatly contradicted by the New Testament. It was Jesus who said: "All men should honor the Son as they honor the father, he that does not honor the Son, does not honor the father who hath sent him" (John 5:25). All the "I Am's" of Jesus testify to the uniqueness of Jesus as the only Revealer of the true God, and to the necessity of accepting His claims. Jesus assertion in John 14:6: "I am the Way, the Truth and the Life, no man can come unto the Father except by me," surely rules out Pelikan's position. Peter's declaration, spoken after Christ's ascension is clear. "Neither is there salvation in any other, for there is no other name under heaven given among men by which we must be saved." This statement of the apostle Peter repudiates the stance and view that Christ is not necessary for eternal salvation. If it does not matter whether or not a person believes in Christ, one must ask: What was the purpose of Christ's incarnation, of His suffering and death? The verse in the Johannine prologue, that Christ lights every person who has come into the world does not mean that all men automatically have received Christ's salvation, but instead means that inasmuch as Christ is the Creator of the Universe and the source for all life in the world every human being, whether aware of it or not, owes his physical life ultimately to Jesus Christ. Vatican II and Pelikan's use of John 1:9 does not support the view that Christ's truth is found in other non-Christian religions by a kind of automatic natural revelation. Paul declared that Christ and Him crucified was foolishness to the Greeks and a stumbling block to the Jews. Simeon had prophesied that the child in Mary's arm would either be for a fall or for an arising in Israel. Jesus is the cause for a fall, when He and His teachings are rejected (Luke 2:32) Simeon predicted that the

Child would be a sign spoken against.

Do Jews and Christians Worship the Same God?

To argue that Christians and Jews worship the same God, "The God of Abraham, Isaac and Jacob," as the pope avers, is just not true! In the Old Testament the Hebrew patriarchs worshipped the true God because they were justified by faith in the coming Messiah, who as the Lamb of God would take away the sin of the world (John 1:19; Is. 53,16). Paul argues in Romans that Abraham was justified by faith (Romans 4:3). Jesus said "that Abraham rejoiced to see my day and he saw it (John 8:56)." The believers in the Old Testament were looking for and expecting the coming of the Messiah. According to Hebrews 11 the Old Testament persons mentioned there are said to have had saving faith and saving faith involves belief in Christ Jesus.

The tragedy that occurred in the first Christian century in Roman Palestine was that the majority of Jews rejected Christ as the fulfiller of the Messianic prophecies of the Old Testament. With the death, resurrection and outpouring of the Holy Spirit a new period in the world's religious history began, because then the great division occurred between Jews and Jews. Those that accepted the atoning work of Christ for the reconciliation of the world separated themselves from those Jews who rejected Christ's claims. With the destruction of Jerusalem in A.D. 70 the rift became even more permanent from that time forward we have the religion centered in the Jew, the God-man of Nazareth, spread by Jewish disciples and apostles throughout the Roman empire. That segment of Judaism that rejected Christ created a large literature, reaching its climax in the Palestinian and Babylonian Talmuds. Thus the life and activities of Christ are the watershed of Jewish history.

In Romans Paul argues that not all descendants of Abraham according to the flesh are the true Israel of God. Every person who is justified by faith is a true son of Abraham. The New Testament Church is in direct lineal descent with the Old Testament people of God. Christianity has a Jewish Bible, both the Old and the New Testaments were composed by Hebrew and Jewish writers. The apostles were Jews. The first Christian congregation was Jewish, in fact, the Jerusalem Church was the mother church of Christendom. Paul the great evangelist for Christ was a Jew, a Hebrew of the Hebrews, and for today's Jews to denounce the Christian faith and attack the reliability of the N.T. is just as anti-Semitic, as it is anti-Semitic for Christians to speak and write against Judaism and its rejection of Christ.

What about "Universality-with Particularity?"

Pelikan has advocated what he has termed "universality with particularly" concept relative to Jesus Christ, namely, that Christ is reflected in heathen faiths generally and particularly in Christianity. But this is setting up a wrong "both and"; but the truth is that it is an "either-or." He claims that as a result of a "curious blend of these currents of religious faith and scholarship with no less powerful influences of skepticism and religious relativism, the universality of Jesus has thus become an issue not only for Christians in the twentieth century, but for humanity" (p. 234).

The Yale professor claims that because of the variety and unity of images of Jesus that there is more than "is dreamt of in the philosophy and Christology of the theologians." For the answer to the question: "Who is Jesus the Christ?" and why did He come and what did He accomplish and what does He demand of all men and women, only the New Testament can answer these questions and their answers are not to be found in a number of misunderstandings and misinterpretations by individuals in the course of the Christian centuries as reflected in a number of erroneous images or inadequate ones of Christ set forth in Pelikan's book.

Examples of Misconceptions about Christ

Pelikan has noted how some early Christian commentators on secular classics like the "The Iliad" reported pre-figurings of Jesus in them in order to keep anti-classical religious zealots at bay. He says that the notion of Jesus as King of Kings fueled the rise and fall of the theory of political divine right in the West. One of the strangest misrepresentations of Christ was committed by the Jesuit Matteo Ricci, who when he came to China as Roman Catholic missionary first dressed as a Buddhist monk, then as a Confucian scholar. Ricci presented the message of Jesus as the fulfillment of the historic aspirations of Chinese culture. Concerning this interpretation of the Jesuit, Pelikan asserted "that it was similar to the way that Jesus had been presented by the early fathers as the culmination of the Greco-Roman faith in the Logos and by the New Testament of the Jewish hope of the Messiah" (p. 223). Again, here it should be noted that the New Testament writers did not ascribe Messianic meaning to Old Testaments text to Christ incorrectly but there was a direct line between Old Testament Messianic prophecies and their fulfillment in Christ: this is the clear assertion of Christ and His apostles.

Another serious misconception about Christ is furnished by the 20th century's attempt to justify liberation theology with Jesus' teachings. Christ was not a political deliverer but a spiritual One. The Christ of Schleirmacher, Harnack, Schweitzer, Thomas Jefferson, and others are caricatures of the real New Testament portrayal of Jesus Christ. Just as Christ was often misunderstood in its century, so the same phenomenon has occurred in the 1900 years since His ascension into heaven.

Questions

1. Pelikan appears to follow the view of New Testament scholars who do not believe that _____.
2. Pelikan seems to be noncommittal toward the historicity of _____.
3. What did Pelikan say about Deuteronomy 18:15? _____
4. Psalms quoted in the New Testament as Messianic are interpreted by critical Old Testament scholarship as _____.
5. Christ's claim to be the Son of God was repugnant to the _____.
6. Is the New Testament anti-Semitic? _____

7. The National Conference of Christians and Jews exonerates ____.
8. The first Christian congregation in Jerusalem was ____.
9. According to Krister Stendahl, the purpose of ____.
10. Nostra Aetate of the Second Vatican Council declared that ____.
11. Rethinking Missions was a frontal attack on ____.
12. What was the stance of Nathan Soederblom? ____
13. Pelikan opposes the clear teaching of ____.
14. Universalism flatly contradicts ____.
15. Every human being ultimately owes his physical life to ____.
16. Do Christians and Jews worship the same God? ____
17. Every person who is justified by faith is a true son of ____.
18. The Apostles were ____.
19. Christ was not a ____ deliverer but a ____ One.
20. The Christ of Schleiermacher, Harnack, Schweitzer, Jefferson are caricatures of ____.

An Analysis of the Vatican's 'Katholischer Kurz Kathechismus'

(A Contribution for Reformation Day and Luther's Birthday)

Christian News, November 10, 1986

The statement has been made that the Roman Catholic Church has changed since Vatican II. In some respects some interesting changes have been made. The liturgy no longer needs to be rendered in Latin, but the vernacular may be used. A new attitude has been manifested toward Jews and the socalled separated brethren. No doubt other changes might also be mentioned. Nuns and priests have not hesitated to criticize and even disagree with the Church. One finds radical movements among Roman Catholic orders in various countries of the world. But the significant question we must ask: "Has Rome changed its theology? Has it given up teachings which Protestants and Lutherans consider as anti-Scriptural?"

As Lutherans celebrate the festival of the Reformation (Oct. 31) and recall that on November 10, 1483 Martin Luther was born, have the differences between Rome and Wittenberg been narrowed and are these differences so insignificant that Lutherans could acknowledge the Pope as the Head of Christendom, as various Lutheran theologians in recent years have advocated?

In this presentation the writer will examine some of the basic teachings of the Roman Catholic Church as reflected in the *Katholisher Kurz Katechismus*, issued by the Directorium Catechisticum Generale, Rome-Vatican 1971. The edition used in this analysis and critique is the German edition Parzeller Fulda, given out by the interdiocesan Catechetical Association.

This *Roman Catholic Short Catechism* was translated into a number of European languages; namely, German, French, Spanish and Danish. The one used in this article is the German version, published in 1975 and has appeared in 13 editions between 1975 and 1982, of which so far 160,000 copies have been printed. The copy we are using is from the 13th edition, issued by Albertus Magnus Kolleg, Haus der Begegnung Koenigstein e. V. This *Short Catechism* has 64 pages and is organized by means of questions and answers with short explanatory paragraphs either proceeding or following a series of questions and answers. The *Catechism* itself has 56 pages and 6 pages of prayers. The *Catechism* is organized around 4 major parts and 93 smaller divisions or sections, with a total of 255 questions and answers. The *Catechism* is clear, precise and easily understandable.

In this analysis and evaluation the writer will translate many of the questions and the answers, so that reader will be able to see what the Roman Church officially teaches in 1986. The major criticism will come after the basic contents has been presented.

Part I deals with God, Creation and Redemption (pp. 1-10) and Jesus Christ is our Master (pp. 11-18).

Part II. The Church and Its Sacraments (pp. 19-25). The life in God. The Seven Sacraments and Prayer (pp. 26-39).

Part III. The Christian Way of Life (Lebensordnung) (pp. 40-52).

Part IV. The Consummation (Die Vollendung) in Glory (pp. 53-56).

Appendix: Prayers (pp. 57-62); Index (pp. 63 -64).

This *Catechism* was printed with the permission of the Church.

PART I

The first question of the *Catechism* reads: Why are we on earth? Answer: We are on earth to know God, to love Him and faithfully to serve him.

The second question reads: Who is the teacher of our faith? Answer: The teacher (Lehrerin) of our faith is the Catholic Church.

Question 3 asks: Who instructs us through the mouth of the Church: Answer: Through the mouth of the Church our Heavenly Teacher instructs us. Comment by writer: This means that the Church is the mediator between Christ and the people. It is not the canonical Word of God but what the Roman Church claims is Christ's teaching because the Church bases its teachings on Scripture and tradition. Christ and the **visible** (emphasis is mine) Roman Catholic Church are identified as one. As Scripture proof 1 Timothy 3: 15 is cited: "The church of the living God is the pillar and foundation of the truth."

Question 4 reads: "Through whom does the Church exercise the teaching office? Answer: The Church exercises (uebt) through the Pope and the Bishops its teaching duty or function. They preserve, proclaim and declare the teachings of Jesus Christ" (p. 3).

Immediately after answer 4 comes the statement: "With the sign of the cross and with the Apostolic Creed we announce and confirm (bekraeftigen) our Catholic faith."

The *Roman Catholic Short Catechism* distinguishes between earthly powers and the Kingdom of God. John 18:16 is cited to support this distinction. In a statement appearing under the caption: "The Message Concerning the Kingdom of God," the *Catechism* states: "We have a part in the Kingdom of God when we recognize Jesus' truth, Law and love and follow them."

Comment: To Nicodemus Jesus said: "Except a man be born again he cannot enter into the kingdom of God (John 3:3)." The Holy Spirit must create a great radical change in the sinner, who is lost and condemned. There are questions and answers in this catechism that agree with what the Bible teaches. Whenever possible this catechism gives Scriptural proof but often cannot. In this presentation the writer is selecting those assertions and those questions and answers where they disagree with Biblical and Lutheran theology. Many of the catechism's statements about God are correct.

In its discussion of the providence of God, question 13 asks: "How does God care for the world? Answer: 1. God sustains the world in its being.

2. God directs men through the moral law (Sittengesetz). 3. God directs the world through the laws of nature. 4. He directs mankind through the Church to salvation. The answer is followed by the following statement: It belongs to the teaching ministry of the Church to explain in a binding manner natural and supernatural law. God gives us all which we need on the way to heaven." The Lord is my shepherd, I shall not want (Ps. 22 in Vulgate).

The *Roman Catholic Small Catechism* teaches that God created the universe out of nothing, but says nothing that He created in six days, a truth set forth in Exodus 20: It was the Devil who tempted man and caused him to sin. The *Catechism* also believes in angels and provide each person with a protecting angel. (Question and answer 29)

Through Adam sin came into the world (Romans 5:12). From Adam his descendants have inherited sin in their origins. Question 33 (p. 10) asks the question: Wherein does original sins consist? Answer: Original or inherited sin consists in this, that because of Adam's sin that we come into this world without the life of grace (ohne das Gnadenleben). Genesis 3:15 does not follow the Vulgate, but the Septuagint. The *Catechism* states that Gen. 3:15 is the first good news about the Savior. An explanation follows this answer, which reads as follows: "With these words God condemned the Devil, who spoke from the serpent. This is the first good news about our salvation. Christ, the eternal Son of God, who became man from Mary, is our Savior. He is the Devil Defeater. Mary was placed by God's grace from the beginning in emnity with the Devil; therefore, she is free from all original sin, was conceived without blemish. This Roman teaching has no basis in Gen. 3:15.

Jesus Christ Is Our Lord

This part has questions 21-32 dealing specifically with the person and work of Christ. Question 33 deals with the mystery of the Holy Trinity.

Question 35 reads: "What did Jesus teach about the Kingdom of God? Answer: He wants to free men from the slavery of the devil and establish God's Kingdom. 2. We must **convert ourselves** (emphasis is mine) and believe His good news." (This would fit in with the Roman Catholic concept that man participated in his conversion.) The *Catechism* follows the second article of the Apostolic Creed and consequently discusses the Virgin Conception and the Virgin Birth.

Questions 44-46 deals with the Virgin Mary in relationship to Christ.

Question 45 asks: "Which advantages characterize Mary?" Answer:

1. She was preserved from original sin and all personal sins. She is full of grace. 2. She always remained a virgin and became the mother of Christ and the mother of God. 3. According to body and soul she was taken alive into heaven.

Question 46 asks: "What does the mother of God mean to us? Answer: The Mother of God is also our mother, our example and intercessor before God." (p. 13) We want to follow Mary's example in faithfully serving Christ (im treuen Christendienst) and dedicate ourselves to her pure heart. We cry to Our dear Lady, the mother of the Church, ask for her intercession in the Ave Maria uttered by the Angel of the Lord, in the rosary and in Mary pilgrimages. (p. 13)

The Death of Christ and the Mass

The ringing of the bells on Thursday and Friday remind us of the suffering and death of Christ which is represented in the Mass, where the priest in an unbloody manner offers up Christ.

The death of Christ is the highest offering He could offer. The highest offering of the New Covenant is the offering of the Mass. Christ's death was complete and is actualized in the Mass.

The Descent of Christ into the Realm of the Dead

The *Roman Catholic Short Catechism* portrays the **descent into Hades** as not being a descent to hell on Easter morning, but as an event that occurred when the soul of Christ after the death of His body went to the righteous , who died before Christ's death , and announced to them redemption. Thus in the Roman Catholic understanding of 1 Peter 3: 13-20 they confess "he descended into the realm of the dead."

In contrast to liberal Protestant and liberal Roman Catholic theologians the R.C. Small *Catechism* believes in the corporal physical resurrection of Christ. Section 29 asserts that the resurrection of Christ is the foundation of the Christian faith. In support it cited 1 Cor. 15:14. The *Catechism* also asserts the fact of Christ's bodily ascension into heaven. Jesus Christ will also return for judgment (Question and Answer 57).

However, the R.C. *Catechism* also declared that Christ comes now to the believer in the Sacrament of the Altar. On Pentecost Jesus sent the Holy Spirit, the third Person of the Holy Trinity. According to question and answer 60, the Holy Spirit proceeds from the Father and from the Son. This Spirit spoke by the prophets (p. 17). It was the Holy Spirit who strengthened the Apostles by enlightening and sanctifying them (Answer to Question 61). Today the Holy Spirit teaches and guides the whole Church and bestows upon it His graces. The Holy Spirit guides and comforts the whole Church but also the individual upon whom He bestows His gifts, His grace. Grace is defined in answer to question 64, "as every inner, supernatural gift, which God gives us for our eternal salvation." Question 65 reads: What does the life of grace bestow upon us? Answer: The life of grace permits us to share in the life of Christ. It makes us children of God, also makes of us brothers and sisters of Christ, temples of the Holy Spirit and heirs of salvation. This is also called sanctifying grace.

This grace is necessary because without it we cannot do anything for our salvation. Each person is provided by God with as much grace as is needed to be saved.

Question 68 asks this significant question: What must we add to God's grace? We must **cooperate** (my emphasis) with God's grace and must not resist it.

Question 69 asks: What does God work through this helping grace? Answer: God helps us through this aiding grace to do the good and to avoid the evil; He enlightens and strengthens us, so that we live as his children and die as such.

The doctrine of the Holy Trinity is said to be an unfathomable mystery (Q. and A. 70).

PART II
The Church and Its Sacraments

We believe in one, holy, catholic and apostolic Church (Credo). By faith and baptism we are members of the Church.

Question 71 asks: How did Jesus begin the founding of the Church? Answer: Jesus gathered the faithful about Himself, and out of them He elected 12 Apostles and made Peter his visible representative on earth. By means of His death Christ has earned life for the Church. Out of the blood and water which came out of the side of Jesus, there came forth the mystery of the Church; she is to transmit eternal salvation (p. 19). Matthew 28:19-20 is understood to refer alone to 12 Apostles.

Question 76 asks: Which offices did Jesus confer upon the Apostles and their successors? Answer: Jesus transmitted (uebertragen) the teaching office, the priesthood, to the Apostles and their successors, the bishops of the Catholic Church.

Matthew 16:16-18 and John 21:15-17 are cited as Biblical authority for the teaching that Peter and his successors are Christ's representatives upon earth. The Pope has the duty to teach, guide and to unify the Church. The Savior completed His Church through the sending of the Holy Spirit on Pentecost. What the soul is to the body, that the Holy Spirit is to the Church on earth. The Holy Spirit guarantees to the Church truth, holiness, sanctity and indestructibility.

Section 36 "The Organization of the Church"

According to Roman Catholic Short *Catechism* the Church is not a democracy, established by the grace or good will of the people, but built by God's grace. The Church is built upon the foundation of the Apostles and their successors, upon the hierarchy. The Church has its power and commission from Christ. Question 81 reads: Who is the successor of Peter? Answer: The successor of Peter is the Bishop of Rome; he is called the Pope or Holy Father, he has the highest governing power. Question 82 asks: Who are the successors of the other Apostles? Answer: The successors of the other Apostles are the bishops, they are supported in their task, especially in the dioceses (Bistum) by priests and deacons. Question 83 asks: What duties do we have over against the Church? Answer: We have the duty: 1. to belong to the Church; 2. to believe what she teaches; 3. to follow what she decides.

The Church Proclaims the Good News

Under this caption the statement is found: Jesus Christ guaranteed the Church for truth for mankind regarding the way to God. "The Church is the pillar and foundation of the truth" (1 Timothy 3:15). We hold as inerrant what the Church has always and in all places believed.

Question 84 asks: Why is the Church infallible? Answer: The Church is infallible because Christ through the Holy Spirit preserves the Church from error. Question 85 reads: What constitutes the teaching office of the Church? Answer: the teaching office of the Church is the Pope and the Bishops who are in fellowship with him. In a statement beneath this answer is the declaration: The teaching office preserves from all error the teaching of Jesus Christ, declares and explains (it). It is necessary to listen to the teaching office, because at all times error endeavors to enter the church.

Question 86 reads: Who has given the Church the gift of infallibility? Answer: The gift of infallibility is possessed by the Pope alone or the Pope and the Bishops in communion with him. This is followed by the significant question? When is the Pope infallible? Answer: The Pope as the chief Teacher of the Church and the totality of the Bishops together with the Pope united, are infallible, errorless when they give decisions concerning matters of doctrine or ethics and they command the Church to accept them. Scriptural proof is adduced: He who hears you, hears me." (Luke 10:16). But nowhere are there Scripture statements, where it is declared that the Apostles and Peter were to pass this power on to their alleged successors! The Encyclicals of the Pope are to be accepted and obeyed (p. 21).

Question 89 asks a very important question: From where does the Church derive its doctrine? Answer: The Church derives its doctrine from Scripture and Tradition. Comment by writer. Much in Roman Catholic doctrine and theology is derived from tradition. True Lutheranism and true Protestantism derive their doctrines only from Scripture.

The Roman Small *Catechism* makes the following statement about the nature of the Tradition: Tradition has developed in the Rule of faith, the liturgy and indifferent life forms (Lebensformen). We prize them highly out of the seed corn (Senftkorn) of early Christian forms a holy tree has developed, not without the gracious guidance of the Holy Spirit.

The next question asks: Why must we accord the Holy Scriptures respect? Answer: We must reverently believe the Holy Scriptures because its authors, writers wrote under the special guidance of the Holy Spirit, who preserved them from all error. However, listen to this qualifier: The teaching office of the Church guards these books and interprets them.

Section 38. treats of "The Command to the Church is Meant for All Believers."

It is the duty of all believers to carry out the Great Commission (Acts 1:8). Questions 95-110 deal with the duty and function of the Church on earth.

Section 38 deals with the fact that the Church sponsors divine services. The Church according to the command and in the power of Jesus Christ offers up the sacrifice of the Body and Blood of Christ, the Holy Mass is the great praise, thank and propitiatory sacrifice of the world Church and of her holy obligations. The holy Mass or Eucharistic celebration crowns and towers over all forms of divine services. For the sanctification of Christians, Christ has instituted the seven sacraments; they mediate to us the grace of Christ.

Question 95 asks? How does the Church exercise her priestly office? Answer: The Church exercises the priestly office when the ordained priest offers the holy Mass, dispenses, consecrates and blesses the sacraments and blesses the people.

The Short *Catechism* claims that Christ continues His work of salvation in the church through the sacraments, when men **cooperate** (emphasis mine) with grace.

Question 96 asks the question: What belongs to a sacrament? Answer: To a sacrament belong: 1. the outward sign; 2. the inner grace; 3. the institution by Christ.

The seven sacraments are the following: 1. Baptism; 2. Confirmation; 3. the Holy Eucharist; 5. Anointing of the sick; 6. Priestly ordination and 7. Marriage.

Question 98 asks: What does the Church do, when she consecrates and blesses.

Answer: When the Church in the name of Jesus consecrates, she designates people and things for God and His service. When the Church blesses, she places men and things under God's special protection.

Section 42 treats of "the One Church."

The Roman Catholic Small *Catechism* informs that there is only one God, one Savior, one true faith and only one Church, which was founded by Christ as the only true one.

Question 99 reads: Why is there only one true Church?

Answer: There is only one true Church, because Jesus Christ only founded one Church for the salvation of all men.

The Roman Catholic *Catechism* recognizes that there are many divisions in Christendom, the cause of which are errors in those denominations which were brought about by pride, desire for power and the sinful life of many Roman Catholics. The divisions and separation from the true Church are a great offense for Christianity, a weakening and contradict the prayer of Jesus: "that ye all may be one" (John 17:21).

How can the unity of the Church be furthered? The catechumen is told in the answer to question 100: 1. When he lives according to the true faith and sets a good example; 2. When he prays for the unity of the Church; 3. When he is tolerant toward the erring ones, but not with their errors.

Which are the marks which Christ has given His Church? (Q, 101). Answer: According to the will of Christ the true Church must be one, holy, catholic and apostolic.

In what must the true Church of Christ be one? Answer: In faith, the liturgy and in its government.

Wherein must the true church be holy? Answer: The Church of Christ must be holy in its commandments, foundational principles, in the seven sacraments and in its numerous saints out of its ranks.

Question 104 asks: Wherein must the true Church be catholic? The true Church must be catholic, i.e. the whole truth of faith must be owned at all times and in all zones.

Wherein is the true Church of Christ apostolic? Answer. The true

Church must be apostolic, i.e. its leaders are followers of the Apostles and possess their authority.

Question 106 asks: Which Church fellowship has the marks of the true Church? Answer: The Catholic Church has the four marks of the true Church of Christ.

This series of questions and answers treating of the marks of the Church is followed by the paragraph: He who is not catholic due not to his own fault, and acts according to his best conscience and is sorry for his sins, will be saved according to the will of Christ by the Church, which prays and offers up for all mankind: for this reason the Roman-Catholic Church is called the only saving (Church). Bad Catholics who are impenitent lose everlasting life.

Section 43 discusses "the Community of the Saints." Just like the various members of the human body constitute a unity, so the Church with all its members constitutes a unity (1 Cor. 12:26). The faithful on earth are united with the saints in heaven and with the departed in purgatory through Christ the head of the Church. The same thought is repeated in question and answer 107.

Section 44 Treats of the Saints the Church Honors

Question 108 asks: Why do we honor the saints. Answer: We honor or reverence the saints because they are friends of God, because God Himself has honored them, and because they are examples and intercessors for us. The next question asks: How do we honor the saints of the Church? Answer. We honor the saints of the Church, in that we observe the saints' festivals, follow their example and hold in esteem their relics and pictures.

Question 109: Why do we especially honor or venerate Mary. Answer: We venerate Mary because she is the mother of God and is our heavenly mother. The "Hail Mary" should be spoken mornings, at noon and in the evening. Also by the use of Mary hymns, Mary pictures, Mary festivals, Marian clubs and pilgrimages to Mary places do we honor her.

Our Life in God. The Seven Sacraments, Prayer

Christ has earned salvation for us. He paid for our sins, Christ's death and resurrection have provided us with a new, supernatural life. Out of his mercy well there issued Baptism. It immerses us into the flood life of the Father, the Son and the Holy Spirit. He who is baptized as an adult must first confess his sins before being baptized.

Normally the priest should perform a baptism, in emergency any Christian can. (Qi and A. 113) No sacrament may be received prior to baptism. According to the Roman *Catechism* water baptism may in cases of necessity be substituted by the desire for baptism (Begierdetaufe) or through the blood baptism of martyrs.

What does Baptism bestow? Answer: 1. Through baptism we become forever Christians; 2. We become children of God and heirs of eternal life. 3. We are free from all sins and sin punishments. 4. There is poured into us three divine virtues: those of hope, faith and love. Baptism is necessary because Jesus said "except a man be born of water and the Spirit he cannot enter the kingdom of God" (John 3:3).

Section 47 Treats "Of the Life of Faith"

Question 117 asks: What must we believe. Answer: We must believe what God has revealed and what the Catholic Church teaches us to believe.

Question 118 reads: Why must we obey God and the Catholic Church? Answer: Because God is the eternal truth, cannot err and does not lie; and 2. Because Christ preserves his Church from error of doctrine.

Question 119 asks: How must our faith be? Answer: Our faith must be catholic, i.e., all encompassing, strong, alive and firm. Although we can be tempted by difficulties concerning our faith, we must not have doubts about the articles of faith.

Section 48 Treats of "The Life of Hope"

Section 49 Treats of "Love to God"

Section 50 Deals With "Love to the Neighbor"

Section 51 Treats of "Following Christ"

Section 52 Deals With "Jesus Teaches Us to Pray"

Questions and answers 117 to 138 are concerned with developing these topics.

The Roman Catholic Small *Catechism* inserts a section on prayer between Baptism and the Sacrament of Confirmation, the second of the seven sacraments taught by Rome.

Section 53 Treats of "Confirmation"

In the holy confirmation Jesus fills us with the Holy Spirit. The Apostles already practiced confirmation; today the Bishop exercises this sacrament for those who are growing Christians. Scripture proof cited is: "Now the apostles laid hands upon them and they received the Holy Spirit" (Acts 8:17).

Question 139 asks: How is confirmation practiced or dispensed? Answer: The Bishop places upon the confirmand the hand, anoints him upon the forehead with Chrisam and utters the words: "Be confirmed through the gift of God, the Holy Spirit."

Question 140 then asks: "What does Christ effect through confirmation? Answer: Through confirmation Christ effects (wirkt) that we are filled with the Holy Spirit, in order that: 1. we confess our faith steadfastly; 2. that we fight against the enemies of our faith; 3. that we participate in the sanctification of the world.

Section 54 Deals With: "Jesus Christ Celebrates In His Church the Holy Eucharist"

Questions and answers 142-155 are introduced by the following statement: Through the celebration of the Holy Eucharist Jesus makes present (Vergegenwaertigt) His cross offering in the Church. Like Mary and John beneath the Cross, we can sacrifice along and thereby receive the sacrificial meal of Holy Communion. 1 Cor. 11:23ff is cited as the words of Jesus by means of which the Lord instituted the Holy Eucharist or the

Sacrament of the Altar.

Question 143 asks: What happens when the words of institution are spoken? Answer: When the words of Jesus are spoken: "This Is My Body," and "This Is My Blood," bread and wine are turned into the Body and Blood of Jesus. Only the appearance of bread and wine remain. The next question asks: Which power did Jesus give with the word: "This do in remembrance?" Answer: With the words "This do in remembrance: He gave the Apostles and their successors the authority to change bread and wine into the Body and Blood." The next question reads. No. 145: Who possesses the power to transform? Answer: The power to transform is possessed by the bishops and priests.

These answers are followed by the statement: The Holy Eucharist is not only holy bread or a sign or a power of Jesus, but Jesus Himself with flesh and blood, with body and soul with Godhead and humanity. The holy Eucharist is not only the proclamation or a remembrance of the Lord, as if He were no longer present.

Question 146 asks: What is Jesus in the Mass? Jesus is both high-priest, the offering and sacrifice or meal. Next Question 147 asks: What does Jesus do in the holy sacrifice of the Mass? Jesus offers Himself, through the priests on the altars totally alive as happened on the cross, except in an unbloody manner. The next question asks: For what purpose has Christ instituted the sacrifice of the Mass? Answer: Christ instituted the sacrifice of the Mass that God is glorified by the ever present sacrifice of the New Covenant; 2. that the Church sacrifices along and receives graces; 3. that Jesus gives us Himself as food. Catholics are obligated on Sundays and feast days to celebrate and attend the Mass. Paragraph 56 deals with The Sacrifice of the Mass and the Everlasting Covenant.

Question 149 reads: "What did Jesus say about the reception of Communion?" John 6:51; 53-54 is quoted which states that the Communicant receives the body and blood of Christ.

Question 150 asks: What do we receive in Holy Communion? Answer: In Holy Communion we receive the Body and Blood of Christ as food for eternal life. Question 151 asks: What does Christ effect in Holy Communion? Answer: Holy Communion unites intimately with Christ, the fountain of all graces, Christ gives new power to live as children of God. Question 152 asks: Who is permitted to attend Holy Communion? Answer: He who is in a state of grace and has the right intention; he may daily attend Communion.

Only believing Catholics are permitted to attend Holy Communion. Venial sins must be confessed by the reception of the Sacrament of Penance. Question 153 asks: How does Paul warn against the unworthy reception of Communion. The Answer cited is 1 Cor. 11:27,29.

Section 57 Treats of "The Worship of the Holy Eucharist"

Before the celebration of the Eucharist the hosts are kept in the tabernacle. The "Eternal Light" burns before it. We owe the Lord in the Sacrament of the Altar our prayers; on Corpus Christi Day the highest honor is bestowed on Christ.

Question 154 reads: How long does Christ remain present in the Sacrament? Answer: Christ remains in the holiest of Sacraments present as long as there remains the appearance of bread and wine.

Question 155 reads: How should we honor the Holy Sacrament? Answer: We should visit Christ often in the Sacrament, we bow entering and leaving the house of God and bend at the end of the Eucharistic blessing, we pray to the present Lord."

Section 58 "From Sin Through Repentance to Jesus Christ"

Throughout a life time the Christian is tempted to sin and sins. Romans 6:23 is cited. Subdivision treats of "Departure from God."

Question asks: Who commits sin? Answer: A sin is committed when a person transgresses a command of God with knowledge and deliberately.

There are mortal sins and venial sins. Question 160 asks: Who commits a mortal sin (Todsuende) Answer: a mortal sin is committed when a person sins in a serious matter, 2. With clear knowledge and 3. at the same time sins with consent. The follow up question reads: What are the effects (or results) of a mortal sin? Answer: Mortal sins robs of the life of grace and all earnings for heaven; 2. brings temporal punishments and eternal damnation. He who fervently prays hardly commits a mortal sin.

Question 162 asks: Who commits venial sins? Answer: A venial sin is committed, when a commandment of God 1. in a smaller matter willingly, 2. or in a grave matter is deliberately transgressed.

The follow up question asks: What are the consequences of venial sins? Answer: the venial sin (laeslichen Suenden) wound and place in danger eventually the life of grace; 2. they darken the love to God and bring temporal punishments.

The Sacrament of Penance

Question 168 asks: With what words did Jesus institute the Sacrament of Penance? Answer: When Jesus on Easter uttered the words found in John 20:22b,23.

Question 169 reads: Why must all sins committed since baptism be confessed? Answer: Because Christ has so ordained; 2. Because the priest as physician of the soul and judge must help and decide. The words of forgiveness or remission of sin, uttered by the priest are the following:

God, the merciful father, has through the death and resurrection of His Son reconciled the world to Himself and has sent the Holy Spirit for the forgiveness of sins. Through the service of the Church He bestows upon you forgiveness and peace. So I pronounce you free from your sins in the name of the Father, the Son and the Holy Spirit. Amen.

Question 170 asks: What does Christ effect through the sacrament of Penance? Answer: 1. In the Sacrament of Penance Jesus remits all confessed sins and eternal punishment; 2. He restores the life of faith or even increases it; 3. He gives us power for the good and helps us to avoid the bad. The next question reads: What is necessary for a worthy reception of the Sacrament of Penance? Answer: 1. We must search our consciences; 2. Arouse repentance; 3. Make good resolutions; 4. Confess our sins; 5. After the pronouncement of forgiveness thank God and carry out the penances (assigned by the priest). Question 172 asks: Who repents unworthily? Answer: Unworthily does he confess, who does not repent of serious sins (Schwere), and will not give them up or deliberately does not confess, or who refuses the necessary means for betterment or will not make good great damages. This is followed by the statement: Venial sins one should confess; mortal sins must be confessed by numbers (Zahl) and important circumstances; Prayer, fasting, alms help as penance to wipe out sins and the punishment for sins, often there remains even though the sins have been forgiven an earthly punishment, and here the Church with its treasury of grace helps along from the meritorious deeds of Christ and the saints. Mt. 18:18 is cited as Scriptural proof for this position.

Question 173 asks: What is released by an indulgence? Answer: Through the indulgence earthly sins are remitted, and for sins that have already been forgiven. This is followed by Question 174: What is necessary to win an indulgence. In order to win or acquire an indulgence one must live the life of grace and perform the prescribed works.

Section 62 Deals With "The Sacrament of Anointing the Sick"
Question 175 reads: What does Holy Scriptures say about anointing the sick? Answer: James 5:14-15 is cited as Biblical proof for this practice. Question 176 asks: Who should be the recipient of the anointing with oil? Answer: When in danger of death: any sickness which is of a serious nature should cause the priest to be called.

Question 177 asks: How does the priest exercise the anointing with oil? Answer: The priest anoints the forehead, the hands of the sick with the oil of the sick and then says:

Through this anointing may the Lord in His mercy help you, may he stand by you in the power of the Holy Spirit. (Answer: Amen). The Lord, who has freed you from sins, save you, and in His grace raise you up. The next question asks: What does the Sacrament of Anointing of the sick accomplish? Answer: The sick anointing remits all confessed sins and the punishment for sins. 2. It often bestows upon the faithful early health; 3. Or it strengthens for the departure to God. People who are extremely ill receive three sacraments: penance for life, anointing with oil for the sick and the Holy Communion.

Stations in the Church
In the communion of the saints there exist different stations in life. Each is assigned specific services and duties. In different sacraments the faithful receive different graces.

Section 63 The Priest and His Ordination
Priests are instruments of Christ between God and the people. They have a special part in the high priestly office of Christ for the honor of God, the service and salvation of the world.

Two questions are devoted to the sacrament of ordination.

Question 179 reads: What does Holy Writ report about the ordination of priests? Answer: Holy Scripture reports that: 1. That Christ bestowed upon or confirmed in the Apostles the priestly authority; 2. That the Apostles conferred this authority by the laying of hands upon others. Question 180 reads: What does Christ bestow through priestly ordination? Answer: Through priestly ordination Jesus bestows: 1. The priestly authority to teach, to sacrifice the Mass, to exercise the sacraments, to consecrate and to bless; 2. The grace to lead a good priestly life and activity. Priests are called clerics, i.e., one removed from this world (therefore wear special clothing). The priest vows celibacy, in order to give himself totally to Christ; 2. To be free for the pastoral work. Baptism, Confirmation and priestly ordination place an immutable seal upon the recipient, therefore these sacraments cannot be repeated.

Section 64 Deals With "The Sacrament of Marriage" (pp. 38-39)

The *Roman Catholic Short Catechism* teaches that marriage was instituted by God in Paradise. Later on, this holy estate was dishonored and destroyed. The Lord has elevated the marriage relationship to a sacrament, by means of which the Lord gives the man and woman who have entered into a lifelong union the graces of Christ and His Church. Marriage is to be between one man and one woman. Question 181 asks: How do bride and groom consummate their marriage? Answer: Catholic brides and grooms declare before a priest and two witnesses that they take each other in marriage. What does Christ effect in marriage? Answer: In the Sacrament of Marriage Christ binds the bride and groom to a holy, insolvable union forever and bestows on them the graces for this station in life: 1. To give them children and to raise them as Christians; 2. To love one another and to help each other; 3. To lead each other to eternal life.

Question 183 asks: What does Jesus say about divorce? The answer is given in the words of Luke 16:18. He who forsakes his wife and marries another commits adultery.

Divorces granted by the state are not valid before God. The Church cannot dissolve a marriage, it can only declare that a marriage was illegal from the very start and never should have been consummated.

Question 184 asks: Why does the church advise against mixed marriages? Answer: Mix marriages constitute a danger for the married couple as well as the children and causes impediments because of the differences of confessions.

Section 65 Deals With "The Celibate Orders"

Monastaries are of inestimable worth for the Church and the world through their prayers and offering, through their works of mercy. The members of an order wish in communion and in following Christ to live on a higher plane, they are a Christ sign for the world and against the enemies of the Church. God and the Church give the virginity of women as preferable to the marriage estate. The evangelical counsels; poverty, God-dedicated virginity and obedience are the guidelines of their lives given as an offering to Christ; they obligate themselves through the vows given the order. Question 185 asks: What do the members of the order pledge? The members of an order pledge personally not to own anything, to live a life of celibacy and for Christ sake to obey their superiors.

PART III
The Christian Life Arrangement (Lebensordnung)

Questions and answers 186-242 are organized around the petitions of the Lord's Prayer. Roman Catholic theology has two pillars on which its rests; philosophy and theology.

Philosophy preceded and sometimes is the basis for Roman theological teachings. This comes out under major Part III.

Section 66 Speaks About "The Arrangement in God's Creation"

All questions of Part III are introduced by this paragraph: God orders the lifeless creatures and the living world through the laws of nature. Creatures without mind are forced to follow the laws of nature. Man has besides natural laws a free will. The free will of man should direct itself according to the natural law of God (Sitteng esetz), in order to live right and attain his eternal destiny.

Question 186 reads: Why is the arrangement of creation holy? Answer: The creation arrangement is holy because it originates with God and by means of which he makes known Holy will.

In Section 67 the *Catechism* discusses the "Responsibility of Human Freedom." Using his reason man is to determine from conscience what the moral law is. The Ten Commandments have been given to elucidate the moral law. The moral law has been especially elucidated in the Sermon of the Mount.

Men and women are obliged to follow their consciences. Question 189 however, recognizes that there can exist an erring conscience. Thus question 189 asks: How must we mold our conscience. Answer; We must examine our conscience and mold it so that it agrees with God's commandments and those of the Church. The Christian way life is organized around the Ten Commandments. Each of the Ten Commandments is treated separately and the appropriate topics discussed.

In connection with the third commandment's instruction there are a number of questions that are significant, because of their theological teaching. Question 203 reads: Which kinds of work are prohibited on Sunday or feast days? Answer: It is forbidden to engage in heavy (gnoebere) physical activities; only in urgent exemptions are they allowed. The great feast days of the church calendar are introduced by fast days.

Question 205 asks: Why does the Church urge us on fast days to bring visible or apparent sacrifices? The Church calls us to make apparent (or visible sacrifices) so that we 1. respond to the sacrifical love of God; 2. that through prayer and fasting we pay penance; 3. that we may help those in need. Question 206 states: Which are the heavy strong fast days? Answer; Ash Wednesday and Good Friday. On those days no flesh is to be eaten (between ages 21-59) and only eat one meal.

Question 207 states: What does the Church demand of us on Fridays?

Answer: On Friday, the death day of Christ, the church requires that a visible sacrifice be made; if it is impossible not to eat meat then there must be abstinence from nicotine, alcohol and prayer must be made.

The next question asks: What blessings does God bestow through fasting? Answer: Through fasting God suppresses sin, elevates our spirit and bestows youthful energy and reward.

Section 75 "Life Under the Order (Ordnung) of the Church"

The Church as the mother of all believers has the right and duty to guide us for Christ. The Church must help guard the faithful against lukewarmness and sin. For this purpose the Church has instituted certain laws to be followed. Question 209 asks: How do the five important church laws sound? Answer: 1. You shall keep the commanded festivals. 2. You should on Sundays and festival days devotionally receive the Holy Mass. 3. You shall observe the commanded fast days. 4. You should at least once a year go to confession. 5. You should at least once a year at the Eastertide receive Holy Communion.

76 Life in Fellowship (Gemeinschaft)

God protects life in community or fellowship in the fourth through the tenth commandments. God has so created us that through fellowship with others we achieve our temporal and eternal destiny.

PART IV
Toward Consummation in Glory Section 88. "Physical Death"

The Roman Catholic *Catechism* teaches that all men must die because through one man sin entered into the world, and death by sin (Rom. 5:12).

Question 243 reads: What happens at death? Answer; At our death the soul separates from the body. The soul immediately comes before the special judgement of God, the body becomes dust. Question 245 asks: Where does the soul go after the special judgment of God? Answer; The soul after the special judgment of God goes either to heaven or into purgatory or hell.

Only by being in the sanctifying grace of God at all times are we able to be prepared for eternity.

89 "Our Concern for the Departed"

Not all departed go at once to heaven; many must be purified before the consummation.

Question 246 asks: Who comes into purgatory? Answer: He comes into purgatory who dies in grace, but still has to pay penance for his sins. This is followed by the paragraph: "Purgatory cleanses the Soul painfully, it matures and sanctifies for the joy of the all holy God. The poor souls for certain will enter heaven. We should help them through prayer, alms, and particularly through the sacrifice of the Mass to remain united to the grace of Christ, that they may attain their goal quickly. The Church does not forget her children, she prays and sacrifices for them, especially on All Souls Day."

Section 90 "Heaven or Hell"

Question 248 asks: Who enters heaven immediately? Answer: He who dies in God's grace and is free of all sins and all punishments for sins enters heaven at once after death.

Question 249 reads: Who enters hell? Answer: He comes into hell who dies in mortal sin.

Question 250 asks: What do the damned suffer? Answer: The damned suffer eternally God's absence and pain.

Section 91 Resurrection of the Dead and World Judgment

In support of a general resurrection of all men, the *Catechism* cites John 5:28-29. After the resurrection of the dead there will occur the world judgment. To those to the right of Him Jesus will utter the words of Matthew 25:34 and to those on the left the words of Matthew 25:41.

Section 92 Speaks of "The Triumph and Completion of the Church"

Question 254 asks: What will happen on judgment day with the visible world? Answer: On the day of judgment the visible world will be changed and a new world created. Then Christ will hand over to the Father a redeemed world.

Critique of the Roman Catholic Short *Catechism*

The theology set forth in this catechism is with few exceptions the same as the theology that was unacceptable to Luther and the Reformers and is evaluated, criticized and rejected in the Lutheran Confessions. In one area particularly has there been an advance in Roman Catholic teaching and that is in the area of Mariology. The Immaculate Conception and the Ascension of Mary, without her body seeing corruption, have become official dogmas of the Roman Church. Functions and accomplishments which only belong to Christ, Mary now has been made to be involved in and gets credit; thus detracting from the saving work of Jesus, the only Savior in the world.

Wherever possible the Roman Catholic Short *Catechism* supports its answers and statements with Scripture passages, but frequently it cannot because the teachings are not found in Scripture but are the product of tradition, human reason, the decisions of church councils. Vatican II again reaffirmed two sources for doctrine: the Bible and tradition. It is presumptious for Rome to claim that it determined the extent of the Biblical Canon and that the world owes the Scriptures to Rome. Furthermore, it is a false claim that Christ made the Roman Church the only interpreter of the Bible. A number of its doctrines rest upon the misinterpretation of the Bible.

The *Roman Catholic Short Catechism* teaches that the Body of Christ which consists of all true believers in Christ who are justified by infused grace and their own works, are the visible Roman Catholic Church. All baptized people are allegedly automatically members of the Roman Catholic Church. Rome makes the mistake of identifying the invisible

1404

Church with the outward physical Church of Rome. The headship of the Church, represented by the Pope, is based on Matthew 16:18, where Peter was told that the Church was to be built on Peter's confession, namely, that Christ was the Son of the living God. Nowhere in Scripture is there any arrangement whereby Peter and the other Apostles were to appoint successors who would wield authority over the Church. In John 21 Peter was merely restored to his apostleship which he had forfeited by his denial of Christ on the Thursday of Holy Week. Paul says that there is one mediator and intercessor between God and man, namely, Christ Jesus. No priesthood is necessary to effect forgiveness of sins or need to have forgiveness of sins.

The *Roman Catholic Short Catechism* leaches that through the sacraments grace is infused into the proper recipients of Rome's seven sacraments. Rome does not understand "grace" as an attitude on God's part, but as something which is infused into the believer, who then is enabled with this new power to meet God's commandments and the requirements of the law and merit justification. Rome confuses the doctrines of justification and sanctification. Christ has become for Rome a new Lawgiver. In its approach to New Testament teaching the Roman Church has introduced many new laws which it is necessary for the believer to keep and thus merit God's approbium. Roman theology is characterized by legalism. The Roman Catholic *Catechism* detracts from the one great sacrifice Jesus made on the cross of Calvary. It is not necessary to crucify Christ anew, every time a Mass is celebrated, except that no blood is shed. The Word of God, furthermore, does not teach that the celebrant at Communion turns bread into the Body of Jesus, and when he blesses the wine, the latter is changed into the Blood of Christ. The transubstantiation teaching violates the Scriptural text which clearly teaches that the recipient receives the Body somehow with the bread, and the Blood somehow with the wine. This is clear from the words of institution.

A comparison of Baptism and the Lord's Supper reveals that a sacrament must have been commanded by Christ, must have an outward sign, and bestow the forgiveness of sins. A comparison of the seven sacraments of Rome shows that only Baptism and the Eucharist meet these three criteria.

Relative to the doctrine of sin, the distinction between venial and mortal is not a distinction the Bible makes. All sins must be considered serious, whether they be in thought, word or deed. Hatred is already murder in God's sight. The immoral desire is already adultery according to Christ. Purgatory is a teaching not contained in Scripture, for once God has forgiven a person his sins, there is no longer the necessity to pay for their allegedly earthly consequences in purgatory. Nowhere in Scripture is there a teaching which asserts that the church has a treasury of good works from which it is possible to bestow them on other individuals.

Luther as champion of the truth protested against the anti-Biblical teachings of the Roman Church. He endeavored to bring the Church back to the pure doctrines of Christ and His Apostles. For his efforts he was excommunicated by the Pope and at the request of the Roman pontiff placed in the ban by the State as a heretic.

The Roman Catholic Church has changed since Luther as far as its doctrinal position is concerned. In fact, it has added other doctrines not contained in Holy Writ;

— In 1854 the Pope proclaimed the dogma of the Immaculate Conception.

— In 1864 the Pope repudiated the democratic principle of the separation of Church and State.

— In 1870 the Pope proclaimed his own infallibility.

— In 1950 the Pope proclaimed the dogma of the bodily assumption of Mary into heaven.

Jesus said: "If ye continue in My Word, then are ye my disciples indeed; and ye shall know the truth, and the truth shall make you free." John 8:31,32.

Questions

1. Have the differences between Rome and Wittenberg narrowed? ____
2. Should Lutherans now acknowledge the Pope as the head of all Christendom? ____
3. The *Catechism* says that God instructs us through ____.
4. The Church exercises the teaching office through ____.
5. Many of the catechism's statement about God are ____.
6. The catechism says nothing about God creating the world in ____.;
7. The catechism teaches that Mary was free from ____.
8. The catechism teaches that man must ____ himself.
9. The catechism says that the Pope is the successor of ____.
10. The catechism says that the gift of infallibility is possessed by ____.
11. Much in Roman Catholic doctrine is derived from ____.
12. True Lutherans and Protestants derive their doctrine from ____.
13. The Short *Catechism* claims that Christ continues his work of salvation through ____.
14. The seven sacraments are ____.
15. Philosophy preceded and sometimes is the basis for ____.
16. The catechism says that at death many must first be ____.
17. "Purgatory" cleanses the soul from ____.
18. The theology set forth in this catechism is the same the theology which ____.
19. The two sources of doctrine which Vatican II reaffirmed were ____.
20. Rome understands "grace" as ____.
21. Christ has become for Rome a new ____.
22. Roman Catholic theology is characterized by ____.
23. All sins must be considered ____.
24. Purgatory is a teaching not found in ____.
25. What teaching has Rome added since Luther? ____

A Critical Response to Plagenz's Essay: "Sorting Out the Word of God"

Christian News, July 15, 1985

Plagens of the Newspaper Enterprise Association wrote an essay entitled, "Sorting out the words of God." The essay is based on the views of G.M. Lamsa, a layman from the East, who, before he died several years ago, had written a number of books pertaining to the interpretation of the Holy Scriptures. Some of his writings were: *More Light On the Gospel*, (Garden City, N.Y. 1968); *Old Testament Light* (1964), *Gems of Wisdom* (1966), *Kingdom on Earth* (1966). The occasion for Plagenz's essay, which appeared in the *Fort Wayne- The News-Sentinel* of June 28, 1985, was the issuing of a paperback version of Lamsa's *The Holy Bible from Ancient Manuscripts*. Containing the Old and New Testaments, translated from the Peshitta *The Authorized Bible of the Church of the East*. The A.J. Holman Company of Philadelphia issued the original, published in 1957. The Bible translation may be said to have been the major **opus** of his life.

In his book *More Light on the Gospel* Lamsa has made the following astounding claim:

> Moreover, the author was educated under the care of today's priests of the Church of the East who knew no other language than Aramaic, and highly educated Englishmen, graduates of Oxford, Cambridge and other English schools. The author, through God's grace, is the only one with the knowledge of Aramaic, the Bible customs and idioms, and the knowledge of the English language who has ever translated the Holy Bible from the original Aramaic texts into English and written commentaries on it, and his translation is now in pleasingly wide use!

Lamsa claimed that the New Testament was not originally written in Greek.[2] According to Lamsa not only were the Gospels originally not written in Greek, but so were also not the Epistles of Paul. Asserted Lamsa: "The Epistles were translated into Greek for the use of converts who spoke Greek."[3] Lamsa was completely ignorant of the spread of Greek as a part of Hellenism which played an important role in first-century Palestine. There is a great deal of inscriptional evidence to support the use of Greek in Roman Palestine.[4]

To support his fictitious claim that Greek was not used by the Jews or first-century Palestine, Lamsa misquoted Josephus as follows: "Josephus states that even though a number of Jews had tried to learn the language of the Greeks, hardly any of them succeeded." (Antiquities of the Jews.XX. 13:2) However, an examination of the context of this quotation will show that what Josephus was asserting was that he had failed to attain a precision in the pronunciation of Greek.

To undergird his theory Lamsa also cited from Rabbinical writings. Thus Lamsa claimed that the rabbis held that the Jews should not study or use Greek. In his introduction to the translation *The Holy Bible from Eastern Manuscript,* he wrote: "Indeed, the teaching of Greek was forbidden by Jewish rabbis. It was said that it was better for a man to give his child meat or swine than to teach him the language of the Greeks."[5] While it is true that the Mishnah, Talmud and the Tosefta ban the teaching of Greek, the correct understanding of the use of Greek by Palestinian Jews has been stated by the Jewish scholar Lieberman:

> In all above mentioned sources, there is no hint of a ban on the study of Greek Wisdom of the Greek language: the injunction involves only the teaching of children. The fear the teaching of Greek may produce or give aid to future informers could be entertained only with regard to children whose development was not yet certain, but not to mature people who seek self-instruction.[6] Yamauchi has pointed out the fact that Rabbi Gamaliel at the beginning of the second century had one thousand students, of whom five hundred studied the Torah and of whom five hundred studied Greek Wisdom.[7] Yamauchi also noted that the Talmudic literature has about fifteen hundred Greek loan-words, which would support the idea that the rabbis had a knowledge of Greek.[8]

Is the Peshita Superior to the Greek New Testament Text?

It is the contention of Lamsa that Syriac was the language spoken by Christ and that the documents of the New Testament were written in Syriac. This would mean that the Greek New Testament was translated from the Syriac into Greek and in this translation process the Greek translators made mistakes and were guilty of mistranslations. Pages xiii-xvi of Lamsa's Holy Bible give examples of such alleged misunderstandings of the Syriac text.[9]

To buttress his theory that Syriac was the language in which the New Testament was originally written Lamas reproduced a letter from Mar Eshai Shimun, Catholicos Patriarch of the East, which asserted:

> ... we wish to state that the Church of the East received the scriptures from the hands of the blessed Apostles themselves in the Aramaic original language spoken by our Lord Jesus Christ himself, and that the Peshitta is the text of the Church of the East which has come down from the Biblical times without any change or revision.[10]

Lamsa's erroneous view which claims that Syriac was the language spoken by Christ and His Apostles rests on a lack of knowledge of the history of the Aramaic language family, one of the important family divisions of the Semitic languages.[11] Aramaic has a long linguistic history. Yamauchi has correctly given the history and relationship of the various dialects of Aramaic to each other as follows:[12]

1. Old Aramaic (925-700 B.C.) Inscriptions from North Syria with Canaanite features.
2. Official Aramaic (700-300 B.C.) Under the Assyrian and Persian Empires. Aramaic was used for official purposes from Asia Minor to Afghanistan. In The Elephantine Papyri, for example and the Aramaic of Ezra belongs to this period.

3. Middle Aramaic (300 B.C. - 200 A.D.) This category would include the Aramaic of the New Testament, the Aramaic of Qumran and Murabba' at Nabatean, Palmyrene.
4. Late Aramaic (A.D. 200-700):
   a. Western: Syro-Palestinian Christian Aramaic. Samaritan. Palestinian Jewish Aramaic
   b. Eastern: Syriac, Babylonian Talmudic Aramaic, Mandaic.
5. Modern Aramaic Aramaic is spoken in isolated villages in the Anti-Lebanon in Syria, in Kurdistan and Azerbaijan, and north of Mosul in Iraq.

Aramaic, the Mother Tongue of Jesus

That Jesus spoke Aramaic has been definitely established by Gutav Dalman and there can be no doubt that our Lord spoke Aramaic as it was in vogue in Palestine in the first Christian century.[13] The New Testament testifies to the use of Aramaic as may be seen from Mark 5:41; Mark 7:34; Mark 15:34.[14] Dalman has shown that a large number of expressions and names found in the New Testament are transliterations of Aramaic. Thus Peter's name Cephas comes from the Aramaic kepha "rock." Thomas is from **toma**, "twin," Matthew is from **Mattay**, **bar** the Aramaic word for "son," is found in such names as Bartholomew, Bar-Jonas, Barsabbas, Barrabbas; Golgotha from **golgolta** "skull." Akeidema from **haqel dema** "bloody ground;" Martha from **mareta**, the expression of Maranatha" (I Cor. 16:22) is from Maran "Our Lord" and **eta** "come."[15]

Were the Gospel Originally Written In Syriac?

A number of years before Lamsa made his Peshitta Bible translation, Charles F. Burney published his *The Aramaic Origin of the Fourth Gospels*, issued by the Oxford University Press, in 1922. It was his theory that the Gospel of John in Greek was based on an Aramaic original. In the 1930 ties a controversy arose between Yale Semitic scholars and the Greek scholars of the University of Chicago over the question as to which was the original language in which the New Testament was written. Charles Cutler Torry, an expert Aramaic and Semitic scholar, was opposed by the Greek N.T. scholar Goodspeed of the University of Chicago. In 1933 Torry published his *The Four Gospels, a New Translation* which was to set forth the view that all four Gospels were written in Aramaic. Torry published his The Four Goepels, a New Testament, which was to set forth the view that all four Gospels were written in Aramaic. Torry based his theory on Semitisms and alleged mistranslations of the Aramaic. Torry reconstructed the Aramaic text and then made his new translation from his reconstructed text. His views found relatively few adherents. Most New Testament scholars were skeptical about his Aramaic background for the New Testament was Matthew Black, who published *An Aramaic Approach to the Gospels* and Acts, first edition 1946 and reissued in 1967 in a third revised edition. Black has utilized Aramaic materials from the Palestinian Targums, the Samaritan Targum, and Christian

Palestinian Aramaic. About this book Yamauchi wrote:

"Though he refers to new evidence of the Targum Neofiti I and the Aramaic materials from Qumran, Black has been criticized for failing to incorporate the new evidence in his third revision."[16]

Syriac Not to Be Identified with
First Century Palestinian Aramaic

A form of Aramaic was spoken around Palestine in the time of Christ and His apostles.[17] Syriac-speaking Christianity had its beginnings around Edessa, which was evangelized between A.D. 116 and 216: It is true that there is the book called Doctrine of Addai which claims to give the correspondence between Jesus and King Abgar V of Edessa.[18] This is considered a fictitious account which no reputable scholar accepts. The Doctrine of Addai is a legendary account of the establishment of Christianity in Edessa.

Syriac, in the light of what is known about the Aramaic dialects, belongs to the Eastern subdivision of Aramaic and not to the Western.[19] In fact, according to Brockelmann, Syriac flourished from the third to the seventh century after Christ.[20] As a language Syriac began to appear as a spoken language toward the beginning of the third century, possibly toward the end of the second. Syriac, therefore, is not suitable as the gauge to be employed in the attempt to reconstruct the language spoken by Christ and His Apostles in the first Christian century.

The Importance of the Peshitta in Textual Criticism

Lamsa has purported to have furnished a reliable English translation of the Syriac Peshitta, one of the three primary translations used by Old and New Testament textual critics in ascertaining the correct text of the Old and New Testaments.[21] The Peshitta was the Bible adopted by Syriac-speaking Christians before the split of the Syriac Church into West Jacobite and East Nestorian branches in the fifth Christian Century.[22]

The Old Testament part of the Peshitta, it is believed was rendered from the Hebrew and Aramaic (Daniel and Ezra) and made possibly by a Jewish person sometime between 150·200 A.D.[23] Before the third century Syriac-speaking Christians adopted it. For purposes of Old Testament textual criticism it has value when used by competent scholars. Later the Old Testament Syriac was brought into line with the Septuagint, a Greek translation of the Old Testament made by various Jewish scholars.[24]

Lamsa contends that the Syriac Old Testament translation is superior to the Septuagint translation, declaring in his introduction to his Peshitta translation that the Septuagint "was never officially read by the Jews in Palestine who spoke Aramaic and read Hebrew."[25] This is not true because Septuagint fragments have been found at Qumran and because of the Greek quotations from the Septuagint round in the New Testament.[26] Furthermore, the text used by Lamsa for His English translation was the important Ambrosianus text in Milan.[27] Various manuscripts of the Peshitta show variant readings which Lamsa ignores claiming that the various manuscript of the Syriac Peshitta text show no variant readings.[28] That such is in no way the case is shown by the fact that for a number of years an attempt is being made to bring out a critical edition of the Peshitta Old Testament text.[29] Yamauchi rejected the idea of Lamsa that "one can revise the Old Testament text on the basis of the ambiguities in either consonants or vocalization of the Syriac Peshitta text is pure fantasy.[30]

The Value of the Syriac New Testament Translation
In opposition to Lamsa, scholars would contend that the New Testament Peshitta is not of great value, because the Peshitta was preceded in time by older texts such as the Sinatic-Curetonian Gospels. Gehman wrote relative to the Peshitta:

Peshitta in Syriac means "Simple" or "Vulgate." The O.T. was made directly from the Hebrew probably in the 2d or 3d century A. D. at a later time it was revised by a comparison with the Greek. It used to be believed the N.T. owed its origin to the effort of Rabbula, bishop of Edessa (411-431), to establish a uniform text by revising divergent copies of the Old Syriac in accord with Greek manuscripts of a later type of text. It is more likely, however, that this work represents an intermediate stage between the Old Syriac text and the final form of the Peshitta. Since the Syrian Church did not accept as canonical the 4 minor Catholic Epistles (II Peter, II and III John, Jude) and the Apocalypse, the Peshitta did not include these five books."

All New Testament textual critics, except those of the Church of the East, believe that the Peshitta is a translation from the Greek and totally reject Lamsa's theory that the Peshitta is the original New Testament and that our present Greek New Testament text is a translation, incorporating translational mistakes because of misunderstanding of the alleged Aramaic original.

The Character of Lamsa's English Translation
Apart from Lamsa's false views about the Peshitta as the original version of the New Testament, his translation of the Peshitta into English might be of service to Bible students. In the opinion and in the evaluation of Yamauchi Lamsa's translation is defective in many respects.[32] According to Yamauchi "in some cases, Lamsa has slavishly copied the King James Version even where the Syriac could be rendered differently."[33] Sometimes Lamsa does offer an original translation, but sometimes he mistranslates as when in Philippians 1:13 he rendered the Syriac Pretorin as "Caesar's court," while actually the Syriac was nothing but a transliteration for the Latin praetorium, the emporer's "pretorian guard."[34] It is especially in the Old Testament where Lamsa engages in speculation and substitutes a different translation from that required by a faithful translation of the Hebrew text. Relative to Matthew 19:24, which according to the Greek, should be translated: "And again I say unto you, it is easier for a camel to go through the eye of a needle. . ."[35] Lamsa rendered: "Again, I say to you, it is easier for a rope to go through the

eye of a needle ..."[35] Kent states: "Camel and needle's eye are meant literally, as attested by a similar Talmudic proverb using an elephant. The simile was intended to show the impossibility by naming the largest beast known in Palestine and the smallest of apertures."[36] To change the Greek **kamelos** "camel" to **kamilos** "rope" or to propose the explanation that Christ was referring to a small gate known as "The Needle's Eye" is to resort to speculation. Jesus may in Matthew 19:24 have used a ridiculous figure to emphasize the point he was making about the danger riches present to people as a stumbling block for entrance into God's kingdom.[37]

The paperback version of Lamsa's Bible often misleads its readers and therefore Plagenz's article to help his readers better to understand the Bible by the examples he gave from the new Harper edition actually do not contribute to a better understanding of God's Word but to a misinterpretation of the Word of Life.

Footnote

1. George M. Lamsa, *More Light on the Gospel* (Garden City, NY, 1968), pp. xx-viii-xxix.
2. George M. Lamsa, *The Holy Bible from Ancient Eastern Manuscripts, Containing the Old and New Testaments.* Translated from the Peshitta, *The Authorized Bible of the Church of the East* (Philadelphia: A. J. Holman Company, 1957), p. i.
3. **Ibid.**, p. xii.
4. **Ibid.**, p. x.
5. **Ibid.**, p. x.
6. Saul Lieberman, *Greek in Jewish Palestine, Studies in the Life and Manners of Jewish Palestine in the II-IV Centuries*, C. E. (New York: The Jewish Theological Seminary of America, 5702-1942), pp. 101-102.
7. Edwin M. Yamauchi, *Greek, Hebrew, Aramaic, or Syriac?, Bibliothee Sacra*, October, 1974, p. 326.
8. **Ibid.**
9. Lamsa, The Holy Bible, **op. cit.**, pp. xiii-xvi.
10. Lamsa, Holy Bible, **op. cit.**, p. ii.
11. Yamauchi, **op. cit.**, pp. 323-327.
12. **Ibid.**, pp. 32-323.
13. Gustav Dalman, *Jesus-Jeshua. Studies in the Gospels.* Authorized Translation by the Rev. Paul P. Levertoff, (New York: The Macmillan Company, 1929), pp. 1-37.
14. **Ibid.**, p.11.
15. **Ibid.**, pp. 12-13.
16. Yamauchi, **op. cit.**, p. 325.
17. Dalman, **op. cit.**, pp. 7-16.
18. Anton Baumstark, *Geschlichte Der Syrischen Literature* (Bon: A. Marcus und E. Webers Verlag Dr. Jur. Albert Ahn, 1922), pp. 27-28.
19. Franz Rosenthal, *Die Aramaistische Forschung selt Noeldeke's Vreoeffent-Lichunge* (Leiden: E. J. Brill, 1939), pp. 179ff.
20. Carl Brockelman, *Syrische Grammatik* (Berlin: Verlag von Reuther & Reichard, 1925), p. 1.
21. Lamsa, Holy Bible, **op. cit.**
22. Yamauchi, **op. cit.**, p. 328.

23. Ernst Wuerthwein, *The Text of the Old Testament*. Translated by Errol F. Rhodes (Grand Rapids: William B. Eerdmans Publishing Company, 1979), pp. 80-81.

24. Bruce K. Waltke,"The Textual Criticism of the Old Testament," in *Biblical Criticism: Historical, Literary and Textual* (Grand Rapids: Zondervan Publishing House, 1978), p. 77.

25. Lamsa, Holy Bible, **op. cit.**, p. ix.

26. J. W. Wevers, "Septuagint," *The Interpreter's Dictionary of the Bible* (Nashville: Abingdon Press, 1962), RZ, p. 276.

27. Yamauchi, **op. cit.**, p. 328.

28. **Ibid.**

29. Cf. Moshe H. Goshen-Gottstein, "Prolegomena to a Critical Edition of the Peshitta," *Scripta Hierosolymitana*, VIII 919621), pp. 26-27.

30. Yamauchi, **op. cit.**, p. 320.

31. "Versions," Henry Snyder Gehman, *The New Westminster Dictionary of the Bible* (Philadelphia: The Westminster Press, 1970), p. 976.

32. Yamauchi, **op. cit.**, pp. 320,330.

33. **Ibid.**, p. 329.

34. Lamsa, Holy Bible, **op. cit.**, p. 1173.

35. **Ibid.** pp. 973-974.

36. Homer Kent, "The Gospel According to Matthew," Charles F. Pfeiffer and Everett F. Harrison, *The Wycliffe Bible Commentary* (Chicago: Moody Press, 1962), p. 964.

37. R. E Nixon, "Matthew," D. Guthrie, J.A. Motyer, A.M. Stibbs and D. J. Wiseman, *The New Bible Commentary Revised* (Grand Rapids: Wm. B. Eerdmans Publishing Company, 1970), p. 841.

Questions

1. Lamsa claimed that the Bible was not originally written in ____.
2. What language did Jesus speak? ____
3. Syriac is not to be identified with ____.
4. What is the Septuagint? ____
5. The Peshitta is a translation from the ____.

The Supreme Court Endorses and Supports Atheistic, Speculative and Unproved Evolution As Factual Science

Christian News, July 20, 1987

On June 19 the U.S. Supreme Court struck down the Balanced Treatment for Creation-Science and Evolution Act of 1981, which required the teaching of creation view of origins alongside evolution in the public schools of Louisiana. The vote was 7-2, Justices Rehnquist and Scalia dissenting against the majority. The decision of this highest court in the United States claimed the Louisiana Law was unconstitutional and that the law was an attempt to advance a religious point of view on the subject of origins.[1] The law it was claimed was endeavoring to present a supernatural being as creator. By seeking "to alter the science curriculum to reflect a religious point of view antagonistic to the theory of evolution," Brennan claimed that the Louisiana Legislature had violated the first amendment respecting an establishment of religion.[2]

Justice Brennen and his colleagues who agreed with him thereby assumed that the evolutionary origins of the universe, the earth and of life is an established and an incontrovertible fact of science. To hold otherwise was to teach myth or even religious teaching. However, these justices believe that evolution, without a God in anyway involved, is so well established so that to oppose it, is to teach the children and young people in high schools and colleges wrong information. In fact, Justice Brennan is so committed to evolution that he claims that to teach creation science side by side with evolution is "a sham."[3] The justices appear to be blissfully ignorant that respected scientists of the past and present have rejected evolution and still do at the present time. That their position on evolution is opposed to the convictions of millions of Jews, Christians, Mohammedans, who believe in the existence of the God concept and that God was the creator of the universe and everything in it, they ignored.

The court's decision reveals the influence of that portion of the scientific community that ever since Darwin's day has ruled the roost of the scientific world and also shows the power which the evolutionary philosophy has wielded over the minds of many in academia and in the various professions, including that of the law, who have been defenders and advocates of what they erroneously have considered to be true science. The influence of science departments in colleges, universities and graduate schools, which has filtered down to the high schools and grammar schools of the nation, is shown also by those who author the textbooks used in various science classes, written by persons who are committed to atheistic evolution.

The Supreme Court decision has been hailed as a great watershed decision and has been proclaimed as a great victory. Some have compared it to the influence of the Scopes Trial of 1929, which it is claimed, was the beginning of the end opposition to the teaching of evolution in public schools in the U.S.A.[4] The news media of the land had articles and editorials ridiculing the Biblical doctrine of creation.[5] Thus Arthur J. Kropp, executive director of People for the American Way of Life, applauded the court's decision and stated: "Even if today's ruling forces the religious right to dig deeper into their bag of creationist disguises, it is a major victory for American public education."[6]

The Inconsistency of the Supreme Court

If God is not to be mentioned and is excluded in the American public schools, then why is it permitted that on the coinage, which even the members of the Supreme Court use, that on it is the slogan: "In God we trust." Each penny, nickel, dime, quarter, half-dollar, coin dollar has upon it the statement "in God we trust." The same assertion is also on the paper coinage of every denomination printed by the United States Treasury. Just what does that slogan mean: "In God we trust." It certainly presupposes His existence and that we place our confidence in Him that He will be with those who believe in Him, act responsibly, morally and ethically. Strange is it not that on millions of coins and bills God may be printed as a belief of America, but in the schools God is not to be even suggested as a possible alternative for answering the important question of origins.

The senate of the United States has a chaplain who opens its sessions with prayer. The serving chaplain may be a Jewish rabbi, a Roman Catholic priest or a Protestant chaplain coming from any number of different churches or some other person believing in God. At the inauguration of a President a number of different clergymen participate in the inauguration ceremonies. Is that not establishing religion and failing to keep church and state separate? How logical is the decision of the Supreme Court in view of a couple centuries of practice? In the law courts of our land, witnesses about to give testimony, are required to take an oath, which reads: "Do you promise to tell the whole truth and nothing but the truth, so help you God?" Is that not introducing the establishment or religion in our legal system?

While the name of God may appear on America's coinage, be used in the halls of congress, be pronounced in the court laws of America, the name of God may not be used in connection with the teaching of creation side by side with the atheistic evolutionary theory.

When the Founding Fathers of the United States of America founded this nation and wrote a constitution for it, they did not intend to banish God completely from American life, but were specifically concerned that no one religion, Protestant or Roman Catholic be established as the religion of the state. The Founding Fathers were either Trinitarians or deists who believed in a Supreme Being who guided the affairs of men. The Declaration of Independence declared that the right of citizens comes from a Creator. No atheism was believed in 1787 or advocated as is now being promoted by the majority Justices when only godless "science" can be taught.

1415

The Views of Humanism and Religion

The denial of the existence of God and his participation in the creation and preservation of the world is the stance of secular humanism, which by an earlier decision of the Supreme Court has been labelled "a religion."[7]

Atheistic associations and the humanist organizations are violently opposed to the belief in God and deny any connection of God with the creation and preservation of the universe and all that is in it. Why should such a form of religion be allowed to be propagated in the name of science, while God as Creator cannot be presented as a viable alternate theory? An injustice is being done to millions of American parents who do not wish their children indoctrinated in a form of science that leaves God out of the picture as Creator. Humanistic views are being promoted in various textbooks, in the social science, geography and science books. This is blatantly being tolerated by the courts in our America.

The Irrationality of Atheistic Evaluation

The atheistic theory of the origins of the universe, the world, the solar system, life and the origin of animal and human life taught in our tax-supported schools is not reasonable and is irrational. To argue for an accidental and haphazard beginning of the universe, the world and life militates against common sense. To claim that matter or energy, from which supposedly the universe and everything in it came and produced itself, is irrational and non-sense. Such a view, hailed as scientific, goes against human experience and human observation. Human experience and observation tell intelligent human beings that a poem has a writer, who created it; a painting is the creation of an artist who purported to convey a message by his painting; a symphony is the product of some gifted musical-talented person. The nation of the United States did not come into existence of itself, but was the creation of representatives from a number of states along the Atlantic seaboard. If a person were to say that a number of years ago a great quantity of lumber, cement, bricks, iron girders, much glass, all manner of electrical equipment deposited itself on the site now occupied by the Empire State building in Manhattan, and further assert that all this conglomerate material created itself into one of the highest architectural structures in the world, that person would be told to see a psychiatrist or even be committed to an asylum for the mentally ill. What is not possible on a limited human scale, however, is predicated by evolutionists as possible on a cosmic scale. While finite minds are required for the production of a book, a painting, a poem, forms of government, buildings, in dealing with our vast and awe-inspiring universe, we are actually asked to believe that there was for it no Creator or Mind that planned it all. Just how rational is that form of "science" that teaches that there was no Planner or Designer or Maker for the millions, if not billions of stars that astronomy has discovered with its powerful telescopes. The origins of the universe, the planetary system, life, man and animals are supposedly the happenings of chance and that the laws of nature and their perpetuation in the universe are simply the result of chance. The great evidences of design in nature and the manner in which the many different laws that control the universe and the creatures that inhabit it, require a "Great Mind," God, to have conceived these laws and imbedded them in the making and sustaining of creation.

Many proponents of evolution take the position that evolution is not a theory but an established fact.[8] George Will, a national syndicated columnist, in an editorial provoked by the Supreme Court's decision, entitled, "Creationism Disguises Dogma As Science," (*Christian News*, July 16, p. 20, Ed.) berated Justice Scalia's dissent and was shocked by his opinion.[9] We cite Will's criticism: "Scalia finds praiseworthy the law's requirements that evolution be taught as a theory, rather than as proven scientific fact." To label evolution as a theory, Will contended, it was misleading because that would: mean that evolution was more less than fact. But, says Will, that would be misleading, for he asserted flatly: "Evolution is a fact about which there are various explanatory theories."[10] Since when does one label facts as theory? We ask Mr. Will.

"Creation science," avers Will, is a theory for people that are only interested in facts that weigh against a rival theory. But no scientist doubts the basic fact that life evolves. However Will does not distinguish between "microevolution" and "macroevolution." That there are variations within the kind is not questioned by those opposed to what has been called the general theory of evolution, which predicates the development of man from the amoeba through all stages of the animal kingdom until development occurred from a monkey, chimpanzee, or an animal common to both animal and man. To blur the picture, Will and others like him, claim that creation science teaches a flat earth and a six day creation. Where are there Christian science teachers who teach the earth is flat? Where are creationists who even predicted large periods of time are referred to by the Hebrew word "yom" or day, or who claim that between Genesis 1, verse 1 and 2 a long period of time elapsed.[12] The opponents of creation science frequently resort to ridicule to make their case seem right and the cause of the opponents appear ridiculous.[13]

In support of the evolution theory it has been stated that seventy-two Nobel Prize winners have come out in support of evolution. There have been periods in the history of science when so-called prestigious scientists held views which subsequent years have shown to be false and incorrect. Since when does one determine truth by numbers?" In philosophy and logic this is known as the **ad honominem argument**.

Is Evolution Really Scientific?

The editor of the Fort Wayne Journal Gazette, a rabid proponent of evolution, wrote an editorial the Sunday after the Supreme Court's decision, entitled, "Religion and science are, not the same thing."[14] Thereby he was implying that evolution was true science, while religion was just belief. Justice Brennan in calling creation science "a sham" was implying that there was a world of difference between evolution as fact and religion as belief and not necessarily true. But just how scientific is the theory of evolution? The American Heritage Dictionary defines "science" as follows:

"Observation, identification, description, experimental investigation and the theoretical explanation of natural phenomena."[15] These four steps, which constitute the scientific methodology employed in scientific facts, have never been employed in the establishment and validation of the evolutionary theory. The whole alleged development of the universe and all that followed that initial beginning in terms of cosmic evolution, biological evolution and cultural evolution is not demonstrable, because this evolutionary process is not reversible and happened once for all. The assumptions that underlie the whole evolutionary theory are not duplicable by experimentation and demonstration. Although divergent views since the days of Darwin have been put forward and no one explanation has been accepted,[16] yet even though the mechanism of evolutionary development has not been determined, still scientists claim that evolution is true.

The problem is that evolution has become the sacred cow before which 480,000 earth and life scientists bow and to which they have pledged their allegiance.[17] To characterize speculation in the name of science as fact these scientists deceive themselves and those whom they attempt to instruct.

Intellectual integrity demands that those who advocate evolution ought to be honest enough to admit that evolution rests on a whole series of assumptions. To advocate evolution as something which has been proved and is well documented, when it is not, certainly must be classified as dogma. In the last two hundred years there have appeared reputable scientists who have not only questioned the conclusion of evolution, but have even rejected them.[18]

Naturalism and Supernaturalism
There are two basic and opposing philosophical positions concerning the universe: naturalism and supernaturalism.[19] The naturalist assumes that the universe is strictly material, while the supernaturalist claims that there are two types of objects in the universe, the natural and the supernatural. The view of naturalism is based on the assumption that there is no such a reality as a Supreme Being or Great designer.[20] The supernaturalist's position is based on certain deductions which can be made by any rational mind. There are certain truths which logically can be deduced from observations of nature. The naturalist accepts only one kind of reality and does not distinguish between matter and spirit. The naturalist assumes that nature is the ultimate reality. The matter of where the universe came from he avoids to answer. The naturalist assumes the impossibility of miracles and of anything interfering with nature. Everything in the natural world must be subject to natural laws, these laws work with regularity and nothing can stop their functioning.[21] This would include the origin of the universe, the origin of the galaxies, the origin of the solar system, as well as the origins of life, of animals and human beings. Naturalism is based on the principle of cause and effect.

Atheistic science must assume that God is irrelevant to the operation of the universe on a day-to-day basis before it can insist that any order is found in the universe. Some scientists argue that law always works the same way, therefore, this consistency proves God has no place in the world. Actually this line of argumentation is what philosophers claim is resorting to tautology.[22] They assume God is not involved in the universe and then since his non-presence is accepted by themselves, they conclude that God does not exist and is not involved in it. But how do they know that God does not exist?

The Origin of the Universe

Cosmogony deals with the origin of the universe. Cosmology theory or philosophy of the nature of the universe. Both of these disciplines are a part of astronomy. Atheistic scientists, who specialize in astronomy, contend that the universe came into existence without God. But from where did the material or energy or the original stuff come from which everything eventually originated in the solar system and on the planet earth? Without taking God into account there are only two options left: the original matter was eternal or is created itself out of nothing. Both positions go against what mankind perceives about reality as it is known. Every effect has been produced by a cause. Cause and effect are a part of the way the universe is constituted. Experience tells human beings that finite things have a beginning. Atheistic evolution begins with a great question mark. How did it happen that such development potential was inherent in this original matter or energy which was the alleged source of everything studied by astronomers, anthropologists, historical geologists, zoologists as physicists? Out of nothing comes nothing, so human experiences teaches mankind. The evolutionary process of atheistic scientists begins with an unsolvable mystery for them and makes an assumption which requires more faith than to believe in an almighty God, who according to Scripture produced the universe by an act of His Will. (Rev.4)

Theories for the Origin of the Universe

Scientists have proposed opposing theories concerning the origin of universe and the solar system. None so far have been able to establish themselves.[23] All thus far have met with opposition. Two theories have been proposed in recent years: the Steady State and the Big Bang.

The Steady State Theory, or as it has also been called the Continuous Creation Theory, has now been abandoned even by its former advocates. Concerning it Henry Morris wrote:

Imagine basing an entire cosmology on hydrogen atoms suddenly appearing out of nothing, coming from nowhere! These imaginary atoms could never have appeared to anyone, since their supposed materialization always was supposed to have occurred as such times and places as never to be detectable. The only reason for postulating such an absurdity was the necessity to escape the creationist implications of the laws of thermodynamics. Hoyle and his followers merely invented what they called the "perfect cosmological principle," stating that the large scale structure of the universe must always be uniform in both space and time. Since the universe appeared to be expanding in space and decaying in time, this principle was held to require continuous creation (actually evolution) of matter or energy out of nothing throughout space and time in order to compensate for these apparent changes thus keeping everything in a "steady state."[24]

But the Steady State Theory was not based upon fact but was metaphysical speculation in order to avoid a confrontation with a Creator.

The Big Bang Theory is the theory presently accepted by atheistic astronomers and those scientists that believe their conclusions. It is looked upon as orthodoxy by the scientific establishment. The Big Bang Theory holds that there was a great primordial explosion which propelled all the matter/energy of the cosmos out radially from its center.[25] The Big Bang Theory is fraught with many problems; it has been shown by scientists that the Big Bang Theory is not the true answer for the way the universe came into existence apart from Gods creative acivity.[26] The primeval explosion is not scientifically observable but the data which are offered in support, of an expanding universe, background radiation, energies available for nucleosynthesis, etc. that appear to give a quasiscientific rationale for the postulation of the Big Bang Theory are against the second law of thermodynamics because explosions create disorder, not order. The primordial super-explosion surely would have produced absolute chaos and the most utter disorder. Thus, Morris asks: "If the universe is indeed a closed system, as evolutionary cosmologists allege, then how in the name of sense and science could this primeval chaotic disorder have possibly generated the beautifully organized and completely ordered universe that we have now? The Big Bang idea, viewed in this light, is as absurd as the Steady State idea."[27]

The Origin of the Solar System

In the opinion of Paul Zimmermann those who have attempted to demonstrate how the solar system could have been formed by chance, without God's creative power have failed utterly to devise a satisfactory cosmological scheme.[28] Thus the world has seen attempts by Immanuel Kant in 1775, who is his *General History and Theory of the Heavens*, made a beginning in endeavoring to answer this question of the origin of the sun and its planets. Fifty years later Laplace's hypothesis was propounded but it was rejected because it was in conflict with the laws of physics. In 1905 Chamberlain and Moulton proposed the first of the encounter theories. Zimmermann has described their view as follows: "These theories postulated that a star came from outerspace and passed very close to the sun. As the result of the gravitational pull between the two bodies a quantity of matter was pulled off the sun. The planets condensed from this envelope of incandescent material."[29] This view was subsequently modified by Jeans, Jeffrys, Lyttleton, and others. Spitzer of Harvard showed the impossibility of this theory. In 1944 C.F. von Weizsacker published what is known as the "dust-cloud" theory. Weizsaker declared that the sun was condensed from a great cloud of interstellar gas and dust. The planets in turn condensed from portions of the cloud far removed from its center. While this theory was acclaimed as the answer for a while, it also was found to have serious flaws. Whipple, Spitzer, Urey, Gamow and others made adaptions of it.

Concerning these and other attempts relative to the origin of the solar system Zimmermann wrote: "All these theories of the origin of the solar system show ingenuity. Some demonstrate great mastery of mathematics but the fact remains that no one yet has proposed a scheme that logically demonstrates how the sun and the planets might possibly have come into being. Much less has anyone proposed any workable theory to account for the universe. Each scheme brought forward has been shown to be defective in one or more vital points."[30]

The Age of the Universe and Age of the Earth

The estimates given by evolutionists who are astronomers and physicists relative to the origin of the universe and the earth are illuminating in showing how speculation characterized and still does in that they do not know when the universe and earth came into existence by development. Edwin P. Hubble, astronomer some years ago at the Mt. Wilson Observatory, Pasadena, California, estimated the age of the sun at seventy billion years. Sir Jeans, the British physicist gave thirty billion years for the age of the earth, and Albert Einstein estimates it at ten billion years. Professor Ernst J. Opik of the Tartu University Observatory in Estonia suggested that the universe was created all at once three billion years ago.[31] Theodosius Dobzhansky in the article on "Evolution" in the most recent 1987 edition of *Encyclopedia American* wrote: "Current estimates of the age of our galaxy are 10 to 15 billion years and of our solar system some 5 billion years. The origin of life on earth is dated at 3.8 billion years.[32] When so-called authorities differ in the billions of years, one gets the suspicion that the whole procedure is pure guesswork after all what is a billion years!

The Origin of Life

Between the inorganic and the organic a great gulf is fixed. How to bridge that chasm evolutionary science has been grappling with for a long time. The view was once held that life originated spontaneously. The Italian physician Rei (1626-1697) showed the impossibility of spontaneous generation by demonstrating that the worms in meat to be fly maggots. However, it was Louis Pasteur who in 1862 removed all doubts by conclusively proving that bacteria can only come from other bacteria.

Those who cannot bridge the difference between inorganic and organic want to eliminate this distinction, claiming it is an unfortunate one. The bridging of the organic and the inorganic, in the estimation of atheistic evolutions, would aid in the promotion of the concept of unlimited transformation of species. Richard Carrington in *A Guide to Earth History* labels the distinction between living and non-living as an unfortunate dualistic understanding of nature which was thwarting the solving of the question as to the origin of life.[33] He contended that, this erroneous understanding of nature has led religion to "charming legends" in which God is depicted as creating different classes of living beings by a kind of celestial conjuring stick.[34]

Science is endeavoring to attack this problem of the origin of life by assuming that the difference between living and non-living is less a difference in kind than in complexity. Thus Carrington complained: "We nowadays accept without question the evolution of higher forms of life from lower, but somehow resist that the idea that life itself would have evolved just as logically from an earlier form of non-living state."[35] Genetics has established that all living organisms have protein, and since proteins are composed of amino acids some scientists believe that under certain circumstances amino acids and other compounds can be produced. Carrington postulated that carbohydrates were long ago formed in the area of the sea and that they combined with nitrogen thus forming amino acids and protein.[36] Phosphorus probably acted as a chemical agitator and thus the process we now call life is supposed to have come into existence. Evolutionists claim that life must not be defined as unique and distinct, but should be regarded as a particular way of organizing certain basic materials of the universe. There are scientists who believe that man will never be able to create life. If life were to be created synthetically in some laboratory, it still would not prove that life originated of itself in that manner.

The Evolution of Plants and Animals

From amoeba to man, that is the history of man's rise from an unicellar animal. Organic evolution concerns itself with the progressive change which plants and animals in association with changing physical surroundings are supposed to have undergone throughout geological time. The term "evolution" as employed in biology is defined as an imminent process whereby the whole universe of life and being was being evolved through the self-differentiation of lower into higher forms. Many scientists who have fought for the principle of evolution have claimed that the progress from amoeba to man has been accomplished by an imminent process apart from God. They believe that evolution is perfectly explainable on a naturalistic basis.

William Matthew, III defined organic evolution: As "a process of cumulative change characterized by the progressive development of plants and animals from more primitive ancestors. A study of evolutionary processes indicates that the plants and animals of today have reached their present state of development as a result of gradual, orderly changes that have taken place in geologic past."[37] The same scientist declared: "Most of the evidence of evolution, **though indirect** (bold face the writers) clearly indicates that all organisms developed from a few simple ancestors to the multitude of diverse and complex organisms that inhabit the earth today."[38]

Proofs for organic evolution are supposed to be supported by both geologic and biological studies. Comparative anatomy, comparative physiology, taxonomy, geographic distribution, controlled breeding, and especially fossil data are cited as furnishing indubitable evidence for the assertion: "Life has developed from amoeba to man."

To accept the evolutionary theory of man's development ultimately from the amoeba presupposes believing in what scientists call microevolution which led to macroevolution. According to Goldschmidt, microevolution is confined to historical changes within species. On the other hand, macroevolution deals with those changes at the level of genus and the higher categories. Goldschmidt claimed that there is a great deal of difference between the two; microevolution does not lead to macroevolution.[39] According to the biochemist Kerkut: "The general theory of evolution rests on a series of fundamental assumption which are not capable of scientific experimentation."[40] In his book, The Implications of Evolution, Kerkut stated: "In effect, much of the development of major groups of animals has to be taken on trust. There is a great deal of circumstantial evidence but much of it could be argued either way." He claimed that the theory "that all living forms in the world have arisen from a single source which itself came from inorganic form should be called the general theory of evolution."[41] In Kerkut's opinion the evidence is so insufficient that at best it can be considered a "working hypothesis."[42]

A number of scientists have claimed that from the beginning the fossil record indicates the existence of major groupings of life which began as fixed patterns. Oswald Spengler in *The Decline of the West* argued that the evidence of paleontology does not support the monophyletic origin of animals.[43] Austin H. Clark claimed that between the major groupings of animals there never were any intermediates. He wrote: "If we are willing to accept the facts, we must believe that there never were such intermediates, or in other words, that these major groupings have from the very first borne the same relation to each other that they have today."[44] The continuity of the development from amoeba to man is not supported by the geological record and is clearly not in harmony with the Scriptures.

The Origin of Man

Text books on zoology and anthropology are proclaiming as a fact the descent of man from animals. William F. Qillian, Jr. claimed that "among competent scholars there are no longer any serious questioning of the evolutionary development of man."[45] The position taken by Darwin in his *The Descent of Man* and of Thomas H. Huxley in *Man's Place in Nature* in the nineteenth century as fact is repeated in the twentieth.[46] Ashley Montagu traced the ancestry of man back to the animal world Relative to man's animal ancestry Montagu wrote: "Man's ancestors will probably be found to have been rather unspecialized creatures possessing a great many primitive characters, creatures that looked very much more like men than they did like apes."[47] The eminent scholar C. F. Weiszaker wrote: "I do not see how we can escape the conclusion that man is physically descended from a being whom biology must clarify as apes."[48] In the century and a half from Newton to Darwin it is asserted by John C. Greene, in his book *The Death of Adam* there was accomplished the death of the Biblical Adam together with administering a final blow to the traditional view of nature.[49] The geologic record is supposed to support a series of fossil men leading back to the animal world.

Henry F. Morris's book *The Biblical Basis for Modern Science* has devoted a number of pages to an examination of the alleged anthropological evidence for man's animal ancestry and has refuted the fictional nature of the claims of the evolutionists.[50] Morris and Gary E. Parker in *What Is Creation Science?* have examined the fossil evidence for man's animal ancestry and have shown how false it is and how one fossil find after the other was either shown to be a fraud or a case of misrepresentation or misinterpretation.[51] While the Scripture assigns to man a high origin, in that God by a special act created man and later Eve, and made them in his own image and likeness, evolution's men and women once were wild animals living in the jungle.

When Justice Brennan labelled creation science as "a sham" and opted for atheistic evolution, he was supporting a theory that in all areas dealing with the question of origins was engaging in wishful thinking and in the construction of a system that was based on a great series of assumptions, on the basis of which conclusions then were drawn which were wrong. When columnist Will was complaining that actually evolution was resting upon facts, he was either writing as a committed believer of evolution, thus assuming that it must be right or he was writing from ignorance of the facts.

The argument has been made by the proponents that only committed evolutionists can criticize the theory of evolution and point out its problems and weaknesses, but its opponents dare not do so. To this, the writer reports, that if its proponents find their evidence faulty and unwarranted, why cannot its opponents use the same evidence for the rejection of aspects of the evolutionary theory? In every other sphere concessions and admissions of weakness and failure are utilized to refute the position of those people whom they oppose. What grants the evolutionists special consideration?

The Biblical View about Creation of the Universe, the Earth and Life Itself

It is however not sufficient to believe just in a Supreme Being, a Great Designer, and a Great Law Giver. The Word of God was spoken on the same questions, for which evolutionary atheistic science has attempted unsatisfactory answers. The author of Hebrews declared: "For every house has its builder, but he who builds all things is God (3:4)." This view complies with common sense, for all buildings and all creations presuppose a maker or creator. David in Psalm 19 argued: "The heavens declare the glory of God and their expanse is declaring the work of his hands. Day to day pours out speech, and night to night reveals knowledge. There is no speech, nor are there words. Their voice is not heard. Their line is gone out through all the earth." (Psalm 19:1-4).[52]

David declared in Psalm 14:1: "The fool hath said in his heart there is no God." The rejection of God in turn leads to moral corruption. Atheism, or for that matter agnosticism goes against the evidence given by nature itself, because nature could not have self-produced itself. The apostle St. Paul claimed in the opening chapter of the letter to the church at Rome that the heathen, the Gentiles, were without excuse because there are certain facts that can be inferred about God from His creation. Thus Paul wrote: "What can be known about God is clear to them because God has made it clear to them. Ever since He made the world, they have seen the unseen things of God-from the things He made they can tell He has everlasting power and is God. Then they have no excuse. They knew God and didn't honor Him as God or thank Him but their thought turned to worthless things, and their ignorant hearts were darkened. Claiming to be wise, they showed how foolish they are, when for the glory of God who cannot die, they substituted images of man who dies and of birds, four footed animals and reptiles" (1:19-23).[53]

A philosophy of life needs to answer three basic questions: "Where did I come from? (Whence?), why am I living (why?), and what will be the end of my life? (whereto?). Those who do not recognize God and His divine Revelation are like a rudderless ship adrift upon the ocean. The evolutionary philosophy (if it can be called a philosophy) has no satisfactory answers for all three. For one thing everything is said to be due to forces that march inexorably forward and life is determined further by chance. The Word of God has the only satisfactory answer for all three questions, and it all begins with the doctrine of creation as originally related in Genesis.

The Triune God, the Creator of the universe and of all in it, made man by a special creative act and gave him the command: "Have children and multiply and fill the earth and control it; and rule over the fish of the sea, the birds in the air, and everything that moves and lives on earth." Man was placed in charge of the Garden of Eden and was given permission to eat of every tree of the garden, but of the tree of the knowledge of good and evil he was not to eat." Man disobeyed God and became subject to all form of death: physical, spiritual and eternal. But in his mercy God gave Adam and Eve the promise of the coming Savior through whom God would be reconciled to mankind through His Son's vicarious substitutionary death. (Gen. 3:15) Those believing this truth would become God's children and inherit eternal life. Now mankind knows where it came from, why it is living and what its terminal destiny can be, namely everlasting life with the Triune God in heaven or life in a new earth where all death and sorrow are banished. Without God the Father, Son and Holy Ghost there cannot be no hope for any human being in this life on earth and especially in eternity. Godless evolution cannot contribute one iota to the happiness of mankind, but makes life meaningless, especially if people belong to the unfortunates of society.

The Doctrine of Creation in Holy Writ

While teaching pupils and individuals that there must be a Supreme Being is a start on which it is possible to build a true and satisfactory doctrine of God, as Creator, Sustainer, Redeemer and Sanctifier, it must be emphasized that the Holy Writ has given many clear assertions about God's activity as Creator and Sustainer, ascribed to each of the Persons of the Triune Godhead. In opposition to atheistic and materialistic evolution the Scriptures affirm that the universe is the result of the creative activity of the Triune God. This revelation is found in the first chapter of Genesis. A study of the entire Judeo-Christian Bible reveals that it presupposes divine creation. The doctrine of creation is not only the starting point of revelation, but all subsequent Scripture recognizes it and builds upon it. There are some 75 passages in the books of the Old and New Testament that speak of the creative work of God. Here are some of them: Gen. 1:1-3 (cf. 1:1-31; 2:1-15); Ex. 20:11; 1 Sam 2:8; Neb. 9:6; Job 12:8,9; 26:7-13; 28:24-26; 37:16,18; 38:4,7-10; Ps. 8:3; 19:1,2; 33:6-9; 65:6; 74:16,17; 78:69; 89:11,12; 90:2; 95:4,5; 102:25; 103:22; 104:2-6; 119:90; 124:8; 136:5-9; 148:5; Prov. 3:19; 8:26-29;30:4; Eccl. 3:11; 11:55; Is. 40:12,26,28; 42:5; 44:24; 44:24; 45:7-12,18; 48:13; 51:13; 66:2; Jer. 5:22; Jer. 5:22; 10:12; 27:5; 10:12; 27:5; 31:35; 32:17; 33:2; 51:15-16; Amos 4:13; 5:8; 9:6; Jonah 1:9; Zech. 21:1; Mal. 2:10; John 1:3-10; Acts 14:15; 17:24; Rom 4:17; 11:36; I Cor. 8:6; 2 Cor. 4:6; Eph. 3:9; Col. 3:9; Col 1.16,17; I Tim. 6:13; Heb. 1:2-10; 2:10; 3:4; 11:3; Rev. 4:11; 10:6; 14:7.

The doctrine of creation is the first doctrine revealed by God in His Word. The Holy Scripture teaches that the work of creation is the product of the activity of the Triune God Father, Son and Holy Ghost. Creation is the work of the undivided Godhead and not the work of the Father, or Son or Holy Spirit alone or separately. The creation of everything brought into existence is ascribed to the essence of God or hypostases in God. It is indeed, ascribed in the Scriptures to each of the three person of the Godhead, to the Father (1 Cor. 8:6), to the Son (John 1:3 and to the Holy Spirit) Gen. 1:2; Ps. 104:3). But the three persons and not to be regarded as separate causes for creation nor as associated causes. For as the three persons are only one God, so there is only one cause for creation. Creation, however, is ascribed by preeminence to the Father.[55] Thus we confess in the Apostles' Creed, that God is "the Father Almighty, Maker of heaven and earth," and in the Nicene Creed, that He is "the Maker of all things, visible and invisible." The latter creed also declares that all things were made by or through the Son, and that the Holy Spirit is the Giver of life. Luther phrased it this way: "The world was made through Christ and by the Father and in the Holy Ghost."[56]

While the doctrine of the creation and preservation of the world and all in it is very prominent in the Old Testament, the New Testament doctrine of creation is not merely a repetition of facts of the Old Testament, but it is in some respects a fuller development and a further unfolding of the Old Testaments truths concerning the creative work of the Triune God. It is as Unger has said: "As the heavenly and spiritual world come more closely into view in the New Testament, along with this comes more clearly the declaration that all spiritual beings, outside of God, owe their origin to Him."[57] The prologue of John's Gospel contains the most complete delineation of the principles underlying the creation and amplifies the first verse of Genesis. In John 1:3 the LOGOS, translated as the

"Word," is Jesus Christ who is portrayed as creating all things. That the Logos was Christ is established from chapter 1:14, where it is stated "that the Word became flesh and tabernacled among us." St. Paul, in another Christological passages asserted: "In Him all things were created, in heaven and on earth, visible, whether thrones, or dominions or principalities, or authorities — all things were created through Him and for Him. He is before all things, and to Him all things hold together" (Col. 1:16-17) Again St. Paul asserted "Yet there is one God, the Father, from whom are all things and for whom we exist, and one Lord, Jesus Christ, through whom are all things and through whom we exist." (1 Cor. 6:8). A similar teaching Paul expresses in Ephesians 3:4: "And to make all men see what is the fellowship of the mystery, which from the beginning of the world hath been hid in God, who created all things by Jesus Christ."[57] With what Paul taught the author of Hebrews agrees: "In these last days God has spoken to us by a Son, who He appointed the Heir of all things, through whom He also created the world (Heb. 1:2)."

For the work of the Holy Spirit in creation is set forth especially in the Old Testament. In addition to Genesis 1:2 the Spirit's creative work is spoken of in Psalm 104:30: "Thou sendest forth the Spirit, they are created." Many commentators believe that in Psalm 33:6,8,9 there is a reference to the creative work of the Son and Holy Spirit: "By the word of the Lord the heavens were made and all host by the breath of His mouth... Let all the earth fear the Lord. Let all the inhabitants of the world stand in awe of Him. For He spoke and it came to be; He commanded and it stood firm."

The Six Days of Creation

In Exodus 20:8-11 God's Word asserts: "For in six days the Lord made heaven and earth, the sea, and all that in them is, and rested on the seventh day." Six 24-hour days followed by another 24-day of rest alone can furnish a consistent analogy for the six days and taking a seventh day of rest. Keil and Delitzsch asserted: "If the days of creation are regulated by the occurring interchange of light and darkness, they must be regarded not as periods of time on incalculable duration, of years or thousands of years, but simply earthly days."[59] Exodus 31:17 supports the same interpretation.

The Purpose of Creation

What does the Word of God teach? From many passages in Holy Writ it seems that creation was to testify to the glory of God. "The heavens declare the glory of God, and the firmament showeth His handiwork" (Psalm 19:1). Isaiah declared: "Even everyone is called by My name, for I have created him" (43:7). Paul at the end of the doctrinal section of Romans exclaimed. "For of Him, and through Him, and to Him are all things; to Whom be glory forever (Romans 11:36)." Let man the creature, including evolutionists, never in any way detract from the glory and honor of almighty God who gave form and life to all things. Let us humbly accept the doctrine as found in the Holy Scriptures. Let us never ascribe to blind chance what God has told us He Himself has wrought with His mighty Word and limitless power!

Footnotes

1. High Court decision settles latest battle over evolution," *Fort Wayne Journal,* June 21, 1987, p. 1A, written by Stuart Taylor, Jr., of *The New York Times.*
2. *Fort Wayne News Sentinel,* "Fallout starts in creation decision "June 20, 1987, page i.
3. "Memories of the Monkey Trial," *Time Magazine,* Monday, June 29, 1987, p. 54.
4. "The Day Darwin and God Met in Court," *U.S. News World Report,* June 29, 1985.
5. George Will, "Creationists disguise dogma as science," column appearing in *Fort Wayne News-Sentinel,* June 25, 1987, p. 11A.
6. *Fort Wayne News-Sentinel,* June 20, 1987, "Praise and Scorn great creationism verdict," page 8A, continued from page 1.
7. Fr. Virgil Blum, "Is Secular Humanism a religion?" *Catholic Twin Circle,* Sunday, May 10, 1987, reprinted in *Christian News,* June 8, 1987, p. 14.
8. Craig Klugman, "Religion and science are not the same thing," Editorial in *The Fort Wayne Journal Gazette,* Sunday, June 21, 1987, p. 12A.
9. George Will, **op. cit.,** p. 11A.
10. Will, **op. cit.,** p. 11A.
11. Cf. the for the difference. cf. John Klotz, *Studies in Creation.* (St. Louis: Concordia Publishing House, 1985), p. 83.
12. Cf. Raymond F. Surburg, "In the Beginning God Created," in Paul Zimmermann, editor, *Darwin Evolution and Creation* (St. Louis: Concordia Publishing House, 1959), pp. 57-58.
13. Stuart Taylor, "High Court decision settles latest battle over evolution." *New York Times,* reprinted in *Fort Wayne Journal Gazette,* June 21, 1987, p. 11A.
14. Craig Klugman, "Religion and Science are not the same thing," **op. cit.,** p. 12A.
15. *The American Heritage Dictionary of the English Language* (New York, Boston/Atlanta: American Heritage Publishing Co., and Houghton Mifflin Company, 1970), p. 1162.
16. Cf. Wilbert H. Rusch, "Darwin, Science and the Bible," in Zimmermann, **op. cit.,** pp. 15-35.
17. "Keeping God out of the classroom," *Newsweek,* June 29, 1987, p. 23.
18. The Creation Research Society has at least 700 scientists with credentials who reject the evolutionary theory as not being scientific.
19. Josh McDowell, Don Stewart, *Reasons Skeptics Should Consider Christianity* (San Bernadino, California: Campus Crusade for Christ, 1981), p. 194.
20. Vergilius Ferm, "Varieties of Naturalism," Vergilius Ferm, editor, *A History of Philosophical Systems* (New York: Philosophical Library, 1950), p. 430.
21. **Ibid.**
22. McDowell-Stewart , **op. cit.** p. 195.
23. Henry A. Morris, *The Biblical Basis for Modern Science* Grand Rapids: Baker Book House, 1984). pp. 146-151.
24. **Ibid.,** p. 151.
25. **Ibid.,** p. 150.
26. **Ibid.,** pp. 151-152.
27. **Ibid.,** p. 151.

28. Paul A. Zimmermann, "The Evidence for Creation," in P.A. Zimmermann **op. cit.**, p. 88.

29. **Ibid.**

30. **Ibid.**, p. 89.

31. Data given in Joseph P. Free, *Archaeology and Bible History* (Wheaton, Illinois: Scripture Press Publications, 1962), p. 22.

32. Theodosius Dobzhhansky, "Evolution," *Encyclopedia Americana*, 4, p. 746.

33. Richard Carrington, *A Guide to Earth History* (New York: The New American Library, 1956), pp. 71-73.

34. **Ibid.**, p. 75.

35. **Ibid.**

36. **Ibid.**, p. 75.

37. William H. Matthews, III, Fossils. *An Introduction to Prehistoric Life* (New York: Barnes & Noble, 1962), p. 156.

38. **Ibid.**, p. 156.

39. Richard B. Goldschmidt, *The Material Basis for Evolution* (New Haven: Yale University Press, 1940). pp. 6-7.

39. G. A. Kerkut, *Implication of Evolution* (New York: Pergamon Press, 1960), p. 7.

40. **Ibid.**, p. 154.

41. **Ibid.**, p. 157.

42. **Ibid.**, p. 157.

43. Oswald Spengler, *The Decline of the West* (New York: Knoph and Company, 1952) 1 II, p. 32.

44. Austin H. Clark, *The New Evolution* (Baltimore: William and Wilkens, 1930), p. 189.

45. William F. Quillian, Jr., "Evolution and Moral Theory in America," *Evolutionary Thought Today in America*, Stow Persons, ed. (New Haven: Yale University Press, 1950), p. 416.

46. Thomas H. Huxley, *Man's Place in Nature* (Ann Arbor: The University of Michigan Press. 1959), pp. 188.

47. Ashley Montagu, *Man: His First Million Years* (New York: The American Library, 1958), p. 41.

48. C. F. Weizsaker, *The History of Nature* (Chicago: The University of Chicago Press, 1949), p. 157.

49. John C. Greene, *Evolution and Its Impact on Western Thought* (New York: The New American Library. 1959). p. 382.

50. Henry M. Morris. *The Biblical Basis for Modern Science* (Grand Rapids: Baker Book House, 1984), pp. 390-413.

51. Henry M. Morris and Gary E. Parker, *What Is Creation Science?* (San Diego: Creation-Life Publishers, 1982), pp. 106-129.

52. *New American Standard Bible.* (New York: Thomas Nelson, 1975), Old Testament, p. 702.

53. William F. Beck, *The Holy Bible - An American Translation* (New Haven: Leader Publishing Company, 1976), New Testament, p. 191.

54. Cf. Raymond F. Surburg, "In the Beginning God Created," in Zimmermann, editor, *Darwin, Evolution and Creation*, **op. cit.**, p. 39.

55. Joseph Stump, *The Christian Faith* (New York: The Macmillan Company, 1932). p. 63.

56. Luther's Saemmtliche Werke (St. Louis: Concordia Publishing House. 1920), 12, p. 137ff.

57. Merrill F. Unger, "Creation" in *Unger's Bible Dictionary* (Chicago: Moody Press, 1957), p. 225.

58. *Holy Bible. The New King James Version* (New York: Thomas Nelson Publishers, 1982), p. 1143. NU-Text omit through Jesus Christ

59. C. F. Keil and F. Delitzsch, *A Commentary to the Old Testament* (Edinburgh: T. & T. Clarke, 1875), I, p. 56.

Questions

1. What did the U.S. Supreme Court rule in 1981? ____
2. Who dissented against the majority? ____
3. The justices were ignorant that ____.
4. What has filtered down to the high schools and grammar schools of the nation ____.
5. What is on U.S. coinage? ____
6. The Founding Fathers were either ____ or ____.
7. The Declaration of Independence declared that the rights of citizens came from ____.
8. Humanistic views are being promoted in ____.
9. George Will insisted that evolution is ____.
10. Is truth determined by numbers? ____
11. Evolution has become the ____.
12. In the last two hundred years there have appeared ____ who reject evolution.
13. Cosmogony deals with ____.
14. Out of nothing comes ____.
15. The rejection of God leads to ____.
16. What are the three questions a philosophy of life must answer? ____
17. Godless evolution makes life ____.
18. God's Word says that God created the world in ____.

The Contribution of Karl Friedrich Keil To Nineteenth Century Biblical Studies

Christian News, April 25, 1988

(An Essay Commemorating the Centennial of His Death)

The Life of K.F. Keil

On May 5, 1988 Karl Kriedrich Keil will be dead exactly one hundred years. Many pastors and today's students know of and possibly use and consult the Keil-Delitzsch commentary of the Old Testament. Keil was one of the conservative Lutheran theologians of the nineteenth century who fought the higher criticism which was promoted and developed by Lutheran theologians and exegetes in Germany and by Protestant scholars in other European countries. Together with Hengstenberg, Keil was one of the chief opponents of negative higher criticism in Germany. Because of his conservative Biblical hermeneutics which he employed in the interpretation of the Bible, he has been ignored by liberal Lutheran and Protestant scholars when discussing the theological developments of the nineteenth century. Before embarking on a presentation of Keil's contributions to Biblical science and also portraying the effects of his literary efforts on his own time as well as on people of the twentieth century, it will be helpful to give a sketch of his life and also enumerate his literary efforts produced during his eighty-one year life.

Karl Friedrich Keil was born at Lauterbach near Oelnitz (25 miles southwest of Zwickau), Saxony, February 16, 1807. He was the son of Johann Friedrich Keil and his wife Sophie, born Modes.[1] Because his family was poor, it was determined that after his elementary schooling that he should learn a trade and he chose that of cabinetmaker (German: Tischler). He went to St. Petersburg, Russia, where his brother had a business making tables and cabinets. His apprenticeship did not work out, because his body was too small and not suited for the joiner's bench and so he returned from St. Petersburg and was sent to the school of Peter, where he distinguished himself so that with the help of the empress Maria Feodorowna, sister of Kaiser Wilhelm I, and wife of Nicholas I he was able to study theology at Dorpat and Berlin.[2] Privately Keil studied the old languages and became proficient in them. In his early home surroundings he was exposed to piety but in his youth he was subjected to rationalism.[3]

During his students days he was greatly influenced by Sartorius and Strauss in Berlin, whose fatherly friendship he was permitted to enjoy.[4] It was especially through Sartorius that Keil came to a knowledge of the true understanding of the Scriptures and of traditional church doctrine, at least in its fundamentals.[5] He wrote a prize essay in which he contrasted rationalism and Scriptural and church doctrine. In 1830 he successfully passed his theological examination and Sartorius recommended that he prepare for an academic career. With the aid of his supporter the

queen, he studied in Berlin, where he entered into an intimate association with Hengstenberg.[6] In 1832 he became licentiate in theology. In
1833 he received a call as private docent at Dorpat and in 1838 moved
into the vacant professorship for Old and New Testament Exegesis,
which he occupied for twenty five years.[7] Keil also served as professor of
Oriental languages. With Sartorius, Busch, later Philippi, Theodosius
Harnack and Kurtz he educated for the Baltic states a generation of pastors and preachers who faithfully adhered to the theology of the Lutheran
Confessions as found in the *Book of Concord* of 1580.[8] In 1859 he settled
in Leipzig, where he devoted himself to literary work and to the practical
affairs of the Lutheran Church. Keil was interested especially in missions
and served for thirty years on a missionary board.[9]

After twenty-five years of teaching he was pensioned by the Russian
government and spent his pension in Saxony. For twenty-eight years he
lived as a private person in Leipzig where from early morning till late at
night he worked on his **Kommentar zum Allen Tesament**. Between
1860·1872 he worked with Franz Delitzsch on **Biblischer Kommentar
zum Alten Testament**. Of the fifteen volumes comprising this commentary ten flowed from the pen of Keil. To this commentary he contributed
the books from Genesis to Esther inclusive, Jeremiah, Ezekiel, Daniel
and the Twelve Minor Prophets.[10] In 1887 he retired to Roedlitz, contin-
uing his literary efforts to the time of his death, which occurred on May
5, 1888.

Keil's Literary Works

The following were the products of Keil's pen between 1833 and 1885
and for over fifty years he was active as author of Biblical and exegetical
books. Among his earlier literary productions were: **Apologetischer
Versuch ueber die Buecher der Chronik und die Integritaet der
Buecher Ezra** (Apologetical attempt concerning the Books of Chronicles
and the Integrity of the Books of Ezra, 1833); **Ueber die Hiram-
Solomonische Seefahrt nach Ophir**, 1834 (Concerning the Sea Journey of Hiram-Solmon to Ophir, 1834); **Der Tempel Salomos,** 1839 (The
Temple of Solomon, 1839); **Kommentar ueber die Buecher der
Koenige,** 1845 (*Commentary on the Books of Kings*, 1845); **Kommentar
ueber das Buch Josual**, 1847 (*Commentary on the Book of Joshua*,
1847); Haevemick's Einleitung In AT. 3 Teile, 1849, *Haevernick's Introduction to the Old Testament. 3 Parts, 1849*; Einleitung In die Kanonis-
cben Schriften des AT. (*Introduction to the Canonical Books of the Old
Testament*, 1853, 3rd edition, 1873); *Biblische Archaeologie*, 1857, 2nd
edition, 1870 (*Biblical Archaeology*, 1857; 2nd edition, 1875.

In the Keil-Delitzsch **Biblischer Kommentar ueber das Alte Testament** the following volumes: were Keil's **Genesis und Exodus,** 1861;
3rd ed. 1878; Leviticus, Numbers and **Deuteronomium**, 1862, 2nd ed.
1870; **Josua, Richter und Ruth,** 1863; 2nd ed. 1874; **Samuelis,** 1865,
2nd ed. 1875; **Koenige,** 1866, 2nd ed. 1876; Chronik, Esra, Nehemla, Esther, 1870; **Jeremia,** 1872; **Ezechiel,** 1868, 2nd ed. 1881; **Daniel,** 1869;
Kleine Propheten, 1867; 3rd ed. 1888.

1432

He also wrote **Kommentare ueber die Makkabaerbuecher**, 1875. As a professor of the New Testament he had taught many books. He wrote commentaries on **Matthaeus**, 1877; **Marcus und Lucas**, 1879; Johannes, 1881; On certain of the Catholic Epistles: **Petrus und Jacobusbriefe**, 1883; **Hebraerbrief**, 1885. In all these commentaries he abstained from using the historical method.

The Theological Viewpoint of Keil's Exegetical Writings

Keil was one of a number of nineteenth century Lutheran scholars who refused to bow before the golden calf of negative literary criticism. He rejected the rationalistic presuppositions of higher criticism and the results which flowed from its application to Biblical literature.

The type of higher criticism promulgated in the nineteenth century was that known as literary. When this was applied to the Old and New Testaments it resulted in a denial of statements that writings made about their authorship, time of composition, unity and integrity.[12] Form criticism, **Sachkritik**, tradition criticism, redaction criticism and structuralism were later developments the product of the twentieth century. Keil took the position, rejected by rationalism and deism, that the books of the Old and New Testaments were the inspired Word of God and thus historically and scientifically reliable.[13] The rationalistic attacks upon the canon of the Bible, its reliability and trustworthiness were rejected by him.[14] Since the rationalism of the eighteenth century served as the basis of the historical-critical method as it was developed and practiced by nineteenth century scholars and exegetes, Keil completely disassociated himself from it as being a sound methodology in Biblical exegesis and rejected the method. Franz Delitzsch, who at first was a very orthodox Lutheran and an author of a number of commentaries in the Keil-Delitzsch, **Biblischer Kommentar zum Alten Testament**, caved into the demands of historical-criticism, as practiced by the Erlangen Lutheran school, of which he was a member.[15] Delitzsch began his academic career as an orthodox Lutheran in the 1830's who was in agreement with the Missouri Synod and its theologians, like Pieper, Stoeckhardt and others.[16] To preserve his reputation as a scholar and not wishing to be labelled a reactionary theologian, he gradually began to adopt higher critical positions on the matters of Biblical authorship and time of a book's origin. Thus he gave up the Mosaic authorship of the Pentateuch, the unity of Isaiah, postulating a number of authors for Isaiah and also surrendered the second century B.C. date for Daniel.[17]

In contrast to Delitzsch, Keil held that the Old Testament Exegete should have as his objective under the guidance of the Holy Spirit assuming the homogeneity of the total Scriptures, to set forth the meaning of the various Biblical writings. This exegetical work was to be done by the use of the best philosophical and linguistic and historical science. The Biblical books were to be interpreted according to their literary character, taking into account the historical background. The exegete was especially to keep in mind and enunciate the plan of salvation as set forth in Holy Writ. The exegete of the Old Testament Scriptures, operating under the influence of the Holy Spirit, was to elucidate the text of the Old Testament in the interest of the Gospel, which was the power of God unto salvation.[18] An Old Testament exegesis should aid and promote the proclamation of salvation and the building up of the Kingdom of God on earth.

A theological discipline which did not keep these objectives in view was for Keil as a Christian and as a theologian, inimical to the purpose of Christian theology as set forth clearly in God's Word, the Bible. Those scholars who attacked the reliability and trustworthiness of Holy Writ, he criticized and strongly disagreed with them. According to one of Keil's sons, who contributed his biography for **The Schaff-Herzog** *Encyclopedia* **of Religious Knowledge** and also for the **Realencyklopadie fuer protestantische Theologie und Kirche** Karl Keil did not as a rule engage in polemics with scholars who were practitioners of the higher-critical method, because he considered such argumentation fruitless.[19] In Keil's opinion his opponents were not employing a true scientific method, a method, if it were sound, would not impose its ideas and theories upon the Scriptural text and deal with the latter as if it were unreliable. He was opposed to higher criticism because it did not foster faith in Jesus Christ and set forth the plan of salvation. Keil firmly believed that by presenting the data of the Scriptural text as reliable and true and not questioning the authoritative statements of Christ about the Old Testament, he was sharing with thousands of students and pastors who used his commentaries a means for sound Biblical teaching and preaching. Concerning the direction and development of Old Testament studies between 1830 and 1888, he was convinced that in due time the church would learn to recognize that it was on an erroneous path and would see the error of promoting a double-creating and soul-destroying method. In his youth Keil had been subjected to rationalism with its faith-destroying views. He believed their higher criticism was spiritually unsatisfying.

Keil and the Hengstenbergian School of Lutheranism

Keil's personal friend, Franz Delitzsch, in the third edition of Keil's *Commentary on The Twelve Minor Prophets*, wrote: "With Keil the last great representative of the school of Hengstenberg has departed."[20] Ernst William Hengstenberg died nineteen years before Keil. The latter had found, by personal study of the Bible, Christ as his Savior, and he saw in the Confessions of the Lutheran Church the clearest expression of true Biblical theology. The *Concordia Cyclopedia* asserted about Hengstenberg: "By his work of interpretation and defense of the Old Testament he became the staunch defender against rationalism, unionism and the mediating theology of his day. As a mouthpiece of his testimony for the truth he founded the Evangelische Kirchenzeitung, a most powerful organ in defense of the truth and attacking error without fear. For forty-two years he was identified with this paper and was the chief contributor."[21] Because of his orthodoxy he was disliked by the authorities in Berlin, who tried everything in their powers to silence his testimony and see him leave Berlin. However, Hengstenberg successfully frustrated all attempts to silence him and have him transferred out of Berlin. He was subjected to violent slander and insult because of his Bible doctrine and his attack upon error. In the opinion of Heick, Hengstenberg and his Sshool were proponents of "the theory of repristination."[22] According to Heick: "This school sought not only to restore the treasures that had been cast aside by rationalism but also attempted to revive the scientific method of Lutheran theology of the sixteenth and seventeenth centuries."[23]

A member of this school was August Vilmar, professor of Marburg University, who asserted the purpose of theology to be: "Theology must know nothing that she has nothing new to say, nothing new to discover, but that her task is to preserve the spiritual treasure that has been given in Holy Scriptures and received by the church in such a form that it may be transmitted to future servants of the church undiminished, certain and in its most useful form."[24] C.P. Caapari (1814-1892) was a pupil of Hengstenberg and called himself "a hardboiled Hengstenbergian." He was the author of a useful Arabic Grammar, translated into a number of European languages. He was also a pioneer in the science of comparative symbolics.

F. A. Philippi, a convert of Judaism to Christianity, was professor at Dorpat and Rostock, who published the extensive and widely employed **Kirchliche Glaubenslehre** (6 vols. 1854 ff.), in which he endeavored to elucidate the ideas of the old Lutheran teachings with a refutation of the attacks made upon them by later times. However, unfortunately Philippi claimed "the source from which dogmatics must draw is the dogmatician's reason enlightened by revelation." He began his Dogmatics with a declaration concerning the inerrant, inspired scriptures, in which he taught an inspiration not of words but of the Word. While claiming that the Bible was inspired he denied the verbal inspiration. He embarked on bitter controversy about the doctrine of the atonement, concerning which Von Hoffman held erroneous and unBiblical ideas.

Although this school purported to return to Lutheran orthodoxy, it did not escape being influenced by ideas emanating from the thought world of the nineteenth century. Kliefoth and Loehe are also said to have belonged to this school but both men exhibited tendencies which were called Roman Catholic, because of their liturgical views. The last member of this school, so Delitzsch claimed, was Keil, who wished to be faithful to the view of Scripture held by Luther and the Lutheran Confessions.[25]

Footnotes

1. Pastor Keil, "Keil, Karl Fridrich," Albrt Hauck (ed), *Realencyklopaedi fuer protestantische Theologie und Kirche* (Tuebingen: J. C. Hinrische Buchhandlung, 1901), X, p. 197.

2. *Die Lutherdt'sche Kirchenzeitung*, 1888, No 39, as cited by Stoeckhardt, *Lehre und Wehre*, 44:371, 1888.

3. **Ibid.**

4. Hauck, *Realencyclopaedie*, **op. cit.**, p. 197.

5. **Ibid.**

6. **Ibid.**

7. W.J.A. Keil, "Keil, Jhann Friedrich Karl," *Schaff-Herzog Encyclopedia of Religious Knowledge*, edited Loettchert, 1955, reprint of 1913 edition (Grand Rapids: Baker Book House, 1913), VI, p. 305.

8. *Die Lutherdt'sche Kirchenzeitung*, **op. cit.**, p. 370.

9. **Ibid.**

10. *John M'Clintock and James Strong, Cyclopedia of Theology and Ecclesiastical Literature* (New York: Harper & Brothers, 1875), V. p. 28.

11. As given in the biographical sketch by Pastor Keil on "C. J. K. Keil," in Hauck, **op. cit.**, pp. 197-198.

12. This is clear from his *Manual of Historico-Critical Introduction* (Edinburgh: T. & T. Clark, 1870. 2 volumes).

13. Pastor Keil in Hauck, **op. cit.**, p. 197.

14. Georg Stoeckhardt, "Karl Friedrick Keil", *Lehre und Wehre*, 44:371, 1888.

15. Cf. Raymond F. Surburg, "The Influence of the Two Delitzsches on Biblical and Near Eastern Studies," *Concordia Theological Quarterly*, 47:228, July, 1983, p. 3.

16. **Ibid.**, pp. 227-228.

17. Hans Joachim Kraus, Geschichte der Historischen-Kritischen Erforschung von der Reformation bis zur Gegenwart (Neukirchen Moers: Verlag der Buchhandlung des Erziehungsverein, 1956), pp. 211-218.

18. Pastor Keil, in Hauck, **op. cit.**, pp. 197-198.

19. **Ibid.**

20. Carl Friedrich Keil, *Biblisher Kommenatar ueber das Alte Testament* (Leipzig: Doerffling und Franke: 1883), Vorwort, p. viii.

21. "Hengstenberg, Ernst Wilhelm," L. Fuerbringer, Th. Engelder and P. E. Kretzmann, *The Concordia Cyclopedia* (St. Louis: Concordia Publishing House, 1927), p. 317.

22. Otto W. Heick, *A History of Christian Thought* (Philadelphia: Fortress Press, 1966), II, p. 200.

23. **Ibid.**

24. August Vilmar, *Die Theologie der Tatsachen wieder die Theologie der Rhetorik, p. 16 as cited by Heick*, **op. cit.**, p. 200.

Questions

1. Keil fought ____.
2. In Berlin Keil entered into an intimate relationship with ____.
3. Keil was especially interested in ____.
4. Between 1860-1872 Keil worked with Franz Delitzsch on ____.
5. Keil refused to bow before ____.
6. Franz Delitzsch caved into ____.
7. Delitzsch gave up the ____.
8. Keil believed that higher criticism was ____.
9. In the opinion of Heick, Hengstenberg and his school were proponents of ____.
10. Kliefoth and Loehe exhibited tendencies which were ____.

February 2, 1989

Dear Brother Otten:

Enclosed please find a response to the article by Larson in *The Journal of the Historical Review*. Relative to another copy of the article "The Holy Spirit in the Book of Acts" I unfortunately did not make an xeroxed copy. I have a handwritten manuscript which I would need to retype but did not keep the footnotes so I would have to look them up or even research them. I could have the article in your hands sometime in the beginning of March, in plenty of time for the Pentecost of *Christian News*. If so, let me know.

Cordially,

Raymond Surburg

"A Reply to 'What Ever Happened to the Dead Sea Scrolls?' By Dr. Martin A. Larson"

Christian News, February 13, 1989

Dr. Martin A. Larson published in *The Journal of Historical Review* an article, entitled: "What Ever Happened to the Dead Sea Scrolls?"[1] This is an essay originally delivered at the 1981 Revisionist Conference (Editor's note: This entire essay was reprinted in the February 13, 1989 *Christian News*). The author himself questioned the pertinence of his topic at a Revisionist Conference.

His rationale or justification for its delivery was stated as follows: "But it does deal with an historical distortion and cover-up of the first magnitude and I hope you will find it interesting and constructive."[2] The obvious parallel which he makes was the following: Just as the Jews and Zionists during the past forty years have been fabricating facts about the activity of the Germans and Germany relative to the treatment of Jews, so the Jews are withholding data they possess about the Dead Sea Scrolls, for their publication would be damaging to the Jews and possibly also Christians.

Larson's Interest in the Dead Sea Scrolls

In the beginning of his article Larson described how he happened to take an interest in the Dead Sea Scrolls (hereafter abbreviated as DSS). After he had written his Ph.D. thesis at the University of Michigan in 1927, which deal with Milton's theology – particularly his Trinitarian concept — he had to devote his time to business pursuits and ceased reading in the field of religion which up to the time of the completion of his doctorate he had pursued avidly. However, after retiring from business he resumed his study of religion. When the DSS were found in 1947, he plunged headlong into research on them and published a book, called *The Essene Heritage.* (1967) "And so the authors of the Scrolls, the Essenes, their writings and their impact on history has been a subject of consuming interest for me."[3]

The writer of this response also has been greatly interested in the DSS since their discovery in 1947, having used information from them in various courses taught on the seminary level. A reading of Larson's article and its views remind the writer of some of the same views that were set forth by authors such as A. Powell Davies,[4] Edmund Wilson,[5] Dupont-Sommer,[6] Allegro,[7] and Charles Potter.[8] The view and theories of all these men were rejected by the majority of scholars who read the same documents Davies, Wilson, Allegro and Dupont-Sommer had, because they were based on a misinterpretation of the DSS.[9]

The main reason for this response to Larson's article is that he has made claims about the Essenes and their relationship to Christianity. Thus Larson stated on p. 121 that *The book of Enoch* and *The Testament of the Twelve Patriarchs* were Essene writings which were recognized as canonical by the early Christians.[10]

In the determination of what constitutes Biblical Christianity the writer accepts the sixty-six books of the Old and New Testaments and any person familiar with the Biblical canon knows that The Testament of the Twelve Patriarchs of The Book of Enoch were never considered canonical by the Roman Catholic Church, the Greek Orthodox Churches or the various churches that constitute what has been called the Protestant Churches. That *The Testament of the Twelve Patriarchs* and *The Book of Enoch* were historically considered canonical is not true. Both of these books are generally classified as belonging to the pseudepigraphical writings, in contrast to the books considered by Protestant as apocryphal, to which fourteen books are assigned, of which ten are regarded as deutero-canonical by the Roman Catholic Church.[11]

The Testament of the Twelve Patriarchs and *The Book of Enoch* were not regarded as canonical by first-century Christians nor by The Synod of Jamnia held in A.D. 90, sponsored by Jews to discuss the Hebrew Old Testament canon. These two books belong to a type of literature that was popular at the beginning of the first century A.D. In this category fall such books as the Legends of Moses, the Book of Adam and Eve, the Ascension of Isaiah, the Apocalypse of the Prophets.[12] Old Legends of the Patriarchs and prophets were assembled and new ones composed. In some Christian lists those writings appear among the canonical and apocryphal books. The orthodox Jews never recognized these books as Holy Writ nor do the writings of the New Testament refer to apocryphal and pseudepigraphical books as authoritative Word of God, even though New Testament writers were aware of their existence.

In his *Introduction to the Intertestamental Period*, the writer wrote: "The value of the Testament of the Twelve Patriarchs is not to be found in the apocalyptic nature but rather in its ethical character, because it contains mostly exhortations and encouragements to follow the God-fear-ing way and gives warnings to eschew sin. The name of the book is derived from the 12 sons of Jacob, who are depicted as speaking with their father Jacob before his death and who, in turn, receive from him a moral testament and ethical injunctions. Even though the work is undoubtedly Jewish and penned in Hebrew, yet there are many Christian interpolations which contain references to an incarnated Messiah. The Testament of the Twelve Patriarchs is a collection of 12 little books, one for each patriarch probably written by one person is as evident by the uniformity of format."[14]

The book contains Messianic predictions. In regard to the messianic predictions, it has been noted, that instead of the Messiah coming from Judah, He would descend from the tribe of Levi. The descent from Levi has been explained by the fact that the greatest part of The Testament of the Twelve Patriarchs was composed during the reign of Hyrcanus as the Chosen Messiah and promised him a glorious future. Therefore, the Messiah had to come from Levi, for John Hyrcanus as a priest. But when he broke with the Pharisees and their descendants and the Maccabeans were far from being the ideal rulers the Jews were looking for, later additions to Testimonies of the Patriarchs announced that the Messiah would come to the tribe of Judah.

According to Robert Pfeiffer, *The Testament of the Patriarchs* was written in Hebrew about 140-110 B.C., probably in the last days of the rule of John Hyrcanus (135-104 B.C.).[15] J.H.E. Thomson disagrees with Charles[16] and others and argues for a dependence of the book on the New Testament and not vice versa.[17] This would mean that it is the product of the last part of the first century and would have no relationship to the Essenes, who were wiped out by A.D. 70. There is no evidence of Essene authorship and that this is an Essene document.

The Book of Enoch Essene?

The Book of Enoch, which according to Larson, is supposed to have affected Christianity, is said by him to be an Essene book.[18] However, scholars who have studied and written on this pseudepigraphical book claim it is a collection of various dates, composed by different authors at different times. In it are found a diversity of religious ideas not always in harmony with each other, but bound together by a unity of tone and outlook. Parts of ten Aramaic fragments of Enoch have been found at Qumran Cave IV. The choice of Enoch as the supposed author of this heterogeneous collection of writings has been explained that the interpretation of the words in Genesis 5:24: "And Enoch walked with God and was no more, because God took him," means that Enoch was translated into heaven during his lifetime, a belief current among the Jewish people at an early time. The visions in the book are portrayed as having been received by the patriarch Enoch. While the book is mainly eschatological in character, there is a fine section 91-104 which contains excellent examples of wisdom literature of the Jews. In the opening part of the book, Enoch saw the vision of the holy One in the heavens, "which the angels showed me, and from them I understood as I saw, but not for this generation, but for a remote one which is for me to come."[19]

Charles' Division of Enoch

Charles divided the Book of Enoch into the following parts: chapters 1-36 (before 170 B.C.); chapters 37-71 (94-64 B.C.); chapters 72-82 (about 110 B.C.); chapters 83-90 (166-161 B.C.; chapters 91-105 (104-95 B.C.).[20] In contradiction, H.H. Rowley placed the earliest parts of Enoch in the Maccabean, not in the pre-Maccabean age, shortly after Daniel.[21] Other chapters are placed after the Maccabean Period but before the Roman conquest of Palestine in 63 B.C. (chapters 85-90; 91-105) exclusive of the Apocalypse of Weeks (chapters 37-71, 1-5, 83-84 and 108), while chapter 12-36; 81-182 and 72-82 are dated before 150 B.C. These latter passages it is assumed were known by Jubilees 4:17-19.

The authentic authors of Enoch are thought to belong to the Pharisaic circles, although some think they should not be taken in too a narrow sense. The fact that in 82:2 the celibate is praised has been taken as evidence for Essene influence, but the rejection of marriage was not a fundamental doctrine of the Essenes. The original language was Hebrew, and only a small portion was written in Aramaic. Later a translation was made into Greek by Alexandrian Jews, and from Greek into Ethiopic. The book was lost, except for extractions made by the chronographer George Syncellus. In the last quarter of the 18th century, Bruce, an Abyssinian traveler, brought copies of the Ethiopic Enoch to Europe.

In the Canonical New Testament book of Jude there is a reference to the Book of Enoch. In verse 14 it is written: "It was to these, that Enoch, the seventh in descent from Adam prophesied, saying: LO! the Lord is come with myriads of his saints to execute judgment upon all, and to convict all the ungodly of all the ungodly deeds which in their ungodliness they have committed, and of all the hard things they have spoken against Him, ungodly sinners that they are." Kretzmann asserted about this verse: "His (namely Jude's) quotation, ascribed by himself to Enoch, the seventh from Adam, may without hesitation be considered as having been taken from the Book of Enoch, for the possibility of the Lord's having acknowledged a fact recorded in an apocryphal book is not excluded. Still it may also have been transmitted to the apostles in some other manner, very likely by the Lord Himself, in one of His discourses on the end of the world. Matt. 24:3-26; Luke 21:5-36."[23]

The Essenes and Christianity

Larson believes that Christianity was influenced by the Essenes, that John the Baptist may have been an Essene, that Christ was acquainted with them and that Christians took over the Teacher of Righteousness (TR) who anticipated the crucifixion and resurrection of Jesus.[24] To properly discuss the view that Christianity is another version of Essenism, it will be advantageous to briefly discuss the Dead Sea Scrolls (DSS) and the relationship of the Essenes to the Scrolls that were found in the wilderness of Judeah.

Since 1947 phenomenal discoveries of great importance to Judaism and Christianity have been made in caves of wadies entering the western side of the Dead Sea. The manuscripts found in five different places not only throw remarkable light on the community at Qumran itself but also give valuable information about the thinking of those times, the character of the Biblical text, the background to the ferment of Biblical interpretation in which Jesus preached and in which the Christian church was formed and various aspects of that time. One find in the Judean desert has shed light on the Persian period of Palestinian history.[25]

The term "Dead Sea Scrolls" is a generic term for various documents found since 1947 in several caves near the Dead Sea. In 1952 a group of scraps were found at Khirbet Mird, the ruins of a Byzantine monastery. The materials were in Greek, Hebrew, Palestinian Aramaic and Arabic, and had their origin between the 5th and 8th centuries. Another group of materials from the Dead Sea were found in 1951 in two caves of Murabba'at, a portion of the Wadi Darajh which enters the Dead Sea at Hebron. They come from the Roman period; most of the documents come from A.D. 131-135 and relate to the Second Jewish Revolt, led by Bar Kochba. In 1963 and 1964 The American Schools of Oriental Research conducted excavations in a cave 8 miles north of Jericho, at Mugh9ret Abu Sinjeh. From the Taamirah tribe 40 documents from this cave were bought and they were papyri placed in this cave when the patriarchs of Samaria fled before the soldiers of Alexander the Great in 331 B.C. The date of these papyri range between 375-335 B.C. They were composed in Aramaic; two were in Paleo-Hebrew. These papyri are the earliest found in Palestine. A fourth group of documents were found by the Hebrew University archaeologists at Nahal Hever and Nahal Se'elim in the Judean desert near En Gedi. Documents were found in caves in these two places which take the world back to the days of Bar Kochba, 131-135 A.D.[26]

Unrelated to those just enumerated are manuscript materials, usually referred to as the Dead Sea Scrolls. Over 40,000 fragments have been found in 11 different caves out of 259 caves examined by Arabs and archaeologists between 1957 and the present. The Qumra or Ain Feksha, MSS are of leather, papyri and copper.

Frank Cross classified the MSS from Qumran as coming from three distinct periods: (1) the archaic period, ca. 200-150 B.C. (2) the Hasmonean period, ca. 150-130 B.C.; (3) and the Herodian period, ca. 30 B.C. to A.D. 70.[27] The caves with Qumran MSS were probably sealed between 50 B.C. and A.D. 70.

Were the Sectaries of Qumran the Essenes?

The community who was responsible for depositing the MSS in the near Wadi Qumran have been identified with the Essenes by many scholars. The Essenes, a religious sect who were known from the writings of Philo, Josephus and Pliny appear to have beliefs and features in common with the sect of people at Qumran.[28] The Roman writer Pliny had described a "city" of the Essenes located in the wilderness between Jericho and En Gedi, near the shore of the Dead Sea.

Comparing the Essenes and the Qumranite sectaries, Yamauchi wrote:

"Both the Essenes and the Qumran sectaries were aseptic groups. Both had probationary periods of their initiates, ranked members, held property in common, practiced repeated immersions, partook of a common meal, refused the use of oil, held apart from the sacrifices of the temple, stressed God's predestination, and were intolerant of outsiders."[29]

While the Essenes were a celibate community, the Qumran cemeteries have revealed the graves of women and may have been the remains of married Essenes, who according to Josephus lived in villages. Scholars believe that the Essenes were pacifists, who however, believed in a great eschatological conflict portrayed in the War Scroll. Josephus claimed that some Essenes were involved in the First Conflict against Rome (A.D. 66-70).[30]

The sect at Qumran were students of the Old Testament Scriptures. In Caves 1,4,10 many Biblical MSS, one complete MS of Isaiah, two thirds of Habakkuk, and fragments of all books of the Hebrew Old Testament were found, with the exception of the Book of Esther. Ten members of the Qumran community were to be studying the Old Testament Scripture all day long.[31]

In addition to the large quantity of Biblical material, MSS of books that heretofore have been classified as "apocryphal" or as pseudepigraphical were also found in the various caves, some titles and samples for the first time. The languages were Hebrew and Aramaic. The following are some of the Apocrypha (called Deutero-canonical by Roman Catholic scholars); Tobit (Cave 4), Judith, Wisdom of Solomon (Khirbet Mird), Ecclesiastes (Cave 2), Baruc (place not identified) and 1 and 2 Maccabees.[32]

The following Pseudepigraphical books were discovered: Enoch (Caves, 1 and 4), those in Cave 4 were all in Aramaic, Jubilees (Caves 4, 2, 1) altogether five MSSs in Cave 4, two from Caves 2 and 1, which are also called "The Little Genesis." Also found were the Book of Noah (Cave 1), the Testament of Levi (Cave 4, and 1), Testament of Naphthali (Possibly in Cave 4), the Sayings of Moses, the Genesis Apocryphon (Cave 1). The latter was originally called the Lamech Scroll. Another MS was The Prayer of Nabonidus.

Distinctive Manuscripts of the Qumran Sect

A number of MSS have been found which were produced by the Qumran Sect itself, most of which were not known before 1947. One document, the *Damascus Document* found at Qumran in a number of copies was known, since it was found in the Genizah of a Jewish synagogue in Old Cairo. At least nine MSS of the Document were discovered in the Qumran caves.

The Manual of Disciple was one of the seven DSS found in Cave 1. This work gives instructions concerning the entrance requirements of the Sect." Also in Cave 1 was found the *Hodayot* or Thanksgiving Hymns.[34] These include some thirty hymns, probably composed by a single individual, the Teacher of Righteousness. From the same cave also comes *The War Scroll* which sets forth the tactics and equipment that the Sons of Light will use in defeating the Sons of Darkness.[35]

Commentaries called *Pesharim* have been found dealing with Psalm

37, Isaiah, Hosea, Nahum and Habakkuk, most hailing from Cave 4.[36] The Habakkuk Commentary, one of the original MSS of Cave, gives us important information about the Teacher of Righteousness and the Wicked Priest. The Nahum Commentary makes reference to historical persons of the second and first centuries B.C.[37]

Other sectarian documents found in Qumran are the Mishmarot. An MS describing the course of the priests adjusted to the Sect's solar calendar. *Testimonia* is a collection of texts related to the Messiah, and are similar to the Messianic texts which could have been used by New Testament writers as they include composite quotations of a type cited in the New Testament. Also found were a *Messianic Horoscope* and a *Cryptic Document* indicating that the Sect was not opposed to astrology.[38]

One of the most recent finds was the Sectarian MS known as The Temple Scroll, whose discovery was announced in 1967 after the June War. The scroll is the longest to come from Qumran and is over 26 feet long, thus longer than the famous complete Isaiah Scroll. According to Yadin it deals with four subjects: 1) religious rules concerning ritual cleanness; 2) sacrifices and offerings; 3) statistics of the king and army; and 4) a detailed description of the temple. The author passes off this scroll as a decree from God.[39]

In Cave 3 there was discovered in 1952 a unique Copper Scroll and it tells of the location of fabulous amounts of gold and silver hidden at 60 places. John Allegro published it together with a translation. Allegro is of the opinion that it contains a map of treasures taken from the temple as the Zealots were fleeing before the Roman armies. A survey by Allegro resulted in finding no treasures.[40]

Influences of Qumran on Christianity

Not only do the Qumran discoveries have importance for the textual criticism of the old Testament text and for the enlargement of our understanding of the Interestamental Period, but scholars have also attempted to show that the views of the Sect influenced the New Testament and postapostolic Christianity, as earlier stated.[41]

Larson quoted De Quincy to the effect that the Essenes were simply Christians gone underground, that the Essenes were not a separate organization for "otherwise we would have to accept the blasphemous conclusion that there were two independent, yet almost identical revelations at the same time and in the same place."[42]

Larson's Interpretation of the Qumran Scrolls

According to Larson in the period from 192-69 B.C. the Essenes produced a great corpus of literature under the inspiration of leaders from generation to generation as the Teacher of Righteousness who was called the Holy Great One and was given other titles signifying revelatory powers as direct conduits of direct messages from the Supreme God of the universe, who by the way was something quite different from Jehovah, the tribal god of the Jews.[43] The latter assertion is made without any proof from the DSS.

Jesus an Essene?

In his article on pages 124-125 Larson put forth the idea advanced over thirty years ago that Jesus was an Essene, in fact he claimed that Jesus "was a full-fledged member of the Essene Order, and that Jesus was persuaded that He was the personage foretold in their Scriptures who would be empowered to establish the kingdom of Righteousness, and, that therefore he broke his vow of secrecy and preached the doctrine of the Order in the highways and byways of Galilee."[44] Larson further-more contended that John the Baptist was an Essene as "were the core of men who established Christianity."[45] Larson also suggested that when the Romans destroyed the headquarters at Qumran, many Essenes be-came an integral part and decisive element in the formation of the Chris-tian movement. Asserted Larson: "There was, in particular, one segment known as the Ebionites, or the Poor Men, who recreated in detail in their own literature, the doctrines, teachings, and the discipline of the Essene communities. Actually, the three Synoptic Gospels are studied with state-ments in complete harmony with the cultic teachings, as in the so-called Sermon on the Mount in Matthew. The more we study the Dead Sea Scrolls and the Christian Scriptures, the more striking are the parallels which become evident. We have already noted that two documents were widely accepted by early Christian converts as genuine Scriptures of their own. Perhaps these converts had previously been Essenes."[46]

No examples are cited by Larson for all his claims of Essene influence on the Gospels. Since both Christianity and the Qumranites had their roots in the Old Testament, it would not be surprising that they held sim-ilar religious and ethical ideals. But in essentials they were miles apart. We have already shown that under no circumstances did the Christian Church accept The Testament of the Twelve Patriarchs or The Book of Enoch as inspired Word of God. They certainly are not found among the twenty-seven canonical books of the New Testament, recognized by Roman Catholic, the Eastern Orthodox Church or the various Protestant denominations.

Were the Ebionites True Christians?

Larson cited the Ebionites as Christians influenced by Essenism. The Ebionites were a second-century sect of Jewish Christians that generally preferred loyalty to the torah, which was reinterpreted to harmonize with tenets of varying groups that apparently practiced severe asceticism. Jesus was regarded as the Messiah but was not divine nor born of the Virgin Mary; Paul was rejected and they used a Hebrew Gospel based on St. Matthew.[47]

Christianity Is Not Essene in Character

It has been claimed by certain scholars, repeated by Larson, in his ar-ticle, that the Teacher of Righteousness (TR) anticipated Jesus' crucifix-ion and resurrection.[48] Professor Andre Dupont-Sommer of the Sorbonne gave a lecture on May 26, 1950 in which he claimed that the Teacher of Righteousness had probably been crucified and had been raised from the dead, and had appeared to judge Jerusalem when the Roman general Pompey entered the Holy City in 63 B.C. Thus he asserted that "Jesus appears in many respects as an astonishing reincarnation of the Teacher of Righteousness."[49]

The journalist Edmund Wilson, author of a best seller called *The Dead Sea Scrolls* in 1955, put forth the view that Jesus may have spent his childhood among the Essenes. However, this contradicts the New Testament that Jesus was obedient to his parents Mary and Joseph and went and lived with them in Nazareth.[50]

In 1956 John Allegro of the University of Manchester propagated similar views as those suggested by Dupont-Sommer. Allegro was a member of an international team of eight scholars entrusted with the official publication of the Qumran materials. He also expounded his views on the BBC broadcasts, he made remarks that were rejected by five of the scholars on that team. Allegro published an article in *Harper's Magazine* in 1966, in which he expressed even less restrained allegations. He claimed that Jesus himself was an Essene, and indeed a "magician."[51] This he did, claimed Yamauchi, by means of "linguistic legerdemain." Other assertions of Allegro were that Jesus was a companion of "harlots" and "publicans" and that these names were disguised titles for Essenes. This is not scholarship but irresponsible guesswork. He even suggested that the "tomb of Absolom" in the Kidron Valley was the Tomb of Jesus, which goes against all archaeological evidence.

In 1956 A. Powell Davies, pastor of All Souls Church in Washington, D.C, published: *The Meaning of the Dead Sea Scrolls*. In this volume, whose chapters have many scholarly footnotes, Davies not only related the exciting story of the DSS discovery, but also purported to give a scholarly interpretation of Holy Scriptures and also of the origins of Christianity. In chapter 3 he discussed "The Sect of the Essenes;" in chapter 4: "The Scrolls and Christian Origins" and in chapter 5: "The Scrolls and Jesus."[52] In these chapters Davies sponsored views similar to those of Wilson, Allegro, Dupont Sommer. What Larson has depicted as historical fact is nothing but a restaurant of theories and allegations written about over thirty years ago.

Misconceptions about the Teacher of Righteousness

The writers just mentioned in the previous paragraph misunderstood the facts about the Teacher of Righteousness, who only appeared about twenty years after the community had even been groping like blind. He may have been the author of the *Hodayot*, the *Thanksgiving Hymns*, which help students of the DSS to get an insight into the Sect's views about salvation. Our knowledge about the leader of the Qumranic Sect is based on the *Habakkuk Commentary*, the *Commentary on Psalm 37* and the *Damascus Document*. From these documents students of the Sect learn that the TR was a priest in Jerusalem. Although the Sect looked for the coming of two Messiahs, a priestly and a kingly one the Teacher of Righteousness was not looked upon as the Messiah.

The History and Historical Background
of the Teacher of Righteousness

Scholars have given different interpretations about the TR's time of existence and the place of the TR in the Sect as well as his relationship to the Wicked Prince (WP). Yamauchi have outlined the different views about the TR and the WP on pages 162ff of *The Stones and the Scriptures*.[53] During the span of 175-62 B.C. Rowley and Black would claim that the TR was the Zodokite High Priest Onanias III and the Wicked Priest was Jason or his brother Menelaus. In the time span 152-139 B.C. Milik, Sutcliffe, de Vau, Vermes, Bruce and Cross believe the Teacher of Righteousness is an unknown person and the Wicked Priest either Jonathan or his brother Simon. In the time period 75-63 B.C. Dupont-Sommer would make John Hyrcanus II the Wicked priest. During the time of the Great War with the Romans in A.D. 66 Roth and Driver tend to identify the TR with the Zealot Menahem and the WP as Eleazar, the son of Ananias the High Priest.[54]

Dupont-Sommer and Allegro on the Teacher of Righteousness

Both Dupont-Sommer and Allegro have used two passages for their comparison of the Teacher of Righteousness with Jesus of Nazareth. They are found in Nahum Commentary II 13b and Habakkuk Commentary XI; 4-8.

The first of these passages reads:

The explanation of the concerns of the furious Young Lion (who...took vengeance on those who seek smooth things - he who hanged living men (on wood . . . which was not formerly (done) in Israel; but he who was hanged alive upon (the) wood. . . (Note items in ( ) represent restorations of gaps).[55]

The Young Lion was identified by Allegro with Alexander Janneus, which is a possibility. The reference to hanging may refer to crucifixion. Jannaeus is known to have crucified 800 rebels. Although the Qumran texts nowhere suggest this, that was Allegro's interpretation inasmuch as the Teacher of Righteousness was an enemy of the Wicked Priest. Other scholars disagree with Allegro. "Those who seek smooth things" of the Nahum Commentary may be the Pharisees who were the enemies of Alexander Jannaeus. H. H. Rowley, a very competent Semitic scholar, rejected Dupont-Sommer's views totally. Rowley remarked: "It would surely be passing strange for the crucifixion of the Teacher of Righteousness — which Allegro thinks to have been of comparable significance for the sect with the crucifixion of Jesus for Christ or Christians — to be unmentioned in the text which has with so much — publicity been declared to record it."[56] In the Habakkuk Commentary XI: 4-8 the following passage was cited by Dupont-Sommer to buttress his sensational arguments about the Teacher of Righteousness.

The explanation of the concerns of the Wicked Priest who persecuted the Teacher of Righteousness, swallowing him up in the anger of his fury in his place of **exile**. But at the same time of the feast of rest of the Day of Atonement he **appeared** before them to swallow them up and to cause them to stumble on the Day of Fasting, their Sabbath of rest, (bold face not in text).[57]

Yamauchi has pointed out that the word rendered "exile" Dupont-Sommer translated **gitw**, which means to strip, and associated with crucifixion.[58] Two other DSS experts. Burrows and Kuhn favored "exile" which Dupont-Sommer adopted in his 1962 translation. However, the crux of the passage concerns "he appeared,"[59] Dupont-Sommer believed that the subject of that verb is the Teacher of Righteousness but older scholars are convinced it refers to the Wicked Priest.[60]

What Can Really Be Asserted
About the Teacher of Righteousness?

The following can be asserted with more or less certainty:

1) The Teacher of Righteousness was persecuted by the Wicked High Priest;
2) He may have been crucified, but there is no explicit statement to this fact in the Qumran documents;
3) The claim that the Teacher of Righteousness was resurrected is a forced interpretation rejected by the majority of scholars who are expert in the Qumran literature.

A Comparison Between the Teacher
of Righteousness and Jesus

Professor Brownlee has listed ten differences between the Teacher of Righteousness and Jesus:[61]

1. Unlike Jesus, the Teacher of Righteousness was a confessed sinner who gratefully acknowledged his dependence upon "the forgiving grace of God." This point and those that will follow are based on the assumption that TR is the author of the Thanksgiving Hymns.
2. "Unlike Jesus, he must suffer in order to be purified from sin."
3. "Unlike Jesus, the Essene Master founded a community vowing hatred toward his enemies. Jesus taught: 'Love your enemies.'"
4. "Both teachers founded a church — but only built a church which the powers of death could not overcome."
5. "Unlike Jesus, the Teacher of Righteousness called his followers out of the world," but Christ commanded His followers "to make disciples of all nations" (Matt. 28:20).
6. "Unlike Jesus, the Teacher of Righteousness does not appear to have been 'a friend of sinners and publicans'."
7. "Unlike Jesus the Essene Master performed no works of healing, nor in other ways did he engage in acts of compassion among the needy." In contrast to Jesus who welcomed the sick and deformed, all people with physical defects were excluded from the Qumran sect.
8. "Unlike Jesus, he was not a prophet, not a redeemer."
9. "Unlike Jesus the Teacher of Righteousness was simply preparing a greater than himself."
10. "Unlike Jesus the Teacher of Righteousness founded a community

enmeshed in legalism." So fanatical were the (Qumranites that they even considered the Pharisees lax. This fact occasioned Stauffer to remark: "I contend that had Jesus fallen into the hands of the Wilderness sectarians, they would have murdered him as ruthlessly as did the Pharisees."[62]

What Happened to the Dead Sea Scrolls?

This is the title of Larson's article in *The Journal of Historical Review*. Since much of the material in caves 1 and 4 consisted of thousands of fragments, it was the task of the international team of eight scholars to figure out which pieces belonged together and constituted a manuscript. The 500-800 scrolls, most in fragmentary condition, have not all been published. The original Qumranic materials was either in the hands of the State of Israel and before the 1967 Six-Day War in the hands of Jordan, deposited in the Museum in Jerusalem.[63] Then obviously Israel also took control of the Jerusalem Museum, controlling practically all of the Qumran finds. Since 1967 publication of DSS materials has ceased. Scholars[64] and Larson have wondered why nothing more has become available. Thus Larson wrote: "One question continued to occupy my interest: what had become of the scrolls? Why were none of them published for so many years? Sometimes I wondered whether they would survive or ever be made available to the public. However, we should note that even in custody of the Jordanians, they were held in the strictest secrecy — and why? I could only surmise that extreme pressure had been exerted by both Christian and Jewish sources: from the former, because it would not be beneficial to them should it be established that this faith grew out of a Jewish cult and was, therefore, not an original revelation: nor would the Israelis wish the Scrolls released since they were filled with fierce denunciations of Jewish religious leaders and civil authorities."[65]

In a column in the New York Times, Tuesday, April 29, 1980, which discussed the views of Dr. Golb of the University of Chicago who at the Proceedings of the American Philosophical Society, 1980, claimed that the Scrolls were taken hurriedly from libraries in Jerusalem and hidden in the Qumran caves in A.D. 70, shortly before the Romans besieged Jerusalem. Golb contended that the scrolls "describe a wide variety of practices, beliefs and opinions . . . spiritual currents which appear to have characterized Palestinian Judaism between the writing of the Old and New Testaments. He does not believe that celibacy characterized the Sect of Qumran."[66] Professor Cross disagreed with Golb's interpretation.[67] Milt Freudenheim, the author of the column in the New York Times, stated that he has contacted Cross by phone, who told him that the unpublished Qumranic material included 14 copies of the Manual of Discipline and "a small body" of local material, including "contracts," lists of names, personal documents on pottery and papyrus. Quite a large number of Biblical commentaries found, "bear the marks of autographs."[68]

Hershel Shanks in a report of a 1980 Biblical Archaeology Congress, held in Jerusalem, devoted the last part of his article "Jerusalem Rolls Out Red Carpet for Biblical Archaeology Congress,"[69] to what Larson complains of in his article that a goodly number, about half of the materials have not yet been published. Twenty-five years have elapsed and nothing is published. However, various scholars complaining about this situation, blame it on various other scholars who had the assignment to study and publish what was entrusted to them. According to the Shank's report it is not Israel which is responsible for the non-publication of the remaining Dead Sea Scroll documents, but the scholars entrusted with their translation, decipherment and publication.[70]

As far as Larson's charge is concerned that both Israel and Christians have much to gain from the non-publication of the remaining finds, this writer would contend that Christianity as reflected in the New Testament is **sui generis** and differs greatly from the theology of the Pharisees, Sadducees, Essenes, the Qumran sectaries, the Zealots or whatever religious views might be found at Qumran in the future. Judaism in its long religious history since the rejection of Jesus as the promised Old Testament Messiah has spawned many divergent types of religiosity, much far removed from the Old Testament Scriptures. Even Marxists who deny the existence of Yahweh are considered Jews! What kind of religious views can surpass the denial of God's existence? Psalms 14 and 53 assert: "Only the fool hath said in his heart: There is no God." (Ed. See "Response to Larson", p. 306).

Footnotes

1. Dr. Martin Larson, "What Ever Happened to the Dead Sea Scrolls?", *The Journal of Historical Review,* Volume Three, Number Two, Summer, 1982, pp. 119-128.
2. **Ibid.**, p. 119.
3. **Ibid.**
4. A. Powell Davies, *The Meaning of the Dead Sea Scrolls* (New York: The New American Library, 1956), pp. 43-132.
5. Edmund Wilson, *The Scrolls from the Dead Sea Scrolls* (New York: The Meridian Books, 1959). CF. also his article in the May 14, 1955 issue of *The New Yorker*.
6. Dupont-Sommer, *The Dead Sea Scrolls* (Oxford; Basil Blackwell, 1952). Also Dupont-Sommer, *The Essenes Writings from Qumran* (Cleveland: World Publishing Company, 1961).
7. John Allegro, *The Dead Sea Scrolls* (Baltimore: Penguin Books, 1956).
8. Dr. Charles Francis Potter, *The Lost Years of Jesus Revealed* (New York Fawcett Books, 1962), pp. 9-154.
9. Edwin Yamauchi, *The Stones and the Scriptures* (Philadelphia and New York: J. B. Lippencott, 1972), pp. 136-145.
10. Larson, **op. cit.**, p. 121.
11. Cf. H. T. Andrews, *An Introduction to the Apocryphal Books of the Old and New Testaments*, Revised and Edited by Charles F. Pfeiffer (Grand Rapids: Baker Book House, 1964), pp. 53-55; 69-72.
12. John E. Steinmueller, *A Companion to Scriptural Studies*. Volume I. General Introduction to the Bible (New York City: Joseph F. Wagner, Inc., 1941), pp. 49-81.
13. Aage Bentzen, *Introduction to the Old Testament*. Volume II, *The Books of the Old Testament* (Copenhagen: G. E. C. Gads Verlag, 1948), pp. 236-252.
14. Raymond F. Surburg, *Introduction to the Intertestamental Period* (St. Louis: Concordia Publishing House, 1975), pp. 128-129.
15. R. H. Pfeiffer, "The Literature and Religion of the Apocrypha," George Buttrick ed., *The Interpreter's Bible*. (Nashville, Abingdon Press, 1952), p. 421.
16. R. H. Charles, *Religious Development between the Old and New Testaments* (Lon-

don: Oxford University Press, 1956), p. 228.

17. J. E. H. Thomson, "Apocalyptic Literature," *International Standard Bible Encyclopedia*, I. 176b.

18. Larson, **op. cit.**, p. 121.

19. CF. Surburg, **op. cit.**, pp 143-144.

20. R. H. Charles, *The Book of Enoch* (London: SPCK, 1952), pp. 31, 56, 95, 129. Cf. also pp. xxxxviii).

21. H. H. Rowley, *The Relevance of Apocalyptic* (London, 19), p. 54.

22. Bentzen, **op. cit.**, pp. 243-244.

23. P. E. Kretzmann, *Popular Commentary of the Bible. The New Testament* (St. Louis: Concordia Publishing House, 1924), II, p. 587.

24. Larson, **op. cit.**, p. 124.

25. Frank Moore Cross, "Papyri of the Fourth Century B.C. from Daliyeh," David Noel Freedman and Jonas C. Greenfield, *New Directions in Biblical Archaeology* (Garden City, N.Y.: Doubleday Company, 1969), pp. 45-69.

26. Cf., Surburg, **op. cit.**, pp. 80-82.

27. Frank F. Cross, "The Oldest Manuscripts from Qumran," *Journal of Biblical Literature*, 74:164, September, 1955.

28. James A. A. Sanders, "The Dead Sea Scrolls - A Quarter Century of Study," *The Biblical Archaeologist*, 36:15-19, December, 1973.

29. Yamauchi, **op. cit.**, p. 136.

30. **Ibid.**

31. Millar Burrows, *The Dead Sea Scrolls* (New York: The Viking Press, 1956), pp. 227-246.

32. Cf. the convenient listing of all DSS manuscripts and where found in H. Wayne House, *Chronological Background Charts of the New Testament* (Grand Rapids: Zondervan Publishing House, 1981), pp. 82-83.

33. Cf. the translation in G. Vermes, *The Dead Sea Scrolls in English* (Baltimore: Penguin Books, 1962), pp. 72-94; The Hebrew and German translation, Eduard Lohse Die Texte aus Qumran (Muenchen: Koesel-Verlag, 1971).

34. Hebrew and German translation, Lohse *Die Texte aus Qumran*, **op. cit.**, pp. 112-175.

35. Vermes, **op. cit.**, pp. 122-148.

36. Eduard Lohse, *Die Texte aus Qumran*, **op. cit.**, pp. 227-244; 261-280.

37. Edwin Yamauchi, "Dead Sea Scrolls," Pfeiffer, Vos and Rea, *The Wycliffe Bible Encyclopedia* (Chicago: Moody Press, 1975), I, p. 438.

38. Mentioned by House, **op. cit.**, pp. 82-83.

38. Cf. Yigael Yadin, "The Temple Scroll - The Longest and Recently Rediscovered Dead Sea Scroll," *The Biblical Archaeology Review*, Vol. 10, October, *No.* 5, 1984, pp. 33-49. Cf. also Menahem Mansoor, *The Dead Sea Scrolls* (Second Edition: Baker Book House, 1983), pp. 194-203.

39. Yigael Yadin, "The Temple Scroll," in Freedman and Greenfield, *New Directions in Biblical Archaeology*, **op. cit.**, pp. 156-166.

40. Cf. John Marco Allegro, *The Treasure of the Copper Scroll* (Garden City, N.Y.: Doubleday & Company, 1964), 186 pp.

41. Frank Moore Cross, *The Ancient Library of Qumran and Modern Biblical Studies* (Garden City, N.Y.; Doubleday & Company, 1961), pp. 195-238.; Charles F. Pfeiffer, *The Dead Sea Scrolls*, Revised. (Baker Book House, 1962), pp. 85-88. Jean Danielou, *The Dead Seas Scroll and Primitive Christianity* (New York: The New American Library, 1958), pp. 87-125.

42. Larson, **op. cit.**, p. 124.

43. **Ibid.**, pp. 121-122.

44. **Ibid.**, p. 124.

45. **Ibid.**

46. **Ibid.**, pp. 124-125.

47. Clarence Tucker Craig, "Ebionism," Vergilius Ferm ed., *The Encyclopedia of Reli-*

gion (New York: The Philosophical Library, 1945), p. 241.

48. Larson, **op. cit.**, p. 124.
49. A. Dupont-Sommer, *The Dead Sea Scrolls*, **op. cit.**, p. 99.
50. As cited by Yamauchi, **op. cit.**, p. 140.
51. John M. Allegro, "The Untold Story of the Dead Sea Scrolls," *Harper's Magazine*, CCXXXIII, August, 1966.
52. Davies, **op. cit.**, pp. 43-131.
53. Yamauchi, **op. cit.**, pp. 141-142.
54. **Ibid.**, p. 142.
55. A. Dupont-Sommer, *The Essene Writings from Qumran*, **op. cit.**, p. 269.
56. H. H. Rowley, "Nahum and the Teacher of Righteousness," *Journal of Biblical Literature*, LXXV: 190, 1956.
57. Dupont-Sommer, *The Essene Writings*, **op. cit.**, p. 266. 58. Yamauchi, **op. cit.**, p. 143.
59. **Ibid.**
60. H. Silberman, "Unriddling the Riddle, A Study in the Structure and Language of the Habakkuk pesher," *Revue de Qumran*, XII, 158-159, 1961.
61. William Browlee, *The Meaning of the Qumran Scrolls for The Bible* (Oxford University Press, 1964), pp. 141-142.
62. Ethelbert Stauffer, *Jesus and the Wilderness Community at Qumran* (Philadelphia: Fortress Press, 1964), p. 21; Oscar Cullmann, "The Significance of the Qumran Scrolls for Research into the Beginnings of Christianity," ed. Krister Stendahl, *The Scrolls and the New Testament* (New York: Harper & Brothers, 1957), pp. 31-32.
63. Larson, **op. cit.**, pp. 127-128.
64. Biblical scholars mentioned by H. Shanks, "Jerusalem Rolls out the Red Carpet for Biblical Archaeology Congress," *Biblical Archaeology Review*, Vo. X. 1984, August, No. 4, pp. 17-18.
65. Larson, **op. cit.**, p. 127.
66. Milt Freudenheim, "Scholar Stirs Debate on Dead Sea Scrolls," *New York Times*, Tuesday, April 29, 1980, C3.
67. **Ibid.**
68. **Ibid.** Shanks, **op. cit.**, pp. 127-128.
70. **Ibid.**

Questions

1. Martin Larson delivered his essay on the Dead Sea Scrolls at ____.
2. When were the Dead Sea Scrolls discovered? ____
3. Larson believes that Christianity was influenced by ____.
4. Larson claimed that Jesus was an ____.
5. Professor Andre Dupont-Sommer claimed ____.
6. Edmund Wilson wrote that Jesus sent his childhood among ____.
7. John Allegro claimed ____.
8. What Larson depicted as historical facts is nothing but ____.
9. What are some differences between Jesus and the Teacher of Righteousness? ____
10. Larson suggested Christians wanted the Dead Sea Scrolls suppressed because ____.
11. Christianity as reflected in the New Testament differs greatly from ____.
12. The fool has said in his heart____.

A Response to Nicholas Carter's 'The Counterfeit Gospel'

Christian News, March 12, 1990

Carter's article is found in the February issue of *Liberty Bell*, a magazine published by Liberty Bell Publications.[1] (Ed. "The Counterfeit Gospel" by Nicholas Carter, published in the February 1990 Liberty Bell, was photographed in the March 12, 1990 Christian News). Liberty Bell Publications claims to be advocates of freedom of speech, freedom of thought and freedom of expression. *Liberty Bell* has been published monthly since September 1973. One of the objectives is supposed to be the "exposure of Ideas suppressed by the controlled news media of this country. The intent of *Liberty Bell* is also to give free reign to ideas, because it is through ideas that the world is ruled. Another objective is to replace institutions or government by men by the replacement of new ideas which are subjected to evolution, change or replacement by the will of the people."[2]

Carter's "Counterfeit Gospel" is an attack on Christ's historicity, His life, death, crucifixion, resurrection, ascension and session at God's right hand. Furthermore, it attacks the first century existence of the 27 books of the New Testament canon. The article completely misrepresents the origin of the Christian Church as well as its development and growth. The knowledgeable reader will find nothing new in this vicious attack on one of the great and influential religions of the world, numbering over a billion members. The reader will find nothing new in this articles, except the cocksureness with which the author has explicated his erroneous views. Former pantheists, agnostics, skeptics, and humanists have set forth similar views in the last three hundred years, especially since the Age of Rationalism. The only works which the author seems to have consulted were those in sympathy with his views. He never takes into account the evidence of Christian apologists who were well acquainted with the pronouncements of Carter and his ilk. Carter claims that Christianity is "a counterfeit Gospel." *The New American Dictionary of the English Language* defines "counterfeit" as "a fraudulent imitation or fascimile."[3] The word "Gospel" is a Biblical word and means "good news," a legitimate designation for the preaching and teaching of Jesus Christ. Paul only claimed to proclaim the "good news" as it had been delivered to him by transmission and also by direct revelation from Christ. Carter's use of a term he rejects is, therefore, strange as a title for his article which denounces the "good news" as presented in the New Testament. One may ask Carter: "what previous 'good news' did later Christians counterfeit?"

The Alleged Secret of the Origins of Christianity
Despite the so called scholarly studies of Couchoud, Guignebert, Klaussner, Schweitzer, et al., Carter claims that students "are still in the dark regarding the actual origins of one of the most famous religions of

our time."[5] Why does Carter have that problem? Because he rejected the first-century origin of the 27 books of the New Testament and, consequently, there is a void for him between the first century B.C. and the second century A.D. Millions of people who have lived since the death and ascension in A.D. 30 till the present time have had no difficulty with the question: "How did the Christian Church originate?" The Gospels and the Book of Acts answer this and similar questions clearly and uneqivocably. The Gospels and Acts state that Christ founded the Christian Church, that He selected 12 men as apostles, trained them and gave them the command to make disciples of all nations (Matthew 28:18-20).

Evolution and Christianity

One of the erroneous presuppositions of *Liberty Bell* Publications and of Carter is their commitment to the theory of evolution and its application as determining factor in the development of religion and of a people's theological convictions. Thus Carter wrote in his article: "What is now recognized is that the Christian religion developed in and through myth. The historical principle that all religions are simply phases of continuous evolution is irrefutable."[6] He further argues: "It is logical to conclude, therefore, that no system of belief inculcated on the dangerous and malevolent territory of faith is rational, and, indeed, there is no evidence to sustain a single one. It logically follows, too, that ALL Christian traditions are myths, and that brings us to the problem: When the facts of reality are woven into 'the antique fables and fairy toys' of transcendentalism, they are never easy to unravel."[7]

Carter and Nineteen Century Theories On Religious Origins

Carter has resorted to arguments advanced by the members of the school of the "Comparative Study of Religion," a term invented by Ernst Renan.[8] This school of the study of religion has been with the world for over a century. Many scholars have written on the subject of the origin of religion.[9] Father Schmitt, a recognized anthropologist, made a thorough study of primitive religions as well of the origin of the idea of God among primitive peoples.[10] His major work on this subject is entitled: *Der Ursprung der Gottesidee* and in this multivolume work he has shown as an anthropologist that in pre-animistic thought that the concept of a High God is the oldest religious conception. Members of the school of religion from whom Carter has drawn his views were all anti-supernaturalists. When the evolutionary school became widespread, it was proclaimed that all religions were evolutionary products. This meant that no one religion could be considered better than the others.[11] This was held to invalidate the claim made by the Christian religion of being the only saving faith. For the students of the school of comparative religions there was no one religion which could lay claim on absolute truth. One writer wrote as follows concerning this issue:

"The depreciation of the Old Testament and the exaltation of the Sacred Books of the East and of other faiths have sometimes gone hand in hand. The words of Christ: 'I came not to destroy but to fulfil'

1453

have been wrested out of their context and made to mean that he came to fulfil the Bhavagita, the Analects of Confucius, and even the Koran, while the Old Testament is designated 'mere folklore' and so often goes by the board."[12] Already Max Mueller (1823-1900) revolted against the evolutionary pattern of the development of religion, which started with animism, followed by polytheism, then followed by henotheism and was climaxed in monotheism. Here was Mueller's final view: the original henotheistic nature worshipped degenerated into polytheism, sank into Fetishism, and then rose in some cases to a new form of Pantheism or Theism.[13] The writer does not agree with this understanding of religious development.

According to the Bible there was in the beginning Biblical theism, the belief in the true God, Yahweh and as people more and more forsook the God of revelation they became polytheists and supporters of animism. Polytheism followed monotheism and not vice versa. This was also the conclusion of Stephen Langdon, professor of Assyriology at Oxford University, who in his study of Semitic Mythology asserted that monotheism was the original religion of the Semites.[14] The same position was also taken by the Egyptologist Flinders Petrie relative to the religion of Egypt.[15]

Myth and Christianity

According to Carter Christianity is built upon the "Christ Myth." Thus he wrote: "The Christ myth is the foundation upon which the Gentile church is built, the orthodox teaching of Christianity."[16] "Catholicism includes in it the principal myth of the Western World, which is not God or the Mother, but what we call the Christ the myth, that is found with practically all ancient peoples of the deliverer, the savior, and with all but the Jews,[17] the sacrificial offering on the fructifying tree.'"[18]

Soulen in his *Handbook of Biblical Criticism* defined "myth" in this way: "In popular usage the term myth connotes something untrue, imaginative or unbelievable, or in older parlance, a purely fictitious narrative usually involving supernatural persons, actions, or events (DED)."[19] Carter claimed that the simplest terms, myths are "life lies that enable us to believe that supernaturalism exists in a natural world . . . and that non-materiality exists in a material world . . . . and that miracles and prophecies exist in a world in which there is no scientific proof of supporting miracles and prophecies."[20]

Carter has drawn his erroneous views about myth from the works of German and European scholarship, from men who were unbelievers and detractors and enemies of Biblical Christianity. The author of "Counterfeit Gospel" went to scholars labelled by W.G. Kuemmel as proponents of "radical criticism." Bruno Bauer and a number of Dutch, French and English scholars questioned the existence of Christ and that Paul was the author of many epistles ascribed to him in the introductions of his letters.[20a]

There were many other scholars like Gunkel who postulated myths in the Old Testament Scriptures. In his introduction to his commentary on

Genesis he labelled the whole book of Genesis as consisting of a number of different types of myths, thereby rejecting the historicity of many occurrences before and after the time of Abraham.[21] Many other portions of the Old Testament were designated as mythological. Hugo Gressman adopted the hermeneutical principles of Gunkel, and Gunkel and Gressman exerted a great influence on the first two editions of the prestigious *German Theological Encyclopedia*, known as Die Religion in Geschichte und Gegenwart.

In the New Testament area Carter had the inspiration for his soul-destroying views in the writings of Rudolf Bultmann, who claimed that the New Testament needed demythologization.[22] In order to get at the real meaning of a text, it needed to be stripped of its myth. While Bultmann did not deny the historical existence of Jesus, he still claimed that the crucifixion and burial of Jesus were the end of Him. All miracles and supernatural events associated with the life of Christ were jettisoned.[23] The views of radical Old and New Testament scholars were fodder for Carter's canon and constituted the essence of his theological repertoire.

It should be pointed out to Carter with regard to myth that as Soulen has noted: "There is no agreement upon definition, whether in terms of its form, that is, its relationship to fairy tales, sagas, legends, tales, epics etc. or in terms of its content and function."[24]

Carter and Psychoanalysis

That the writer of "The Counterfeit Gospel" should make use of psychoanalysis in his discussing the Biblical doctrine of the atonement is not surprising, believing that by Christ's sacrifice the wrath of God has been appeased, the sinner is said to palm off on someone else the accumulated transgressions and misfortunes which he shrinks from bearing himself. "This trick" avers Carter is "pscho-analytically speaking, the soul seeks to deceive God and evade the claims of justice. This desire leads to the creation of an elaborate scenario in which salvation deity dies for the benefit of his worshipers. The Son, the divine sacrifice for all, submits to the Father and is then slain, which results in a sense of guilt, the need for self-punishment to relieve it, and a mystery of salvation based on the suffering and death of the savior-god conceived as redemptive."[25] And what is the outcome of such thinking? Here is Carter's psychoanalytical explanation. "And thus is that myth esoterically reinforces that most subversive of psychological defenses — the denial of reality."[26]

Many psychologists and psychiatrists reject psychoanalysis as a valid method for dealing with men's minds and actions. Freud contended that the laws governing physical processes are applicable to human personality. Orville Waters evaluated the Freudian system as "basically materialistic and deterministic, with no place for human freedom. If the motive power of personality derives from a pleasure-seeking id, a subjectivistic, hedonistic ethic follows."[27]

Carter portrays mythology as a person or force that creates all manner of "fetishes, legends, parables, poems, songs, prayers, moral apothegms and widely exaggerated exploits."[28] It is not mythology which makes human beings do and write certain things, but human being who makes decisions and create beliefs. Carter places the cart before the horse!

The Inability of Skeptics and Agnostics To Comprehend the Bible

Carter is an outspoken opponent and hater of the contents of Holy Writ. That should not surprise Christians, because Paul warned that the man by nature, that is the unregenerate person, is not able to comprehend the things of the Spirit of God, for they are foolishness to him.[29] The Gospel was a stumbling block to the Greeks, who claimed to be the intellectuals of Paul's time, so in centuries past the same Word of God is intellectually unacceptable to men, who have adopted materialism and made man the end of all things. Instead of believing in God they rationalize about "Mother Nature" or that mythology acts as a womb that produces all manner of odious things, such as miracles, prophecies and belief in the deity of Christ.[30]

Myth and the Historical Christ

One of the main points of contention of Carter's "The Counterfeit Gospel" is that Christendom has wrongly propagated the harmful myth that Jesus of Nazareth lived in Palestine from circa 6 B.C. to A.D. 30. That is the great lie and deception foisted on the world, and that he wishes to help straighten out the thinking and belief of people in general and especially the Christian Churches, who for nearly two thousand years have been deceived by this piece of mythology. Carter calls it "the Christ myth." There is no reliable evidence—thus Carter reasons for such an untenable belief. He tries to muster various types of evidence to show the absurdity of the statement with which the article in the *New America Encyclopedia* dealing with Jesus Christ begins, namely that Jesus of Nazareth was born in Bethlehem in 6 BC.[31] To a person answering Carter to the effect that the four gospels explicitly say so, he would retort that the 27 books of the New Testament only originated in the second half of the second century and in the years beyond.

Relative to Carter's denial of Christ's ever having lived, F.F. Bruce, former Rylands professor of Biblical criticism, asserted: "Some writers may toy with the fancy of Christ's myth, but they do not on the grounds of historical evidence. The historicity of Christ is as axiomatic for an unbiased historian as the historicity of Julius Caesar, it is not historians who propagate the Christ myth theories."[32] Otto Betz in his book. *What Do We Know About Jesus?* concluded his study about Jesus and His life with the words that "no serious scholar has ventured to postulate the non-historicity of Jesus."[33]

What are the sources for the historicity of Jesus? In response to this question, Dr. John Warwick Montgomery wrote: "What then, does a historian know about Jesus? He knows, first and foremost, that the New Testament documents can be relied upon to give an accurate portrait of Him. And he knows that this portrait cannot be rationalized away by wishful thinking, philosophical presuppositions, or literary maneuver-

ing."[34]
Christ's Historicity Established by Non-Biblical Evidence
Carter tries his utmost by misrepresentation of historical data of heathen sources and by the omission of facts inimical to his views to show that the first-Christian century and the first half of the second century A.D. knew nothing of Jesus Christ,[35] who according to the New Testament lived in Palestine, that he was born in Bethlehem of Judea, that He lived for about thirty-three years, that he preached for about three years, was condemned to death by the Romans under Pontius Pilate and was buried."

Alleged Evidence by Carter For Christ's Non-Existence
The author of "The Counterfeit Gospel" claims that Yadini's find on Masada, which included fragments from the Book of Numbers and the Book of Psalms, 64 documents all told, that they do not once refer to Jesus.[36] We retort that this is a good example of the argument from silence. Bar Kochba, who claimed to be the "star of Jacob," foretold in Numbers 24:17, and who was recognized as the Messiah by a number of Jews, would hardly mention Jesus of Nazareth because at that time (A.D. 70-73) he was waging a war with the Romans and not with the Christians. The fact that these Dead Sea Documents area documents do not mention other historical personages of that era, does not prove that those people did not live. Furthermore, Philo of Alexandria, who left fifty writings, never mentions Jesus Christ, and this omission is considered by Carter as evidence for his contention that Christ never lived and is a creation out of "the womb of mythology."[37] Philo lived from Circa 25 B.C. to A.D. 45. Most of his life was lived before 30 or 33 A.D. when Jesus was crucified. It was only after A.D. 45 that Christianity was beginning to be spread throughout the Roman empire, especially that it was being proclaimed on Cyprus, Syria, Asia Minor, Greece, Italy and later Spain. Here again the argument from silence is being appealed to.

Josephus and Jesus Christ
There are two passages in the Greek writings of the Jewish historian Josephus that refer to Jesus Christ. The first has been challenged as a Christian interpolation. Some scholars reject it an unauthentic; others consider it genuine. The first reads as follows:

> Now there was about this time Jesus, a wise man, if it be lawful to call him a man, for he was a doer of wonderful works, a teacher of such men as receive the truth with pleasure. He drew over to him both many of the Jews, and many of the Gentiles. He was the Christ, and when Pilate, at the suggestion of the principal men among us, had condemned him to the cross, those that loved him at first did not forsake him; for he appeared unto them alive again the third day; as the divine prophets had foretold things concerning him. And the tribe of Christians so named from him are not extinct to this day.[38]

Josephus, who at first served as general in Galilee in the Great Jewish War of A.D. 66-70, later went over to Rome and consequently was looked upon by his countrymen as a traitor; could have possibly written this description of the life and ministry of Jesus. Carter, like other scholars, rejects the account of the *Jewish Antiquities*.[39]

There is a second reference to early Christian history in the *Jewish Antiquities* that need not be a Christian interpolation. It relates to James, the brother of Jesus. It reads as follows:

> But the younger Ananus who, as we said, received the high-priesthood, was of a bold disposition and exceptionally daring, he followed the party of the Sadducees, who were severe in judgment above all the Jews, as we have already shown. As therefore Ananus was of such a disposition, he thought he had now a good opportunity, as Festus was now dead, and Albinus was still on the road; so he assembled a council of judges, and brought before it the brother of Jesus the so called Christ, whose name was James, together with some others, and having accused them as law-breakers, he delivered them to be stoned.[40]

Carter claims that it is mystifying that the Christ's existence is not found anywhere in "ANY of the writings of the first century of the Common Era (as Jews prefer to render the period A.D.) aside from the New Testament — which cannot be under any circumstances be accepted as a factual historical document."[41] Roman literary figures as Seneca, Pliny the Elder, Juvenal, Martial, Quintilian, Epictetus, Plutarch and others appear not to have a single reference to the Nazarene or Christ. Conveniently, Carter does not mention Roman writers which do refer to Christ or Christianity. He does not mention Tacitus as a reliable writer, for the latter in his historical work the Annals, XV,44 refers to the historical fact that during the reign of Tiberius that a man called Cregtus (Christ) was crucified and to the fact of the existence of Christians in Rome.[42] Tacitus (A.D. 52-59) was a Roman historian who in A.D. 112 was Governor of Asia and was the son-in-law of Julius Agricola who was Governor of Britain A.D. 80-84 and so a reputable writer of events of which he had had first-hand knowledge.

Then there is the testimony of Lucian, a satirist of the early second century who spoke scornfully of Christ and the Christians. He connected the Christians with the Jewish synagogue in Palestine and among other things said this about Jesus that he was a man "who was crucified in Palestine because he introduced this new cult into the world. . . Furthermore their first lawgiver persuaded them that they were brother one of another after they have transgressed once for all by denying the Greek gods and by worshipping that crucified sophist himself and living under his law."[43]

It is thus clear that there were some Roman writers who knew of Christ and referred to him in the beginning of the second century A.D. But those writers who did write about Jesus and His followers are repudiated by Carter because they show the falsity of his erroneous views. The following is the way he dismissed the first-century Roman evidence against his theory: "It is neither logical nor scholarly to accept the second century speculations, the conclusions of Tacitus, Suetonius, Pliny the Younger and others, who were motivated to write off Christian legends that were being developed as if they were rooted in historical fact rather than in evolutionary folklore and evangelical fancy."[44] How convenient! When writers agree with Carter's position, they are considered as writing history, when they disagree they report fanciful stories and give statements that cannot be labelled factual.

The Historicity and Reliability of the New Testament Writings

The most important source for the life of Jesus and of the apostolic church are the writings of the New Testament, twenty seven books all told. These primary sources Carter rejected as written in the second half of the first century.[45] They are creations of other individuals than those set forth in various writings of the New Testament themselves, as well as the belief of the individuals who ascribed the four gospels to the Apostle Matthew and the Apostle John and Mark, a friend of Peter and of Luke, a friend of Paul. Scholars who are specialists of the Greek New Testament completely disagree with the contention of skeptics who deny the first century origin of the canon of the New Testament.[46] There exist writings by individuals that were born in the last half of the first century and who lived into the first half of the second century who knew a number of the New Testament writers.[47] Anyone doubting the first-century origin of the New Testament should read the study of F.F. Bruce, The New Testament Documents: Are They Reliable? This scholar averred: "The importance of evidence lies in the link with the apostolic age and his ecumenical associations. Ireneus was brought up in Asia Minor at the feet of Polycarp, the disciple of John and Ireneus became Bishop of Lyons A.D. 180."[48] His work refer to the canonical recognition of the Fourfold Gospel, to Acts, Romans, 1. and 2 Corinthians, Galatians, Ephesians, Philippians, Colossians, 1 and 2 Thessalonians, 1, 2, 3 John and Revelation, 1 and 2 Timothy, Titus and 1 Peter.

Clement of Rome (A.D. 95-97) has references to Matthew, John, Romans, Ephesians, 2 Thessalonians, Titus, Hebrews, James, 1 and 2 Peter. These are books which appeared before Clement of Rome and consequently are from the first Christian Century.[48a] This explodes the theory of Carter that there were no New Testament writings prior to the latter half of the second century after Christ's birth.

Orthodox Christianity and Judaism

An argument against the historicity of Jesus advanced by Carter is the non-acceptance by orthodox Judaism of Jesus Christ and his teachings. He claims that the Jews did not accept the claim of Jesus' messiahship, predicted in the Old Testament.[49] However, the Book of Acts reports that the Christian church began its existence in Jewish Jerusalem and that many Jews and proselytes constituted the first Jerusalem congregation formed by Peter and John, two Israelites. Later on the Jerusalem congregation, which initially numbered 3,000 grew to 5,000 in the city of Jerusalem and that many priests even joined the first Jerusalem church.[50] Congregations established later by Paul were comprised both of Jews and Gentiles.

Christianity and Fable or Myths

Carter has accused Christianity as being a product of "the womb of mythology" and that the Christian faith is built upon the foundation of a non-existence person and believes the writings of the Apostles, men who also never existed.[51] However, Peter, often referred to in the Four Gospels and also in the Book of Acts, disclaimed the contention of people, who like Carter today, claimed that the faith that emphasized "Christ and Him crucified" was only built on fables. Thus Peter wrote to a group of congregations in Asia Minor: "We did not follow any clever *myths* when we told you about the power of our Lord Jesus Christ and His Coining. No, with our own eyes we saw the majesty (I Peter 1:16).[52] (Italics are the authors)

Christianity's Supposed Derivation from the Mystery Cults

Carter has revived an old discredited theory that New Testament Christianity derived its teachings from the mystery religions, which sprang up during the Hellenistic era. The *Liberty Bell* writer claims that many erudite scholars of the 19th and 20th centuries were utterly convinced that Christianity and its alleged founder did not exist. Men such as Volney, Depuis, Kulicher, Bauer and Kalthoff, are the authority for the contention of the fact that Christ never lived.[53] However, these individuals are all skeptics and opponents of historic Christianity. Since Christianity was not founded in the first century and originated at a much later time, so it was the mystery religions which furnished the basic tenets used by Catholicism in its construction of the "Christian myth." J. Gresham Machen in his book, *The Origin of Paul's Religion* has devoted chapter vi, entitled: "The Religion of the Hellenistic Age" and chapter vii, "Redemption in Pagan Religion and in Paul." About the supposed borrowing from the mystery religions.[54] In this book Machen, formerly of Princeton Theological Seminary, has totally rebutted the claims of Reitzenstein and others relative to the alleged borrowing by Paul of baptism and the doctrine of salvation. Here is Carter's contention: "From time immemorial the death and resurrection of a salvation-deity considered to be both human and divine have been the primary tenets of the gospels of the many Gentile mystery (meaning secret) religions within Hellenistic Asia and the Far East."[55]

The Rejection of Carter's Claim Relative to the Non-Historical Character of Jesus

According to Carter, Jesus of Nazareth never lived and all that the Gospels and the rest of the New Testament assert about Him is A LIE. The foundation of the Christian religion rests upon the lies of deceivers. Is it not strange that for nearly two thousand years millions of people have believed in Jesus Christ, the God-man, and have suffered martyrdom for Him? Every encyclopedia of American and European publishers has treated Jesus of Nazareth as a person who lived in the first century A.D. The present *Encyclopedia Britannica* has a biography of Jesus covering 11 pages, with two columns to a page, and contains about 22,000 words.[56] All secular encyclopedias treat the New Testament as the primary source for the life of Christ.

The description of Jesus of Nazareth received more space than that given to Aristotle, Cicero, Alexander the Great, Julius Caesar, Buddha, Confucius, Mohammad or Napoleon Bonaparte in the Britannica.

Carter's Denial of the Existence of the Twelve Apostles

Since the three Gospels contains a listing of the Twelve Apostles and the Book Acts also gives a listing the Christian Church has never doubted their existence and activity in carrying out the Great Commission to evangelize the whole world.[57] Such a denial on the part of the *Liberty Bell* author is consistent with his rejection of the entire New Testament as containing a library of first-century books. Second and third century authors have recorded traditions about all Twelve Apostles, which House has conveniently collected in the book, *Chronological and Background Charts of the New Testament.*[58]

Here is Carter's view about the Twelve Apostles: "What of the famous Twelve known as the Apostles? The word apostle means 'to send' or 'commission.' The apostolic implication is that the Apostles were with Jesus and commissioned by Jesus to go forth and preach. But just as there was no Jesus, there were no Apostles commissioned by Jesus."[59] Carter submits the following as to how there were twelve men supposedly commissioned by a non-existent Christ: "Why twelve? The number betrays a symbolic intention. With twelve helpers Joshua passed through the Jordan, Jason went after the golden fleece with twelve helpers. The sun wanders through the twelve signs of the zodiac. And so it was that Jesus wandered through the Holy Land with twelve disciples. In the religion of sun worshipers the twelfth month is the betrayer of the sun that sickens and dies at the winter solstice. Ergo, Jesus is betrayed by the twelfth disciple."[60] Can a rational person find a more weird hermeneutic than that practiced by Carter?

The fact that Eusebius does not mention the Twelve Apostles, so claims Carter, is another proof that the Twelve never existed but "the Apostles" were all chosen long after the first century by different people in different times — which makes it nearly impossible to determine "who's who and whose what in the apostolic circle."[61]

Carter's Ignoring of the Book of Acts and St. Paul's Epistles

Carter rejects all documents that deal with the first century A.D. as well as the 13 Epistles of Paul. All reputable scholars consider the Book of Acts as containing reliable history. In Acts, Luke has given an account of the spread of the Gospel beginning with the Ascension in A.D. 30 or 33 until Paul's first Roman imprisonment (A.D. 60). There was a time in the 19th century when the historical value of Acts was questioned by critical scholars, especially by the Tuebingen School.[62] Scholars like Ramsay,[63] Harnack, A.T. Robertson and others rejected the views of Tuebingen school. Harnack, not exactly known as conservative scholar, earlier in his career questioned the reliability of Acts. But later Harnack in his Acts of the Apostles wrote as follows: "The book has now been restored to the position of credit which is its rightful due. It is not only, taken as a whole, a genuine historical work, but even in the majority of its details it is trustworthy . . . Judged from almost every possible standpoint of historical criticism it is solid, respectable, and in many respects an extraordinary work."[64]

Intertwined with the Acts of the Apostles is the life of Paul as well as his letters to congregations and to individuals which constitute the Pauline corpus of inspired writings.[65] Carter completely ignored this important body of literature and speaks of a first century void of evidence for the growth and development of the Christian church throughout the Roman world.

Carter and the Apostolic Fathers

Again Carter complains about lack of historical material for the first half of the second century. And yet there is a group of writings, known as The Apostolic Fathers.[66] Because they believe in Christ and employ the New Testament writings they are not to be trusted for they totally negate his theory of "Christ-myth." Their references to Christ and the teachings of Christianity are no doubt frauds and if they lived and were active, they simply perpetuated the fraud of the Christ-myth.

Early Church History According to Carter

On pages 51-53 Carter sets forth his views about the development of Christianity into a mystery cult. Church history begins for him with the second half of the second century. He gives his views about the origin of the Four Gospels. He claims that the bubble must be broken that the Gospels were by individual people. The Epistles of Paul postdate him by several centuries. *The Muratorian Canon* is for Carter the beginning of the development of the beginning of the Gospel. According to Carter, Jesus of Nazareth never lived and all that the Gospels and the rest of the New Testament assert about Him is A LIE.

The fact is that the *Muratorian Fragment* shows that many books of the New Testament were in existence in the late second century.[67] He argues from the earliest copies of New Testament, which in turn were made from previous copies, that the Book of the New Testament did not originate earlier than the copies which have been found, which would be the Chester Beatty Papyri. He does not tell his readers that New Testament textual critics have a fragment of the Gospel of John dated on paeleographical grounds as coming from A.D. 135.[68] The *Bodmer Papyrus II* has been dated as coming from A.D. 150-200 and it contains nearly all of John Gospel.[69] For many Greek writers our oldest MSS are a thousand years removed from the time of writing and in the case of a number of Roman writers 300 years, yet these authors and their works are not rejected by classical scholars, why should the New Testament be questioned as to its first century origin when the textual evidence exists in great abundance in copied MSS as compared with classical authors.[70] Carter claims that the New Testament was the result of a long Christianizing process that occurred during the period of time when "the Gentile Fathers of the Church were determining what religion was supposed to be." But Carter does not tell his readers who these people were who periodically changed the contents of the New Testament and how this was accomplished. If any of our readers has any doubts about the history of the New Testament canon, he should read chapter 4, "The Reliability of the Bible," in McDowell, *Evidences That Demand A Verdict*.[71]

The presentation of the history of the Church from A.D. 30 till 430 given by Carter will not find any support in reputable church history books. Let the reader compare Carter's interpretation of the growth and development of the church as found in Latourette, A History of Christianity, pages 3-236,[72] or Qualben's *A History of The Christian Church*, pp. 1-136.[73]

Contradictions in the New Testament

In order to discredit the contents of the New Testament, like other skeptics, he brings up a number of the alleged contradictions supposedly contained in the Gospels. Carter has problems with what occurred at Easter morning. The contradictions enumerated have been answered by William Arndt in his book *Does the Bible Contradict Itself?* and in *Bible Difficulties*.[74]

Carter's Confusion of Apostolic Christianity With Catholicism

Carter depicts the Christian doctrine as borrowed and developed by the Roman Catholic Church. However, there is a difference in some requests as to what the infallible and inspired New Testament taught, and the interpretations of postapostolic and pre-Nicene writers as to what they believe the Bible to say. There were developments in the second and third centuries which should not be attributed to New Testament Christianity. The great problem with Carter is that he starts his discussion of church history nearly two centuries too late.

Carter's Materialism and Other Philosophical Systems

Carter appears to be a materialist, holding to the idea that reality is of one kind: namely material.[75] Even people who are not Christians reject materialism as wrong, preferring idealism as the ultimate reality. The truth is that Christianity teaches both the reality of matter and of spirit. Christianity teaches the dualistic character of reality.[76]

The Inability of Mankind to Accept the Myth of Christianity

Carter's disappointment with Christians is stated thus: "With what are we left? Without doubt, the most intriguing 'Who done it?' in the history of Western world — a world, incidentally, that seems to be patently indifferent to this remarkable mystery. Not that this cavalier attitude is surprising, considering the fact that virtually no one today is aware of the radical conclusions cited above."[77]

It is the duty of people to accept the harsh judgments of Carter's essay and yet he tells Christians they then must "to decide how to save what is worthwhile in their religion."[78] But we ask, if it is true that there is no God, that He has given no revelations of his will and plan of salvation, that MIRACLES AND PROPHECY AND THE SUPER NATURAL ARE IMPOSSIBLE, IF Christ never lived, that the New Testament does not come from the first century A.D., if the Christian religion is built upon lies and is a fraud, what precisely is there to save? Carter's essay might be termed a defense of Karl Marx's famous dictum: "Religion is the opiate of the people."

Footnotes

1. Nicholas Carter, "The Counterfeit Gospel." (*Liberty Bell*, 17:42-53, February 1990).
2. **Ibid.**, p. 3.
3. William Morris, Editor, *The American Heritage Dictionary of the English Language* (Boston and New York; Houghton Mifflin Company, 1970), p. 303b.
4. G. Abbott-Smith, *A Manual Greek Lexicon of the New Testament* (New York: Charles Scribner's Sons, 1929), p. 184.
5. Carter, **op. cit.**, p. 42.
6. **Ibid.**
7. **Ibid.**
8. Samuel M. Zwemer, *The Origin of Religion* (Nashville: Cokesbury Press, 1935), p. 33.
9. Cf. "Religionsgeschichtliche Schule," in Richard N. Soulen, *Handbook of Biblical Criticism Atlanta*: John Knox Press, 1931), pp. 166-168.
10. Cf. W. Schmidt, *The Origin and Growth of Religious Facts and Theories*. Translated by H.J. Rose (New York, 1931); W. Schmidt L'Origine de l'idee de Dieu. Articles in Anthropos, Vienna, 1908-1910.
11. Cf. Raymond F. Surburg, "The Influence of Darwinism," in Paul A. Zimmermann, *Darwin, Evolution and Creation* (St. Louis: Concordia Publishing House, 1959), pp. 201-202.
12. As cited by Zwemer, **op. cit.**, p. 42.
13. According to Zwemer, **Ibid.**, p. 33.
14. Stephen Langdon, *Semitic Mythology* (New York: Cooper Square Publishers, 1964).
15. As stated by Halley in his *Bible Handbook* (Grand Rapids: Zondervan Publishing House, 1975), p. 62.
16. Carter, **op. cit.**, p. 43.
17. **Ibid.**
18. **Ibid.**, p. 43.
19. Soulen, **op. cit.**, pp. 14-125.
20. Carter, **op. cit.**, p. 44.
20a. Elgin Moyer, *Who Was Who in Church History* (Chicago: Moody Press, 1962), p. 32.
21. Herman Gunkel, *The Sagas of Genesis* (New York: Schoken Books, 1901), p.
22. Rudolf Bultmann, "New Testament and Mythology," in Hans W. Bartsch, Kerygma and Myth. Revised translation by Reginald Fuller (New York: Harper and Brothers, 1961).
23. Rudolf Bultmann, Kerygma and Myth, **op. cit.**, pp. 33-39, also Rudolf Bultmann, *Jesus and the Word* (New York: Scribner's and Sons, 1938).

24. "Myth, Mythology," in Soulen, **op. cit.**, p. 125.

25. Carter, **op. cit.**, p. 44.

26. **Ibid.**

27. Orville Waters, "Psychoanalysis," in Carl F. Henry, Baker's *Dictionary of Ethics* (Grand Rapids: Baker Book House, 1973), p. 555.

28. Carter, **op. cit.**, p. 44.

29. 1 Corinthians 2:14.

30. Carter, **op. cit.**, p. 44.

31. *The Americana Encyclopedia*, 1988 edition, cf. the articles on "Jesus Christ."

32. F.F. Bruce, *The New Testament Documents: Are They Reliable* (Downer's Grove, IL: InterVarsity Press, 1972), p. 119.

33. Otto Betz, *What Do We Know About Jesus* (London: SCM Press 1968), p. 9.

34. Dr. John Warwick Montgomery, *History and Christianity* (Downers Grove, IL: Inter-Varsity Press, 1964), p. 40.

35. Carter, **op. cit.**, pp. 46-47.

36. **Ibid.**, p. 44.

37. **Ibid.**, p. 45.

38. Flavius Josephus, *Jewish Antiquities*, xviii, 33 (Early second century)

39. Carter, **op. cit.**, p. 45.

40. Flavius Josephus, *Jewish Antiquities*, xx.

41. Carter, **op. cit.**, pp. 45-46.

42. Ornelius Tacitus, Annals, xv, 44.

43. Lucian, *The Passing Peregruis*, as cited by Josh McDowell, *Evidences That Demand An Answer* (San Bernardino" Crusade for Christ, 1972), p. 84.

44. Carter, **op. cit.**, p. 46.

45. **Ibid.**, pp. 47-43.

48. Everett Harrison, *Introduction to the New Testament* (Grand Rapids: Wm. B. Eerdmans Publishing Company, 1964), pp. 91-114. Dolad Guthrie, New Testament Introduction (Downers Grove, IL: InterVarsity Press, 1974), pp. 1-984.

48a. H. Wayne House, *Chronological and Background Charts of New Testament* (Grand Rapids: Zander Van, 1931), p. 22.

49. Carter, **op. cit.**, p. 46.

50. Acts chapter 2:41; 2:47;6:1,7.

51. Carter, **op. cit.**, p. 50.

52. William Beck, **The Holy Bible, An American Translation**. (New Haven: Leader Publishing Company, 1976), New Testament section, p. 296.

53. Carter, **op. cit.**, p. 47.

54. Gresham Machen, *The Origins of Paul's Religion* (New York: The Macmillan Company, 1928), pp. 211-292.

55. Carter, **op. cit.**, p. 43.

56. "Jesus Christ," in *The Encyclopedia Britannica*, Macropedia section, 10; 12 145a-155b.

57. The following are the lists as given in the Gospels: Matthew 10:2-1; Mark 3:13b-19a; Luke 6:13-16; In Acts 1:13. Cf. Adam Fahling, *The Life of Christ* (St. Louis: Concordia Publishing House, 1936, pp. 250-261.

53. House, **op. cit.**, pp. 133-134.

59. Carter, **op. cit.**, p. 50.

60. **Ibid.**, pp. 50-51.

61. **Ibid.**, p. 61.

62. F.F. Bruce, *The Acts of the Apostles, The Greek Text* (Grand Rapids: Wm. B. Eerdmans Publishing Company, 1951), pp. 15-18.

63. "Ramsay vs. The Tübingen School," in Edwin Yamauchi, *The Stones and the Scriptures* (Philadelphia and New York: J.B. Lippencott Company, 1972), pp. 92-97.

64. Adolf Harnack, *The Arts of the Apostles*, pp. 298ff. as quoted by A.T. Robertson, "Acts of the Apostles," in James Orr, (General Editor, The International Standard Bible *Encyclopedia* (Grand Rapids: Wm. B. Eerdmans Publishing Company, 1939), I, p. 45b.

65. Geo. W. Clark, *Harmony of the Acts of the Apostles and Chronological Arrangement of the Epistles and Revelation and Explanatory Notes and Valuable Tables*, Designed for Popular Use (Philadelphia: American Baptist Publication Society, 1897), 408 pp.; Frank J. Goodwin, *A Harmony of the Life of St. Paul According to The Acts of the Apostles and the Pauline Epistles* (New York; American Tract Society, 1913), 240pp.

66. Cf. description of "Apostolic Fathers in Jerald C. Brauer, *The Westminster Dictionary of Church History* (Philadelphia: The Westminster Press, 1971), pp. 48-49.

67. **Ibid.**, 580.

68. Bruce Metzger, *The Text of the New Testament, Its Transmission. Corruption, and Restoration* (New York: oxford University Press, 1964), pp. 38-49.

69. Cf. chapter 4, "The Reliability of the Bible," in McDowell, **op. cit.**, p. 49.

70. **Ibid.**, p. 48.

71. **Ibid.**, pp 42-79.

72. Kenneth Scott Latourette, *A History of Christianity* (New York: Harper & Brothers, 1953), 1516 pp.

73. Lars P. Qualben, *A History of the Christian Church* (New York: Thomas Nelson and Sons, 1940), 644 pp.

74. Both volumes at first separately published in 1962 and 1932 are now combined in William Arndt, *Bible Difficulties and Seeming Contradictions,* Edited by Robert G. Hoerber and Walter R. Roehrs, (St. Louis: Concordia Publishing House, 1937), 263 pp.

75. Carter, **op. cit.**, p. 44.

76. Leander S. Keyser, *A System of Christian Evidence* (Burlington, Iowa: A System of Christian Evidence, Fifth revised edition; Lutheran Literary Board, 1939), pp. 192-198.

77. Carter, **op. cit.**, p. 47.

78. **Ibid.**

Questions

1. The intent of *Liberty Bell* is to ____.
2. Nicholas Carter claims that Christianity is ____.
3. How did the Christian church originate? ____
4. *Liberty Bell* and Carter are committed to the theory of ____.
5. Carter said that ALL Christian traditions are ____.
6. Carter has resorted to arguments advanced by ____.
7. According to the Bible there was in the beginning ____.
8. Polytheism followed ____.

9. Carter derived his views from____.
10. What did Rudolf Bultmann claim? ____
11. Many psychologists and psychiatrists reject ____.
12. Carter portrays mythology as ____.
13. According to Dr. John Warwick Montgomery, what does the historian know about Jesus? ____
14. Does the Jewish historian Josephus refer to Christ? ____
15. Who was Tacitus? ____
16. The most important source of the life of Jesus are ____.
17. Congregations established by Paul were comprised both of ___ and ____.
18. Peter wrote that "We did not follow any ____."
19. What did J. Gresham Machen rebut in *The Origin of Paul's Religion?* ____
20. What does Carter say about the twelve Apostles? ____
21. What did Harnack say about Acts? ____
22. The Bodmer Papyrus II has been dated as ____.
23. Readers who doubt the history of the New Testament canon should read ____.
24. What does William Arndt answer in his *Does The Bible Contradict Itself* and *Bible Difficulties?* ____
25. Carter's essay might be termed ____.

Carter's Counterfeit Gospel

March 24, 1990

Dear Brother Otten:

Regarding your request that I shorten my article: "The Response to Carter's Counterfeit Gospel," I would suggest to drop the footnotes, because Carter did not give any proof for his assertion.

Carter's article was 12 pages in *Liberty Bell*, all but 2 pages have over 400 words to a page. I would estimate that Carter's article had between 4900 and 5000 words. By omitting my introduction and my footnotes I would have about 4900 words.

On February 19, I was involved in a serious auto accident; a woman drove right into the flow of traffic and due to the woman's carelessness, I was nearly killed and my car was demolished, necessitating the purchase of a new car. I sustained injuries to my legs, knees, and ribs. My arthritis was also worsened by the traumatic shock.

You could send what you published minus the introduction and footnotes. Right now I am not in a position to retype my response. I teach 7 hours a week (two short of a full teaching load for the spring quarter) at the seminary.

May you and your congregation have a blessed Lenten and joyous Easter season.

Yours in Christ,

Raymond Surburg

Carter Answers *CN* —
Says Christianity Is a Delusion
Man Created Christ for Himself

Christian News, June 3, 1991

"God did not create the Christ for Man — Man created the Christ for himself," writes Nicholas Carter in a response to a challenge from *Christian News.*

CN has reprinted several articles by Carter in *Liberty Bell*, which attack Christianity and the Bible. *CN* has published responses by Dr. Raymond Surburg and Charles Provan and suggested that *Liberty Bell*, a revisionist publication, also publish the responses since *Liberty Bell* claims to be willing to publish both sides of issues. The March 11 *CN* said in an editorial titled "Who Are The Bigots — CHRISTIANITY BASED ON EVIDENCE:"

"The Christian is not a bigot. He bases his opinions upon solid evidence and not majority opinion.

"*Christian News* has at times quoted from various publications often referred to as 'revisionist.'

These publications are correct when they insist on basing opinion upon solid evidence rather than majority opinion. However, when it comes to matters pertaining to the Bible and historic Christianity some of the revisionists are as bigoted as the 'exterminationist' scholars they rightly accuse of bigotry for not studying the evidence for their exterminationist theory that the Germans exterminated six million Jews during WWII. The revisionists correctly observe that the exterminationists in their publications refuse to give the revisionists an opportunity to present their case.

Liberty Bell (Box 21, Reedy WV 25270) is a revisionist publication which has published articles showing that the Germans did not execute six million Jews during WWII. *Liberty Bell* says that it 'strives to give free reign to ideas.' Unfortunately, scholars who defend historic Christianity and refute the attacks upon Christ published in *Liberty Bell*, have not yet been able to get their articles and refutations of *Liberty Bell's* attacks upon historic Christianity published in *Liberty Bell.*"

CN wrote to Carter on March 3: "Your articles indicate that you are one of the best informed of the critics of historic Christianity. We reprinted your material to show our readers what the critics of Christianity are saying. Many of our college students must face the kind of arguments you raise against Christianity and we want to show them that there are valid arguments which show you are mistaken. You claim that the Bible contains many errors. In his response, Dr. Surburg noted: 'The contradictions enumerated have been answered by William Arndt in his *Does The Bible Contradict Itself and in Bible Difficulties.*' A few years ago these books were reprinted by Concordia Publishing House, 3558 South Jefferson Ave., St. Louis, Missouri 63118. We have often recommended these books to answer the critics who claim the Bible contains errors. The March 4 *Christian News* reprinted Dr. Robert Dick Wilson's *Is The*

Higher Criticism Scholarly? Dr. Arndt, best known for the Arndt-Gingrich Lexicon, was a leading Greek scholar. Wilson was one of the greatest Old Testament scholars of this century. He knew some 26 languages, If you can show the errors in Arndt's *Does The Bible Contradict Itself* and *Bible Difficulties* we will publish them together with a response.

"Robert Dick Wilson in his *Is The Higher Criticism Scholarly?* includes several questions for skeptics, who reject the inerrancy of the Bible and the Mosaic authorship of the Pentateuch, to answer. Some of Wilson's questions are included in our 'Open Letter to the Great Scholars at Concordia Seminary, St. Louis' which originally appeared in the June 25, 1973 *CN* and is reprinted on p. 10 of the March 4, 1991 *CN.*

"Why not have *Liberty Bell* reprint Wilson's '*Is The Higher Criticism Scholarly?*' from pages 11-14 of the March 4 *CN* together with your answer to Wilson's questions. *CN* will publish your answer to Wilson's questions. We can give you up to 10,000 words to respond to Arndt and to answer Wilson's questions.

"Some *Liberty Bell* readers are disappointed with *Liberty Bell* for not publishing a response from a well-informed Christian to your attacks upon Jesus Christ and historic Christianity."

"We have not hesitated to let our readers know where such literature attacking Christianity is available. Unfortunately, it appears as if *Liberty Bell* and other revisionists publications, which have attacked Christianity, have not told their readers where they might get books written by scholars, such as Arndt and Wilson, who defend Christianity. You fellows remind us of the liberal professors we had in various graduate schools and during our seminary days who told us all about the works of liberal scholars and skeptics but little, if anything, about the writings of Bible believing scholars who refuted the works of such liberals as Loisy, whom you mention in your letter.

"It is our contention that Christianity is the only religion based on real historic fact, it has withstood the sharpest attacks of the critics for the last 2,000 years and it will stand until the end of time. Our prayer, Mr. Carter, is that you will carefully study the works of such scholars as William Arndt, Robert Dick Wilson, J. Gresham Machen, Francis Pieper, C.F.W. Walther (particularly his Law and Gospel) and some of the many other scholars whose writings appear in our *Christian News Encyclopedia.* If you would like a free copy of this encyclopedia to review in *Liberty Bell* we will be glad to send you the four volumes. It has many articles on the Bible and its critics. Some of the articles are listed on page 23 of the March 4, 1991 *CN.*

"Christianity, contrary to some revisionists, has nothing to hide. We believe in putting 'all the facts on the table.'

"May God the Holy Ghost lead you to recognize that the Bible is indeed God's directly revealed Word and that Jesus Christ is your Savior from sin and only hope for eternal salvation.

"Sincerely yours,
Herman Otten"

"P.S. The enclosed March 5,1990 *CN*, which has an editorial on 'Revisionists and the Dead Sea Scrolls,' has as its lead story 'Jericho Wall Did Crumble — ARCHAEOLOGY AGAIN SUPPORTS BIBLE.' It features the work of Archaeologist Bryant T. Woods. Nelson Glueck in his Rivers in The Desert says that nothing archaeologists have ever unearthed has disproven anything in the Bible. Glueck was a leading Jewish scholar and some years ago was featured on the cover of Time. If you can refute what Woods has written in the Biblical Archaeologist or if you know of some facts unearthed by archaeologists, which refute anything in the Bible, perhaps you can include this in any possible response article you might send *Christian News*."

Carter wrote to *CN*:
Nicholas Carter
12319 Manor Drive No. 208
Hawthorne, CA 90250

Dear Pastor Otten,
I am amazed to see that you are still whining about the fact that *Liberty Bell* did not publish a response to my writing by Dr. Raymond Surburg.

I have a copy of the Surburg article, and I have never seen a sloppier or more unprofessional piece of writing in my life. The editor of *Liberty Bell* wrote Mr. Surburg and asked him to not only shorten it, but to recast it in a professional manner. To my knowledge, Surburg did not respond.

You are not telling your readers WHY the article wasn't published. Instead, you are implying that Dietz is not honest enough to publish a rebuttal to an article in the Bell. I was hoping that you were an honest man. I am now beginning to doubt that.

You published a letter of mine to *CN* in the March 11th issue. I had to learn about it from someone who wrote to me. You didn't even send me a copy.

I am attaching a response to your "challenge." Since you seem to place such a high value on "honesty and courage," will you be honest enough and courageous enough to publish it?
Very truly yours,
Nicholas Carter
Nicholas Carter
12319 Manor Dr. No. 208
Hawthorne, CA 90250
April 8, 1991
213-679-8758

In continuation of Pastor Otten's usual policy, we would like to say to Nicholas Carter that we welcome his response to our attack on his propositions, and hope that he has enough honesty and courage to either prove us wrong or admit his error. We look forward to one or the other for future issues of *CN*.

(Ed. This is a quotation from an article by Charles Provan answering Carter's attack on Christianity.)

Dear Pastor Otten,

Prove that you are wrong? Pander to your irrationality? Nonsense! The burden of proof always falls upon those who claim that the moon is made out of bleu cheese, or that there are little green men on Mars.

With regard to your metaphysical culpability, however, the very nature of the universe proves that you are wrong, it is metaphysically impossible for the universe to be both natural and supernatural . . . impossible for universal laws to be both immutable and arbitrary . . . impossible for burning fire to be hot on one day and cold on another . . . impossible for there to be virgins births, or for people to be able to walk on water, and for truly dead creatures to return to life.

My universe includes this earth physical reality, the human senses, perceptions, reason, and science. Caught up in causeless, arbitrary, contradictory, and inexplicable beliefs — that are to be believed because they are unbelievable — you (the witch doctors for Christ) spend your lives denying reality in order to make room for faith — a faith that destroys the ability to separate truth from error.

The dedication of your consciousness to the non-rational, is not blindness, but a refusal to see because of indoctrination; not ignorance, but a refusal to know the truth, because of the fear of reality, it isn't going quietly into that good night that's tough. It's going without easy answers. And this simple fact gives every stone-age witch doctor and New Age Guru the hook he needs to ensnare his share of the human species.

What sets your curiously coiled creed apart from the real world is the fact that the entire metaphysical system is poised upon the single stem of miraculism. There isn't one orthodox tenet of Christendom — from the magical sacrament of baptism to the miraculous conception — that can be scientifically demonstrated to have been a part of the real world. There isn't a miracle, or a prophecy, or any other wondrous event that can be shown by a process of reason based on perceptual observation to have occurred in the real world. There isn't a single non-natural Christian event that can be measured by natural standards.

Why the miraculous, the wondrous, the stupendous? Who on earth would believe in a half-man, half-God creature — a freak of SUPER nature — if he hadn't supposedly fulfilled in detail, predictions supposedly made by prophets hundreds of years before? . . . or if a virgin birth wasn't a part of the story? . . . or if there was no alleged resurrection from the grave?

Trapped in a world of your own making — a non-natural, non-rational non-scientific metaphysical realm — you desperately try to justify supernaturalism in a natural world. You try to fit supernatural results into a natural framework of logic. You try to derive from the facts of reality, and validate by the process of reason, the non-real, and the non-reasonable. And that, in the proverbial nutshell, explains your paranoia — your obsession with my little articles which virtually no one will ever read.

Oh, you are as happy as clams in wet sand for just as long as you are idolized by the "simple faithful" who blindly accept every tag, label, myth, and illusion that you proffer; but, let just one human being, who is afraid of neither reality nor mysticism, declare that the fundamental facts of the Christian religion are no more than delusions, fancies, and hallucinations, and it hits you where it really hurts — right in the old sub-conscious, where that little voice is always whispering, "I don't want to hear that!"

No one knows better than you that you don't have a rational leg to stand upon; and you are not alone. Metaphysical inferiority attacks all slaves of mythology and superstition in the face of rationalism, it is for that very reason that, in terms of human suffering, the evil that mystics do, can never be measured. Curiously, the poets and philosophers who rhapsodize about how the death of one will diminish all never seem to be existentially aware of the fact that ignorance rooted in the fear of reality diminishes us all. To be anti-reality is to be anti-mind and anti-life.

The human being is not an unnatural phenomenon. He is a contractual being. He has to plan his life long-range; he must make his own choices; he must deal with others by voluntary agreement; he must be guided by his own independent judgement; he must unify his consciousness and clarify his grasp of reality. And yet, if he wishes to be "religious" in the orthodox sense, he must allow himself to be conditioned to believe that his salvation depends upon an acceptance of unnatural phenomena. He must recognize supernatural and wondrous events as facts of reality, and then attempt to integrate them, without contradiction, with the factual events of the real world. He must not look for causes in any metaphysical sense; and he must not question the orthodoxy.

The fact that there are many people who are religious in the orthodox Christological sense means nothing. At one time, the entire family of earthily man believed the world to be flat But that didn't change the truth of the matter. In order to be one of the "simple faithful," a person must submit to a creed. He must sacrifice his reasoning to faith, and subordinate it to the demands of the witch doctor. Faith in the superiority of the witch doctor leads to faith in the super-natural. It is that simple, and — that tragic.

A final thought: God did not create the Christ for Man — Man created the Christ for himself.

Very truly yours,
Nicholas Carter

Questions
1. The Christian basis his opinions on ____.
2. Some revisionists are as bigoted as ____.
3. *Liberty Bell* published articles showing that the Germans did not ____.
4. Christianity is the only religion based on ____.
5. Nelson Glueck in his Rivers in the Desert says ____.
6. Nicholas Carter said that God did not create ____ for Man – Man created ___ for himself.

A Response to Ben Klassen's
"The Greatest" Hoax of All Time"
Raymond Surburg

Christian News, September 24, 1990

"The Greatest Hoax of All Time" by Ben Klassen is one of a number of pamphlets published by Sunshine Publications, Ooltewah, Tennessee. All these pamphlets are anti-Christian and anti-religious. Klassen's pamphlet appeared in the book, Nature's Eternal Religion.[1] The arguments against Christianity are similar to those found in the article in Liberty Bell's "The Counterfeit Gospel," authored by Nicholas Carter.[2] Ben Klassen's views on Essenism are in a few points similar to those of Larsen's "What Ever Happened to the Dead Sea Scrolls?"[3] The articles by Klassen, Carter and Larsen have a number of concepts in common. They are opposed to Christianity and its supernatural character. These three articles are featured by misinformation, misinterpretation and by vitriolic attacks on Christ and His teachings. Christ and the New Testament are called "hoax," "myth," "fabrication," "deception," "lie," "fraud" and "unhistorical."

The Klassen article differs in one respect from the others in that it speaks about a Jewish conspiracy to overthrow the Roman Empire.[4] Thus Klassen wrote: "Never again did the white race shake off the control of the Jews. Never again did the white race regain control of his own thinking, of his own religion, his own finances nor his own government. Unto this day the white race has not remained in control of its own destiny."[5] This statement appears to be strongly anti-Semitic.

Klassen bemoans the fact of the breaking down of Roman civilization, whose contributions to Western civilization have been significant. A system of religious views concocted by Jewish New Testament writers, fabricated during the first and second centuries and adopted at Nicaea, where more fabrication took place, was employed by Jews to topple the great Roman Empire.

The reader will find it difficult to find views about European civilization that are more absurd than the fantasizing of Klassen about the influence exercised by Jews on the course of European history.

The Greatest Hoax According to Klassen

The great hoax that has been perpetrated upon the world was the fabrication of the existence of Jesus Christ, who supposedly was depicted as divine. Klassen claims that there is not one shred of evidence for the historical existence of Jesus Christ. There is not one scintilla of proof for the frequent statements found in the Gospels that Jesus ever was active in Palestine and that He pursued a ministry between A.D. 30 and 33. Not one single portrait of Jesus has been found or a painting of Him, something one would expect had he lived. Thus the world has statues of Cicero, Caesar Augustus, Marcus Aurelius.[6] To this objection this writer would ask: Does the world have statues or paintings of all important men who have ever lived? The Christian movement from the very first was a persecuted movement and so unlikely for people to make portraits or paintings. Further, Christ came to establish a spiritual kingdom and not an earthly one.[7] Christianity only became powerful in the fourth and fifth centuries. Only when Constantine declared Christianity to be a **religio licita** (i.e. a religion recognized by the state) that there was the tendency to come out into the open. Klassen is resorting to the argument from silence, which neither proves or disapproves the historicity or non-historicity of a person or historical occurrence.

Did Herod Kill the Infants of Bethlehem?

Klassen also cites the impossibility of Herod the Great's murder of the babes of Bethlehem, claiming that such an event would have been recorded in some writing of that era.[8] Again, one should note the use of the argument from silence. Josephus, however, has shown that Herod the Great was an ambitious king who even killed those of his own family.[9] The murder of the children of Bethlehem is true to character of Herod revealed by history. While thousands of scholars in the past two thousand years have recognized the historical character of the Gospels, Klassen rejects the reliability of the entire New Testament.

He further claims that the fact that the Christians have placed the birth of Christ in A.D. 1 shows the New Testament to be a fraud, inasmuch as the death of Herod the Great occurred in 4 B.C.[10] In response to this fallacious argument, it should be noted that the New Testament itself gives no such date and that all reputable conservative scholars believe that Christ was born in either 6 or 5 B.C.[11]

Is Christ Mentioned in Roman Secular Sources?

It is the erroneous claim of Klassen that there is not one single mention of Christ in secular literature, which constitutes proof positive that Jesus was a fabrication by those who created the New Testament.[12] This allegation, however, is not true.[13] Klassen is either ignorant of the fact that there are references in heathen literature, indications of the existence of Jesus Christ. Thus Josephus, the Jewish historian who first served as general in Galilee in the Great Jewish War of A.D. 66-70, later went over to the Romans and consequently was looked upon by his Jewish countrymen as a traitor, wrote in his Jewish Antiquities as follows:

Now there was at this time Jesus, a wise man, if it be lawful to call him a man, for he was a doer of wonderful works, a teacher of such men as receive the truth with pleasure. He drew over to him many of the Jews, and many of the Gentiles. He was the Christ, and when Pilate, at the suggestion of the principal men among us, had condemned him to the cross, those that loved him at first did not forsake him; for he appeared unto them alive again the third day, as the divine prophets had foretold these . . . concerning him. And the tribe of Christians, so named from him are not extinct to his day.[14]

Some scholars have rejected this passage as an interpolation, while others have declared it to be authentic.

About a second passage of Josephus there need be no doubt. This second reference is a statement about James, the brother of Jesus. It reads as follows:

> But the younger Ananus who, as we said, received the high priesthood, was of a bold disposition and exceptionally daring, he followed the party of the Sadducees, who were severe in judgment above all the Jews, as we have already shown. As therefore Annas was of such a disposition, he thought he had now a good opportunity, as Festus was now dead, and Albinus was still on the throne so he assembled a council of judges and brought before it the brother of Jesus the so-called Christ. . . whose name was James, together with some others, and having accused them as law-breakers, he delivered him to be stoned.[15]

The Name of Christ in Latin Writers

Klassen claims that the name of Christ is found nowhere in Roman authors. That is not true for Tacitus is a reliable writer, who in his **Annals, XV, 44** refers to the historical fact that during the reign of Tiberius that a man called Chrestus (Christ) was crucified and to the fact of the existence of Christians in Rome. Tacitus (A.D. 52-59) was a Roman historian who in A.D. 112 was Governor of Britain A.D. 80-84 and so a reputable writer of events of which he had firsthand knowledge. Suetonius (A.D. 120), a Roman historian under Hadrian, wrote: "As the Jews were making constant disturbances at the instigation of Chrestus (another spelling of Christ), he expelled them from Rome,"[16] and in another of his works says: "Punishment by Nero was inflicted on the Christians, a class of men given to a new and mischievous superstition."[17]

There is also the testimony of Lucian, a satirist of the early second century who spoke scornfully of Christ and the Christians. He connected the Christians with the Jewi.sh synagogue in Palestine and among other things said about this Jesus that he was a man "was crucified in Palestine. Because he introduced a new cult in to the world furthermore, their first lawgiver persuaded them that they were brothers one of another after they have transgressed once for all by denying the Greek gods and by worshipping that crucified sophist himself and living under the law."[18]

It is thus clear that there were some Roman writers who knew about Christ and his followers. However, the most important source for the life of Christ and of early Christianity is to be found in the Four Gospels and in the rest of the New Testament. Klassen claims that there are no eyewitnesses for the first half of the first century A.D., specifically the years A.D. 30-33, when Christ exercised His ministry.[19] But the fact is that two of the Four Gospels authors were eyewitnesses, because they were a part of the Twelve Apostles who were with Christ during most of His earthly ministry in Palestine. Matthew, once a tax collector, wrote his Gospel from having been with Christ after his call to the apostleship. John, the fourth Evangelist, was also an eyewitness of most of the data recorded by him in his Gospel, which contains 93 per cent of new materials when compared with the Synoptic Gospels. There are second century writers that inform their readership that Mark, the author of the second Gospel, obtained his material from Peter, also one of the eyewitnesses of the activities and sayings of Jesus.[20] The third Evangelist, Luke, tells his readers precisely how he came about to compose his life of Christ. Thus Luke wrote this introduction to his Gospel: "Seeing that many have taken it in hand to draw up an account of those matters which have been fully established among us, just as they reported them to us, who were from the beginning eye-witnesses and ministers of the word, it seemed to me also, after investigating the course of all things accurately, from the beginning, to write to you in order that you make known, most excellent Theophilus, the certainty of the story you have been taught by word of mouth."[21] The Four Gospels are reliable accounts of the historical events recorded regarding Christ's life together with the record of His teachings. The truthfulness of the 89 chapters that constitute the first four books of the New Testament is not affected by the fact that the first Three Gospels were composed between A.D. 50-70. The Fourth Gospel was probably written in the 90's of the first century.

Competent New Testament scholars have shown that the Gospel of John must have been written in the first Christian century for a portion of the Gospel of John was found in Egypt, dated by textual critics as originating around A.D. 125, which in turn must be based on an earlier copy.[22] The Bodmer Papyrus II has been dated as coming from A.D. 150-200 and it contains nearly all of the Gospel of John. For many Greek and Roman authors classical scholarship only possesses copies that are a thousand years older than copies of the Greek New Testament. Yet classical scholars do not question the historicity of these Greek and Roman authors, even though their transmission is late. The genuineness of the classical writers is not doubted, even though the surviving copies are late. Why should the books of the New Testament be rejected as authentic, as first-century documents?[23] Because the New Testament writings, and especially the Gospels, contain miraculous elements and a record of the miracles of Christ does not **eo ipso** mean the historical narratives of the Gospels are fabrications.

The Koran With the Miraculous Not Rejected

The Koran of the Mohammedans contains many references in various surahs of the Koran that claim that the messages Mohammed received from Allah were written in heaven. Scholars who have written on the subject of Mohammedanism have not rejected the Koran as a book delivered by Mohammed to his followers. Why should one not accord the Gospels the same consideration? Incidentally, Mohammed recognized Jesus as a person who lived, as held by all Christians at that time.

F.F. Bruce, *The New Testament Documents: Are They Reliable?*, shows that there were links between writers of the first century Biblical books and the second century: there thus being a continuity with the original autographs and the following centuries. Thus Bruce declared: "The importance of evidence lies in the link with the apostolic age and its ecumenical associations, Ireneus was brought up in Asia Minor at the feet of Polycarp, the disciple of John and Ireneus, became bishop of Lyons A.D. 180. "His work refers to the canonical recognition of the Fourfold Gospel, to Acts, Romans, I and II Corinthians, Galatians, Ephesians, Philippians, Colossians, I and II Thessalonians, I, II, III John, I and II Timothy, Titus, I Peter and Revelation."[24]

Clement of Rome (A.D. 95-97) has references to Matthew, John, Romans, Ephesians, II Thessalonians, Titus, Hebrews, James and II Peter. There are books which appeared before Clement of Rome and consequently are from the first Christian century.[25] This explodes the theory of Klassen that there were no New Testament writings before the latter half of the second century after Christ's birth.

The Alleged Contradictions of the Gospels

The fact that the Gospels contain numerous contradictions according to Klassen is positive proof for him of the fabrications of the materials found in the first four New Testament books.[26] Klassen claims that there are so many errors that he is not interested in citing any cases. Klassen is completely ignorant of the fact that the New Testament, which includes the Four Gospels, has been the subject of intense study since the earliest days of its usage as the Bible for all branches of Christendom. The teachings of the New Testament have been under attack throughout the centuries. The Four Gospels, especially, have been attacked by the enemies of the Christian faith in the early Christian centuries, for that reason there exists a group of writers, called by church historians as the Christian Apologists, who defended Christianity against attacks from both Judaism and Roman paganism.[27] Then beginning with the Age of nationalism readers of church history will find that Christianity was subject to attacks by rationalists and those opposed to the supernatural character of the Christian faith as expressed in the writings of the New Testament.[28]

The Gospels Thoroughly Scrutinized

No other books of the New Testament have been so thoroughly examined as the Four Gospels have been. Its alleged contradictions and difficulties have been minutely examined by any number of scholars who have written on the life of Christ. In most cases there is no problem in answering and solving them. At this juncture the writer would mention books like the following: Fahling, *Life of Christ*,[29] Ylvisaker, *The Four Gospels*,[30] Arndt, *Does the Bible Contradict Itself?*[31] Arndt, *Bible Difficulties*,[32] Shepard, *The Christ of the Gospels*,[33] Roney, *Commentary in the Harmony of the Gospels*,[34] and Haley, *Bible Difficulties*.[35]

The Problem with agnostic or atheistic humanists is that they are so hostile toward Christ and Christianity that they have no knowledge of this vast apologetic literature, which adequately answers their objections. To do justice to any writing, a reader must approach it with sympathy and with the purpose of fairly assessing a piece of writing. There are certain moral and spiritual qualifications that an interpreter needs to possess, especially, when dealing with the Bible, God's Word. When the Gospels are labelled a "fraud," "a hoax," there should be adequate data to substantiate such charges. Klassen simply resorts to pontification and makes derogatory statements without bringing adequate proof.

The evidence he does adduce is completely wrong. Thus Klassen claims that the Sermon on the Mount, which Jesus allegedly never spoke, was taken over literally from the writings of the Essenes.[36] This is an assumption for which he does not cite real proof. That the New Testament and the Qumran writings had certain ideas in common is due to the fact that both had their origins in the Old Testament. Any individual who has really studied the Qumran writings and read all the statements of Christ will readily notice a world of difference between the two. A true analysis of the Sermon of the Mount will reveal that it is unique and that Christ has gone beyond the Old Testament understanding of the law. Klassen seems to believe that the Sermon on the Mount is the sum and substance of Christ's teaching. However, the truth is that the Sermon on the Mount, is only one of five different sermons spoken by Jesus around which Matthew has organized his Gospel.

Misrepresentations of the Dead Sea Scrolls

A good portion of Klassen's diatribe against Christianity deals with the Dead Sea Scrolls. There are a few views that are similar to those expressed by Larson in his article in *The Historical Review* entitled: "What Ever Happened to the Dead Sea Scrolls?"[37] The Essenes were a religious Jewish sect that lived in the area of the Dead Sea at a site called Qumran, near Ain Feksha. Many scholars have identified this religious association with the Essenes. Pliny, Philo and Josephus have written about them. Between 1947 and 1956 a number of Biblical scrolls as well as other documents setting forth the organization, religious practices and theological views of this sect, probably the Essenes, were found. Although not all scholars are willing to make this identification, but simply call them the Qumran sect, or the Qumranites. The Essenes are described as not marrying and so would be a celibate male community. However, in the territory associated with the sect, a cemetery was found having the bodies of children, which therefore raises some problems with the ascetic theory.

Klassen has a few of the arguments advanced by Larsen about Christ's dependence on Essenism. The uniqueness of the New Testament is ascribed to borrowing from the Essenes. Baptism and the Lord's Supper are supposedly taken from the Essenes.[38]

In distinction to other opponents of the Christian faith alleging dependence on Essene teachings, Klassen has come up with some very strange views about the use of the Essenes as being employed by Judaism. Thus he wrote: "However, we want to go further into the teachings which were pounced upon by the Jews to be formulated into a well distilled poisonous brew and then fed to the Gentiles."[39]

When Did the Essenes Cease to Exist?

Klassen claims that the Essenes of Qumran were terminated by the year A.D. 100. It was between A.D. 70-100 that the forgers of the New Testament used ideas from this sect. Furthermore, the Jews knew where the Essenes were to be found but refused to divulge their hiding places. The Jews wanted to prevent them from being destroyed so that they might use their teachings, taken over by Christians, to subvert the Romans and thus wreak revenge on their destroyers.[40]

However, the facts are that the Essenes or Qumranites deposited their sacred books on other writings dealing with the sect into various caves, when the Romans were marching toward Jerusalem to conquer and destroy the Holy City. It appears that the Qumranites never returned to their dwelling places and were conquered and possibly killed in connection with Jerusalem's destruction in A.D. 70. According to Klassen even though the orthodox Jews disagreed with the Essenes, who rejected orthodox Judaism, the Jerusalem leadership as reflected in the teachings as held by the Sanhedrin, supposedly used Essene teaching for purposes of revenge against the Romans. Klassen contends that the first practitioners of Christianity were the Essenes. Klassen's unique theory is that Essenic teaching, found in the New Testament, was employed for purposes of subversion. Thus Klassen has concluded that Christian beliefs as supposedly set forth in the Sermon of the Mount did not originate with Christ at the time that he lived, but actually appeared one hundred years earlier with the Essene sect, who lived near the Dead Sea. Klassen, further, holds, that the Elders of the Sanhedrin recognized the Essene teaching as being deadly and suicidal, that the Jewish leaders further took this doctrine and distilled and refined it into a working creed, which they then with great energy and tremendous amount of propaganda (in Which the (sic) excel), promoted and distributed this poisonous doctrine among the Romans."[41]

Klassen's Praise of Certain Essenic Teachings

Klassen praises the Essenes for some of their positions averring that they were the first ones to abolish slavery as an institution. Furthermore, inspired by the Teacher of Righteousness they promulgated pacifism This was a wonderful religion for the Jews to sell to the Romans, for if the latter could be convinced that they ought to embrace pacifism they would have to submit to others and so the Jews would get revenge for the Romans' sack of Jerusalem.[42]

The author of "The Greatest Hoax of the Ages" gives no evidence for this novel theory, which must be categorized as phantasy. There is no historical evidence to support these speculations. It comes to this writer as a complete surprise that the Essenes were Pacifists, that this is not the case will be recognized by any person acquainted with the Qumranic literature. One of the distinctive scrolls of the Essene sect found at Qumram is "The War Scroll," whose full title is: "The Scroll of the War of the Sons of Light against the Sons of Darkness." According to this writing there will be war between two groups of peoples, a war in which the Sons of Light will participate and in a war which they will help to defeat the Sons of Darkness.[43] The Essenes are thus portrayed as waring. Some pacifism!

Klassen follows the earlier studies made by scholars like Davies,[44] Edmund Wilson,[45] Dupont Sommer,[46] Allegro[47] and Charles Potter.[48] He claims that distinctive New Testament teachings were borrowered from the Essenes who existed already a hundred years before Christ's birth and time. Any person, however, who had a first-hand knowledge of the Qumran documents and also correctly understands the New Testament will readily recognize that there is a tremendous difference between the theology of the Essenes as compared with the theology of Christ and the Apostles as found in the books of the New Testament.[49] Klassen makes the sweeping statement that the Essenes "not only had evolved the same ideas set forth in Matthew, Mark, Luke and John, but the wording, phraseology and sentences were the same and they preceded the supposed time of the Sermon on the Mount by anywhere from 50-150 years."[50] Thus Klassen concludes that there is nothing original or "new" teachings in the instructions of a figure supposedly appearing from heaven in A.D. 1 and who preached between A.D. 30-33.

Klassen joins those scholars who aver that Christian baptism was borrowed from the Essenes who practiced daily washings or ablutions.[51] There is a great difference, however, between these daily washing rites and the institution and practice of baptism. Christian baptism is performed just once, and is not a daily practice.[52] Furthermore, a formula is used at the Christian baptism, namely, that the baptized person is baptized in the name of the Father, Son and Holy Ghost. Christian baptism was also administered to children and infants, a practice unknown to the Essene community. Yet despite these great differences, Klassen concludes: "So the Christian Apostles can hardly be credited with having instituted the ritual of baptism."[53] This is a direct denial of Christ's command in Matthew 28:18-20.

Klassen claims that people have been misled when they are taught that Christianity was the teaching of Christ and the inspired Apostles, when in actuality the Essenes were the first practitioners of Christianity. Klassen's shallow and wrong interpretations reveal the fact that he does not really understand the theology of the Essenes nor does he comprehend in the least the true nature of New Testament teaching. One wonders whether he has ever read the Dead Sea Scrolls and the New Testament in the original Greek.

There is a world of difference between the theological system of the Qumran sect and the requirements of Christianity. The teachings of the Essenes was a religion of work righteousness. It had a wrong conception of the Old Testament Messiah because the people of Qumran expected the coming of two different Messiahs, one from the house of Aaron and the other Messiah was to be from the house of David.[53a] Christianity knows of but one Messiah, whose coming was fulfilled in the birth and life of Jesus of Nazareth. The Qumran theology emphasized the keeping of the law and also did not offer up sacrifices as was required by the Mosaic law. It was also a very exclusivist society which was therefore not missionary minded as was orthodox Judaism. Certainly Christ commanded his disciples to make disciples of all nations (Matthew 28:20). Christ demanded of all people that they accept Him as Savior and Redeemer. There is no doctrine of the vicarious and substitutionary atonement in Essenism as is in the Gospels and the Pauline and Petrine Epistles. The doctrine of justification by faith, the central doctrine of Christianity, is totally absent in the Qumran writings. That the Essenes were the first practitioners of Christianity is a deduction completely founded on speculation and an idea created out of thin air. Klassen makes the claim that the Christian Lord's Supper was borrowed like baptism from the Essenes. Just as others of the same ilk, Klassen assumes that the Lord's Supper or the Eucharist was suggested by the Essenes' practice of having common meals. The Christian's Last Supper, however, is quite different because each participant in the Holy Meal receives the bread and wine individually and the words: "Given and shed for you for the remission of sins," "Take eat this is my body," "take drink this is the blood of the new covenant," are spoken. There is nothing like this in the Qumran.

Klassen's Erroneous View About Christianity

Klassen's entire article promotes misconceptions about Jesus Christ and the Christian faith. One strange argument advanced against the authenticity of the New Testament is his assertion that inasmuch as Christ and Paul spoke in Hebrew that the New Testament should have been written in their language and not in Greek as is the case with the New Testament in its original form.[54] Both Jesus and Paul spoke Aramaic and not Jewish. That Jesus and Paul did not know Greek and did not speak it cannot be proved. There is much evidence that Jewish people in the first century knew and used Greek as is evident from many Jewish inscriptions which are in Greek. The inscription upon the cross, setting forth the reason for Christ's crucifixion. Was in three different languages, namely Aramaic, Greek and Latin. People in Palestine in the first century were bilingual, if not polyglottal. The reason for the books of the New Testament being composed in Greek was that Greek was the universal language of communication in the Roman Empire. Latin was the language of the courts but Greek the language of culture and normal communications.

According to Klassen's wrong portrayal of the origins of Christianity, the New Testament was produced by the Jewish hierarchy and was a conspiracy coordinated by Jews and this Jewish conspiracy had many members and co-workers. Thus Klassen wrote: "It (namely, the New Testament) was not written at the time of Christ at all, but the movement was given great promotion by the combined efforts of the Jewish nation. As they organized and promoted ideas further, these were reduced to writing considerably later than the years 30-33 A.D. when Christ supposedly came out with these startling and 'new' revelations."[55] According to Klassen: "The conclusions are that they were written by Jewish persons whose identity we shall never know and were written collectively by many authors, were revised from time to time and not only in their original formation and formulation but have been revised time and time again throughout the centuries to become more effective and more persuasive propaganda."[56]

The New Testament informs its readers that Paul was the author of his 13 letters, Peter the author of two, James, Jude each of one and John the author of three epistles. The same John is the writer of The Apocalypse and according to late first-century and early second-century Christian authors, Matthew, Mark, Luke and John wrote the Four Gospels. The only book whose author is unknown in the whole New Testament is the writer of the Epistle to the Hebrews. The method described by Klassen as to the manner in which the New Testament came into existence is contrary to what the early church knew about its New Testament and modern conservative Biblical scholarship has a very good idea about the history of the writings of the twenty-seven books of the New Testament.[57]

All writers, with the exception of Luke, were Jews, but Jews who accepted Christ as their Savior and Lord. They were not just Jews, but believing followers of Christ. Klassen cannot furnish one piece of evidence that the various writings of the New Testament were constantly tampered with to meet a certain agenda. Klassen is expert in making assertions which have no basis in reality. Never does he furnish evidence for his radical statements.

Pages 1-5 and the 15 lines on page 6 are printed in typewriter print and this is followed on pages 6-8 in smaller printer's type and has the caption:

The 'Christianity Hoax" No Spooks in the Sky No Christ, No Historical Evidence

A good portion of this Addendum after his article "The Greatest Hoax of All Time," has three sections, each separated by the printing device x x x x.

In addendum 1 Klassen claims that there is no evidence for the basic claims of Christianity, in the three sections or portions of the Addendum some of the same arguments and theories appear as in his hoax article. The claim "for a life in the hereafter," the claim for the existence of a hell, the claim for the existence of a heaven, the claim for spooks in the sky and even the historical existence of Jesus himself have no basis in fact. "The Gospels were not written by Matthew, Mark, Luke and John but by individuals who wrote these garbled self-contradictory concoctions."[58] The materials in the Gospels are put together from myths.[59]

The guarantee for the authenticity of the Gospel was the Roman Catholic Church, a product of the first three hundred years of the Christian centuries. The Catholic Church backed its authority by Matthew 16:18: "Thou art Peter and upon this rock I will build my church and the gates of hell will not prevail against it," is criticized by Klassen as using the argument in a circle.

Superstition and Gullibility

The Christian Churches, so Klassen, contends, fosters superstitions

1482

and condemns those who insist religion should make sense, is logical. They do not allow people to think for themselves and develop their own religious ideas. Only when people are gullible do they accept the authoritative belief and dogmas of the churches. Thus Klassen claims "as we have pointed out earlier, the combination of gullibility and superstition have wreaked havoc on the White Race, and the Jewish mind manipulators have exploited those two human weaknesses to the utmost to our detriment and to their benefit."[60]

Part 2 of the Addendum

In Part 2 of the Addendum (page 7) answers the questions where did Christianity come from. Did a Jew, a fanatic found it as the New Testament describes?

In this part Klassen repeats many of the same misconceptions he has in his article: "The Greatest Hoax of All Time." The Bible was patched together out of fables, myths, bits and pieces of other religions, until finally there was a movement going that pulled down the Roman Empire brought about by Constantine.[61] In this part there is no mention of the Essenes as furnishing the ingredients used in the brew the Jews employed to overthrow the Roman Empire.

In Part 2 Klassen informs his readers that it was Constantine, a Roman Emperor, a murderer of his wife and killer of thousands, who in A.D. 313 put Christianity in business.[62]

In his article "The Greatest Hoax etc." Klassen claimed that it was the Ecumenical Council of A.D. 325 that determined the Creed of Orthodox Christianity and not the Essenes.[63] Klassen avers that Constantine made the Christian Religion the exclusive religion of the Roman Empire. Latourette, however, states that the Christian religion up to that time outlawed, was given official standing as a religion allowed to be practiced.[64]

Klassen has put forward the strange and totally erroneous view that it was the Council of Nicaea which created the New Testament and then combined it with the Old Testament, thus producing "the Bible." Thus he wrote: "The final package that emerged from the Council of Nicaea was called the New Testament, a contradictory, demented conglomeration of far-out nonsense."[65] Constantine forced this Bible upon people. To this we respond by asserting that there is no evidence for forcing people either militarily or legally to accept the new Bible.

Church historians and New Testament specialists that deal with the canon of the New Testament depict a completely different account as to the coming into existence of the New Testament.[66] The **Muratorian Canon is proof against the stance of Klassen that the New Testament was determined by the Council of Nicaea. The Canon Muratori** is a fragment (85 lines) of a Latin treatise on the Bible canon, giving a list of books of the New Testament accepted as canonical in Italy about the latter half of the second century, it mentions the Gospel of Luke (which it calls the third) John, the Acts, Paul's epistles to the Corinthians, Ephesians, Philippians, Colossians, Galatians, Thessalonians, Romans, Philemon, Titus, Timothy, Revelation, Jude, two Epistles of John, the Wisdom of Solomon, and doubtful the Revelation of Peter."[67]

The Nicene Council According Church History

Klassen has totally misrepresented the purpose of the Nicene Council of A.D. 325. The Creed of Nicaea was written for the settlement of the Arian controversy. The Creed of 325, which culminated into the Niceno-Constantinapolitan Creed of A.D. 381, grew out of the immediate necessity of guarding the apostolic teaching about the deity of Christ against the heresy of the inferiority of the Son. The Arians claimed that Christ was of similar essence with the Father but not of like essence.[68]

The two Greek words involved were homouos versus homoios, the difference was one letter, namely, the Greek letter yot. The Nicene Council did not create the New Testament which had been in existence since the end of the first century A.D.

Confusion and Illogical Thinking
Characterizes Klassen's Views

Klassen portrays a confusing picture as to the real cause for the downfall of the Roman Empire. In his article "The Hoax etc." and in the Addendum he gives three different contradictory answers as to the reason for Rome's fall. On page 3, column B of his essay it was Essenic teaching which was used to distill a poisonous brew that was fed the Romans and caused their demise.[69] In his Hoax article he claims "that the Elders of the Sanhedrin recognized this teaching as being deadly and suicidal, that they further took this doctrine and refined it into a working creed, the Jews then with a great deal of energy and tremendous amount of propaganda (in which they excelled) distributed this poisonous doctrine among the Romans."[70] Yet in his Addendum Constantine is depicted as the promoter of the Christian faith and the Roman Emperor who forced people to abandon the pagan Roman religion.[71] One must raise the question here: "Whom does Klassen claim to be the cause for Rome's downfall?"

In the final paragraph of the Addendum Klassen expresses the view that it was the Jews, scattered throughout the Roman Empire, who resorted to mind manipulation and found a relatively unimportant religion called the Essene to poison the Roman mind.[72]

Real Reason for the Fall of the Roman Empire

The reasons for the fall of the great Roman Empire were many. The moral decay of Roman society together with the fact that the Roman gods failed to satisfy the real religious needs of the people of the Roman Empire, are the true causes. Despite state-sponsored persecutions of the third century by a number of emperors, ultimately Christianity conquered Rome because of its soul-satisfying message and hope of a future life.[73]

Christ a Real Historical Person

Klassen, Carter and other opponents to Christianity assert that the historical existence of Christ is a myth and a hoax. They know that, if Jesus Christ had never lived, that the claims of Christianity would fall and the Christian faith would have no foundation whatever. The New Testament which mentions Jesus Christ in every New Testament book would be a fraud. They well know that Christianity is Jesus Christ.

Klassen, Carter and the haters of Christianity should explain to the world how such a hoax should have been believed by millions of people during the nearly two thousand years of its existence. How can one explain the fact that more biographies, books and monographs have been written about Jesus Christ than any other person in history. All the major encyclopedias of America and Europe give the impression that Jesus Christ lived and scholars wrote rather long articles about Jesus revered by millions as both God and man. Thus *The Encyclopedia Britannica* begins its life of Jesus Christ with the assertion that Jesus was born in 6 B.C.[74]

Relative to the issue of Christ's historicity, F. F. Bruce, former Rylands professor of Biblical Criticism, asserted: "Some writers may try with the fancy of Christ's myth, but they do not on the grounds of historical evidence. The historicity of Christ is axiomatic, for an unbiased historian as the historicity of Julius Caesar. It is not historians who propagate the Christ myth theories."[75] Otto Betz, in his book: What *Do We Know About Jesus?* concluded his study of the life of Jesus with these words: "No serious scholar has ventured to postulate the non-historicity of Jesus."[76]

What Are the Sources for Christ's Historicity?

In response to this question Dr. John Warwick Montgomery wrote: "What then, does a historian know about Jesus? He knows, first and foremost, that the New Testament documents can be relied upon to give an accurate portrait of Him. And he knows that this portrait cannot be rationalized away by wishful thinking, philosophical presuppositions, or literary maneuvering."[76]

In opposition to Klassen the judgment of history is that those so called scholars who deny Christ's history are really pulling the greatest hoax of the ages.

Footnotes

1. May be obtained from Box 400, Otto, North Carolina 28763.
2. Nicholas Carter, "The Counterfeit Gospel," Liberty Bell, 17:14-53, February 1990.
3. Martin A. Larsen, "Whatever Happened to the Dead Sea Scrolls?, *The Journal of Historical Review*. Number Two, Summer, 1982, pp. 119-123.
4. Ben Klassen, "The Greatest Hoax of All Time," printed as a pamphlet, pp. 5-6.
5. **Ibid.**, p. 6.
6. Klassen, **op. cit.**, p. 1, column B.
7. John 18:36.
8. Klassen, **op. cit.**, p. 2, column A.
9. Flavius Josephus, *Antiquities of the Jews*, XVII, v. 3-7.
10. Klassen, ep. cit., p. 3, column A.
11. Harold W. Hoellner, *Chronological Aspects of the Life of Christ* (Grand Rapids: Zondervan Publishing House, 1977), pp. 11-23.
12. Klassen, p. 1, column B, cf. also Addendum, p. 6.

13. cf. Josh McDowell, *Evidences That Demand a Verdict*, 1, (San Bernardino, California: Campus Crusade for Christ, 1972), pp. 83-39.

14. Flavius Josephus, *Antiquities of the Jews*, XVIII, 33.

15. **Ibid.**, p. XX.

16. Suetonius, *Life of Claudius*, 25.4.

17. Suetonius, *Lives of the Caesars*, 26.2.

18. Lucian, The Pass as cited by McDowell, **op. cit.**, p. 84.

19. Klassen, **op. cit.**, p. 4, column A.

20. cf. The quotation on Eusebius as printed in Graham Scroggie, *A Guide to the Gospels* (London: Pickering and Inglis, 1948), p. 12.

21. Helen Barrett Montgomery, *The New Testament in Modern English* (Philadelphia; The Judson Press, 1934), p. 147.

22. Bruce Metzger, *The Text of the New Testament: Its Transmission and Composition and Restoration* (New York: Oxford University Press, 1964), pp 38-49.

23. Cf. McDowell, **op. cit.**, 1, chapter 4 "The Reliability of the New Testament."

24. F.F. Bruce, *The New Testament Documents: Are They Reliable?* (Downers Grove, IL: Inter-Varsity Press, 1972), p. 119.

25. Everret Harrison, *Introduction to the New Testament* (Grand Rapids: Wm. B. Eerdmans Publishing Company, 1964), pp. 91-114; *Donald Guthrie, New Testament Introduction* (Downers Grove, III. Inter-Varsity Press, 1974). pp 1-960

26. Klassen, **op. cit.**, p. 2, column B.

27. Lars P. Qualben *A History of the Christian Church* (New York Thomas Nelson and Sons, 1940), pp. 82-84.

23. Kenneth Scott Latourette, *A History of the Christian Church* (New York: Harper & Brothers, 1953), p. 587; pp. 1002-1095.

29. Adam Fahling, *The Life of Christ* (St. Louis: Concordia Publishing Company, 1936), 742 pp.

30. Joh Ylvisaker, *The Gospels* (Minneapolis: Augsburg Publishing House, 1932), 700 pp.

31. William Arndt, *Does the Bible Contradict Itself?* (St. Louis: Concordia Publishing House, 1926), 112 pp.

32. William Arndt, *Bible Difficulties* (St. Louis: Concordia Publishing House, 1933), Note: These two Arndt volumes have been reissued as one volume under the title *Bible Difficulties and Seeming Contradictions*. A revised edition done by Robert G. Hoerber and Waiter R. Roehrs in which the revisers sometimes changed Arndt's views.

33. J. W. Shepard, *The Christ of the Gospels* (Grand Rapids: Wm. B. Eerdmans Publishing Company, 1946), 650 pp.

34. Charles P. Roney, *Commentary on the Harmony of the Gospels* (Grand Rapids: Wm. B. Eerdmans Publishing Company, 1948), 573 pages.

35. Alvin Haley, *An Examination of the Alleged Discrepancies of the Bible*, (Nashville: Gospel Advocate Company, 1967) a reprint of 1869 edition, having 473 pages.

36. Klassen, **op. cit.**, p. 5, column B.

37. Cf. footnote 3.

38. Klassen, **op. cit.**, p. 4, column A.

39. **Ibid.**

40. **Ibid.**, p. 5, column B.

41. **Ibid.**, p. 4, column B.

42. Klassen, **op. cit.**, 4, Column B.

43. Mehahem Mansoor, *The Dead Sea Scrolls* (Second Revised edition; Grand Rapids: Baker Book House, 1983), pp. 57-68.

44. A. Powell Davies, *The Meaning of the Dead Sea Scrolls* (New York: New American Library, 1958), pp. 43-132.

45. Edmund Wilson, *The Scrolls From the Dead Sea* (New York: The Meridian, Cf. also Wilson's articles in The New Yorker, May 14, 1955.

46. Dupont Sommer, *The Dead Sea Scrolls* (Oxford: Basil Blackwell, 1952). Cf. also Dupont Sommer, *The Essene Writings From Qumran* (Cleveland: World Publishing Company, 1961).

47. John Allegro, *The Dead Sea Scrolls* (Baltimore: Penguin Books, 1958).

48. Charles Francis Potter, *The Lost Years of Christ Revealed* (New York, Fawcett Books, 1962), pp 9-154.

49. Edwin Yamauchi, *The Stones and Christianity* (Philadelphia: J.P. (Lippencott, 1972), pp. 136-148.

50. Klassen, **op. cit.**, p. 3, Column A.

51. **Ibid.**, p. 4, Column B.

52. Edmund Schlink, *The Doctrine of Baptism*. Translated by Herbert J.A. Bouman (St. Louis: Concordia Publishing House, 1972), p. 109.

53. **Ibid.**, p. 3, Column B.

53a. Charles F. Pfeiffer, *The Dead Sea Scrolls and the Bible* (Grand Rapids: Baker Book House, 1969), pp 126-132.

54. Klassen, **op. cit.**, p. 4, Column A.

55 **Ibid.**

56 **Ibid.**

57. Cf. Theodore Zahn, Grundriss der Geschichte des Neutestamentlichen Kanon (Leipzig: A. Dieterische Verlaghandlung Nacht, 1904), pp 1-92.

58. Klassen, **op. cit.**, p. 6.

59. **Ibid.**

60. **Ibid.**, p. 7.

61. **Ibid.**

62. **Ibid.**, p. 7.

63. **Ibid.**, p. 5, column A.

64. Latourette, **op. cit.**, p. 92.

65. Kiassen, ep. cit., p. 7.

66. Merrill C. Tenney. *The New Testament - A Historical and Analytic Survey* (Grand Rapids; Wm. B. Eerdmans Publishing Company, 1953), pp. 417-428.

67. Jerald A. Brauer, *The Westminster Dictionary of Church History* (Philadelphia; The Westminster Press, 1971), p. 580.

68. "Nicene Creed," *The Concordia Cyclopedia* (St. Louis; Concordia Publishing House, 1927), p. 547.

69. Klassen, **op. cit.**, p. 3, Column B.

70. **Ibid.**, p. 4, Column A.

71. **Ibid.**, p 5, Column B.

72. **Ibid.**, p. 3.

73. Qualben, **op. cit.**, pp. 103-104.

74. "Jesus Christ," *The Encyclopedia Britannica*, Propaedia, X, 145.

75. F. F. Bruce, The New Testament Documents: Are They Reliable? (Downers Grove,

IL; Inter-Varsity Press, 1972), p. 119.
76. Otto Betz, What Do We know About Jesus? (London; SCM Press, 1963), p 9.
77. Dr. John Warwick Montgomery, History and Christianity (Downers Grove, Inter-Varsity Press, 1964), p. 40.

Questions

1. Ben Klassen appears to be strongly ____.
2. What is the Greatest Hoax according to Klassen? ____
3. Klassen rejects the reliability of ____.
4. All reputable conservative Christian scholars believe that Christ was born in either ____ or ____.
5. Two of the four Gospel writers were ____.
6. Mark obtained his material from ____.
7. What did Luke write in his introduction to the Gospel of Luke? ____
8. For many Greek and Roman authors classical scholarship only possesses copies that are ____.
9. Clement of Rome (A.D. 95-97) has references to ____.
10. Who were the Christian Apologists? ____
11. The Sermon on the Mount is only one of ____.
12. Who were the Essenes? ____
13. Klassen contends that the first practitioners of Christianity were ____.
14. Klassen follows the earlier studies made by scholars like ____.
15. The teachings of the Essenes was a religion of ____.
16. Jesus and Paul spoke ____ and not ____.
17. The subscription on the cross was in ____.
18. The reason the books of the New Testament were written in Greek is ____.
19. Klassen is an expert at ____.
20. The *Canon Muratori* is ____.
21. The Creed of Nicea was written for ____.
22. The Arians claimed that Christ was ____.
23. The two Greek words involved were ____ and ____.
24. ____ ultimately conquered Rome because of ____.
25. What did Dr. John Warwick Montgomery write about Christ's historicity? ____
26. Who is pulling of the greatest hoax of the ages? ____

The Misuse of the Bible to Falsely Support Homosexuality and Lesbianism

Christian News, January 14, 1991

Marshal Alan Phillips, a member of the Hate Violence Reduction Committee of the Los Angeles Human Relations Commission, wrote an article for *The Los Angeles Times*[1] in which he indicates his support for homosexuality and lesbianism. This article was reprinted in *The Fort Wayne Journal Gazette* on Tuesday, December 11, 1990.[2] In this essay he marshalled all the new alleged evidence made by recent literature to support these two lifestyles, believed by him to be unfairly discriminated against. (Ed. "New Bible sensitive to gays" by Marcel Alan Phillips was photographed in the January 14, 1996 *Christian News*).

In his opinion the 1990 revision of The Revised Standard Bible of 1952 and 1959 has taken a step forward in removing one of the religious objections to homosexuality. The 1959 RSV was alleged to be a revision of the King James Bible of 1607-1611. He specifically called the new rendering of I Corinthians 6:9, which in the 1959 RSV read: "Do you not know that the unrighteous will not inherit the kingdom of God? Do not be deceived; neither the immoral, nor idolators, nor adulterers, nor homosexual, nor thieves, nor the greedy, nor drunkards, nor revilers, nor robbers will inherit the kingdom of God." In the King James Version it rendered "arsenokoitai" and "malakoi" as "effeminate" and "abusers of themselves with mankind." The new revision of the 1990 *Revised Standard Version* has substituted "male prostitutes" "sodomites." "This new rendering, it is claimed by Phillips, "to be an improvement, but slightly off if one looks closely at the Greek." How it is supposedly not faithful to the Greek Phillips does not tell his readers.

The essay for *The Los Angeles Times* also takes up the cudgel for homosexuality and lesbianism by challenging the story of the destruction of Sodom as having been misinterpreted in claiming that the people in Sodom were not punished by God for homosexuality, for deviate sexual behavior, but because they would not extend hospitality to Lot's visitors.

Phillips refers to the book of John Boswell, *Tolerance and Homosexuality*, in which the chairman of the Yale University History department, claimed that he spent ten years in research into the legal, literary, theological, artistic and scientific records of Greek writers contemporaneous with Paul and concluded that the words "malakoi" and "arsenokoitai" did not refer to homosexuality or to the practitioners of this sexual practice. The term "malakoi" supposedly means "sexually loose," and "arsenokoitai" merely means male prostitutes, men who have sexual intercourse either with male or female persons. It is Phillips conclusion, as stated in the beginning of his defense of homosexuality and lesbianism: "But a new Protestant version of the Bible is more tolerant of and sensitive toward **homosexuality, removing one more obstacle** to full societal acceptance of homosexuality, including legal marriage and its associated benefits."[4] (emphasis authors)

The Bible Condemns Homosexuality and Lesbianism

By no stretch or the imagination can any responsible scholar, who takes the Scriptural text the way it is written claim that either the Old or the New Testament support homosexuality and lesbianism. The Bible denounces and condemns sexual relations of men with men, of women with women. The Book of Leviticus, which discusses sexual intercourse with whom an Israelite was not to have sexual relations state in chapter 18:22: "You shall not lie with a man or with a woman: it is an abomination."[5] In Leviticus 20:13 there is this prohibition against homosexuality or sodomy: "If a man has intercourse with a man as a woman, they both shall be put to death. They both commit an abomination. They shall be put to death; their blood shall be on both of their heads."[6]

St. Paul in Romans Condemns Homosexuality and Lesbianism

Paul in his great theological letter, Romans, composed in A. D. 58, in the first chapter has depicted the corruption of the Roman world of the first century A.D. Verses 18-32 portray to what low moral depths the Roman Empire had sunk. Paul in chapter 1:18-3:20 has powerfully set forth the need for the world for the Gospel, specifically, the need for justification by faith, because all men. Jew and Gentile have sinned and are under the wrath and judgment of God.[7] The need for justification by faith in Christ's atoning work is shown by the many sin of which mankind was guilty in the Roman Empire. In Roman 1:19-32, Paul has listed a large catalogue of sins engaged by many in the Roman Empire. The different sins may be classified into impiety (1:19-23) and sins of iniquity (1:24-32). Wrath, guilt, and liability to punishment are appropriate, because men know the truth and suppress it. What is known about God has been made clear to them (Acts 14:17). The eternal attributes of omnipotence and deity, though invisible for a man not to worship God. The race refusing to glorify and thank became stupid, so stupid as to fall to the level of idolatry. They worshipped birds, beasts and even reptiles.[8]

Results of Ignoring God

One of the results of ignoring God in creation and one might appropriately add is, that man and woman consider themselves wise, and they became fools. The New English Bible of 1970 rendered Romans 1:24: "For this reason God has given them up to shameful passions. The women have exchanged natural intercourse for unnatural, and their men in turn, giving up natural relations with women, burn with lust for one another, male behaving indecently with males and are paid in their own person the fitting wages of such perversions."[9]

Phillip advanced in support of homosexuality the views of Father John McNeil, professor Robbin Scruggs of Union Theological Seminary, as well as the position of the fundamentalist Maury Johnstone, all who claim that homosexuality is not forbidden in the Bible. One wonder what Bibles they are reading! What passages could be clearer than the passages from the Hebrew Pentateuch or the latter part of chapter one of Romans! Then there is the punishment visited upon homosexuals and sex perverts de-

1490

scribed as occurring in the days of Abraham and Lot.

The Ruin of Sodom as Descriptive of Homosexuality Punishment

According to *The New American Dictionary of the English Language* the word "sodomy" is defined as "Anal copulation of one male with another 2. In some legal usage anal or any copulation with a member of the opposite sex or any copulation with an animal. (Middle English from Old French, sodomite: from Late Latin Sodoma, Sodom).[10]

The same dictionary has this to say about the city "Sodom:" "a city that with Gomorrah was destroyed by fire because of the depravity or its inhabitants." The verb "to sodomite" is defined as "one who practices sodomy."[11]

The most pathetic attempt to support homosexuality is the claim that Sodom was not destroyed because of its homosexuality but because the people of Sodom refused to extend hospitality to the guests of Lot. "Phillips claimed that since 1955 Derrick Sherwin Halley's publication of *Homosexuality and the Western Christian Tradition* scholars have concluded that Sodom was destroyed not for its practice of homosexuality but for its treatment of visitors.

What Does Genesis 19:1-19 Teach About Homosexuality?

Genesis 19:25 states: "Then the Lord rained brimstone and fire on Sodom and Gomorrah. So he overthrew those cities, all the plain, all the inhabitants of the cities and what grew on the ground."[13] "The next morning Abraham went early in the morning to the place where he had stood before the Lord. Then he looked toward Sodom and Gomorrah, and toward all the land of the plain, and he saw, and behold, the smoke of the land which went up like the smoke of a furnace."[14] "And it came to pass, when God destroyed the cities or the plain, that God remembered Abraham, and sent Lot out of the midst of the overthrow, when He overthrew the cities in which Lot had dwelled."[15]

The question which Phillips and those he mentions need to answer is: why were Gomorrah and the cities near Sodom wiped out when they were not guilty of refusing hospitality to Lots' visitors? Where in the Old and New Testaments is there a statement that if a person or persons or a city who do not how hospitality that they are to be put to death?

The True Teaching for Sodom and Gomorah's Destruction

The interpretation which Halley advanced, accepted by Phillips, goes against the context of Genesis chapters 18 and 19. Chapter 18 of Genesis records two important historical facts. After wailing for twenty four years, God tells Abraham that Sarah will become pregnant and give birth to a son, which was the primary consideration for Abraham becoming a great nation and inheriting the land of Canaan and especially that through one of his Descendants the nations of earth would be blessed.[15a] One of the three visitors who visited Abraham was God Himself, who further informed him that the cities of Sodom and Gomorrah were to be destroyed. Thus Moses wrote in 18:20: "And the Lord said, 'Because the out-cry against Sodom and Gomorrah is great and because their sin is very grievous. I will go down and see whether they have done according to the outcry against it that has come to Me, and if not, I will know (v. 23).'" This announcement was followed by the first intercessory prayer of the Old Testament and also of the Bible, in which Abraham prevailed upon the Lord to promise that if even ten righteous people could be found in Sodom, He would not destroy the city.[16]

The Real Sin of Sodom

When Lot received two angel visitors, he brought them to his home and offered them hospitality. Before the angels went to sleep the men of Sodom, both old and young coming from every quarter of the city, called to Lot: "Where are the men who came to you tonight? Bring them out to us that we may know them carnally."[17] *The 1970 New English Bible* rendered verse 5 this way: "Bring them out to us so that we may have inter-course with them."[18] Beck, in his An American Translation rendered this verse: "Bring them out to us that we may rape them."[19] Theophile Meek, in the University of Chicago's An American Translation rendered this crucial verse: "Bring them out to us that we may have intercourse with them."[20] The verb used in Hebrew is "yadah," used in Genesis 4:1 of Adam who ("yadah") his wife, that is bad sexual intercourse and as a result of it, Eve, the mother of all living, gave birth to Cain. The Chicago's An American Translation rendered Genesis 4:1: "The man had intercourse with his wife Eve, so he conceived and bore Cain."[21]

The three primary translations of the Old Testament: Namely the Sep-tuagint, the Peshitta and the Vulgate all three render this verse that the men of Sodom wanted to have relations with the Angels, who looked like men to them. The LXX has in its Greek text the words: "Eksagage' autous pros humas, hina suggenometha autois."[22] In English: "bring them out to US' that we may know them. *"The Clementine Vulgate* transates verse 6: "educ illos huc, ut cognoscrmaseos."[23] Bring those out, in order that we may know them." Knox, in his translation based on the Vulgate, ren-dered this verse this way: "Bring them out here to minister to our lust."[24] The Syriac reads: "Aphek enon lan weyadah enon," which means: "Bring them out to us that we may know them."[25] George Lamsa, in his English translation of the Peshitta, gave this translation: "Bring them out that we may know them."[26]

By reason of hospitality rather morality Lot offered his unbetrothed daughters instead. When these men decided to proceed with their rape intentions, they were struck with blindness, so that they could not carry out their deviate sex on Lot's guests. For unnatural sex practices Sodom and Gomorrah and other nearby cities were destroyed for practices which go against nature or God's order.

How responsible scholars can completely misread and interpret clear texts as found in the Old and New Testaments is only explainable in that for the sake of popularity and to be men pleasers, texts in Genesis, Leviti-cus, Romans, and 1 Corinthians are interpreted in such a way as to condone homosexuality and lesbianism. The texts, which condemn what goes contrary to nature and God's intended use of sex are clear leaving no room even for doubt. The Old Jewish commentators of Rashi, Ibn, Ezra, and Rashbam considered the request of the people of Sodom to have sexual relations with Lot's visitors as sinful, also the fact that Lot would offer his two daughters as persons with whom to have sex, the same commentator considered reprehensible.[27]

The Word Homosexuality Not Used In Bible

The argument advanced by Phillips that the word "homosexuality" is not used either in the Old or New Testaments,[28] is a very weak one and surprises a person that such a kind of argument supposedly is said to allow for sexual relations of man with a man or women with each other. The history of the vocable "homosexuality" is of Medieval linguistic origin.[29] However, everything involved in modern homosexuality and lesbianism is clearly set forth and condemned in the Bible. The word "trinity" is not used in the Bible, but the teaching of God's oneness or unity and that this same God has manifested Himself in three distinct Persons is clearly set forth in both Testaments of Holy Writ.[30]

Sodom and Gomorrah as Examples or God's Punishment of Great Sin in Subsequent Old Testament Book[31]

Moses warned his people that if they would break the Sinaitic Covenant and commit the abominations of the land of Canaan that Israel's land would become "brimstone, salt, and burning; it is not sown nor does it bear, nor does any grass grow there, like the overthrow of Sodom and Gomorrah Admah, and Zeboim, which the Lord overthrew in His anger and wrath" (Deut. 29:23).[32] The prophet in the Great Assize chapter where Isaiah depict the rejection of Judah and Jerusalem, calls them: "You rulers of Sodom give ear to the law of our God," (1:10).[33] The people are addressed as "the people of Gomorrah." (1:10). In Isaiah 1:9 Isaiah reminds the Judeans that: "If the Lord of hosts, had not left us a few survivors, we should have become like Sodom and Gomorrah."[34] Isaiah in his prophecy predicting the destruction of Babylon, announced: "And Babylon theology of kingdoms, the beauty of the Chaldean pride, will be as when God overthrew Sodom and Gomorrah" (1:19).[35] The prophet Jeremiah in his prediction or the destruction of Edom declared: "Edom shall become a horror; everyone who passes by it will be horrified and will hiss because all of its disasters, as when Sodom and Gomorrah and their neighbor cities were overthrown, says the Lord, no man shall sojourn in her" (49:18).[36]

Ezekiel compared the Northern Kingdom of Sodom and the Southern kingdom by implication to Gomorrah. Both of these divided kingdoms are guilty of harlotry, both physical and spiritual. Ezekiel Chapter 16 sheds light also on the immoralities which were committed at Sodom and Gomorrah of Lot's day and shows no relationship between failure to receive guests and the vile sexual practices, perpetrated in daily individual lives and the participation in vile sexual practices at the high places of

Canaan. Amos, Isaiah's contemporary, rebukes the religious practices of the Northern Kingdom and listed among punishments already visited upon the Israelite nation: "I overthrew some, of you, as when God overthrew Sodom and Gomorrah and you were as a brand plucked from out of the burning" (Amos 4:11).[37] In the Davidic psalm, number 11, David probably has Sodom and Gomorrah in mind, when he declared: "On the wicked he will rain coals of fire and brimstone, a scorching wind shall be a part of their cup."[38]

It is apparent from references to Sodom and Gomorrah, the cities of the plain (Gen. 19: 29), "Admah and Zeboim (Deut. 29:23; Hos. 11:8) and possibly Zoar that all these cities were destroyed because of heinous sins, committed against natural law as well as the laws God gave Israel.

The Meaning of the Greek "Arsenokoitai" and the Greek "Malakas"

Phillips began his article: "New Bible Sensitive to Gays" in this way: "Large numbers of Christians believe that God condemns homosexuals and uses the Bible to justify continued civil discrimination. But a new Protestant version of the Bible is more tolerant of and sensitive toward homosexuality, removing one more obstacle to full societal acceptance of homosexuals including marriage and its benefits."[39] As already indicated in the beginning of this retort to Phillips the new 1990 RSV uses "Male prostitutes" and "sodomites," in place of the 1959 versions of "homosexuals." This raise, the question about the translation of "arsenokoitai" and "malakas."

In classical Greek according to Benselers *Griechisch Deutsches Woerterbuch* the word "arsenokoitai" means paederest, "Knabenschaender,"[48] which also is the meaning which Souter gives in his *Pocket Lexicon of the New Testament*.[41] The Roman Catholic scholar Francis Spencer, rendered "Malaks" as "Effeminate" and "azrsenokoitai" as "sodomites."[42]

The Opinion of Greek Scholarship on "Arsenokoltai" and "Malakos'

The Expositor' Greek New Testament comments on "arsenokoitai" as follows: "Whose sin of Sodom as widely and shamefully practiced by the Greeks; cf. Rom. 1:24ff. written from Corinth."[43]

The great New Testament Greek scholar A.T. Robertson says in his exposition of the New Testament relative to I Corinthians 6:9: "Do not be led astray by plausible talk to cover up sin as mere animal behavior. Paul has two lists in verses 9 and 10, one with repetition of **oute**, neither (fornicators idolaters, adulterers, effeminate or malakos) abusers of themselves with men or **arsenokoitai**, or sodomites as in 1 Tim. 1:10, a late word for this horrid vice, thieves, covetous), the other with **ou** not (drunkards, revilers, extortioners). All these will talk short of the kingdom of God. This was plain talk to a city like Corinth."[44]

Rienecker, *A Linguistic Key to the Greek New Testament*, wrote about "malakos" as follows: "soft effeminate, a technical term for the passive partner in homosexual relations," for which he cited such scholars as Barrell and Conzelmann.[45] Adolf Peissmann in his book, *Light from the Ancient East*, p. 164 makes the same point based on extra-Biblical linguistic data.[46] For further substantiation one should consult Theodore Klauner, Reallexicon fuer Antike und Christentum.[47] Relative to "arsenokoitai" Rieneker claims the singular refers to a male who has sexual relations with a male, homosexual.[48] The Jewish punishment for the sin of homosexuality was stoning.[49]

The one-volume Hailings Dictionary of the Bible has a seven-line statement on homosexuality which read thus: "The term homosexual is found only in I Cor. 6:9 (RSV, RV "abusers of themselves with mankind, or with men"). "The practice is condemned in connection with male prostitution."[50]

Wilbur Gingrich defined "arsenokoites" as "a sodomite" or "paederes" and "malakas" a "effeminate" or "homosexual."[51]

G. Abbott-Smith defined "arsenokoites" as a "sodomite" and "malaka" as "effeminate" and in I Cor. 6:30 as probably in obscene sense.[52]

Phillips Misrepresents Paul's Ethical Position and Teachings

Paul trained at the feet of Gamaliel surely knew his Old Testament and chapters 18 and 20 of Leviticus and its repetition in Deuteronomy. He certainly would never have approved of homosexuality and even refused to condemn deviant sexual practices found in use in the immoral Roman empire. He certainly denounced both homosexuality and lesbianism as he did in Roman's 1:18-33. Nothing in Paul's writings could be construed as favoring them or even taking a neutral stance over against them. The passages in I Corinthians 6:9 and I Timothy 1:10 should be only explained within the whole context of Paul's announced position on these forms of sexual aberrations. It is absolutely wrong to interpret an author against himself, especially when he has so clearly denounced homosexuality and lesbianism.

Homosexual and Lesbian Marriages?

Phillips in his article defends and advocates marriages in which two men are married (?) to each other and likewise two women live together as married people with all the rights normally guaranteed to married people.[53] Here the liberal establishment, humanists, atheists; agnostics are emptying the historic concept of what is involved when a man (male) and a woman (female) live together in marriage to procreate and raise children and also have legitimate sexual intercourse. To speak of two men as married to each other or two women as married completely empties the word "marriage" of its meaning. God instituted the marriage estate as a divine institution and said to the first human couple, Adam and Eve, "Be fruitful and multiply and fill up the earth (Gen. 1:28)."

To define the living together of two men who practice anal intercourse as natural is the height of absurdity. The same testament also applied to two women as "married" when they cannot become one flesh, which only can be effected by a man and a woman who are physically equipped for normal sexual intercourse.

It is an insult to God to expect a Christian pastor to sanctify with word and sacrament such heinous God-forbidden sexual relations. It is a disgrace when Protestant churches accept homosexuals and lesbians as students to prepare to teach God's Word and then upon graduation ordain them into the holy ministry. The spiritual state of many churches today is reminiscent of the Old Testament church which was guilty of syncretism and the adoption of pagan religious views, all of which are most severely condemned in the Old Testament.

Footnote

1. Marshall Alan Phillips, "New Bible Sensitive to Gays."
2. *The Fort Wayne Journal Gazette*, December 11, 1990, p. 9A.
3. Phillips, **op. cit.**, p. 9A, column I.
4. **Ibid.**, p. 9A column 1.
5. *The Holy Bible, The New King James Version* (Nashville, Thomas Nelson Publishers, 1982, p. 120.
6. *The New English Bible With Apocrypha* (Oxford University Press and Cambridge University Press, 1970), p. 132.
7. Paul's *Epistle to the Romans* 3:27-28.
8. Gordon H. Clark, "Romans" in Carl F. Henry, *The Biblical Expositor* (Philadelphia, A.J. Homan Company, 1960), III. p. 239.
9. *The New English Bible of 1979*, **op. cit.**, New Testament, p. 192.
10. *The American Heritage Dictionary of the English Language*, William Morris (Editor), I Boston, New York: American Heritage Publishing Co., Inc., and Houghton Mining Company, 1970), p. 1227a.
11. **Ibid.**
12. Phillips, **op. cit.**, p. 9A, column 3.
13. Translation of Hebrew Text, Genesis 19:24-25.
14. Genesis 19:28, based on Hebrew text.
15. Genesis 19:27-28, *New American Standard* (Philadelphia; A.J. Holman Company, 1973), Old Testament, p. 13.
15a. Genesis 3:15.
16. Genesis 18:19-33.
17. William F. Beck, *The Holy Bible - An American Translation* (New Haven: Leader Publishing Company, 19761), Old Testament, p. 18.
18. Genesis 19:6, *New King James Version*, **op. cit.**, Old Testament, p. 16a.
19. Genesis 19:5, (The New English Bible of 1970), Old Testament, p. 9.
20. *The Holy Bible - An American Translation*, J.M. Powis Smith and Edgar J. Goodspeed (Chicago: The University of Chicago Press, 1931), Meek's translation, p. 28.
21. **Ibid.**, Old Testament, p. 8.
22. *The Septaugint Version of the Old Testament* (London: Samual Bagster and Sons, Limited, no date), p. 21.
23. Biblical Sacra. Juxta Vulgatam Clementinam (Romae, Tornai et Parisiis: Desclee ey Socii, 1938), p. 17.
24. *The Holy Bible, A Translation from the Latin Vulgate in the Light of Hebrew and Greek Orginals* (New York, Sheed and Ward, 19561, p. 14. Translation of Bishop Knox.

25. *Katheba Kadishe*, (London: Trinitarian Bible Society, 1913), p. 10. column b.

26. George M. Lamsa, *The Holy Bible From Ancient Near Eastern Manuscripts, Containing The Old and New Testaments*. Translated from the Peshitta (Philadelphia: A.J. Holman Company, 1957). Old
Testament), p. 24.

27. A. Cohen, *The Soncino Chumash, The Five Books of Moses with Haphtorah* (London: The Soncino Press, 1956). pp. 93-94.

28. Philipps, **op. cit.**, p. 9A, column 3.

29. *The American Heritage Dictionary of the English Language*, **op. cit.**, p. 1227.

30. John Theodore Mueller, *Christian Dogmatics* (St. Louis: Concordia Publishing House, 1955), p. 147.

31. G. Henton Davies. "Geneis," *The Broadman Bible Commentary* (Nashville: Broadman Press, 1969), I, p. 189.

32. Quoted from *The New King James Version*, **op. cit.**

33. **Ibid.**, p. 664.

34. Isaiah 1:9, according to *The Revised Standard Version of 1959*.

35. Isaiah 13:9, according to *The Revised Standard Version*.

36. Jeremiah 49:17-18, according to the 1959 *Revised Standard Version*.

37. Amos 4:11, according to the *Revised Standard Version*, 1959.

38. Psalm 11:6, according to the *Revised Standard Version*.

39. Phillips, **op. cit.**, p. A, 9, column 1.

40. *Benseller Griechisch-Deutches Schulwoerterbuch*. Fierzehnte Auflage, bearbeited vo Adolf Kaegi (Leipzig: and Berlin: Druck und Verlag von B.G. Teubner, 1940), p. 121. Kaegi claims that was also the usage in New Testament.

41. Alexander Souter, *A Pocket Lexicon to the Greek of the New Testament* (Oxford: At the University Press, Clareton, 1917), p. 38 gives a paederest.

42. Francis Alosyius Spencer, *The New Testament of Our Lord Jesus Christ*, Edited by Charles J. Callahan and John McHugh (New York: Macmillan and Company, 1944). p. 434.

43. G.G. Findlay, *The Expositor's Greek New Testament* (Grand Rapids: Wm. B. Eerdmans Publishing Company, 1924), III, p. 817.

44. A. T. Roberson, *Word Pictures of the New Testament* (Nashville: Broadman Press, 1931), IV, p. 119.

45. Fritz Rienecker, *A Linguistic Key to the Greek New Testament* (Grand Rapids: Zondervan Publishing House, 1980), II, p. 56.

46. Adolf Deisman, *Light From the Ancient East*, Translated by Leonard Strahan (London: Hodder and Stoughton, 1927), p. 164.

47. Theodore Klauser, (Editor) *Reallexicon Fuer Antike und Christentum* (Stuttgart: Anton Hiersemann, 1976), IV, pp. 620-650.

48. Rienecker, **op. cit.**, II, p. 56.

49. Strack and Billerbach, *Kommentar zum Neuen Testament aus Mishnah und Talmud, III*, p. 70f.

50. James Hastungs, *Dictionary of the Bible*, Revised Edition by Fredrick C. Grant and H. H. Rowley (New York: Charles Scribner's Sons, 1963), p. 394.

51. F. Wilbur Gingrich, *Shorter Lexicon of the Greek New Testament* (Chicago: The University of Chicago Press, 1975), p. 28 and 130.

52. G. Abbott Smith, *A Manual Greek Lexicon of the New Testament* (New York: Charles Scribner's Sons, 1929), pp. 60, 277.

53. Phillips, **op. cit.**, A 9, column 3-4.

Questions

1. Marshall Alan Phillips says that the ____ translation of the Bible has taken a step forward in supporting ____.
2. How does the English Standard Version translate 1 Corinthians 6:9? ____
3. Sodomy is defined as ____.
4. Why was Sodom destroyed? ____
5. The Jewish punishment for the sin of homosexuality was ____.
6. What is the height of absurdity? ____
7. The spiritual state of many churches today is ____.

Hengstenberg, a Nineteenth Century Lutheran Apologete and Defender of Biblical Christianity

Christian News, July 15, 1991

One of the important conservative Old Testament scholars in Germany in the nineteenth century was Ernst Wilhelm Hengstenberg, born nearly one hundred and ninety years ago (1809). His influence was considerable in stemming the tide of higher criticism in Germany and in other West European countries, as well as in the United States. *The Concordia Cyclopedia* gave him this evaluation: "By his word of interpretation and defense of the Old Testament he became the staunch defender against rationalism, unionism, and the mediating theology of his day."[1]

The Life of Hengstenberg

Ernst Wilhelm Hengstenberg was born at Froedenberg (a village of Westphalia, near Hamm, 22 miles north northwest of Arnsberg) October 20, 1802.[2] He was born into a family which had a tradition of producing men for the ministry of the church. Because of health reasons Hengstenberg did not attend a public school but his father, a Calvinist conservative pastor, instructed his son so well that the son was qualified to enter the newly-founded university of Bonn.[3] Here he prepared himself for the ministry by a good grounding in philology and philosophy, under Freytag and Giessler he studied Old Testament exegesis and church history. In addition, Hengstenberg received a thorough and complete course in classical philology. He gave special attention to Aristotle's philosophy and he especially concentrated on Arabic. The results of his Aristotelian philosophical studies resulted in a German translation of Aristotle's Metaphysics (Bonn, 1824), while as a result of his Arabic studies he issued an edition of Amrul l' Kair's *Moallakah* (Bonn, 1825) and by means of the latter obtained his doctor of philosophy degree.[4]

Hengstenberg wished to begin a thorough study of theology, but lack of money prevented this objective from being fulfilled immediately after the reception of the Ph.D. However, at Freytag's recommendation he was appointed assistant to Staehlin at Basel where he participated in Staehlin's Oriental investigations. At Basel he had leisure time which he utilized for serious study of the Scriptures.[5]

Hengstenberg's Theological Career as a Lutheran

Through private study of the Bible he had found Christ his Savior and as a result of the study of the Lutheran Confessions he came to the conviction that they contained the clearest exposition of the Scriptures. In 1824 he became Privatdocent (lecturer) and in 1825 a licentiate in theology and in 1826 professor extraordinary. His thesis for the baccalaureate in theology was a defense of Protestantism and a severe rebuke of criticism employed by rationalism when interpreting the Bible. At Berlin

1499

Hengstenberg became the head of a seminar on Old Testament problems. Because of the manner in which he conducted his seminar at Berlin, his fame in Germany caused many students to come to sit at his feet and as guide and counselor he exercised a profound and beneficial influence on many in Europe and in America in the nineteenth and to some degree in the twentieth century.[6]

Hengstenberg's Friends

Hengstenberg's friends included such scholars as Neander, Friedrich Strauss, Therenius, and many younger clergymen of Berlin. His association with these men as well as the growing of his orthodoxy brought upon Hengstenberg the opposition of the Berlin authorities. Attempts were made to remove him from Berlin under the guise of promotion to other cities. Minister Von Allenstein made a number of attempts to induce him to take positions which were supposed to be advancements.[7] However, Hengstenberg was convinced that God wanted him to serve in Berlin. In July, 1827 he became the editor of *Die Evangelische Kirchenzeitung*, a periodical which was to exercise a great influence far and wide on the religious life of the nineteenth century.

Hengstenberg's Battle Against Rationalism

Bachmann wrote about Hengstenberg's effort against rationalism as follows: "Once convinced that his proper field lay in the career thus opened to him, Hengstenberg entered with vigor on a task that he was to carry-on under great discouragement for forty-two years. No man of our time has been exposed to more opposition and enmity, ridicule and slander, open and secret denunciation than the editor of Evangelische Kirchenzeitung."[8] Kahnis, an opponent of Hengstenberg, asserted the opinion: "The opinion of the world during the last forty years has associated with Hengstenberg's name all that it finds condemnatory in the revival of a former faith, Pietism, a dead orthodoxy, obscurantism, fanaticism, Jesuitism, sympathy with every influence or retrogression."[9]

In his *Die Evangelische Kirchenzeitung* Hengstenberg consistently attacked rationalism and its spirit. Rationalism as an abstract system was repudiated. Individuals, whoever they were, who exhibited the rationalistic spirit came under attack. Rationalism as a principle objected to and rejected the supernatural. Hengstenberg attacked all who denied the deity of Christ, all who exalted matter over spirit or who adored or paid too much attention to human reason. In opposing rationalism and its proponents Hengstenberg quoted Holy Writ and the creeds of the church.[10]

The Writings of Hengstenberg

It was especially in the Old Testament field that Hengstenberg was active and at home. In his major writings he took issue with negative higher criticism as applied especially to the Old Testament, as it was developed in Germany by De Wette, Eichhorn, Ewald, Bleek and K . H. Graf and later by the once very conservative Franz Delitzsch, who as a member of the Erlangen faculty was constantly adjusting his views to the latest higher critical theories, as is evident from a study of his later commentaries.[11]

The Christology of the Old Testament

A major work of Hengstenberg was his *Christologie des Alten Testaments*, originally published in three volumes in Berlin, 1829-35. It appeared in four volumes, translated by T. Meyer, (T & T Clark, 1854-58). A reprint of the English translation by R. Keith with some abridgment by T. K. Arnold was published in 1956 (Zondervan Publishing House, Grand Rapids). Every critically-oriented book that this writer has consulted has denounced the book as a return to what is called "repristination theology." Heick made the following assertion about it: "He defended the genuineness of all biblical writings according to the understanding of the churchly and rabbinical tradition. The Old Testament was looked upon as a prophecy that contained all essential Christian doctrines."[12]

Not only have critical scholars objected to Hengstenberg's christological interpretation of the Old Testament, and they have especially faulted him for following the New Testament's interpretation of the Old Testament of finding predictions about the coming Messiah.[13] The concept of prophecy and fulfillment, so clearly set forth as a hermeneutical principle of primary importance, in the Bible, his enemies labelled this stance of Hengstenberg and the Jewish synagogue as practicing an allegorical method, supposedly now completely discredited.[14] It is interesting to note that the Essenes of Qumran believed in the coming of possibly two or even three different Messiahs.[15] The Gospels clearly show as historical documents the fact that at the time of Christ the Jewish people were looking for the coming of the Messiah.[16] The problem with historical critical scholars is that they have a veil over their eyes when they read the Scriptures, especially the Old and New Testaments.[17] The book of Joachim Kraus, *Geschichte der Historischen-Kritischen Erforschung des Alten Testaments* has devoted four solid pages to a critical analysis of Hengstenbergs' understanding of the Old Testament and has severely denounced his understanding of the Scriptures of the Old Covenant as outmoded and as a return to "repristination theology,"[18] of allegorizing, of reading his dogmas into the Old Testament. Hengstenberg's colossal mistake was that he failed to allegedly see the truth of critical scholars who now have succeeded after 1800 years of church history to make the Old Testament truly understandable. Kraus is utterly amazed that Hengstenberg believed in the unity of Isaiah and that chapter 53 of the latter book speaks about Christ.[19] Critical scholars of the nineteenth century hated Hengstenberg for persuading many pastors and laymen for accepting the Mosaic authorship, the unity of Isaiah, the unity of Zechariah, the sixth century date of Daniel, the unity of the Pentateuch,[19a] the direct action of Yahweh in history, the many miracles that God did though his servants and the fact that the canon of the Old Testament was complete by around 400 BC, a position rejected by higher critics.[20]

The Commentary on the Psalms

Der Kommentar ueber die Psalmen, four volumes, treated the 150 psalms and this major work was published between 1842-1847 in Berlin. It follows the method of interpretation employed in the ancient church as well as the hermeneutic that characterized Luther and the authors of the Lutheran Confessions. In a number of Psalms he adopted a typological method of interpretation, which was not in agreement with the interpretation of these psalms as seen in the writings on the Psalms in *Luther's Works* or the interpretation in the Lutheran Confessions. Instead of following a rectilinear interpretation he weakened the Messianic meaning of those psalms, wrongfully interpreted in a typological manner. For example in Psalms 2, understood as being directly predictive about the person and work of Christ, and so cited in the books of the New Testament, Hengstenberg treated typologically, Walter A. Maier, Sr. in his *Notes on the Hebrew Psalms* made this a specific criticism of Hengstenberg's *Commentary on the Psalms*.[21] The apostle Peter in his Pentecost sermon Acts 2 called David a prophet and quotes from Psalm 16 and Psalm 110 and specifically declared that in those psalms the second king of Israel that he spoke of not David but of Christ. When Peter was released from prison and appeared to the praying members of the Jerusalem church, they all lifted up their voices and quoted Psalm 2:1-2 as being a prediction which had just been fulfilled by the release of Peter and John.[22] John Goldingay in his *Old Testament Commentary Survey*, while he mentions a number of commentaries produced in the nineteenth century, as those of Alexander and Kirkpatrick does not allude to this major commentary from the pen of Hengstenberg.[23] Childs, *Old Testament Books for Pastors and Teachers* completely ignored Hengstenberg's *Commentary on the Psalms*.[24]

The Reverend W. B. Pope, principal of Wesleyan College, Didsbury, Manchester in his "Essay on the Life and Writings of Hengstenberg" said about Hengstenberg's *Commentary on the Psalms*, "The Christology of the Old Testament was followed by a Commentary on the Psalms, which for a long time was both in Germany and in the English translation in this series, the leading book for the preacher's use, combining in a remarkable way the results of adequate learning in the unfolding of the structure, and the materials of profitable application."[25]

Other Books Dealing of Old Testament Subjects

A monograph on *Balaam and His Prophecies* appeared in 1842; this was followed by a commentary on *Canticles* and one on Ecclesiastes. A more extensive work came from Hengstenberg's pen on *The Prophet Ezekiel* (2 volumes, 1849-1851; 2nd edition, 1861-1862. From notes left after his death, there was published posthumously *The Book of Job* (2 volumes, 1870, 1875).

A History of the Old Testament

Hengstenberg wrote a two-volume history of the kingdom of God in the Old Testament under the German title: "*Geschichte des Reiches*

Gottes unter den Alten Bund (zwei Baende, Berlin, 1869-1871, which appeared as *The History of God Under the Old Testament*. He also authored a book, *Ueber den Tag des Herrn*, Berlin, 1852, English translation as *The Lord's Day*, Edinburgh, 1853. A book was also published by Hengstenberg which dealt with the sacrifices of the Old Testament, which appeared in Berlin as *Die Opfer der Heiligen Scrift*. Sixteen years prior to the latter publication, Hengstenberg penned his *Aegypter und die Buecher Moses*, which was translated into English as *Egypt and the Books of Moses*, Edinburgh, 1843.[26]

Hengstenberg and New Testament Studies

Although Hengstenberg was mainly an Old Testament exegete and expositor, he did venture into the New Testament field by publishing a three volume *Commentary on the Gospel of St. John*, 1st edition: 1861-63; 2nd edition: volume 1, 1867.

Because there were over four hundred references in the Revelation of St. John to the Old Testament, Hengstenberg was drawn to write a commentary on Revelation, 2 volumes, 1st edition: 1849-1851; 2nd edition: 1861-1862. He also endeavored to show the Old Testament background of the Gospel of John. *Vorlesungen ueber die Leidungsgeschichte, Lectures on the History of the Passion*, was issued posthumously in Leipzig, 1875. That Hengstenberg was well versed in the New Testament Scriptures is not a surprise because of the close connection he saw between the Old Testament prophecies of the Messiah and their fulfillment in the life of Christ and in the establishment of the Christian church.[27]

Hengstenberg and the Discipline of Old Testament Theology

The editor of Die Kirchliche Schriftzeitung did not write a work entitled: "A Biblical Theology of the Old Testament." The interested student could go to the many writings of Hengstenberg, especially his Christology of the Old Testament, The History of the Kingdom of God in the Old Testament, his commentaries on the Psalms, Ezekiel, Ecclesiastes, Song of Songs, Job and other writings and construe a Biblical theology of the Old Testament.

Payne in his *The Theology of the Other Testament* devoted chapter 2 to the history of the development of Old Testament theology as a separate discipline,[28] usually considered to have been inaugurated a little over two hundred years ago when John Philip Gabler, in his inaugural address, "Concerning the Correct Distinction Between Biblical and Dogmatic Theology," insisted that in the future he demanded that Biblical theology be treated as the "religious ideas of Scripture as an historical fact, so as to distinguish the different times and subjects, and so also different stages in the development of these ideas."[29] Such a distinction requires the separation of the Old Testament from New Testament theology. The first Old Testament theology that carried this out was that of Lorenz Bauer in 1796.[30] What was not at once realized was the fact that Gabler's demand sprung from the latter's own rationalism and distrust of the supernatural. Bauer refused to equate the thought of the Old Testament writers with true theology nor would he accept the supernaturalistic events and phenomena recorded in the Holy Scriptures. In the opinion of Payne, "it was E.W. Hengstenberg who demonstrated the value of Old Testament theology in his monumental *Christology of the Old Testament*, published in 1829-35 and revised in 1854-57. Liberal writers have indeed criticized Hengstenberg's work as a reading of New Testament Christology back into the Old Testament. At most points, his exegesis is careful and represents the revelations which God actually made. The presuppositions of disbelievers simply prohibits them from acknowledging such revelations as historical."[31] In the last half of the nineteenth century Gustav Oehler's *Old Testament Theology* was one of the last conservative Old Testament theologies to be written. He lived from 1812 to 1872. He was thus a contemporary of Hengstenberg. He was professor of Old Testament theology at a number of institutions and even declined the call to the University of Erlangen to become the successor of Franz Delitzsch. In 1845 he published *A Prolegomena to the Theology of the Old Testament*. It was only after Oehler's death that his complete notes on the Old Testament were published as *Die Theologie des Alten Testments*. Oehler was recognized throughout Germany as a competent Old Testament scholar. Oehler took cognizance of the various commentaries of Hengstenberg and other of his major writings. There are over forty references to *The Christology of the Old Testament, The History of the Kingdom of God* and his exegetical commentaries. Since both were Lutheran, who like other Christians accepted the Bible, Old and New Testaments as God's Word and regarded as correct the New Testament's interpretation of Old Testament events, were opposed to the radical theories being advocated and promoted by liberal Lutheran scholars, it is not surprising that these two Lutheran Old Testament scholars would basically agree on most matters involving the understanding of the Old Testament. Hengstenberg was also a contemporary of von Hofmann and Keil and Delitzsch. In the opinion of Pope von Hofmann and Oehler must have had some influence on Hengstenberg.[35] Thus this English evaluator of Hengstenberg's efforts in Old Testament exegetics, wrote: "There can be no doubt that Hoffmann's original and most suggestive volumes on prophecy written on principles diametrically opposed to those of Hengstenberg, had the effect of making him more careful in his realization and exposition of the facts interwoven with all predictions."[36] Relative to Oehler's influence. Pope wrote: "Nor can it be doubted that the works of another very eminent writer, Oehler, especially his *Prolegomena to the Biblical Theology of the Old Testament*, (1841) influenced considerably his estimate of the progressive character of revelation from Mosaicism to Hebraism and prophetism. But Hengstenberg, like all voluminous writers, must be judged by his works in their final form."[37]

Hengstenberg and "the Awakening"

In the beginning of the nineteenth century there occurred a revival movement which can be classified as "restoration theology." By the latter term is meant "the tendencies to attain theological goals primarily by turning back to older prerationalistic tradition. Die Erweckung was an heir of the older Pietism, above all in its Wuerttemberg form but it was also related to the newly awakening religious interest in Romanticism and German idealism. Its proponents were pleased to claim Luther as a patron (the Reformation jubilee of 1817 served to actualize Reformation ideas), but no importance was attached to the differences between Luther and Reformed theology."[38]

Hagglund listed Ernst W. Hengstenberg and Friedrich Adolph Philippi (died 1882), Kirchliche Glaubenslehre, 1854-71, as the major spokesmen for "repristination theology."[39] This school of thought aimed at a resuscitation of old Protestant theology. This theological school looked upon old Lutheranism as adequate for the evangelical church. It rejected the use of modern science in its presentation of the interpretation of Holy Writ and the employment of the Lutheran Confessions. The repristinationists rejected the idealistic transformation that had occurred and was being promoted by Hegel and Schleiermacher. This school of theological thought exercised considerable influence of the church life and had many fruitful consequences for good. Hagglund evaluated repristination theology in his History of Theology. Here is his evaluation: "Its achievements were limited, however, by its lack of attention to Luther's theology, and also by reason of the fact that its leaders seemed to ignore the differences between the intellectual assumptions of the old Protestant period and of the 19th century. They instinctively accepted the world view and thereby fell into contradiction. A genuine repristination proved to be impossible. The older Lutheran position was denied at several points. The distinction between Lutheranism and Calvinism was considered unimportant."[40]

The Concordia Cyclopedia while praising the efforts of Hengstenberg as an apologete of the Christian faith and also of Lutheranism, has this demurer: "It must be said, however, that in the end he remained within the 'Union' ('What God hath put together, let no man put asunder') and refused to break with the rationalists in the Church. Sternly opposed to rationalism, he yet bespoke a certain means of freedom for theology. In his later years he adopted a Romanizing view of the doctrine of justification."[41]

In his *Commentary on the Apocalypse* he taught a millennium which was not consonant with the rejection of Millennialism in the Lutheran Confessions (Augsburg Confession, Article XVII).

The Critical Theological Climate in Germany
During Hengstenberg's Lifetime

When Hengstenberg began his public career in Berlin, the character of theology controlling Germany was bad and its religious outlook was gloomy. Pope, as a Britoner, evaluated the German religious situation as follows: "The expectation excited some dozen years before by the fresh tide that had been poured into German thought and life by the war of independence had to a great extent disappointed."[42] It was hoped that the tide of rationalism had been checked and better days were ahead for the evangelical Church in Germany.

1505

It was further hoped that the Union of Lutheran and Reformed Churches, brought about more or less by force by King Frederick William was to improve the religious situation. By the Tercentary of the Lutheran Reformation, (1817) it was hoped that the Prussian Union would improve the religious situation and Protestantism would flourish. The plan of the King's Union proved very divisive with Calvinism battling Lutheranism and Lutheranism fighting Reformed theology.[43]

Rationalism Not Thwarted

Rationalism was not stopped or hindered but flourished because its proponents occupied important university chairs and had as representatives first-rate scholars who appeared to have scholarship on their side. In fact, the old vulgar rationalism of Roehr and Bretschneider was in the ascendency. In Halle, there were 800 students subjected to the rationalism of Gesenius and Wegschneider. The professors who taught future pastors were skeptical about the reality and trustworthiness of the Bible and challenged and destroyed the great Christian verities taught in the Ecumenical Creeds of Christendom.[44]

Hengstenberg lived till 1869, and thus he lived through nearly seventh-tenth of the nineteenth century, which witnessed some remarkable theological developments, inimical to historic Biblical Christianity and hostile to a sound conception of Lutheran doctrine. Between the year of his birth (1802) and the year of his death, Hegel, Schleiermacher and Negative Biblical Criticism appeared, whose views affected untold numbers of people in Germany, Holland, France, England and America.[45] Two years after Hengstenberg's birth Immanuel Kant died, who lived from 1724 till 1804. During and after Hengstenberg's liefetime, Kant exercised an important influence on modern philosophy.[46] *In Kritik der reinen Vernunft* he tried to show that the transcendent world, the existence of God and the immorality of the soul were unknowable to pure reason. Then in *Kritik der praktischen Vernunft* he endeavored to undo the views expressed in *der reinen Vernunft*. Freedom of man, the immortality of the soul, the existence of God (the three great principles of the "Enlightenment") could be held. These three are postulates which could be established by the practical reason, i.e. of conscience. Prominent in Kant's ethics is what he called the "categorical imperative." In his *Religion in-errhalb der Grenzen der blossen Vernunft* he declared that morality is the essence of religion. Saving faith he claimed was identical with leading a god-pleasing life. Hengstenberg had also to contend with the Kantian view of religion which negated basic Biblical doctrines.[47]

Wilhelm Hegel's Influence

Hegel and Hengstenberg were contemporaries in Berlin. After 1818 Hegel was the main proponent of Absolute Idealism in modern philosophy.[48] Everything that exists is the result, ultimately, of the development of one absolute thought or idea, or, expressed in terms of religion, the world including nature and humanity, is only the self-manifestation of God. Though Hegel claimed that his philosophy was in agreement with

Christian doctrines and was hailed by many as the most rational explanation of Christianity, reconciling theology and philosophy perfectly, it was still in reality pantheism and amounted to the complete negation of Christianity. Hegel did not believe in a concrete, historical Jesus, and his Neo-Hegelian school of philosophy led to a destruction of the historical foundations of Christianity.[49]

Schleiermacher's Influence

Another opponent of apostolic Christianity was Friedrich Daniel Schleiermacher (1768-1834).[50] In 1796 he issued his *Reden Ueber die Religion*, written against the prevailing "Enlightenment." In this work he set forth his conception of religion and the Church. Schleiermacher defined "religion is the taste and feeling for the infinite." In 1810 he became dean of the faculty of the University of Berlin. At the same time he was the preacher of the Dreifaltigkeitskirche in Berlin. Schleiermacher's chief work was *Der Christliche Glaube nach den Grenzen der evangelischen Kirche im Zusammenhalt dargestellt* (1821-1822). Here he defined religion "as feeling of absolute dependence upon God, who is the highest causality." Christ's supernatural birth, resurrection, ascension and second coming are discarded. In Christ was supposedly found the highest degree of God-consciousness. Redemption for man was said by Schleiermacher to consist in this high consciousness which Christ showed man.[51]

Although Schleiermacher believed that he was attacking and refuting rationalism, his views negated every basic doctrine of the Christian faith. He completely failed to do justice to the Christian faith. He laid the foundation for a highly subjective character of present day theology. He was both a rationalist and a pantheist. By no stretch of the imagination can Schleiermacher's religious views represent apostolic Christianity. Hengstenberg opposed him vigorously just as he had Hegel.[52]

Higher Criticism in the Nineteenth Century

During Hengstenberg's teaching and preaching lifetime, the documentary hypothesis was being gradually formed and developed by a trial and error method.[53] Gottfried Eichhorn (1752-1827) elaborated Astruc's documentary hypothesis which distinguished a number of different sources in Genesis, but did not question the Mosaic authorship.[54] By extending Astruc's principles to the whole Pentateuch, Eichhorn in the last edition of his Einleitung in das Alte Testament (1823-1826) came to the conclusion of the existence of sources in the Pentateuch which he claimed were later than Moses, although he concluded that some source might go back to the time of Moses.[54] It took nearly a century for what became known as the Final Documentary Hypothesies to be developed, as it was popularized by Julius Wellhausen (1844-1918) in his *Prologemena to the History of Israel*.[55] During

Hengstenberg's later years the critical approach to the Old Testament produced a number of incompatible theories relative to the alleged composition of the Hebrew Torah, the Pentateuch of Moses. At first De Wette (1778-1843) accepted the view of the Roman Catholic priest Alexander

Geddes, who claimed the Pentateuch was a fusion of a number of fragments and for a while the "frangmentary theory" held the field.[55] According to this view a Redactor at the time of the Exile (586-539 B.C.) combined the many fragments into a book, later called by the Jews "the Torah." It was a theory that was soon discarded but out of it came the position that Deuteronomy was not of Mosaic origin. The present book of Deuteronomy was believed to hold a core, denominated "the Deuteronomic Code" (chapter 12-26) and came from the time of Josiah (640-609 B.C.). That Deuteronomy was of sixth century B.C. origin became one of the five pillars on which the Final Documentary Hypothesis rested.[56]

The supplementary theory followed the fragmentary, with the former assuming a **Grundschrift** (a foundational document) which in the course of time was supplemented by other sources.[57] This appeared in the 11th or 10th centuries B.C. and thus Mosaic authorship was impossible. The most important proponent of this view was Georg Augustus Ewald (1803-70), a German Orientalist and Old Testament scholar.[58]

Three years before Hengstenberg's death, there appeared a work by Graf (1866) in which the documentary hypothesis was rather clearly delineated. The Final Four Source Hypothesis is often referred to as the Graf-Wellhausen theory. The latter assumed that the religion of Israel developed from fetischism to polytheism into an ethical monotheism at the time of the eighth century prophets.[59]

Hengstenberg's Influence

Critical scholars rejected Hengstenberg's hermeneutics, theology, isagogics and exegesis. Hahn in his book, *The Old Testament in Modern Research* never mentions Hengstenberg, although he refers to Graf and Wellhausen some thirty times.[60] Hahn claims that Graf-Wellhausen laid the foundation for the critical studies of the twentieth century.

However, whether liberal scholars like it or not, Hengstenberg did especially influence the theological outlook of the century in which he lived. According to Pope, Hengstenberg influenced English theological thought for many of his writings were translated into English and were a standard part of T. & T. Clark's Theological Library. Thus Pope wrote concerning Hengstenberg's influence in England: "Hence no German name is so familiar in England as that of Hengstenberg; certainly none is so thoroughly identified with all that is sound and honest and loyal in German theology. And this prominence among ourselves corresponds very fairly with his prominence in the Contentnt."[61]

Pope further claimed that Hengstenberg has been "the foremost in Lutheran Protestantism for nearly a generation."[62] While Franz Delitzsch was constantly changing his mind and adjusting to the latest view in Biblical criticism, he did assert about Hengstenberg that "he brought Old Testament exegesis back to a churchly basis."[63]

Edward Young made this judgment about Hengstenberg: "Ernst Wilhelm Hengstenberg, H. Ch. Haevernick, and C.F. Keil voiced protests against the critical treatments of the Old Testament. These men were believing scholars who wrote with high regard for the integrity and trustworthiness of the Bible. Their writings have had a great influence in England and America."[64]

Hengstenberg's Influence on Conservative Old Testament Scholarship in the Twentieth Century

Between the years 1831 and 1836 Hengstenberg produced his Beitraege zur Einleitiung ins Alte Testament. The first volume of this series dealt with Die Authentic des Daniel und die Integritaet des Sacharjah.[65] The Authenticity of Daniel and the Integrity of the Book of Zecharia published 1831, the second volume in this series of entitled: Die Authentic des Penteuches (The Authenticity of the Pentateuch). The latter work comprised 502 pages.[66] Twentieth century defenders of the Christian faith as Ludwig Fuerbringer,[67] W. Moeller,[68] Unger,[69] Young,[70] Archer,[71] Harrison[72] and others read Hengstenberg's writings and were no doubt influenced in their thinking by his isogogics, hermeneutics and Christological understanding of the Old Testament.

Footnotes

1. L. Fuerbringer, Th. Engelder, P. E. Kretzmann, *The Concordia Cyclopedia* (St. Louis: Concordia Publishing House, 1927), p. 313.

2. Joh. Bachmann, *E.W. Hengstenberg nach seinem Leben und Wirken: Guetersloh*, (1875), I.

3. Johannes Bachmann, "Hengstenberg, Ernst Wilhelm," *Realencycklopedie der Protestantischen Theologie und Kirche*, 1899, vii, p. 670.

4. W. B. Pope, "Essay on the life and Writings of Hengstenberg," in prologue to evaluation of the life and work of Hengstenberg in *History of the Kingdom of God under the Old Testament* (Edinburgh: T. & T. Clark, 1872), II, p. x.

5. Bachmann, E.W. "Hengstenberg" in *Raelenencyklopedie*, **op. cit.**, p. 671.

6. **Ibid.**, p. 672.

7. "Hengstenberg, Ernst Wilhelm," *Concordia Cyclopedia*, **op. cit.**, p. 317.

8. Bachmann, **op. cit.**, p. 672.

9. Pope, **op. cit.**

10. Hans Joachim Kraus, *Geschichte der Historischen-Kritischen Erforschung*, (Neu Kirchen Kreis Moers: Verlag der Bucbhandlung des Erziehungsvereins, 1956), p. 212.

11. Pope, **op. cit.**, p. xx.

12. Otto W. Heick, *A History of Christian Thought* (Philadelphia, Fortress Press 1966), p. 201.

13. Kraus, **op. cit.**, p. 220.

14. **Ibid.**

15. Charles F. Pfeiffer, *The Dead Sea Scrolls and the Bible* (Grand Rapids, Baker Book House, 1969), pp. 130-132. Menahem Mansoor, *The Dead Sea Scrolls* (Grand Rapids: Baker Book House, 1983. Second Edition, p. 157.

16. Matthew 11:3; Matthew 26:63; Luke 2:21; Luke 20:41; John 1:41; John4:25. Acts 3:18.

17. 2 Corinthians 3:14-18.

18. Kraus, **op. cit.**, pp. 2-3-207.

19. **Ibid.**, p. 215.

19a. *Beitraege zur Einleitung ins Alte Testament, Die Authentic des Daniel und die Integritaet des Secharjah* (Berlin, Ludwig Gehmigke, 1830).

20. Emil Kraeling, *The Old Testament Since the Reformation* (New York, Harper & Brothers, 1955. Completely ignores life and Old Testament contributions of Hengstenberg).

21. Walter A. Maier, Sr. *Notes on Selected Psalms* (St. Louis, Concordia Mimeo Company, 1929 of Psalms 2,16,22.

22. Acts 4:23-32.

23. John Goldingway, *Old Testament Commentary Survey* (Madison, Wis., Student Fellowship - A Division of Inter-Varsity Christian Fellowship, 1981), pp. 29-30.

24. Bernard S. Childs, *Old Testament Survey for Pastors & Teachers* (Philadelphia: The Westminster Press, 1977), pp. 59-63.

25. Pope, **op. cit.**, p. xxvii.

26. Information taken from Bachmann, **op. cit.**, p. 673.

27. **Ibid.**, p. 673.

28. J. Barton Payne, *The Theology of the Older Testament* (Grand Rapids: Zondervan Publishing House, 1962), pp. 25-43.

29. **Ibid.**, p. 27.

30. **Ibid.**, p. 28.

31. **Ibid.**, p. 27. .

32 Gustav Friedrich Oehler, *Theology of the Old Testament.* Edited posthumously and it was preceded by his *Prolegomena to the Old Testament Theology*, 1845. The complete theology was translated by George F. Day (New York, Funk & Wagnall, 1883). 594 pp.

33. Adolph Spaeth, "Oehler, Gustav Friedrich," H.E. Jacobs and John A. Haas, Editor, *The Lutheran Cyclopedia* (New York: Charles Scribner's Sons, 1899), p. 352.

34. **Ibid.**

35. Pope, **op. cit.**, xxviii.

36. **Ibid.**

37. **Ibid.**

38. Bengt, Hagglund, *History of Theology*, Translated. Gene J. Lund (St. Louis, Concordia Publishing House, 1968), p. 363.

39. **Ibid.**

40. **Ibid.**, p. 363.

41. *Concordia Cyclopedia*, **op. cit.**, pp. 317-318.

42. Pope, **op. cit.**, p. xiv.

43. **Ibid.**

44. Lewis W. Spitz, "Lutheran Theology After 1580," E. Lueker, *The Lutheran Cyclopedia* (St. Louis: Concordia Publishing House, 1954), p. 639.

45. E. H. Klotzshe, *The History of Christian Doctrine* (Burlington, Iowa, The Lutheran Literary Board, 1945), pp. 310-320.

46. Horst Stephan, *Geschichte der Evangelischen Theologie* (Berlin, Verlag Alfred Toepelmann, 1938), pp. 23-31.

47. Jerald C. Brauer, Editor, *The Westminster Dictionary of Church History* (Philadelphia, The Westminster Press, 1971), pp. 470-471.

48. Elgin Moyer, *Who Is Who in Church History?* (Chicago, Moody Press, 1962), p. 190.

49. Hegel, "Georg Wilhelm Friedrich," *The Lutheran Cyclopedia*, **op. cit.** p. 452.

50. Klotzsche, **op. cit.**, pp. 311-314.

51. *The Lutheran Cyclopedia*, 1954, **op. cit.**, p. 953.

52. **Ibid.**

53. Edward A. Young, *An Introduction to the Old Testament* (Grand Rapids: Wm. B. Eerdmans Publishing Company, 1969), pp. 119-122.

54. Merrill F. Unger, *Introductory Guide to the Old Testament* (Grand Rapids: Zondervan Publishing House, 1951), pp. 242-243.

55. **Ibid.**, p. 243.

56. **Ibid.**, p. 244.

57. Gleason L. Archer, *A Survey of Old Testament Introduction* (Chicago: Moody Press, 1975), pp. 83, 86.

58. R. K. Harrison, *Introduction to the Old Testament* (Grand Rapids: Wm. B. Eerdmans Publishing Company, 1969). p. 15.

59. **Ibid.**, 19ff.

60. Herbert F. Hahn, *The Old Testament in Modern Research With a Survey of Recent Literature by Horace D. Hummel* (Philadelphia, Fortress Press, 1954), p. 31.

61. Pope, **op. cit.**, p. vii.

62. **Ibid.**, p. viii.

63. Hengstenberg in *Lutheran Cyclopedia*, 1899, **op. cit.**, p. 218.

64. Young, **op. cit.**, p. 22.

65. Ernst Wilhelm Hengstenberg, Die Authentie des Daniel und die Integritaet Sachar jab Beitraeg zur Einletung ins Alte Testament Band I (Berlin: Bel Ludwig Oehmigke, 1931.

66. Ernst Wilhelm Hengstenberg, *Die Authentie des Pentateuches. Beitraege zur Einlietung ins Alte Testament*, Zweiter Band (Berlin: Bei Ludgwig Oemigke, 1936.

67. Ludwig Fuerbringer, *Einleitung in das Alte Testament* (St. Louis, Concordia Publishing House, 1923), pp. 15-24.

68. W. Moeller, *Die Einheit und Echtheit der 5 Buecher Mosis* (Bad Salzusien: Selbverlag des Bibelbundes, 1931), 480 pages.

69. Unger, **op. cit.**, pp. 213-276.

70. Young, **op. cit.**, pp. 42-61.

71. Archer, **op. cit.**, pp. 54-178.

72. R. K. Harrison, *Introduction to the Old Testament* (Grand Rapids, William B. Eerdmans Publishing House, 1969), pp. 171, 424, 504, 1040a.

Questions

1. When was Hengstenberg born? ____

2. Hengstenberg came to the conviction that the Lutheran Confessions contained ____.

3. Hengstenberg became the editor of ____.

4. The once conservative Franz Delitzsch was constantly ____.

5. A major work of Hengstenberg was his ____.

6. When was a reprint of the English translation by R. Keith published? ____

7. The problem of historical critical scholars is that they ____.

8. Why did critical scholars of the nineteenth century hate Hengstenberg? ____

9. What criticism did Walter A. Maier, Sr. make of Hengstenberg's comments on Psalm 2? ____

10. Who was Gustav Oehlers? ____

11. What was "restoration theology?" ____

12. "Repristination theology" aimed at ____.
12. *The Concordia Cyclopedia*, while praising Hengstenberg, said that in his later years he adopted ____.
13. Immanuel Kant tried to show ____.
14. Hegel did not believe in ____.
15. Scheichermacher's views negated ____.
16. Who popularized the Final Documentary Hypothesis? ____
17. The Graf-Wellhausen theory assumed that ____.
18. What did Edward Young say about Hengstenberg? ____

A Comparison of Traditional Roman Catholic Ethics Versus Historical Lutheran Ethics

Christian News, Octotober 21, 1991

On October 31st the Lutheran Church will observe the 474th anniversary of the nailing of the ninety-five theses on the church at Wittenberg (generally held to mark the beginning of the Protestant Reformation) as well as the 508th anniversary of the birth and baptism of Martin Luther on November 10th and 11th respectively. By baptism Luther became a member of Christ's kingdom, as well as a member of the Roman Catholic Church and also a member of the invisible Christian Church.

In the past it was customary in connection with Reformation Sunday to recall the necessity of the Protestant Reformation or Revolt as Rome calls it, and to emphasize the restoration of New Testament apostolic Christianity to the world.[1] However, in recent years, under the influence of ecumenism many Protestants and Lutherans have become convinced that the reformation by Luther was a great and tragic mistake and that the Protestant Revolt divided Christendom unnecessarily and that the differences between Rome and Wittenberg are not that important inasmuch as Protestant and Rome Catholics recognize Christ as Savior, Redeemer and Helper; doctrinal differences are just different ways of looking at truth from different angles.

Unfortunately, those who hold it does not matter what one believes and that doctrinal latitude is permissible, show a great ignorance of the New Testament and often do not actually know the doctrines of either the Roman Catholic Church and of the Lutheran Church.[2]

Jesus, in connection with the giving of the Great Commission, instructed his disciples to make disciples of all nations by baptism and instructing them in all things whatsoever Jesus had commanded them. Christ also promised His disciples the Holy Spirit, who would guide them into all truth (John 16:13). All the books of the New Testament were written by apostles or by persons close to the apostles, so that the entire New Testament, like the Old Testament, is the Word of God. God's Word in the New Testament contains numerous passages that call upon believers to avoid false prophets and to contend for the truth once delivered to the saints. Indifference to God's truths and teachings is condemned by Christ and his New Testament writers.[3]

No Radical Changes in Rome Despite Vatican II

Although certain administrative changes were made by the participants of Vatican II (1965), no real radical doctrinal changes have been made by the teaching magisterium of the Roman Catholic Church.[4] Its doctrinal system which ultimately controls Christian living is still the same. The time-honored teachings which were a stumbling block in the days of the Reformation are still the same as they were in 1517 and in

1580, when *The Book of Concord* was published, containing confessions that clearly explicate the differences between Rome and Wittenberg.

In this presentation the ethical differences between Rome and historic Lutheranism will be set forth and it will be shown that theologically and ethically great differences do exist between traditional Roman Catholic teachings and ethical practices flowing from such doctrinal positions.

The Difference Between General and Christian Ethics

Ethical theory and ethical views can be traced to pagan antiquity.[5] Butler defined general ethics in this matter: "The science which investigates individual conduct. It inquires into the moral sentiments, intentions, the good and the right, the criteria of obligations, conscience and the summum bonum or highest good, and its freedom."[6] Another definition reads like this: "General ethics is the science which treats of the sources, principles and practice of right and wrong."[7] Christian ethics can be described "as the science which treats of the sources and principles and practice of right and wrong in the light of Holy Scripture, in coordination with nature and reason, especially when they do not contradict God's Word."[8]

The Separation of Christian Ethics From Christian Dogmatics

For many centuries Christian ethics was treated as a part of Christian dogmatics.[9] It was in the post-Reformation age that Christian ethics was separated from Christian dogmatics and treated as a distinct discipline.[10] In the last few hundred years a large literature has grown, dealing with ethical principles and practices and there exist many different schools of interpretation.

Christianity, unfortunately, is divided into many particular churches and denominations and the fact is that there are significant differences between individual churches and denominations.[11] Naturally these differences are reflected in their respective traditions, manifested in their moral and ethical practices.

Rome's Ethics Versus Lutheran Ethics

The eminent twentieth-century Lutheran Scholar Reu claimed: "The difference between Roman Catholic and Lutheran Ethics is particularly pronounced and of fundamental importance."[12]

Roman Catholic Views on Religion and Morality

Christian ethics distinguishes between religion and morality. The Roman Catholic Church does not distinguish between religion and morality.[13] Stump claimed that Rome misconceives the two concepts because its teachings regard faith as merely intellectual assent to the teachings of the Church and morality as obedience to the commands of the Church, as well as to the law of God.

The Church of Rome interposes itself between the soul and God and makes man's relationship dependent upon the Church instead of dependence upon Christ.[14]

One of the significant contributions of the Lutheran Reformation was to give mankind the true conception of both what is involved in religion and in ethics and the Reformation stressed the vital and true connection between them.

The Sources of Roman Catholic Ethics

Roman Catholic ethicists derived their ethical principles from the Bible (which includes the Apocrypha), the decisions of church councils, encyclicals of various Roman Pontiffs,[15] and from philosophy. By contrast, Lutheran ethics is derived from Holy Writ or from the deductions which can be made based on other ethical and doctrinal teachings.

Thomas Aquinas and Roman Catholic Ethics

Eeningberg claimed that "Roman Catholic ethics and moral theology today are dependent in their mainlines, upon the massive structure of moral theory in the thirteenth century."[15a] While contemporary Catholicism displays a variety of opinion in moral theory,[16] all of it can be measured in distance from Thomas. The main body appears to be close to the Angelic Doctor in form or substance.[17] Eeningberg contended that "the main traditional Thomistic accents conserved are those summed up in the insistence upon the rational, legal, and therefore objective character of the moral order approved by God. These have been the characteristic marks of Catholic moral theory for centuries."[18]

The Great Difference Between
Roman Catholic And Lutheran Ethics

Also in the realm of ethics the confessional character of an ethical system is apparent. Central to Rome's ethics is the morality of work righteousness. Rome never truly comprehended the teaching of Paul. "Therefore we conclude that man is justified by faith without the deeds of the law." (Romans 3:28) This would make Roman Catholic ethics appear to be legalistic, slavish and mercenary.[19]

Reu claimed that Lutheran ethics is the ethics of the justified man, justified and therefore made good by the grace of God for Christ's sake. The Lutheran Christian is impelled by an inner necessity, in grateful love for this grace received to obey the Father's will, and to obey it freely, not as a series of individual tasks to be performed but as the uniform governing principle of his life.[20]

The Influence of Pagan Philosophy on Roman Catholic Ethics

As was already stated, Thomas Aquinas had a determining influence on Roman Catholic ethics. Aquinas was greatly influenced by Aristotle. The latter dominated medieval ethics, and was highly regarded in all seats of learning. Aristotelian ethics was not a matter of man's inner disposition, but of external actions. Aristotle taught that man became morally good by doing good, much as the same way a harpist achieves control over his instrument by constant practice. The Greek philosopher believed that the law was the best teacher of morality, because the law insisted that a person do good. Semi-Pelagianism controlled the Medieval Church, which meant that natural man was not totally depraved. This latter concept of natural man was joined with Aristotelian ideas. Aristotle was revered as the master of all former disciplines, such as philosophy, logic, rhetoric and grammar.[22]

Luther in his conflict with the theology and ethics of the Roman Catholic Church from the start rejected the pagan and anti-Scriptural views of Aristotle. Reu averred: "In the field of formal sciences Luther was willing to accord him due honor and Melanchthon made deliberate use of him in some of his textbooks. But his influence with respect to matters of morality and religion was opposed by Luther most emphatically." "Let philosophy remain in its own sphere," he said, "for which it was given by God." Luther completely rejected the stance of Aristotle when the latter taught "by doing good we become good."[22]

Roman Catholic Theology's Wrong View of Man's Original Nature

Unfortunately, the Roman Catholic Church held and still holds the position that man by nature, the natural man, possesses all the powers necessary to do God's will and that these powers need only be stirred up by the Holy Spirit in order to enable man not only to do all that God demands, but also to perform the works of super-arrogation which may be applied to the credit of other persons.[23]

In contrast to this teaching the Reformation denied the possession by man of powers needed to live an ethical life[24] but that the natural man needed to be born again (John 3:5) and by God's grace had to be regenerated. It is as Stump has pointed out: mankind must distinguish between the natural man's powers to do good the realm of **justitia civilis,** (civil justice) and the regeneration to do good works in the realm of **justitia spiritualis.**[25] Luther[26] and other Reformers repudiated the idea that Christian ethics is merely a supplement of natural ethics, and contended that Christian ethics alone supplies for mankind the ethical goals demanded by God.[27] In the Lutheran Church the distinction between General Ethics and Christian Ethics was made by Calixtus (1586-1656), who made the regenerated person the subject of his Christian ethics.

The Doctrine of Justification by Faith in Lutheranism

When Luther discovered the doctrine of justification by faith a new world opened for him.[28] It became clear to Luther that the most important factor in religion and ethics was this teaching, totally obscured by Rome, namely that religion was a personal relationship between God and the individual, and not as Rome taught about man's relationship to the Church. Reu expressed that position in this way: "Now there is 'peace without end' between man and God. Because in justification man has been declared righteous and because the old has been done away and with faith and the Holy Spirit new powers have entered into him, man can now do what is really good for he himself has now become good. Indeed, now he must do good, for the new Spirit within him. Will not suffer him to remain idle but constantly urges him to do it."[29]

The Christian will perform good works not out of fear of the law which demands them, but out of love for what a gracious God has done for him.[30] Furthermore, the regenerated child of God will not perform works in order to merit justification. No longer do Christians have to wonder whether they have done enough good works to merit salvation.

Reu also claimed that a major difference between Roman Catholic ethics and Lutheran ethics was the latter's distinction between a higher and lower form of reality.[31] Thus he wrote: "The higher form involves a renunciation of the use of the rights and gifts of nature and culminates in the three vows of chastity, poverty and complete subjection to ecclesiastical superiors. The lower form permits the unrestricted use of all things natural."[32] By contrast Lutheran ethics only postulates one form of morality, which is obligatory for all, and regards the things and ordinances of the natural world to be divinely established when God created the world and that mankind was given the right and power to use them.

The Pagan Sources for the Distinction Between Lower and Higher Morality

This distinction between a higher and lower morality used in Roman Catholic ethics can be traced to two roots. Here Roman Catholicism shows the influence of pagan stoicism.[33] Pagan Roman stoicism classified humanity in three classes: the wise, the fools and the **proskoptontes** (i.e. "aspirants") or such who worked at being wise. Fools were said to have no virtue at all, thus there were only two classes left: the common or lower virtues of the aspirants and the higher virtues of the wise. Cicero in **De Officias** gave a detailed discussion of this teaching. Ambrose adopted this distinction in his book **De Officiis Ministrorum** and he was the popularizer in the Church and it highly claimed that while heathens were capable of putting forth the ideal of the wise man praising his virtues, the Church alone had such wise men. Ambrose focused on those who practiced the lesser virtues, kept the commandments of the decalogue, but also those who practiced the evangelical counsels **(consilia evangelica,** Matt. 19), love their enemies, give their goods to the poor, fast voluntarily, show mercy and practice celibacy. This was the beginning of the later distinction between monastic and lay ethics.[34]

The second root can be traced to the wrong concept, distinctive of Medieval theology, concerning the Christian relationship to the natural and purely human. Medieval theology distinguished sharply between nature and grace. Nature in the course of time was identified with the earthly reality, carnal and sinful. It was against marriage, the possession of wealth, the use of worldly goods, government, war, barter and trade, commerce and industry, all considered sinful. The truly ethical was considered to be found in the renunciation and avoidance of all earthly things. For church members to employ and use all these earthly things required the Church to make special concessions regarding them. Of this teaching Reu made this judgment: "It is beyond estimation what embarrassment and confusion, what violations and coercion of conscience, and what obstruction of progress in sound living have been caused by this doctrine."[35]

1517

Lutheran Opposition to Higher Morality Concept

The Roman Catholic Church taught the higher existence of special classes within the Church, based on the teaching of a higher and a lower morality. Over against this ethical teaching Luther placed Galatians 3:28 and II Corinthians 5:17.[36] In these verses Paul did not allow for racial, cultural and social differences as affecting the believers relationship to God. Before God all justified persons stand in the same relationship and in the same place, whether they be clergy or laity.[37] If this is true, how can there be different levels of morality, as Rome taught? Luther completely rejected this erroneous identification of nature and sin. All cultural orders are a part of God's creation. How can marriage,[38] government[39] and other legitimate endeavors be part of a lower moral order? How can life instituted for the common welfare of mankind be classified as sinful? Against Roman Catholic ethical misconceptions the Wittenberg Reformer advanced I Timothy 4:2-4 and Colossians 2:16-18 as well as Galatians 5:1 and I Corinthians 3:23. Marriage, family, the government, the arts and crafts were for Luther holy orders and occupations and there was nothing morally wrong with being involved in these God-instituted institutions.

Major Differences Between
Roman Catholic and Lutheran Ethics

The Nature of Man's Constitution

Both Roman Catholic and Lutherans taught that all members of mankind are sinners and commit many different kinds of sins. However, the Roman Church makes a distinction between venial and mortal sins.[40] In themselves considered, all sins are mortal and can lead to eternal destruction, unless repented. According to the Sermon on the Mount evil thoughts, evil words can lead to destruction, unless repented. The Roman Catholic distinction between venial and mortal sins is one the Catholic Church has created but has no valid grounds in Holy Scripture.[41] **The Katholischer Kurz-Katechisms** in question 33 asks this: "What is meant by original sin? Answer: Original or hereditary sin consists in this, we come into the world without the life of grace."[42] Historic Lutheranism teaches that each person being totally depraved by nature and an enemy of God has no morally good in him.[43] Man by nature is a lost and condemned sinner. The teaching on original or hereditary sin has important implications for the realm of Christian Ethics. It will be helpful to be clear on the teaching of original sin and determine why people are sinful and guilty, and are in need of forgiveness of sins. God, strictly speaking, does not forgive the sin but the sinner who needs forgiveness.

Roman Catholic Ethics and External Positive Law

The Roman Catholic Church, opposed by Luther, does not hold that faith is an operative power, working through love, but is held as a belief of the Church and that morality consists in doing what the Church commands. In historic Lutheranism faith is considered both as a receptive and operative power and holds that the moral life is the outgrowth of love, which is the resultant of true justified faith.[44]

The Roman Catholic Concept of Love

In the Ancient Church, on the basis of I Corinthians 13:13, faith, hope and love are considered the three outstanding theological virtues. This stance was also a common belief of the Middle Ages and was taken over by the Roman Catholic Church. It is still the belief of the Church of Rome, which in its doctrine of **fides informis** and **fides formata,** gives to love a place above faith. But faith, hope and love are not really coordinated virtues. On the basis of the New Testament, historical Lutheranism postulated an essential difference between faith on the one hand, and love and hope on the other. Faith stands in a class by itself and strictly speaking is not a virtue.[45] Faith is the means by which there comes into existence a relationship between God and man, a relationship alone which makes love or hope or any other Christian virtue possible. On Scriptural grounds historic Lutheranism would ascribe to faith the preeminence over love and hope inasmuch as faith is the originating and vital principle, with hope and love being the resultant from faith. Love is indeed the essence of the Christian life and ethics, but faith is its prerequisite.[46]

Lutheranism Rejects the Socalled "Evangelical Counsels"

Since the ethical conduct of Christians is a unit, controlled by one principle of love and comprising obedience to God's will as the sole norm, the Roman Catholic Church distinguishes between two orders of morality, as previously pointed out, namely, the lower order consisting of obedience to the commandments of God and the Church and the higher one consisting of obedience to the evangelical counsels; this distinction is not biblical and has no foundation in God's Word.[47]

Roman Catholic theology made the assumption that by infused grace transmitted in the Sacraments, Christians can perform good works which are meritorious and that obedience to the evangelical counsels of perfection are not regarded as optional conduct when they are performed, and result in the accumulation of benefits available for other individuals who are in need of good works.

Luther and the Lutherans objected to the evangelical counsels as being works of super arrogation and as being unable to help any person to qualify for heaven. It is as Stump has correctly evaluated this teaching and practice of Rome: "In ascribing to good works a meritorious character and coordinating them with faith as the instrumental cause of salvation the Roman Catholic Church sets the merits of Christ in the background."[48] Luther and his followers maintained that good works have nothing to do with obtaining salvation, but that they are done and can alone be done by individuals who are already saved by grace through faith.[49]

Differences on the Concept of Prayer

During the Middle Ages the Roman Catholic Church considered prayer not as communion with God, but rather as a meritorious act along with fasting and almsgiving.[50] The mystics looked upon prayer as a means of absorption into God. The Reformation restored the meaning of prayer as communion with God by resorting to the doctrine of justification by faith.

Asceticism Taught by Rome Rejected by Luther

Asceticism, a phenomenon found in many heathen religions, developed in Church History very early and has persisted to modern times.[51] When asceticism was adopted, the concept was promoted that living an ascetic life was a great religious work and had meritorious value. Asceticism was engaged in by monks and hermits and often monks tried to outdo each other in almost suicidal treatment of themselves. In fact, these severe ascetic practices were even considered expiatory.[52] Such asceticism is greatly opposed to the Biblical teaching that sinners are only justified by faith and not by works.

To this day celibacy and asceticism are still practices in the Roman Church. In the past they have constituted a source of strength for the promotion of Roman Catholic faith and in the promotion of varied activities in which the Roman Church has been engaged in for centuries.

Difference in the Concept of the Family

The Roman Catholic Church espouses a different view of the family than does Lutheranism. God instituted holy matrimony and the proper use of sex. Adam and Eve were joined together by God and they were to remain together until death would disrupt this relationship. The family occupies a prominent role in the Old Testament. The families of Noah and Abraham stand out at the beginning of two different historical periods; in the beginning of the human race, and again after the Great Flood. Israel as a nation came from Jacob's family. The world's Savior was born of the house and lineage of David. As a growing child and adolescent, Jesus was obedient to his foster father Joseph and to his mother Mary. Christianity found its way into families and into its societies. Paul and Peter both taught the importance of the family (Ephesians 5:22; Colossians 3:18; 1 Peter 3:11). The Roman Catholic Church by its exaltation of the monastic life and the celibate state deprecated the family and its place in society, but Luther and the Reformers restored the value and dignity of the family. Roman Catholicism exaltation of celibacy is based on the misinterpretation of Holy Writ and its wrong theological teaching on the value of ascetitism.[53]

Clerical Celibacy Versus Married Clergy

In 1073 Gregory VII forced the greatest divorce in history in that married clergy had to separate and ever since priests have been forbidden to marry.[54] Even though Peter was married and Paul asserted that he like Peter could have had a wife and although[55] Paul specifically permitted a bishop to be married to one wife,[56] the Roman pontiffs have ever since forbidden the clergy to become married, creating a great problems for the clergy as well as for nuns and brothers with regard to the sexual urge, given by God for the procreation of children. Many sexual deviations have been shown to characterize the lives of men and women prohibited from finding legitimate release for a strong sexual urge.

Clerical celibacy was rejected by Luther who married the nun Catherine from Bora and thereby established the evangelical parsonage.

Difference of Divorce Between Roman Catholic Moral Theology and Lutheranism

The Roman Catholic Church has pronounced marriage a sacrament and teaches that a special kind of grace is infused into the married couple.[57] Only a priest is authorized to perform the sacred nuptial ceremony and bestow the grace needed to live a God-pleasing married life. There is no teaching in the Bible that supports the teaching that marriage is a sacrament and bestows sanctifying grace.[58] Reu stated on the other hand: "Lutheranism, however, holds that marriage is a part of the natural order established in creation and governed by God's providence."[59]

Different Teachings on Divorce

When the LORD God established marriage in the garden of Eden, He told Adam and Eve: "What God has joined together, let not man put asunder." The Roman Catholic Church denied the possibility of divorce. In the New Testament Jesus repeated this same prohibition. The Roman Church denies the possibility of divorce at all, and allows only a separation a **mensa et thoro** (bed and board).[60] It however, does make allowances for the fact that certain marriages, according to Church law should not have occurred, that they may be annulled and thus in actuality never took place. Other Catholics have argued that in view of the fact that Mark's and Luke's accounts of the Lord's words about divorce (Mark 10:11,12, and Luke 16:18) no exceptions can be permitted, even if adultery has been committed by one of the marriage partners. However, Lutherans would respond that Mark and Luke do not contradict Matthew 5:32, 19,9, for they merely set forth the morally unbreakable relationship of marriage, while Matthew states in addition that adultery as well as death represents an ipso facto break of the marriage tie.[61]

It has been the general position of Lutheran theologians that on the basis of I Corinthians 7:15 willful and malicious desertion constitutes additional valid grounds for a divorce by making the conjugational relationship impossible.[62]

The Nature of the Church in Roman Catholicism and in Lutheranism

A proper conception of the nature of the Church is essential to a formulation of the duties of those who constitute the Roman Catholic communion. The Roman Catholic Church conceives the Church as the external hierarchy and knows knothing of the idea that the true Church is invisible, the communion of all true believers in Christ.[63] The right relationship to Christ by the believer is supposed to be conditioned on the right relationship to the external Roman Catholic Church, which has claimed that outside of the Roman Church there is no salvation. Accepting the doctrinal system of the Roman Catholic Church constitutes the way to salvation for its members.

The Reformer rejected this concept of the Church and found the essential principle of Church membership in justifying grace. All those who from the heart believe in Christ's atoning blood are by the virtue of a Holy Spirit created faith, members of the communion of saints or the invisible Church. In its essence the Church is not an external organization but a spiritual fellowship with Christ as its head. The visible church has in it true believers and make-believers, called in Scripture hypocrites. Rome claims to be the visible Body of Christ and whatever laws and decrees passed by the teaching magisterium of the Church are to be obeyed as if Christ has spoken or given them. The concept of the Church as conceived and taught by Roman Catholicism and the Biblical Lutheran teaching accounts for far reaching differences in the ethical practices of both communions.

"In ascribing to good works a meritorious character and coordinating them with faith as the instrumental cause of salvation the Roman Catholic Church sets the merits of Christ in the background."

Footnotes

1. L. Buchheimer, *The First Gospel and Other Sermons*, (St. Louis: Voelkening 1913), pp. 606-611; R. E. Golloday, *The Challenge of a New Day* (*Columbus: Lutheran Book Concern*, no date), pp. 348-353. Larimer, Seeger and Bowers, *Gospel Preaching For Today* (New York: Falcon Press, 1933), II, pp. 240-250.

2. Cf. *The Split Between Roman Catholicism and Christ* (New York: No Date), Sixth Printing, 40 pp.; Theodore Hoyer, *Why I Am Not A Roman Catholic* (St. Louis: Concordia Publishing House, 1953), 80 pages James Portland, Catholic Beliefs (Franciscan Publications, 1984), 63 pages.

3. Romans 16:17; Luke 21:17; I Tim. 6:3,5; Titus 3:10; 2 John 10-11; Gala. 3:29; Jude 3.

4. Paul Emm, *The Moody Handbook of Theology* (Chicago: Moody Press, 1989), pp. 601-609.

5. Arthur Butler, *A Dictionary of Philosophical Terms* (London: George Rutledge & Sons, no date), pp. 33-34.

6. **Ibid.**, p. 33.

7. Leander S. Keyser, *A Manual of Christian Ethics* (Burlington, Iowa: The Lutheran Literary Board, 1926), p. 9.

8. **Ibid.**, p. 10.

9. Joseph Stump, *The Christian Life - A Handbook of Christian Ethics* (New York: The Macmillan and Company, 1930), p. 11.

10. **Ibid.**, p. 13.

11. Cf. F.E. Mayer, *The Religious Bodies of America* (St. Louis: Concordia Publishing House, 1956, cf. especially Part II, "The Holy Catholic Church," pp. 25-124.

12. Johann Michael Reu, and Paul H. Buehring, *Christian Ethics* (*Columbus: The Lutheran Book Concern*, 1935), p. 36.

13. Stump, **op. cit.**, p. 7.

14. **Ibid.**, p. 7.

15. J. J. Bourke, "Ethics," *New Catholic Encyclopedia*, 5, p. 571.

15a. Elten Eenigenburg, "Roman Catholic Ethics," in Carl F. Henery, Baker's *Diction-*

ary of Christian Ethics (Grand Rapids: Baker Book House, 1973), p. 593b.

16. Charles L. Manske, *World Religions* (Irvine, California: The Institute for World Religions, 1991), Appendix "Six Types of Christian Ethics" taken from the Notes of H.R. Niebuhr.

17. J. J. Bourke, "Ethics," *New Catholic Encyclopedia*, V. (1967), p, 571a.

18. Eengenburg, **op. cit.**, p. 593b.

19. Gordon H, Clark, "Aquinas," *Baker's Dictionary of Ethics*, **op. cit.**, pp. 32-33.

20. Reu, **op. cit.**, pp. 38-39.

21. **Ibid.**, p, 37.

22. Stump, **op. cit.**, p. 11.

23. Ludwig Ott, *Fundamentals of Catholic Dogma* (St. Louis: Herder, 1957), pp. 103-106.

24. **Ibid.**, p. 112.

25. Stump, **op. cit.**, p, 11.

26. *The Weimar Edition of Luther's Works*, 30 I, 178f.

27. Reu, **op. cit.**

28. E. G. Schwiebert, *Luther and His Times* (St. Louis: Concordia Publishing House, 1950), p. 2.

29. Reu, **op. cit.**, p, 39.

30. *Weimar Edition of Luther's Works*, 27, 380f.

31. Reu, **op. cit.**, p. 41.

32. **Ibid.**,p. 41.

33. Bourke, **op. cit.**, V. p. 573.

34. Reu, **op. cit.**,p, 41.

35. **Ibid.**, p. 42.

36. *Weimar Edition of Luther's Works*, 11, 415.

37. **Ibid.**

38. *Weimar Edition of Luther's Works*, W-T, no. 433.

39. *Weimar Edition of Luther's Works*, 10, III, i, 626.

40. F. Tower, "Actual Grace," in G, D, Smith, *The Teaching of the Catholic Church* (New York: Macmillan Company, 1954), I, p. 593.

41. Francis Pieper, *Christian Dogmatics* (St. Louis: Concordia Publishing Company, 1954), I, pp. 567-568.

42. *Katholischer Kurz-Katechismus* by Interdiocesan Catechetical Association, 1975, issued by Directorium Generale of the Vatican in Rome, Italy, (Koenigstein e vi Albertus-Magnus Kolleg, 13, Autlage, 1982), p, 10.

43. *Apology of the Augsburg Confession*, Article II, Theodore Tappert, *The Book of Concord* (Philadelphia: Muhlenberg Press, 1959), p, 100.

44. *Weimar Edition of Luther's Works*, 10, I, 1, 114.

45. **Ibid.**, 10, II, p, 323.

46. F. Godet, *Commentary on St. Paul's First Epistle to the Corinthians* (Edinburgh: T. & T. Clark, 1893), I, p. 262.

47. Reu, **op. cit.**, p. 44.

48. Stump, **op. cit.**, p. 112.

49. *Weimar Edition of Luther's Works*, 12, 557f.

50. E. Tower, "Sanctifying Grace," *The Teaching of the Catholic Church*, **op. cit.**, I, p. 572; Stump, **op. cit.**, p. 184.

51. Lars P. Qualben, *A History of the Christian Church* (New York: Thomas Nelson & Sons, 1940), p. 156. Konna, "Gregory VII," *New Catholic Encyclopedia*, VI, 773. Konna, "Gregory VII, Pontificate," *New Catholic Encyclopedia*, VI, p. 773.

52. *The Apology of the Augsburg Confession*; Tappert, **op. cit.**, p. 192, Par. 68.
53. *Apology of the Augsburg Confession*, Tappert, **op. cit.**, p. 244.
54. First Lateran Council, 1123, Canon XXI.
55. Cf. Mark 1:30; I Corinthians 9:5.
56. 1 Timothy 3:2.
57. Ott, **op. cit.**, pp. 460, 467.
58. **Ibid.**
59. Reu. **op. cit.**, p. 267.
60. Stump, **op. cit.**, 218.
61. Edward A. Koehler, *A Summary of Christian Doctrine* (Detroit and Oakland: By L. A. Koehler and A. W. Koehler, 1952), p. 291.
62. Ott, **op. cit.**, pp. 276-277.
63. Henry Nash, in Samuel Macauley Jackson, *The New Schaff-Herzog Encyclopedia of Religious Knowledge* (Grand Rapids: Baker Book House, 1950), IV, p. 188.

Questions

1. Under the influence of ecumenism many Protestants and Lutherans have become convinced that the Reformation by Luther was a ____.
2. God's Word calls upon believers to avoid ____.
3. Were radical changes made at Vatican II? ____
4. Christian ethics can be described as ____.
5. Rome makes man's relationship dependent upon the ___ rather than ___.
6. Lutheran ethics is derived from ____.
7. Central to Rome's ethics is ____.
8. Who had a determining influence of Roman Catholic ethics? ____
9. Aristotle taught that man became materially good by ____.
10. What was Luther's attitude toward Aristotle? ____
11. The Reformation denied the possession of man ____.
12. What teaching did Rome totally obscure? ____
13. The Christian will perform good works not out of ____ but out of ____.
14. How does Rome show the influence of pagan stoicism? ____
15. Before God all justified persons stand ____.
16. What did Luther advance before Roman Catholic ethical misconceptions? ____
17. The Roman Catholic Church makes a distinction between ___ and ___ sins.
18. Rome teaches that morality consists in doing what ____.
19. Rome gives to ___ a place above faith.
20. Why did Lutheranism reject the "evangelical Counsels?" ____
21. The Reformation restored the meaning of prayer as ____.
22. What is Asceticism? ____
23. How did Rome depreciate the family? ____
24. In 1073 Gregory VII forced the greatest ___ in history?
25. Does the Bible teach that marriage is a sacrament? ____
26. On the basis of 1 Corinthians 7:15 willful and malicious dissertation ____.
27. The true Church is ____.

An Anti-Biblical Cult Total Bahaism Not Completely Represented

Christian News, September 13, 1993

The News Sentinel of Fort Wayne (the evening paper) on Monday, August 23, 1993 (photographed in the September 13, 1993 *Christian News*) had as its lead article on page 15 the following entitled: "Bahai faith is unifying spark for all religions."

Bob Caylor, a reporter for the *News Sentinel* reproduced the contents of an interview with an apostate Lutheran, formerly of New Haven (city adjoining Ft. Wayne), who had embraced Bahaism. His name was Menking who relayed to Caylor how he and later his two wives had accepted Bahaism.

Menking described the features of Bahaism which would appeal to the average person who was interested in world peace, general education for all, equality of the sexes and the promotion of democratic principles. However, Menking, as reported by Caylor, did not give other aspects of Bahaism which would not be received by Mohammedans, Roman Catholics, and Protestants.

The Origins of Bahaism

The Penguin Dictionary of Religions states about Bahaism as follows: "A faith arising out of the Babi movement in Persia. Baha'u'llah (1817-92) was originally a Babi who in exile acquired the conviction that he was the prophet foretold by the Bab. His faith of Baha'ism developed subsequently from an authoritarian, post-Shi'i sectarianism into a universalist religion of humanity."[1]

Although Bahaism developed from Islam it, however, is quite un-Islamic.[2] In 1819 a Shiite Muslim was born who claimed to be the Bab(gate) who became the forerunner of Baha'u'lah, the half-brother of the Babi, who claimed to be "the greatest of God's manifestations."[3] The name Baha'u'llah means "Glory of God." His followers, called Bahis, believe that the last dispensation of the world began May 23, 1844. Baha'u'llah is considered the fulfillment of prophecies of the Old Testament.[4] Baha'u'llah was exiled from Persia for 40 years because of his teachings and died later in prison in 1892.[5] His successor was his son Abdu'l Boaha, who died in 1921. He spent months propagating the Bahai faith in the United States. Abdul Baha named as his successor—Shogi Effendi, known as "the Guardian of the Baha'i Faith." Shogi Effende died in 1957 and he administered the expansion and strengthening of the organization. The current worldwide organization is governed by nine men from the Universal House of Justice, Haifa, Israel. The Bahai National Assemblies elects members every five years.

The rulings of the assemblies are deemed "infallible".[6]

The Unique Position of Baha'u'llah

Baha'u'llah is spoken of in Bahai literature as a divine manifestation, also his writings are quoted frequently and are considered divine. Baha'u'llah declared that he was the long expected educator and teacher of all people, the channel of wondrous GRACE that would transcend all previous outpourings, in which all previous forms of religion would be merged as rivers in the ocean.

Baha'u'llah's religious system has been summarized by Manske as: (1) Independent investigation of truth; (2) the essential harmony of science and religion; (3) the equality of men and women; (4) the elimination of prejudice of all kinds; (5) universal compulsory education; (6) a spiritual solution to the economic problems of the world; (7) a universal auxiliary language, and (8) universal peace upheld by a world government.[7]

Previous Revelations and Religions
before the Advent of Baha'ism

According to Bahai theology, God is supposed to have manifested himself through nine prophets: Moses, Buddha, Zoroaster, Confucius, Jesus Christ, Muhammad, Krishna the Bab and Baha'ullah. The nine different prophets changed their views throughout Babi history. Any person who has studied the religious views of these nine religious personalities knows that they do not have much in common; they certainly gave divergent and contradictory answers to life's three great questions:

1) Whence did I and the Universe come from; 2) Why am I living, and 3) Where am I going? Is there a life beyond death? It would appear that the founders of Bahaism selected these "prophets" in order to make their novel religious views appealing to the major religions of the world. The Shiite view that the Bab was the twelfth Iman the Shiites were expecting to come would have an appeal for one of the major sects of Islam. The selection of Buddha would find favor with the millions of Buddhists found in India, Southeast Asia and China and Japan. The selection of Krishna would appeal to millions of Hindus. Listing Zoroaster would appeal to the Parsees of Iran and especially in India. The selection of Jesus Christ should find favor with Roman Catholics, the Eastern Orthodox and Protestant churches. Since it was only in the early nineteenth century that this new and final revelation was made, Mirza Ali Muhammad (1819-50) and the Bahis had to create some continuity with the past.[8]

Major theological views of Bahaism

Mayer in his discussion of Bahaism has summarized some of the basic theological concepts as follows: The world is a continuous divine emination. Man is a product of a long process of evolution. Death is merely the new birth, through which the soul enters into a larger life, there to continue its growth until it reaches such perfection as to enter into the presence of God. The body serves only as a temporary housing for the soul. These basic principles are usually presented under various forms of unity.[9]

The Concept of Unity as Perceived by Bahaism

Mayer has discussed the concept of unity under: (1) The "unity of thought"; (2) the unity of truth and prophecy; (3) the unity of religion; (4) The unity of mankind; (5) the unity of two worlds; and (6) the unity of religion anti-science.[10]

It is in connection with these "unities" that the real religious views of Bahaism are really advocated and the article in The News Sentinel did not give them and so it really was misleading for readers learning about the religion of Bahaism.

The Bahai religious data treating of "unity of thought" have revealed that it espouses pantheism, and operates with an impersonal God, who is the source of everything since God is the only "all in all" there is no room for evil. Evil is but the absence of good or the lesser degree of good, but never a positive thing.[11]

According to Bahaist teaching the unity of truth and prophecy, there can be only one truth, which is living and progressive. The Old Testament Moses and the New Testament prophet, the Christ, had not attained a full and complete knowledge of truth. In the early nineteenth century the full truth had been given to Baha'u'llah, son of Abdul Baha. His teaching must he accepted as correct.[12]

The unity of religion has been expressed in the nonagon temple at Wilmette, Illinois. In Baha'i symbolism "nine" is the number of perfection, just as they have nine different prophets and at present nine elected officials run the affairs of this world-wide organization. Relative to the unity of mankind Mayer wrote: "that unity of mankind is based upon the premise that a spiritual power breathed in this age into the soul of humanity shall remove every cause of difference, misunderstanding, discord and disagreement."[13] To achieve this wonderful unity of mankind, there should be compulsory education for all, poor, and rich and the equality of the sexes must be insisted on, there must also be the creation and adoption of a universal language. A world court should be established and a league of all nation should be formed to run the government of the world.[14]

The concept of the unity of two worlds represents a unique teaching of Bahaism. The latter holds that the embodied and disembodied human beings are united and that these are in constant and inevitable communion between the two. The living are to love, help and pray for the dead, those who are advanced in the unity of two worlds, are to help the undeveloped souls.

Regarding the unity of religion and science Bahaists are taught that truth is one, and one should not find a conflict between science and religion.

Bahaist Ethics and Religious Practices

Bahais are expected to live a high and moral life. They are to live in accordance with Baha'u'llah's message and laws to pray and meditate daily, to give to the Bahai Fund and to teach others about Baha'u'llah's messages to mankind. Alcoholic beverages, narcotics, or other habit-forming drugs are prohibited. They have no clergy and at present are a lay-directed movement. Worship is practiced on the first day of each of the nineteen months that make up the Bahai calendar. It is also the day for conducting business and socializing. The holy days of Bahaism are the festival of Ridva, the Declaration of Bab, the Ascension of Baha'u'llah, the Martyrdom of Bab, the Birthday of the Bab, the Birth of Baha'u'llah, the Ascension of Abdul Bab and New Ruz (New Years).[15]

All writings of Bab, Baha'u'llah and Abdul Baha are regarded as Bahai sacred texts. The works of Shogi Effendi and The Universal House of Justice are considered infallible. Manske listed their world membership as being 4,691,890 or one per cent of the world population, with 319,819 living in America.[16]

An Evaluation of Bahaism

Since Bahaism appeared only in the early nineteenth century, most of its religious ideas were taken from religions existing before or during the Bab's and Baha'u'llah's lifetimes. Like many other cults of today the Bahai originators and leaders are tantamount to being deified. Bahaism's claim that it unifies all religions is not true. Actually it is an amalgam of ideas taken over from other faiths which preceded it by over a thousand years. Bahaism is a religious system that all major religions repudiate. In every respect it contradicts Judaism and Christianity at its most crucial points. Evaluating it from a Biblical perspective, its cosmology, its doctrine of God, its concept of sin, its soteriology, its ecclesiology, its teaching on sanctification and its eschatology are totally unacceptable and must he rejected as anti-biblical.

Footnotes

1. "Baha'is," *The Penguin Dictionary of Religion*, edited by John H. Minnells (London: Penguin Group, 1984), p. 60b.
2. Irvine Robertson, *What the Cults Believe* (Chicago: Moody Press, 1979), p. 143.
3. Paul E. Johnson, "Bahaism," in Vergilius Ferm, *An Encyclopedia of Religion* (New York: The Philosophical Library 1945), p. 52b.
4. F. E. Mayer, *The Religious Bodies in America* (St. Louis: Concordia Publishing House, 1956), p. 562.
5. Charles L. Manske, *World Religions* (Irvine, CA: The Institute For World Religions, 1991), No. 5.03 "Bahai Faith."
6. **Ibid.**
7. **Ibid.**
8. "Baha'i," William H. Geentz, *The Dictionary of the Bible* (Nashville: Abingdon Press, 1986), p. 160.
9. Mayer, **op. cit.**, 562b.
10. Ibid.
11. J .E. Esslemont, *Baha'u'llah and the New Era* (New York: Bahai Publ. Committee, 1957), pp. 261 ff.
12. Mayer, **op. cit.**, p. 542.
13. **Ibid.**
14. "Bahai, Gentz, *Dictionary of the Bible*, **op. cit.**, p. 100.

15. Manske, **op. cit.**,
16. **Ibid.**

Questions

1. What is the origins of Bahaism? ____
2. How did Charles Manske summarize the Baha'u'llah's religion system? ____
3. What are life's three greatest questions? ____
4. Bahaism espouses ____.
5. What is the world membership of Bahai? ____

Response to Larson

August 19, 1993

Dear Pastor Otten:

Enclosed please find a response to Larson's article in the March-April issue of *The Journal Historical Review*. I must apologize for my typing, with which I have my difficulties. If you cannot read it, discard the article. Please under no circumstances send my typed efforts to the editor of *The Journal of Historical Review*.

At the end of May I had cataract surgery on my left eye and am still waiting for my glasses.

I was sorry to hear about the water damage to your camp and trust the water has been receding. I imagine you had a hectic summer because of the big flood of 1993.

Wishing you and your work God's blessings, I am,

In Christ,

Raymond F. Surburg

Christianity Founded on Historic Fact Not Myth

A Response to Larson's Alleged Borrowings by Christianity

From Essenism and Mithraism
Christian News, October 4, 1993

The Journal of Historical Review in its March-April issue of 1993 (reprinted in the October 4, 1993 *Christian News*) has in a section called "Ancient Religions," in which Dr. Larson discusses in a LIMITED FASHION "The Essene Origins of Jesus' Teachings," and also "Mithraism: Formidable Rival of Christianity." The article is based on Larson's book, The Religion of the Occident, published in 1954 and on a revised edition appearing under the title "The Story of Christian Origins." Larson claims that "in this book I explain how an obscure Middle East cult could become the basis for Christianity, and how this dynamic new faith was able largely to replace the existing religions and become one of history's most powerful and durable institutions."[1]

Larson began his religious interpretation with a study of the pre-Christian mystery religions,[2] of which Shirley Case claimed there were the following: 1) the different Greek mystery cults, celebrated at different places in Greece; 2) the Phrygian mysteries centered about the mother goddess Cybele, with whom the youthful male youth, a god Attis had become associated at an early time; 3) In Syria there had existed a similar cult as described in Phrygia. The Hellenistic period saw Aphrodite and Adonis become gods chiefly connected with the same mysteries that had emanated in Phrygia; 4) The Egyptian mysteries of Isis and Osiris-Serapis who rapidly spread throughout the Roman Empire; 5) Persian Mithraism, the last mystery religion to spread throughout the Roman Empire.[3]

Larson labelled all these mystery religions as "savior cults." He claimed: "All these were based upon belief in a great demigod who is born of a virgin, dies a sacrificial death and after arising from the grave, returns to the realm of glory. The blood and flesh of these saviors, which were symbolically ingested by their adherents in mystical rituals, were believed to confer blessed immortality upon the partakers."[4]

The Alleged Essene Origins of Christianity

According to Larson an obscure sect was founded in Palestine about 180 B.C. Eventually this sect separated itself from the Sadducees and Pharisees. The members of this religious group lived in the desert around Qumran and around 140 B.C. are supposed to have become radical in their religion regarding law and practice.[5] "In fact," averred Larson, "over the years they gradually absorbed elements from Zoroastrianism, Pythagoreanism and Buddhism."[6]

The sect supposedly was the Essenes, or the Holy Ones. With the 1947 discoveries at Qumran the scholarly world has learned much about the Essenes, available in what is now known as *The Dead Sea Scrolls* CDSS). These documents claims Larson have broadened and deepened mankind's knowledge of the Essenes and also have made apparent influences of the Essenes upon the writers of the New Testament, and especially upon Luke.[7] This is Larson's erroneous interpretation of the Dead Sea data. Larson also believes that Jesus was an Essene before his thirtieth year. Later on Jesus is said to have proclaimed many of the order's doctrines in Palestine. The Essenes had a strong in influence on passages in the Synoptic Gospels (of) Mark, Matthew, and Luke and the three gospels show similar parallels as those that can be duplicated in the Essene documents. Such evidences claims, Larson, are incontestable.

Larson's Anti-Christian Bias

In evaluating Larson's views about the influence of Essenism and Mithraism on the Christian faith of the New Testament and on early Christianity, it will be helpful to ascertain Larson's stance over against the Judeo-Christian faith. In his address before the Institute of Historical Review,[8] Larson stated that he had been raised as a fundamentalist and at the age of twenty he had rejected the teachings and dogmas of the Bible. However, he did not give up his interest in religion as a sociological phenomenon, one which had a great influence on societies. His anti-biblical bias is shown by his quotation with approval of the statement of a certain philosopher who said that men create their gods in their own image, and certain it is that human beings in almost all times have believed a variety of supernatural beings and one scholar declares "the greatest miracle of all is the capacity to believe things for which there is no evidence."[9]

In the same address, presented April 4, 1988, at the Eighth Revision Conference Larson also declared: "Why, then do Christian organizations desire the oblivion of the Scrolls? The reason is that they have always held their creed was a single origin, miraculous and supreme revelation without predecessor or outside contribution."[10] Larson condemns this with these words: "The fact is that nothing is further from the truth; Christianity is a composite of doctrines, teachings and ideologies which have forerunners in previous religion. With proximate source in the Essene cult. Larson charges present day Christians as deliberately ignoring the Dead Sea Scrolls and of being guilty of endeavoring not to see published the remainder of the Dead Sea Scrolls yet in need of publication.[11]

In his article in The Historical Review, March-April, 1993 Larson furnishes no proof for his assertions but simply pontificates and sets forth his position on the alleged influences of Essenism on New Testament religion. However, his views on the alleged influence of Essenism on Christianity have been clearly delineated in his essay in The Historical Review, entitled: "Whatever Happened To the Dead Sea Scrolls?"[12] His extreme views on the supposed influence of the Essenes upon Christianity are in line with those of A. Powell Davies, Edmund Wilson, Dupont-Sommer, John Allegro and Charles Potter, all of whom have denied the originality of the religious views of The New Testament and subscribe to the theory of the influence of Essenism on the New Testament writers.

This writer wrote a response to Larson's article, entitled "Response to Whatever Happened to the Dead Sea Scrolls," which appeared in the *Christian News*, February 18, 1988.[13] What Larson has set forth as historical conclusions is nothing but a restatement of theories and allegations written and propounded over thirty years ago. Larson has engaged in misinterpretation of such books as The Book of Enoch and The Testament or the Twelve Patriarchs, described by Larson as Essene literature and as such affecting biblical Christianity.[14]

The Christians of New Testament times and even of later times never accepted these writings as canonical and they constitute no evidence for Essenic influence. Larson has followed the interpretation of critical scholars that both John the Baptist and Christ had been Essenes. There is not one shred of solid evidence that substantiates that John the Baptist and Jesus Christ were ever in contact with the Essenes. Among the proofs for Essene influence on Christianity Larson asserted: "That many passages in the Synoptic Gospels (of Matthew, Mark, and Luke) are similar is uncontested and his denunciations, of the Scribes and Pharisees are virtually identical."[15]

The idea that Essenism influenced Christianity was already propounded by Renan, the French skeptic, who termed Christianity "an Essenism which succeeded on a great scale." That there are existing certain similarities between the Qumran community life and the life of the early church is hereby acknowledged.[15a] However, as Charles Pfeiffer has warned that there is the danger, however, of overlooking the differences which also exist. "The contrasts frequently appear more significant than the comparisons."[16] An excellent discussion of all aspects of the similarities between Essenism and Christianity has been done by Floyd Filson in his article "The Dead Sea Scrolls and the New Testament."[17] While admitting the usefulness of the Dead Sea Scrolls as furnishing material for the better understanding of the background of the New Testament, Filson wrote: "But there is nothing in the contents of the scrolls or in a careful comparison of them with the New Testament which warrant hasty statements that the Christian Gospel was taken over from the Qumran sector is basically dependent on that sect for its message and way of life. To assert this would exaggerate greatly the importance of the scrolls and minimize the originality and creative significance of Jesus and the New Testament leaders and writers."[18]

Both Christianity and Essenism had a common origin—the Scriptures of the Old Testament. That a number of Essene ideas had parallels in the New Testament is accounted for by the roots of both religious groups being in the canonical books of the Old Testament. Averred Mansoor: "The Dead Sea Scrolls provide us with the setting in which Christianity was born and show us the roots of some of the ideas and also its unique and distinctive character."[19]

Attempt to Connect New Testament Teaching
With the Savior Cults of the Near East

Larson postulated that the idea of a dying and rising Savior came to Christianity through the Essenes.[20] Larson has advocated views about the Teacher of Righteousness (TR) which originally were propounded by Allegro and Du Pont-Sommer. The Teacher of Righteousness appeared in Jerusalem twenty years after the Essene community was groping blind. According to the interpretations of Allegro and DuPont-Sommer the TR was crucified in Jerusalem by the Wicked Priest. Later on the Essenes are alleged to have put forth the belief that TR had been raised from the dead.[21] The Alleged crucifixion and resurrection of the TR was adopted by Christians and ascribed to Jesus. A number of scholars among them H.H. Rowley, have questioned the interpretation of the texts cited, their reading as well as their interpretation. A number of competent students of the Qumranic literature declare that there is no evidence whatever for the TR's resurrection.[22]

Baptism and The Eucharist Supposedly
Borrowed from the Essenes.

The Qumran sect practiced daily ablutions and also engaged in a common meal. These practices are said by Larson and others to be the basis for the practice of baptism and the celebration of the Lord's Supper. Christianity did not obtain its baptism practice from the Essenes, because Judaism already baptized new heathen converts to Judaism. Daily ablutions characterized Essenism, while Christian baptism was performed but one time and had a completely different purpose than the ablutions in Essenism. The Lord's Supper differed considerably from the common meal in which the Qumranites participated The Eucharist was celebrated to commemorate Christ's death and also to receive the body

1533

and blood of Christ under the bread and vine. Christ and his followers associated with the Scribes, the Pharisees and Sadducces and participated in all the festivals and religious ceremonies of Judaism. They had no contacts with the Qumranite community. New Testament Christianity did not borrow from the Essenes as Larson would have his readers believe.

Did Mithraism Influence Christianity?

Just as first-century biblical Christianity is said to have been influenced by Essenism according to Larson, so he also claims second and third-century Christianity is alleged to have been influenced by Mithraism.[23] The later religion has been perceived as the last mystery religion of antiquity, which became especially popular in the Roman Empire. Shirley Case stated that Mithra, an Iranian hero-divinity, "had devoted himself to the service of mankind and after a last meal, celebrating the success of his redemptive labors. He ascended into heaven henceforth he continued to minister help to the faithful in their conflict with Satan and his hosts."[24]

The process of initiation into full membership in the cult was elaborate. The candidate passed through seven grades which prefigured the passage of the soul after death through seven heavens to the first abode of the blessed. Each grade was entered by observing ablutions, sacred meals and other sacramental rites.[26] Mithraism restricted its membership only to men and became very popular among the Roman soldiers. Mithraism became Christianity's chief rival on the frontiers of the Roman Empire during the second and third-centuries. Larson claims that there are numerous parallels between Christianity and Mithraism.

Thus he wrote: "Reflecting common Oriental origins, Christianity and Mithraism shared many similarities in both doctrine and ritual. The followers of each creed shared belief in a great flood and a sacred ark, a last judgment, the resurrection of the flesh and the ultimate triumph of good over evil. Most of these doctrines are found in the Old Testament, in which Christianity had its roots. The story of Noah's Ark and the Great Flood, the doctrine of heaven and hell, the last judgment, the doctrine of the resurrection, the atoning work of the Messiah are all found in the pre-Christian Bible. The teachings of Christ were claimed by Him to be a continuation of the Old Testament, of which Christ said, "that they spoke of Him."[27] Larson also claimed "that in each religion priests presided over rituals that made the use of bells, candles and holy water. Devotees took part in a sacred communion banquet of bread and water (and possibly wine in a ceremony that paralleled the Christian Eucharist.)"[28] However, it should not be forgotten that the use of bells and holy water are not in use in the New Testament and that the canonical books of the New Testament are the source for the establishment of Christian doctrines and for the determination of ethical practices. Larson himself claimed in his discussions of Essenic influences on New Testament Christianity that both baptism and the Lord's Supper were borrowed in the first century from the Essenes. Why should Christians borrow rites from Mithraism when Christians were already employing them in the first century?

Larson averred that Christians and Mithraists were mystical brethren. Each is said to have believed that the founder was the mediator between God and men and that through him alone was salvation possible. Larson further claimed that both taught the doctrine of primitive revelation. Both are supposed to have emphasized the constant warfare between good and evil, required abstinence and self-control and bestowed the highest honor on celibacy.[29]

Most of these parallels or seeming similarities can be found in the Old Testament or in the first-century New Testament and do not need to be assigned after A.D. 100, the close of the New Testament canon. Wrong developments in the post-apostolic period are not to be attributed to the teachings of the New Testament, the real and only source for biblical Christian beliefs. Since Mithraism flourished especially in the second and third centuries, Larson's alleged borrowings of Christianity from Mithraism are not supported chronologically. In fact, it has been suggested that Mithraism borrowed its concepts from Christianity.

Larson has also contended that Christianity later in its history made Sunday, "which the apostolic church had never observed but which had always been sacred to Mithraism,"[30] as their holy day. That claim is not true. In the New Testament one observes that Christians met on Sundays in an article on "Sunday," in The Westminster Dictionary of Worship, it is asserted: "The Sabbath set the pattern of a weekly day of worship, but in the New Testament it was the first day of the week (Acts 20:1) which took the position."[31] In the year A.D. 54 Paul wrote from Corinth: "On the first day of the week let each of you put aside and save etc."[32] This indicates that Christians met on Sunday for worship. McArthur concludes: "By the end of the first century the expression the first day of the week" has been replaced the expression "the Lord's Day."[33] The use of the name Sunday dates from the middle of the second century. In 321 the official recognition by the Roman State of Sunday as the day of Christian worship had been incorporated and as a day of public rest meant that the Sabbath had been incorporated into and transformed by the Christian tradition.[34]

Did Christianity Borrow Its Christmas, December 25 from Mithraism?

Mithraism celebrated December 25 as the birthday of Mithra and supposedly influenced the selection of December 25 as the observance day of Christ's natal day.[35] Duchesne has pointed out that "at Rome, Hippolytus at the beginning of the third century, in his Commentary on Daniel fixed the date at the beginning of the third century, on Wednesday the 18th of December, that according to Hippolytus that was in the forty-second year of the Emperor Augustus."[36] Saint Clement of Alexandria, who died a little more than a century after the death of St. John, speaks of the observance of the celebration of Christ's natal day. Saint Chrysostom, the Bishop of Constantinople, in the fourth century spoke of the feast of the

Incarnation as being of great antiquity.[37] It is sometimes asserted that the Roman Church was influenced in fixing on December 25th for the purpose of turning away the faithful from the excesses of the then ancient festival of the Satunolia by diverting their thoughts to our Lord's nativity. But as the Saturnalia began on December 17th and ended on the 23rd so this theory must be discarded.[38] A Duchesne says that a better explanation is that based on the pagan festival of the Natalis Invicti! The Invictus is the Sun, whose birth coincides with the winter solstice, that is, with the 25th of December according to the Roman calendar.[39]

Gwynne accepts this explanation and states: "And if this be the true date (as there is great probability), it is very significant that it is the third day after the true winter solstice (December 22nd) when the sun, after reaching the lowest point on the human horizon, begins to ascend and to bring back light and life to a darkened and dying world. So also Christ, 'the Light of the World the Sun of righteousness,' arises with healing in His wings."[40]

Christianity's Triumph Over Mithraism

Larson admitted that Mithraism could not compete successfully with Christianity. Mithraism was not a religion for all mankind, for women were excluded, who played such an important role in the life of Jesus while on earth. Women were considered dangerous because they were the source of erotic desire for men. The great difference between Mithraism and Christianity was that the former worshipped the deity Mithra, who had no basis in existence, while Christ came to earth to redeem mankind.

Latourette's Evaluation of Christianity

The reason for the success of Christianity in the third-and fourth-centuries is stated by Latourette as follows: "Christianity was all inclusive. More than any of its rivals it appealed to men and women from all races and classes. In contrast with the philosophies, which were primarily for the educated, Christianity had a message for the simple and the ignorant. It also won some of the keenest minds and most highly trained minds. Membership in the mysteries was expensive and therefore chiefly for the well to do. Christianity was for both rich and poor. Mithraism was only for men. The Gospel was proclaimed to both men and women. Why this inclusiveness? It was not in the parent Judaism."[41]

Latourette showed that in many different ways Christianity made an appeal to those seeking for ultimate fulfillment.[42] The Yale church history professor asked: "Whence came all these qualities which won for Christianity its astounding victory?" Latourette's response was: "Careful and honest investigation can give but one answer, Jesus."[43] "It was faith in Jesus and his Resurrection which gave birth to the Christian fellowship and which continued to be its inspiration and its common tie. It was the love displayed in Christ which was ideally to a marked extent in practice; the bond which held Christians together. The early disciples unite in declaring that it was from the command of Jesus that the Gospel was proclaimed to all regardless of sex, race or cultural background. The new life in Christ might express itself in many forms, but its authenticity was to be proved by high, uncompromising moral qualities as set forth by Jesus. Hence the combination of flexibility and inflexibility. As against the mystery religions those cults which had as much superficial similarity to Christianity, it was partly a theology, a metaphysic which gave the latter its advantage, but it was chiefly against the mythical figures at the heart of the mysteries. Christians point to Jesus, a historical fact."[44]

Machen on Mystery Religions

J. Gresham Machen, one time professor at Princeton Theological Seminary, who wrote a book, *The Origin of Paul's Religion*, in which he, among other things, specifically took up the issue of the influence of the mystery religions on New Testament Christianity. Chapters 5 and 6 of this volume specifically examine this question 45 and Machen shows especially in chapter 6 the impossibility of Christianity being dependent on Mithraism.[46]

An Evaluation of Larson's Hermeneutical Approach to the New Testament

As has already been indicated Larson repudiated biblical Christianity and appears to have become an atheist or an agnostic. He has allied himself with the school of comparative religions, known in German as "religionsgeschictliche Schule." Soulen has described "the history of religion(s)" school as "often connoting an anthropological approach to general religious subjects without concern for historical relativity."[47] The term "religionsgeschichtliche Schule" is applied to a group of Protestant scholars in Germany, who at the turn of the century, sought to understand the religion of the Old and New Testaments within the context of their historical environment, including the other religions of that time and region. Members of this school were Herman Gunkel, Johannes Weiss, Wilhelm Bousset, Wilhelm Heitmueller, Hugo Gressman, Albert Eichhorn and others.[48] Their approach to biblical LITERATURE WAS RADICALLY DIFFERENT from the German idealism of the nineteenth century, during which these scholars lived. Soulen correctly evaluated the presuppositions of this approach when he wrote: "It is their work which first threw light on the immense distance which separate the world of the Bible; its understandings and expectations; from that of our own thus exacerbating the problem with which theologians struggle of making the Bible meaningful to people of today.[49]

Larson has adopted views that were current in the 19th century, beginning as early as 1829. Bousset (1865-1920) applied the method of "comparative religions" to the study of the relationship of Hellenistic religions to early Christianity and Judaism.[50] Bousset is particularly noted for *Die Religion des Judentums im spat hellnistischen Zeitalter* (1903, revised by Hugo Gressman, 1966). Another work of cited is that of Richard Reitzenstein, who wrote a book on hellenistic mystery religions, Gnosticism, and Mandaism and thus to the cultic background of the New Testament gcnerally.[61] His major book is entitled: *Hellenistic Mystery Religions: Their Basic Ideas and Significance.*

Larson in his *The Story of Christian Origins* has followed the insights and conclusions of the "school of comparative religion(s)." For him Christianity is an amalgam of teachings borrowed from other cults and its Bible contains, myths, and for the most port, unreliable religious teachings.

Footnotes

1. *The Journal of Historical Review*, March-April, 1993, p. 33.
2. **Ibid.**
3. Shirley Case, "Mystery Religions." Vergilius Ferm, Editor, *An Encyclopedia of Religion* (New York: The Philosophical Library, 1947), pp. 511-512.
4. **Ibid.**, p. 512.
5. Larson, **op. cit.**, p. 33.
6. **Ibid.**, pp. 33·34.
7. **Ibid.**, p. 34.
8. Martin Larson, "An Update on the Dead Sea Scrolls." *The Journal of Historical Review*, Spring 1988.
9. **Ibid.** The essay was delivered April 11, 1988
10. **Ibid.**
11. **Ibid.**
12. Martin Larson, "What Ever Happened To the Dead Sea Scrolls?" *The Journal of Historical Review*, volume three, Number two, Summer, 1982. Reprinted in *The Christian News Encyclopedia*, V. 3485-3486.
13. Raymond F. Surburg, *The Christian News Encyclopedia*, V, 3485-3486.
14. **Ibid.**, p. 3485.
15. Larson, *The Journal of Historical Review* March-April, 1993, p. 34.
15a. Menahem Mansoor, *The Dead Sea Scrolls* (Second Edition; Grand Rapids: Baker Book House, 1983), pp. 153-154.
16. Charles Pfeiffer, *The Dead Sea Scrolls and the Bible* (Grand Rapids: Baker Book House, 1969), pp. 97-98.
17. Floyd V. Filson, "The Dead Sea Scrolls and the New Testament," In David N. Freedman and Jonas Greenfield, *New Directions in Biblical Archaeology* (New York: Doubleday and Co. 1971) pp. 142-157.
18. **Ibid.**, p. 155.
19. Mansoor. **op. cit.**, p. 154.
20. Larson, "The Essene Origins of Jesus' Teaching," *The Journal of Historical Review*, March-April, 1993, p. 34.
21. Surburg, **op. cit.**, p. 3486, column b.
22. Surburg, **op. cit.**, p. 3486b
23. Larson, "Formidable Rival to Early Christianity," **op. cit.**, p. 34.
24. Case, "Mystery Religions," **op. cit.**, p. 512 b.
25. **Ibid.**
26. Larson, "Formidable Rival to Early Christianity," **op. cit.**, p. 35.
27. John 5:39.
28. Larson, "Formidable Rival to Early Christianity," **op. cit.**, p. 35.
29. **Ibid.**

30. **Ibid.**, p. 35.
31. A. A. McArthur, "Sunday," in J. G. Davies, *The Westminster Dictionary of Worship* (Philadelphia: The Westminster Press, 1972), p. 351b.
32. I Corinthians 16:2.
33. McArthur, **op. cit.**, p. 351b
34. **Ibid.**
35. Larson, "Formidable Rival to Early Christianity," **op. cit.**, p. 35b.
36. Walker Gwyne, *The Christian Year* (New York: Longmans Green and Company, 1915), p. 55.
37. **Ibid.**, p. 55.
38. **Ibid.**, p. 56.
39. As cited by Gwynne, **op. cit.**, p. 57.
40. **Ibid.**, p. 57
41. Kenneth Scott Latourette, *A History of Christianity* (New York: Harper and Brothers, 1953), p. 106.
42. **Ibid.**, pp. 106-17.
44. **Ibid.**
45. J. Gresham Machen, *The Origin of Paul's Religion* (New York: Macmillan Company, 1928), pp. 211-292.
46. **Ibid.**, pp. 255-292.
47. Richard N. Soulen, *Handbook of Biblical Criticism* (Second Edition; Atlanta: John Knos Press, 1976), pp. 167-168.
48. **Ibid.**, p. 168.
49. **Ibid.**, p. 168.
50. Bousset, Soulen, **op. cit.**, p. 35.
51. "Reitzenstein, Richard (1861-1931)," p. 167.

Questions

1. Martin Larson in *The Journal of Historical Review* labeled mystery religions as ____.
2. What was Larson's erroneous interpretation of the Dead Sea data?
   ____
3. Larson charged Christians with ____ the Dead Sea Scrolls.
4. Christianity did not borrow from ____.
5. Was the use of bells and holy water in use in the New Testament? ____
6. Christianity was both for the ____ and the ____.
7. Latourette said that Christians point to ____, a historical fact.
8. What did J. Gresham Machen show in *The Origin of Paul's Religion*?
   ____
9. Larsen appears to have become an ____.

Jerome Biblical Commentary and
The New Catholic Encyclopedia, Rome and Bible

Dear Dr. Surburg,

Here is a response (The response follows Dr. Surburg's "The Radical and Revolutionary Changes Occurring in Roman Catholic Biblical Studies Between 1893 and 1993" on the next page), which I just received to your article on Rome and the Bible. You may want to respond.

Note that the latest book by Father Raymond Brown (featured on the cover of April 4 *Newsweek* and also in the April 1 *TIME*) has Rome's imprimatur. Check the April 4 *CN* for details. Also note the latest statement on the Bible by Rome on p. 1 of the April 4 *CN*. Bishop Williamson claims to be a Roman Catholic bishop. He follows Lefebvre, the Roman bishop the pope excommunicated for defying the Pope's authority. Williamson now teaches at a traditionalist Roman Catholic Seminary in Winona, Minnesota. I regularly receive his newsletters and reports on the seminary.

Conservative Roman Catholics are separately trying to defend the Pope, yet it is obvious that the Pope is now promoting universalism, evolution, and destructive higher criticism.

God's blessings,

Herm

Ed. *The Jerome Biblical Commentary* is edited by Raymond E. Brown, S.S., Joseph A. Fitzmeyer, S. J. and Roland E. Murphy, O. Carm. This 925 page commentary was published by Prentice – Hall, Inc. Englewood Cliffs, New Jersey. The Foreword is by His Eminence Augustin Cardinal Bea, S.J.

It has the Imprimatur of Lawrence Cardinal Shehan, Archbishop of Baltimore June 6, 1968. It is dedicated "To the Memory of Pope Pius XIII Whose Promotion of Biblical Studies Bore Fruit in the Second Vatican Council."

"Catholics Write Liberal Bible Commentary – *The Jerome Biblical Commentary* Says Bible Contains Errors, Fiction, Myths, and Legends," the lead story in the December 9, 1968 *Christian News*, published many sections from the commentary. "New R.C. Commentary Rejects Vicarious Satisfaction of Christ," on p. 1 of the December 16, 1968 *Christian News* showed that the *Jerome Biblical Commentary* attacked a central doctrine of Christianity. An editorial in the December 16, 1968 *Christian News* on the commentary said that "The same anti-Christian doctrines formerly condemned by Rome are now advanced in an officially approved commentary."

In 1967 McGraw-Hill published the 15 volume 15 million word *The New Catholic Encyclopedia*. It was quoted at great length in the May 29, 1967 *Christian News*. "A New Faith", *CN*'s editorial on this massive officially approved work, concluded: "May God raise up leaders within the Roman Catholic Church who will reject the anti-Christian New Faith of this New *Encyclopedia* and begin another Reformation within this church which will turn millions to trust in Christ and not in their own works for eternal salvation."

Through the years, *CN* has found few scholars who have read *The Jerome Biblical Commentary* and *The New Catholic Encyclopedia*.

The Radical and Revolutionary Changes Occurring in Roman Catholic Biblical Studies Between 1893 and 1993

Christian News, March 14, 1994

The Roman Catholic Church has experienced radical and revolutionary changes in the field of Biblical studies in the twentieth century. Between the issuing of the encyclical of Leo XIII, called "Providentissimus Deus," November 18, 1893 and the publication of Pius XII's "Divine Afflante Spiritu" on September 30, 1943 significant changes have taken place in Roman Catholic Biblical Studies, involving the use of a new hermeneutics and radical changes in Biblical interpretation. Within less than fifty years views that were considered heretical and labelled as coming from modernism and were considered anti-Biblical were adopted and thus at least a thousand years of church history were jettisoned.

The climax of these once forbidden positions were finalized in the revised edition of *The Jerome Biblical Commentary*. This commentary purports to give the products of scientific exegesis to which later pontiffs committed the Church. Views and positions that were condemned and denounced between 1893 and 1907 by popes Leo XIII, Piux X and Benedict XV, as well as warnings by Pope Pius XI, were ignored by Roman Catholic savants and were propagated in Roman Catholic seminaries, universities and by the Roman Catholic publication houses. For in 1943, Pope Piux XII issued an encyclical, entitled, "Divina Afflante Spiritu," a document whose wording was interpreted to allow Roman Catholic scholars to use the most scientific method in dealing with Holy Writ, and with it began the adoption of a higher criticism against which at least four popes had warned in various documents. Roman Catholic pundits adopted the liberal higher criticism which liberal Lutheran and Protestant scholars had produced in nearly the past two hundred years.

In this presentation the situation between 1893 and 1907 will first be set forth and then followed with a delineation of radical and revolutionary developments between 1943-1993.

I. Later 19th Century and Early 20th Century Views About the Bible and Its Interpretation

A. Providentissimus Deus

The encyclical "Providentissimus Deus" was issued November 18, 1893 had as its concern to foster correct and edifying study and exposition of Holy Scriptures.[1] Leo XIII held that the canonical books of the Old Testament (the Hebrew canon plus 13 apocryphal books) were written under the inspiration of the Holy Spirit, that they have God for their author and as such had been delivered to the Church.[2] This belief "has been perpetually held and professed by the Church in regard to the Books of both Testaments and there are well known documents of the gravest kind, coming down from the earliest times, which proclaim that God who spake by the prophets, then by His own mouth, and lastly by the Apostles composed also the canonical Scriptures and that these are His own oracles and words—a Letter written by our Heavenly Father and transmitted by the sacred writers to the human race in its pilgrimage so far from the heavenly country. If then such and so great is the excellence and dignity of the Scriptures that God Himself has composed them, and that they treat of God's marvelous mysteries, counsels and works, it follows that the Branch of Moral Theology which is concerned with the defense and elucidation of these divine books must be excellent and useful in the highest degree."[3] Leo XIII warned against those who attempt to defile or corrupt it, either on the part of those who impiously assault the Scripture, or those who are led astray into fallacious and imprudent novelties.

"Providentissimus Deus" was concerned with combating the liberalism afloat at his time in Europe and America and it warned against rationalism and by implication the negative higher criticism which was also affecting certain Roman Catholic scholars.

Pope Piux X in Motu Proprio "Praestantia Scriptura Sacrae," summarized Leo XIII's Encyclical as follows: "After describing the dignity of the Sacred Scriptures and commending the study of it, set forth the laws which govern the proper study of the Holy Bible; and having proclaimed the divinity of these books against errors and calumnies of the rationalists, he at the same time defended them against the false teaching of what is known as the higher criticism, which as Pontiff most wisely wrote, are clearly nothing but commentaries of rationalism derived from a misuse of philology and kindred studies."[4]

Pope Pius X in "Motu Proprio Praesentia Scripturae" issued November 18, 1907 at Rome came out strongly against modernism or theological liberalism. He strongly supported the decision of the Pontifical Council of Commission on Biblical Matters which asserted that Moses wrote the Pentateuch (Or Jewish Torah), but that he might have used sources in penning the Five Books of Moses.[5] This decision supported previous orthodox positions in the area of Biblical Introduction which in the early decades of the 20th century were being ignored or even rejected by Roman Catholic scholars. Pius X even threatened with excommunication those who would not accept the Decree of the Sacred Congregation. The Pontiff clearly implied that in the Roman Church there were those theologians who were promoting and defending heresies[6] "especially when they advocate the error of modernism that is the synthesis of all heresies."

Pope Pius X was engaged in doctrinal matters in a relentless struggle against modernism, the chief tenets of which were condemned by the decree of the Holy Office "Lambentabilis sane exitu" and the encyclical "Pascendi Domini Gregis," both issued in 1907.[7]

Lambentabilis Sane Exitu
This is a decree issued by Pope Piux X, written by the Roman congregation of the Holy Office and signed by Pope Pius X on July 4, 1907.[8] It

listed sixty-five propositions taken from the writings of modernistic scholars, particularly of A. Loisy, but also of G. Tyrrell and E.E. Le Roy.[9] "This papal document condemns as errors the claim that scientific research, even in theological matters ought to be free from ecclesiastical control, the rejection of scriptural inerrancy, religious pragmatism and an evolutionistic conception or the dogmas and institutions of the Church, whose objective character and absolute value was denied by modernism. The decree was followed after an interval of two months by the encyclical 'Pascendi Gregis.'"[10]

Spiritus Paraclitus

On September 15, Pope Benedict XV issued the Encyclical *"Spiritus Paraclitus,"* on the occasion of the fifteenth centenary of the death of Jerome. In this document the pontiff restated the Roman Catholic position with regard to Holy Scriptures.[11] The Bible, composed by men were inspired by the Holy Spirit, and therefore had God Himself as the principal Author, the individual authors of single canonical books constituted his live instruments: In consequence, the Bible as such is said to be absolutely free from error as was already declared in the Encyclical *"Providentissimus Deus"* of Leo XIII. Barrois in summarizing "Spiritus Paraclitus" stated: "Apparent anomalies extant in Scripture may be explained by distinguishing primary and secondary elements of which the latter would not be guaranteed, nor by opposing absolute to relative teaching; nor by the hypothesis of implicit quotations, on which the sacred author would not pass judgment; nor by assuming the existence in the Bible of pseudo historical literature or literary patterns which could seem to give fiction an appearance of reality. These and similar principles of criticism are condemned as being incompatible with the traditional exegesis of the Roman Church."[12]

II. Roman Catholic Biblical Studies Since Pius XII
A Shift In Roman Catholic Circles Toward Modernism

When Steinmueller issued the seventh printing of his *A Companion to Scripture Studies*, three volumes, there was no hint of the use of the historical critical method propounded and used by liberal Lutheran and Protestant Old and New Testament scholars.[13] However, a change occurred with the issuance of Pius XII's *"Divino Afflante Spiritu."*[14]

The Significance of the Encyclical "Divino Afflante Spiritu"

On September 30, 1943 Pius XII issued a document which set forth the best means of promoting the study and diffusion of Holy Scripture. It outlines a complete program of scientific investigation of the Bible, starting from the necessity for the exegete to be thoroughly acquainted with the principles of modern linguistics and methods of textual criticism. It stresses the importance of patristic, ecclesiastical, and profane literature for a theological interpretation of the sacred text. While acknowledging the necessity of taking into account the literary patterns and formulas used by the inspired writers, it recommends some caution in approaching the bearing upon the substance of the Biblical message.[15]

Between 1943 and 1953 great changes occurred in Roman Catholic Biblical interpretation. In 1953 there was published *A New Catholic Commentary*, to which a host of Roman Catholic Biblical scholars contributed.[16] The article on "Higher Criticism" by R. Dyson and R. Mackenzie stated: "The greatest landmark in the modern history of Catholic Biblical criticism was the publication (September 30, 1943) of the Encyclical *Divino Afflante Spiritu,* and it was generally hailed as a sign that the Era of Modernism was officially closed, that the line between what was dangerous in doctrine and what has not been clearly drawn and that now the true freedom of the sons of God had been restored in Catholic exegesis."[17] The words of Pius XI's Encyclical which encouraged critical research are said to be the following: "Let all other children of the Church bear in mind that the efforts of these valiant laborers in the vineyard of the Lord are to be judged not only with fairness and justice, but also with the greatest charity; they must avoid that somewhat indiscreet zeal which considers everything new for that very reason a fit object for attack by suspicion. Let them remember above all rules and laws laid down by the Church are concerned with the doctrine of faith and morals; and that among the many matters set forth in the legal, historical, sapiential and prophetical books of the Bible there are only a few whose sense has been declared by the authority of the Church, and that there are equally few concerned which the opinion of the Holy Fathers is unanimous. There consequently remain many matters, and important matters in the exposition and explanation of which the sagacity and ingenuity of Catholic interpreters can and ought to be freely exercised.... The true freedom of the sons of God, loyally maintaining the doctrine of and at the same time gratefully accepting as a gift of God, and exploiting every contribution that secular knowledge may afford, must be vindicated and upheld by the zeal of all, for it is the condition and source of any real success, of any solid progress in Catholic science."[18]

Changes in Roman Catholic Approach in Biblical Studies

Since 1943 the majority of books dealing with matters pertaining to the area of Biblical introduction and Biblical commentaries have cited the *Encyclical "Divino Afflante Spiritu"* as the justification for following the lead and procedure of liberal Protestant higher criticism. Peter Ellis's *The Men and Message of the Old Testament* makes a great deal of use Pius XII's Encyclical in the second chapter of his book, extensively used in Roman Catholic circles.[19] His analysis of the Pentateuch, used the Graf-Wellhausen views of the origin and development of the Pentateuch.[20] He adopted from the Poly-chrome Bible the idea of using different colors for the various sources woven together by the priest who sometime after B.C. 400 put together the Pentateuch.[21] Roman Catholic scholars in Holland or the Netherlands, produced A. van den Born's *Bijbels Wordenboek*, whose Second Revised Edition appeared 1954-1957 and was translated by a number of Roman Catholic scholars under the direction of Louis Hartman, Executive Secretary of the Catholic Biblical Association of America.[22] This was published with the Nihil Obstat of Peter Ellis and the Imprimatur of Francis Cardinal Spellman. A new Roman Catholic translation of the Bible was undertaken by members of Catholic Biblical Association of America, sponsored by The Bishops' Committee of the Confraternity of Christian Doctrine. In its "Introduction to the Books of the Old Testament" it set forth the validity of the documentary hypothesis.[23]

The Encyclical "Humani Generis" of Pius XII

One wonders whether Pope Pius XII agreed with the interpretation and use made of his 1943 encyclical, for on August 12, 1950 he issued *Humani Generis* which denounced "historicism" which tended to reinterpret Christian doctrines on the basis of an evolutionistic philosophy and existentialism, a system which according to the Pope "neglects the immutable nature of things, while being concerned exclusively with the "existence individual realities." Both ideologies are held as introducing relativism into dogmatics, and for the tying of the essential beliefs with the passing conditions of history, and thereby endangering faith itself.[24] Roman Catholic scholars were encouraged to follow the philosophy of Thomas Aquinas as the best way of establishing a foundation for theology. Catholic scholars were warned to be very cautious in their dealing with non-Catholic scholars. Yes, Roman Catholic exegetes were scrupulously to observe the decree of the Pontifical Commission (June 30, 1909), on the interpretation of the first eleven chapters of Genesis as history and not a myth.[25]

Despite Pius XII warnings the higher critical movement took over the Roman Catholic Church as is clearly evident when consulting a number of very important commentaries and Bible dictionaries. The revolution in Biblical studies in Roman Catholic seminaries and Roman Catholic universities may be seen by consulting *A Commentary of the Holy Scripture,* edited by Orchard, Sutcliffe, Fuller and Russell, which appeared in 1952, to which 43 scholars contributed articles dealing with general Biblical introduction and commentaries on all the books of Bible recognized as canonical by Rome.[26]

Another book showing how Roman Catholicism took over liberal Protestant higher criticism was McKenzie, *Dictionary of the Bible*, written by July 1962.[27] The Liturgical Press of Collegeville, owned by the Order of St. Benedict of Collegeville, issued an *Old Testament Reading Guide* in 31 books. A periodical *The Bible Today* has popularized higher criticism for Roman Catholics. In 1973 *The Bible Today Reader* was published, containing selections from the first ten years of *The Bible Today.*[28] In 1966 there appeared *The New Catholic Encyclopedia* which sets forth very clearly how Roman Catholic Biblical scholarship had adopted this soul-destroying higher criticism. In 1963 McGraw Hill Book Company published *Encyclopedic Dictionary of the Bible*, a translation and adaption of A. Vanden Born's *Bijbels Wordenboek*, second revised edition, 1954-1957; the work of thirty-two Dutch scholars and seventeen American Roman Catholics.[29] Then in 1968 the prestigious *Jerome Biblical*

Commentary, edited by Raymond Brown, Joseph Fitzmyer and Roland Murphy, covering the Old and New Testaments saw the light of day.[30] This commentary was hailed as the epitome of Roman Catholic Biblical scholarship. What a difference between the views of the former old *Catholic Encyclopedia* or Steinmueller's *The Companion to Sacred Scriptures*, 3 volumes dealing with both the Old and New Testaments.[31]

The New Roman Catholicism as Exemplified by Grollenberg

The Dominican Grollenberg in the introduction to his Book, *Jesus,* records how he was brought to change his mind, and this may be cited as an example which was typical of the revolutionary changes which occurred in Roman Catholic studies between 1943 and 1993, and of the Roman Catholic understanding of how the Bible was to be understood. Thus he relates how in 1934 he joined the Dominican order, which he claims gave him a sheltered life within Catholicism. "During our theological education we heard virtually nothing about the intensive criticisms of Christianity dating back to the previous century or about the violent revolutions in the life and thought of the European world. At best these were described to us only to be rejected as nonsense in the triumphant tones of people who knew that they had the truth."[32] In 1946 Grollenberg was sent to Palestine to do Biblical research with the French Dominicans in Jerusalem, where the Dominicans were concerned with historical questions: Who wrote the books of the Bible? In what circumstances? What were they really trying to say? He also worked in summer at archaeological sites. This was followed by introducing young Dominicans into the world of the Bible. He taught the parables of Christ and he saw clearly what brought Christ into conflict with Jewish society. His preoccupation with Jesus resulted, so Grollenberg claimed, "in that I believed less concerned with all kinds of 'truths' from the past: Doctrine of God and the Trinity, of Christ as the God-man, of the means of grace, the infallibility of Christian dogma, biblical inspiration and so on. It is not that these are no longer true for me, they are no longer relevant, and do not fit in with the rest of my thinking. As a result I have grown away from the world of unassailable institutions and divine certainties in which I grew up."[33] The result was that in his book, Jesus, Grollenberg has completely misunderstood Jesus and never asserted that He was the Savior of the world, whom every person needed to believe in for eternal life.

Raymond E. Brown on Roman Catholic Higher Criticism

Raymond E. Brown in 1961 delivered a paper titled: "Our New Approach to the Bible," in which he set forth the reasons for the Roman Catholic embracement of liberal higher criticism.[34] Brown claims that the paper was delivered at a time when the new approach was fighting for its life.[35] In a postscript he stated: "It is a great joy that now a few years later the clouds have lifted and the hopes of the writers for tolerance and acceptance have been granted beyond expectation. Vatican II has adamantly refused to approve any statement on Revelation which would set the biblical movement back. Teachers in Rome who were under a cloud of suspicion have been restored to their chairs of biblical studies."[36]

Higher Criticism Found in Roman Catholic Bible Translations

The Jerusalem Bible in its "Introduction to the Pentateuch" asserted that Christ did not ascribe the entire Pentateuch to Moses.[37] Thus it informs its readers: "Now modern Pentateuchal study has revealed a variety of style, lack of sequence, and repetition narratives which made it impossible to ascribe the whole work to a single author."[38] *The Jerusalem Bible* subscribes to the liberal Protestant theory of JEDP. Relative to other Biblical books it incorporates the same isagogical views as those found in liberal Protestant Bibles.[39] *The New American Bible*, sponsored by The Bishops' Committee of the Confraternity of Christian Doctrine also has embraced the JEDP theory in its "Introduction to the Books of the Pentateuch."[40] Higher critical views on the time and authorship of Biblical books as set forth in higher critical volumes of liberal Biblical scholars are also found here.

The New Jerome Biblical Commentary

In 1990 there appeared a revision of *The Jerome Biblical Commentary* which was designed to apprise interested readers of recent scholarly insights into biblical criticism.[41] *Christian News* of August 1, 1993 on pages 6-8 has listed seventy-eight views and pronouncements held and advocated by *The New Jerome Biblical Commentary*, all of which can be found in liberal higher critical books and commentaries of Protestant proponents.[42] On so many beliefs about the nature and inspiration of the Bible, its isagogics and interpretations, Roman Catholic and liberal Protestants agree. Here is a classification of opinions of Roman Catholic critical scholarship not in agreement with views held from apostolic days till early into this twentieth century.

Old Testament Biblical Introduction

The Jerome Biblical Commentary (hereafter *JBC*) claims Moses did not write the Pentateuch. The Torah or Penteteuch is a product of the early fifth and late fourth centuries put together from various sources written centuries after Moses' time. The major sources were the Jahwist (9th century), E (Elohim 8th century), D (Deuteronomy) and the P (Priestly Code) (p. 4). *JBC* claims that Leviticus was the work of many hands, edited by a Priestly Redactor (p. 6). Deuteronomy was not written by Moses but reached its present form by a long process of formation (p. 61). This theory was already advocated by 1806-1807 by De Wette.[43]

Relative to the Book of Joshua *JBC* teaches that "the book is a kind of historical or theological fiction, whose primary sources was the pre-existent literature of Israel ... various episodes are cautionary tales (p. 112)." Archaeology allegedly has shown that many episodes are not historical (p. 112). The Book of Judges contains material based on fairy tales. The Books of Kings do not present real, accurate history and includes fiction. It is not a work of a political or social history but of theological history.

Kings is not interested in recording historical events (p. 160). Varied sources were employed by the author or authors of I and II Kings, using popular tales (I Kings 3:18-27) and miracle stories (II Kings 20) to archival records (I Kings 4:7-19) (p. 180).

The prophet Isaiah did not write chapters 40-66. (p. 190). The latter chapters were authored by the Second Isaiah, one of the greatest prophets of the Old Testament. The trito-Isaiah is responsible for chapters 55-66, who in turn was supposedly influenced by the Second Isaiah and Ezekiel (p. 191). It was a number of Protestant scholars who came up with the division of Isaiah, 1-39, 40-55 and 55-66 in the eighteenth and nineteenth centuries now accepted by Roman Catholic scholars. Chapter 40-55 and 56-66 are dated as having been written in the exilic and postexilic periods (p. 230). The prophet Jeremiah did not write much of the book ascribed to him (p. 266). Ezekiel, the *JBC* claims, contains imaginative mythological allusions and "wild visions" (p. 305). Two Chronicles has pious suggestions and exaggerations (p. 373). The Book of Daniel was not written by the prophet-statesman Daniel. The *JBC* declares: "Stories about Daniel are clearly haggadic; in their entirety they cannot be taken as history." The author is not interested in history and so he cannot be accused of historical inaccuracies (p. 408), although critical negative Protestant scholarship refers to a number of historical inaccuracies militating against the historicity of Daniel (p. 406).

The Book of Ruth is a short story set into history (p. 557). Relative to the historicity of Esther, *JBC* averred: "Scarcely any scholar would argue for the historical character of the work. The story is now given in the Hebrew Bible as a fictional character" (p. 576). The Esther narratives are supposed to be reminiscent of legends told of Elijah and Elisha in 1 and 2 Kings (p. 589).

The Jerome Biblical Commentary and New Testament Introduction

Critical New Testament scholars claim that Matthew and Mark made use of a source called Q (German for Quelle). *JBC* claims that much material in the Gospels comes from the unknown Q source (p. 598), which nobody has ever found or to which any ancient writer ever refers. The Protestant invented this Q document and Catholic scholarship has adopted it. The *JBC* follows Protestant scholarship in claiming that Matthew copied the whole book of Mark and therefore the First Gospel could not have been written by an apostle, who would have been an eyewitness (p. 630). Matthew contains contradictions or puzzles (p. 631). *JBC* further asserted that the entire Sermon on the Mount was not proclaimed by Christ (p. 640). The Bible was wrong when it has Paul as author of Ephesians and Colossians, but instead was written by a school of disciples which promoted the thinking of Paul (p. 884).

Relative to Jesus Christ, God's Son, the *JBC* has some shocking assertions and conclusions. Thus it is stated on page 1318: "Fundamentalists naively and uncritically equate the Christ of the Gospels with the historical Jesus allowing no place for the developments of Jesus traditions in the early church." Recent liberal Roman Catholic scholarship agrees with Protestant critical scholars as to who the historical Jesus really was. The genealogies of Matthew and Luke are of questionable historicity (p. 1319). However, the Bible Student might note that the gospels portray Jesus of Nazareth as having a mother and also brothers. Does that portrayal not show that Christ was a historic personage? Paul referred to Jesus as "the mediator between God and man, the man Christ Jesus" (I Timothy 2:5). On page 1317 the *JBC* commentator argues: "The fact that all four Gospels are faith documents reflecting the later theology means, however, that John is not to be rejected automatically in favor of the Synoptics. Although the saying traditions have undergone massive reworking some individual dates preserved in John seem even more reliable than parallel materials in the Synoptics. Judgments must be made on the merits of individual cases. The rest of the NT yields few data about the historical Jesus and the Apocryphal gospels may yield at most some sayings" (p. 1317).

The Apostle St. John did not write Revelation or The Apocalypse, because the case for an apostle having written it is not very strong.

The Historical Character of Old Testament Narratives

The Books of Exodus and Psalms teach that the Hebrews crossed through the Red Sea, marching between two walls of water (cf. Also I Cor. 10:1-2). The *JBC* denies these assertions. According to Exodus 14:4 the *JBC* claims that the Hebrews turned toward Baal-Zephon to cause Pharaoh to chase them through the area of Lake Balah and set for the ultimate battle and victory (p. 491). The commentator on Exodus denies that God supplied the Israelites with manna from heaven and sometimes with quails in the Sinai wilderness. The manna did not come from heaven but was picked of the tamarist tree of Palestine and Sinai. What does this do to the testimony of Christ in John 6?

Joshua's account of the falling of the walls of Jericho in the fifteenth century B.C. is not true. Archaeology supposedly has discredited the account which later Scriptures has accepted as true and as historical (pp. 112, 114). The account of the Angel of the Lord killing 185,000 soldiers is a legendary expansion (p. 248). In Daniel, chapter 3:1-97 (Vulgate text) is a haggadic story, the same is true of Daniel in the lion's den, or the affliction of Nebuchednezzar recorded in chapter 4, because no record has been found of Nebuchednezzar's insanity (pp. 412-414). The events in the Book of Jonah are not true, but only a legend (p. 580).

The Jerome Biblical Commentary and
Historical Events in the New Testament

"'The slaughter of the Bethlehem children two years and younger by Herod may not be historical, but bears resemblance to Pharaoh's killing the Hebrew babies in Exodus 1:6 (p. 636). The descent of the Holy Spirit upon Jesus as He was baptized by John the Baptist in the Jordan was not a visible occurrence but merely an inner experience (p. 638). The temptation of Christ in the wilderness was borrowed from the document

"Q" and is a piece of midrash. The account of Jesus conquering the Devil, given in Luke 4:1-13, contains myth. The *JBC* claims it is difficult to assess its historicity (p. 688). Some of the miracles reported in the Four Gospels may not have occurred but much of the data connected with them was added by the Early Church (p. 1317). Opines the *JBC*: "Fundamentalists naively and uncritically equate the Christ of the Gospels with the historical Jesus, allowing no place for the development and reinterpretation of the Jesus-traditions in the early church" (p. 1318). The Virgin Birth has been questioned on scientific grounds but the reader is told to believe what the Church says rather than the Bible (p. 1319). That raises the question: Where did the Church get this belief from, if not from the Holy Scriptures?

The resurrection of Christ was not necessarily an event that transpired in calendar time. The *JBC* accepted the view of Pannenberg that Christ was not raised bodily or physically (p. 1375). The resurrection narratives of the Gospels are contradictory (p. 1375). The Lukan account of the Ascension of Christ is erroneous (p. 1377).

The Jerome Biblical Commentary and Old Testament Prophecy

The *JBC* joins Protestant higher critics in rejecting the traditional views held both by former Roman Catholic scholars and conservative Christian scholars that Christ was foretold in the Old Testament. Thus the writer of Genesis claims that Genesis 3:15 is not a prophecy of the Messiah, but that the word "seed" (Hebrew "tzera") refers to the human descendants of Eve who will regard snakes as their enemies (p. 12). In Genesis 49:10 the *JBC* claims, that Shiloh does not refer to Christ, but rather to the extension of Judah's dominion. This latest of Roman Catholic Commentaries asserts that the word of God mediated through the prophets is not exclusive or eminently predictive (p. 198). Biblical prophecy does not give a photographic picture of the future (p. 1917). Isaiah 7:14 does not predict a virgin conception and later a virgin birth, as Matthew clearly declared in chapter 1:23. In Ch. 19:25-27 Job does not refer to the Messiah as Jerome wrongly has translated (p. 478). Micah 5:2 is not a direct prophecy of Christ's birth occurring in Bethlehem of Judea (p. 253). Psalm 8:10, as Peter asserts in Acts, does not refer to Christ. Psalm 22:2 and Psalm 1:0 do not refer to Christ as Jesus asserts in Matthew. The Bible is wrong, so contends the *JBC*, when it says that David wrote various psalms which spoke about Christ's coming (p. 1311).

The hermeneutics practiced by scholars setting forth the new radical and revolutionary views is not the hermeneutics of the Council of Trent, the Vatican Council of 1870 as well as that of previous Biblical Roman Catholic savants prior to 1942,[44] except those who had already broken with their churches rules of Biblical interpretation. The adoption of negative higher critical views about the nature of the Bible, the doctrines of revelation and inspiration were to have the same devastating effects as those for which liberal Protestants and Lutherans were responsible.

The Jerome Biblical Commentary and
The Inspiration of the Bible

The *JBC* stated: "A serious reading of the Old Testament and the New Testament raises cogent objections against a naive or simplistic theory of verbal inspiration ... The biblical texts themselves clearly suggest that normal human writings processes were at work on the production of Scripture... Such objections have led most contemporary scholars to abandon the theory of plenary verbal inspiration (p. 1028). Steinmueller in *A Companion to Scripture Studies* asserted about Leo XIll's 'Providentissimus Deus' that the Pontiff positively teaches that all the books which the Church receives as sacred and canonical are written wholly and entirely with all their parts as the dictation of the Holy Spirit" (p. EB 109). Hence, inspiration by its very nature embraces wholly and entirely everything written by the hagiographer.[45]

Further, Steinmueller contended that the general extension of inspiration has been settled by the Encyclical "Providentissimus Deus" of November 18, 1893, which repeated the words of the Council of Trent and of Vatican I. The entire books with all their parts (EB 109).[46]

The Greek text of Matthew 5.7 and the text of Luke 6:20-38 ascribe the Sermon of the Mount and the Sermon on the Plain specifically to Jesus, the *JBC* claims that a number of verses were never uttered by Jesus.

The inerrancy of Scripture is questioned by a writer of the *JBC* as follows: "The truth of Scripture lies not so much in that its passages are without error, but in that through them God manifests his fidelity to his people, bringing them in union with himself" (p. 1030). Leo XIII in "Providentissimus Deus" emphasized the absolute inerrancy of the Scripture. This contradicts Vatican II which according to the *JBC* (p. 1030) gave only a sentence to the books of Scripture ... teaching firmly faithfully and without error that truth which God wanted put into the sacred writings for the sake of our salvation, as a result of inspiration means that only certain positions, but not all of Scripture is inerrant.

This significant change in Roman Catholic circles indicates that the theologians have a different understanding of the nature of the Bible and also due to the enlightenment of conciliar directives (p. 1030). At Vatican II Cardinal Koenig pointed out the errors in the Biblical books and asserted that they are deficient in accuracy regarding both historical and scientific matters. In fact, Koenig claimed the Scriptures never claim to be inerrant (p. 1030).

Previous Christian Doctrines
Denied by Roman Catholic Higher Criticism

The creation account as recorded in Genesis 1:1-2:3 is not accurate or true because modern science makes impossible a literal interpretation of the opening three chapters of Genesis (p. 1029). The *JBC* claims the theory of evolution makes it impossible to take seriously the imaginative and theological portrayal of earliest humanity in Genesis 1-3. Adam and Eve were not the first human beings. The *JBC* speculates that the uni-

1551

verse is 4-5 billion years old; a billion years ago simple life forms appeared, to be followed by more complex, leaving a fossil record about 600 million years ago. In Africa fossils of primates occurred 600 million years ago (p. 1220). The oldest hominoid fossils found are those of Ramapithecus, dated circa millions years ago (p. 1241). The Bible clearly teaches that man was Adam fashioned from the earth and Eve was made by God from Adam's body as Roman Catholic theology taught in its dogmatic books.[47]

The Evolvement of God According to
The Jerome Biblical Commentary
God, known in Hebrew as Yahweh, first allegedly appeared in Moses' time, who promoted a practical but not theoretical monotheism (p. 1226). The idea of "God the Father," claims *JBC*, was influenced by the concept of El, the high God of the Ugaritic pantheon (p. 1225). The unicity of God is supposedly first set forth in Deuteronomy-Isaiah in the sixth century (p. 1287). What about Exodus 20:3 spoken around 1446 B.C.? The *JBC* denied the creation of the world **ex nihilo**. (e.i. out of nothing) p. 1293). Liberal Roman scholars contend that the Old Testament does not present a clear doctrine of salvation, but rather records a historical development that supposedly is difficult to synthetize (p. 1308). The Holy Spirit, the third person of the Trinity, is not found in the Old Testament. In the Old Testament the Holy Spirit is not a person but a principle of action (1290).

Scientific exegesis does not support the Virgin Conception and Birth of Christ. However, Roman Catholics are told that they should accept the doctrine on the authority of the Church (p. 1318).

Despite the fact that Muslims, Jews and pagans do not believe in Christ as their Savior and Redeemer, they still worship the same God as do Roman Catholics (Cf. John 3:15; 3:36; Acts 4:12, I Tim. 2:5-6 for a rebuttal).

The Jerome Biblical Commentary
and Biblical Christology
The Christology teachings of the New Testament cannot be trusted. The genealogies of Matthew and Luke of Christ's parentage are of questionable authority (p. 1328). Jesus did not coexist with the Father from all eternity. Asserts the *JBC*: "Only after Jesus' death do we finally have a believing acknowledgement of him as the Son of God (p. 1358). There are miracles ascribed to Jesus which he did not perform (p. 1370). The resurrection of Christ was not physical or corporeal but a spiritual one. The different accounts of the resurrection appearances are contradictory for there are supposedly two different traditions in the New Testament (pp. 1371, 1375).

The Jerome Biblical Commentary and St. Paul
Paul was not consistent in his statements about the future or the personal after life. These statements may be understood as symbolical statements and as such do not have the same authority as doctrinal or

systematic or theological assertions. It is also possible to assume that after his conversion Paul experienced a change of mind in the course of his ministry (p. 1369). The Pauline doctrine of the vicarious atonement is repudiated by the *JBC*. Thus on page 1399 one reads: "It is not Pauline teaching that the Father willed the death of his Son to satisfy the debts owed to God or to the devil by human sinners." Or again: "The bloodshed in sacrifice was not then, a vicarious punishment meted out on the animal instead on the person who immolated it" (p. 1399).

A number of Roman Catholic books, written before and even after 1943, reflect the traditional conservative views of historic Roman Catholicism on the nature and origin of various Biblical books. An examination of Giuseppe Ricciotti's *History of Israel*, 2 volumes,[48] shows that the critical views advanced in Ellis's *The Men and Message of the Old Testament*, Harington's *The Old Testament Record of Promise*[49] are not accepted, but the Mosaic authorship is held and the historical books of Joshua, Judges, 1. and 2. Samuel, 1. and 2, Kings, 1. And 2. Chronicles, Ezra, Nehemiah, Esther are reliable historical books. The concepts of myth, saga and legends ascribed by modern critical Roman Catholic professors to the just enumerated books is not propagated. The three volumes of Steinmueller's *Companion to Scripture Studies* dealing with introductory isagogics and the discussion of both Old and New Testaments follows the historic understanding of Christendom. Steinmueller did not follow or subscribe to the interpretation of Pius XII's encyclical as others had done. The Jesuit Leo-Dufour in *The Gospels and the Jesus of History* rejected the views of Roman Catholic scholars who questioned many historical events in the life of Christ.[50]

The Exegetical and Theological Situation After Pius XII

After Pius XII (1939-1958) there were elected popes such as John XXIII (1958-1963), Paul VI (1963-1978) and John Paul II (1978-) who have fostered and advanced views contrary to their predecessors Leo XIII, Pius X, Benedict XV and Pius XII. Vatican Council II (1962-65) was called under Pope John XXIII, a council designed for updating (Aggiornamento) of the world-wide Roman Catholic Church. Significant changes were made by the council members, which actually contradicted theological of the Council of Trent (1545-63) and Vatican I (1870). The higher criticism which many bishops and cardinals had embraced was not rejected but perpetuated at Vatican II. Out of the new hermeneutics and interpretation also came theological views which were completely contrary to traditional Roman Catholic theology. Bishop Williamson, head of the St. Thomas Seminary in Minnesota, in a September 1, 1993 letter to the friends and benefactors of his seminary, outlined how especially with Pope Paul VI (1963-1978) a new naturalistic theology had been promulgate.[51] Msgr. Montini, later Pope Paul VI, was an admirer of the theology of de Lubac in particular, also of Blondel, von Baltasar, and Teilhard de Chardin.

Bishop Williamson cited from the Si Si No No which has analyzed recent theological developments in Roman Catholicism in the last forty

1553

years. By the time of his death Paul VI had smashed all conservatives and the new theology was in control, aided and abetted by Ratzinger,[52] de Lubac, von Baltasar. These three constitute the think-tank of the current pope, John Paul II, who has embraced philosophical views that are totally anti-Biblical and also anti-traditional Roman Catholic. John Paul II in his first encyclical in 1978 commemorated the statement from the Council Document "Guadium and Spes" (No. 22) "By his incarnation the Son has united himself in some fashion with every man." That statement if true would mean that every man born alive is potentially saved by Christ, but he will have to do something about it to be saved. Father Doermann claims that the pontiff when he holds that all men consciously or unconsciously are in a state of effective redemption by Christ Jesus. So whoever has human nature is in a state of grace, and so all man by virtue that they are men, are saved. Von Baltasar has pointed out: "Hell exists, but is empty."[53]

Williamson has also concluded "that if every man at every moment has grace by merely having a human nature then every man in some way belongs to the Church, so the supernatural Church coincides with natural mankind."[54] The difference between the Church and those not in it, is the degree of difference of their "being in Christ." Thus those people outside of the Church already have a self-awareness and have a primary revelation in themselves by virtue of their humanity. The revelation through Christ is only a secondary revelation. From this it follows that all religions are neither true or false.[55] Hence John Paul II's respect for all religions. This explains the current pope's action at Assisi as well as other statements recognizing Jews and Moslems and the peoples of non-Christian religions as having God and not in need of salvation.

Cardinal Ratzinger, Prefect of the Congregation of the Doctrine of the Faith, according to Si Si No No has been responsible for working together with de Lubac, von Baltasar in promoting modernism and responsible for promoting a new Christology. Ratzinger in his "Introduction to Christianity" in 1968 wrote about the incarnation: "God comes to pass for men through men, nay even more concretely through the man (referring to Jesus) in whom the quintessence of humanity appears and who for that very reason is at the same time God himself."[56] However, the deity of Christ is taught as being eternal. John 1:1-3 clearly teaches the deity of Christ existed before the Word assumed humanity (John 1:14).

Higher criticism has led to the complete change of Roman Catholic theology. The current liberalism in Rome would not have been possible without the adoption of negative Biblical higher criticism.

Footnotes

1. Cf. George Barrois, "Providentissimus Deus," in *The Twentieth Century Encyclopedia of Religious Knowledge* (Grand Rapids: Baker Book House, 1955), II, p. 922; J.A. Steinmueller, *A Companion to Scripture Studies* (New York: Joseph E. Wagner, 1946), I, pp. 395-419. Text: *Leonis XIII, Pontificis Maximi, Acta,* Vol. XII, pp. 326-364.
2. Steinmueller, **op. cit.**, I, p. 395.

3. **Ibid.**, p. 396.

4. **Ibid.**, p. 26.

5. Decision by Pontifical Commission, June 26, 1906.

6. Steinmueller, **op. cit.**, p. 427.

7. Barrois, **op. cit.**, p. 922.

8. Text: Acta Sanctae Sedis, Vol. XI (1940); English translation, Steinmueller, I, 1946, pp. 420-525.

9. Steinmueller, **op. cit.**, I, pp. 419-422.

10. Barrois, "Pius X," *The Twentieth Century Encyclopedia of Religious Knowledge II*, pp. 882-883.

11. Text: *Acta Apostolicae Sedis*, Vol. XII (1920), pp. 381-422; Steinmueller, **op. cit.**, pp. 429-459; Barrois, "Spiritus Paraclitus," *Twentieth Century Encyclopedia*, **op. cit.**, II, p. 1057.

12. Barrois, **op. cit.**, p. 1047.

13. Steinmueller, *A Companion to Scripture Studies* (New York: Joseph Wagner, 1941), I, 478 pages; II, 347 pages; III, 409 pages.

14. Text: *Acta Apostolicae Sedis* (1943), pp. 297-328; J. E. Stenimueller, *A Companion to Scripture Studies,* I, pp. 460-483.

15. Barrois, "Divino Afflante Spiritu," *Twentieth Century Encyclopedia of Religious Knowledge*, **op. cit.**, II, p. 1057.

16. Orchard, Sutcliffe, Fuller and Russell, *A Catholic Commentary on Holy Scripture* (New York: Thomas Nelson & Sons, 1953), 1,312 pages.

17. **Ibid.**, p. 65.

18. **Ibid.**, p. 66.

19. Peter F. Ellis, *The Men and Message of the Old Testament* (Collegeville, Minnesota, 1963), pp. 104-17.

20. **Ibid.**, pp. 4, 51ff.

21. **Ibid.**, pp. 57-62.

22. L. F. Hartman, *Encyclopedia Dictionary of the Bible* (New York: McGraw-Hill Book Company, 1963), 1,634 pages.

23. *New American Bible* (Cleveland and New York: The Catholic Press & The World Publishing Company, 1970), pp. xi, xii.

24. Text: *Acta Apostolicae Sedis*, Vol. XXXII (1950), pp. 561-578; Barrois, "Humani Generis," *Twentieth Century Encyclopedia*, **op. cit.**, I, p. 638.

25. *The Jerome Biblical Commentary* contradicted Pius XII on the historicity of Genesis 1-3.

26. Orchard, Sutcliffe, Russell & Fuller, *A Roman Catholic Commentary*, **op. cit.**, 1,312 pages.

27. John L. McKenzie, *Dictionary of the Bible* (Milwaukee: Bruce Publishing House, 1965), 954 pp.

28. *The Bible Today Reader* (Collegeville: The Liturgical Press, 1973), 424 pp.

29. Cf. footnote 22.

30. Raymond E. Brown, Joseph A. Fitzmyer and Roland E. Murphy, *The Jerome Biblical Commentary* (Englewood Cliffs: Prentice Hall, 1968), volumes I and II.

31. Cf. footnote 13.

32. Louis Grollenberg, Jesus (Philadelphia: The Westminster Press, 1977), pp. 2-3 of the author's introduction.

33. **Ibid.**

34. Raymond E. Brown, *New Testament Essays* (New York: Image Books: Garden City; Doubleday & Company, 1968), pp. 22-25.

35. **Ibid.**, p. 35.

36. **Ibid.**

37. *The Jerusalem Bible*, General Editor Alexander Jones (Garden City, New York: Doubleday & Company, 1968), pp. 3-7.

38. **Ibid.**, p. 7.

39. **Ibid.**, pp. 1124-1125.

40. The New American Bible, translated by members of the Catholic Association (Cleveland and New York: World Publishing Company), 1970, p. xi.

41. *The New Jerome Biblical Commentary*, edited by Raymond E. Brown, Joseph Fitzmyer, and Roland E. Murphy (Englewood Cliffs: Prentice Hall, 1990. No less than 70 different scholars contributed to this commentary having over 1,600 pages."

42. Cf. Bernard W. Anderson, *Understanding the Old Testament* (Third Edition; Englewood Cliffs: Prentice Hall, 1975), 649 pp. John Bright, *A History of Israel* (Philadelphia: Westminster Press, 1972), 519pp.

43. Merrill F. Unger, Introductory Merrill F. Unger, *Introductory Guide to the Old Testament* (Third Edition; Englewood Cliffs: Prentice Hall, 1975), 649 pp. John Bright, *A History of Israel* (Philadelphia: Westminster Press, 1972), 519 pp.

44. Godfried Holberg, *Katechismus der biblischen Hermeneutik* (Freiburg im Breisgau: Herderische Verlagshadlung, 1914), 45 pp.; John E. Steinmueller, *A Companion to Scripture Studies*, **op. cit.**, I, p. 225-249.

45. Steinmueller, **op. cit.**, I, p.22.

46. **Ibid.**, p. 23.

47. George D. Smith, *The Teaching of the Catholic Church* (New York: The Macmillan Company, I, pp. 206-207; Ludwig Ott, *Fundamentals of Catholic Dogma*. Edited in English by James Canon Bastable (St. Louis: B. Herder Book Company, 1957), pp. 94-96.

48. Guiseppe Riccotti, *The History of Israel*. Translated from the Italian by Clement Della Penta and Richard T. A. Murphy. Second Edition (Milwaukee: Bruce Publishing House, 1952), 2 volumes.

49. Wilfrid J. Harrington, *The Old Testament Record of Promise* (Image Books; Garden City, New York: Doubleday and Company, 1976), 240 pp.

50. Xavier Leon Dufour, *The Gospels and the Jesus of History* (Image Books; Garden City, New York, 1970), 312 pp.

51. Father Williamson's entire letter was published in *The Christian News*, September 20, 1993, pp. 6-7.

52. **Ibid.**, p. 6

53. **Ibid.**

54. **Ibid.**

55. **Ibid.**, p. 7.

56. **Ibid.**, p. 6.

Questions

1. The climax of once forbidden positions in Rome were finalized in ____.
2. "Divina Afflante Spiritus" allowed Roman Catholic scholars to ____.
3. "Providentissiumus" had as its concern to ____.

4. Pope Leo XIII warned against ____.
5. Pope Pious X came out against ____.
6. The Pontifical Council on Biblical Matters asserted that Moses ____.
7. Lambentabilis Sane Exitu rejected ____.
8. In Steinmueller's *A Companion to Scripture Studies* there is no hint of ____.
9. What did the Catholic Biblical Association say about the documentary hypothesis? ____
10. *The New Catholic Encyclopedia* set forth clearly ____.
11. *The Jerome Biblical Commentary* was hailed as the ____.
12. Raymond E. Brown set forth ____.
13. What did *Christian News* show about *The Jerome Biblical Commentary*? ____
14. What is the "Q" source? ____
15. According to the *Jerome Biblical Commentary* (JBC) the account of the Lord's Angel killing 185,000 soldiers is ____.
16. The JBC accepted the view of Pannenberg that Christ was not raised ____.
17. The JBC says that Isaiah 7:14 does not ____.
18. What does the JBC say about evolution? ____
19. The JBC teaches that Muslims, Jews, and pagan still worship the ____ god Roman Catholics do.
20. Vatican Council II contradicted ____.
21. What does Cardinal Ratzinger (who became Pope Benedict XVI) say in his "Introduction to Christianity" about the incarnation? ____
22. What has led to the complete change of Roman Catholic theology?____

Catholic Editor Responds to Surburg

Dr. Timothy A. Mitchell
Editor, Pro Ecclesia
63 East 9th St.
N.Y.C., N.Y. 10003
Christian News, April 25, 1994

The major problem in Dr. Raymond Surburg's scholarly look at the "Changes Occurring in Roman Catholic Biblical Studies Between 1893 and 1993" (*CN*, March 14) is that neither hypothesis has been established, despite an acquaintance with the teachings and the scholarship involved.

These two hypotheses are: 1) that three of the last four popes (excluding John Paul I) "have fostered and advanced views contrary to their predecessors Leo XIII, Pius X, Benedict XV, and Pius XII."

And 2) that significant "changes were made by the council members (at Vatican II) which actually contradicted theological (sic) of the Council of Trent and Vatican (1870)."

Both theses can only be proven by contrasting the earlier and later writings. This simply was not even attempted and well it wasn't; for it cannot be done.

Rather, the author ably demonstrated the thesis that some "opinions of Roman Catholic scholarship" as contained in *The New Jerome Biblical Commentary* are "not in agreement with views held from apostolic days till early into the twentieth century."

Actually, these views are not in agreement with views held in Rome today, which is why there is this disclaimer as pointed out in the August 2, 1993 edition of the *Christian News*, namely, that "those who grant the nihil obstat and imprimatur may not agree with all of the opinions expressed."

As to other claims erroneously argued on the basis of a letter by Bishop Richard Williamson of Minnesota (who is not a Catholic bishop), it must be emphatically pointed out that no "theological views" of Vatican II are "completely contrary to traditional Roman Catholic theology."

Secondly, Pope John Paul II "has embraced (no) philosophical views that are totally anti-Biblical ..."

Thirdly, his reference to Jesus uniting Himself "in some fashion with every man" (Guadium et Spes, no. 22) is the start of a scriptural exegesis on the Incarnation and Christ as the New Man. It has nothing to do with the absurd conclusions of Williamson (repeated by Surburg) on potentiality, universal salvation, grace, nature, or the emptiness of Hell.

Fourthly, the Church does not teach "that all religions are neither true or false," as Williamson invalidly concludes. Vatican II is clear in the Decree on Ecumenism, the Fathers teach precisely the opposite, namely, that while all religions are in some sense true, "the very fullness of grace and truth (is) entrusted to the Catholic Church." (para. 3)

In closing then, while the study reveals much scholarship and work, it fails to adequately address both hypotheses, much less establish them, while inadvertently misrepresenting what the Church is teaching today. It did establish, however, the prudence of the censor in disavowing the questionable opinions of the *Jerome Biblical Commentary* and of the Church for separating herself from the nonsensical ramblings of Bishop Williamson.

In a word, Dr. Surburg has presented not one magisterial teaching of the Church since 1943 to demonstrate there has been any "change of Roman Catholic theology," which was the expressed purpose of the undertaking.

Ed. See *The Christian News Encyclopedia* for documentation on the anti-Christian theology which today is promoted within the Roman Catholic Church and tolerated by the Pope. *CN* has often said that about the only Roman Catholics excommunicated today by the Pope are those who challenge the Pope's authority. Bishop Williamson is a follower of the late French Archbishop Marcel Lefebvre. Lefebvre complained about theological modernism being tolerated by the Pope. The Pope excommunicated him but has not excommunicated liberal Roman Catholic theologians who deny such doctrines as the virgin birth and physical resurrection of Christ. Bishop Williamson is the head of a traditionalist Roman Catholic Seminary in Winona, Minnesota. The Pope is an evolutionist who denies the inerrancy of the Bible. He is a universalist who maintains that Jews, Muslims, and other non-Christians can get to heaven without saving faith in Jesus Christ. See the documentation in the *Christian News Encyclopedia*. Conservative Roman Catholics are refusing to face the clear documented fact that the Pope is doing very little about the many liberal theologians within the Roman Catholic Church. The Pope has never voiced any objection to the rank heresy which is in the *Jerome Biblical Commentary* and the *New Catholic Encyclopedia*, which is dedicated to the Pope. Both of these widely publicized works have Rome's imprimatur which at least means that Roman Catholic leaders insist that there is nothing in them contrary to Roman Catholic doctrine. Both works are reviewed in the *Christian News Encyclopedia*. The Pope in his latest encyclical, *Veritas Splendor*, refers to Christ as a second Moses, and maintains that man must at least in part work his own way to heaven. The Pope continues to deny the central doctrine of Christianity, justification by faith alone.

Questions
1. Mitchell was editor of ____.
2. Mitchell claims that Richard Williamson invalidly concluded ____.
3. The French Archbishop Marcel Lefebvre complained about ____.
4. The Pope is an ____ who denies the ____.
5. Has the Pope ever voiced any objection to the rank heresy in the *Jerome Biblical Commentary* and the *New Catholic Encyclopedia*? ____
6. The Pope continues to deny the central Christian doctrine of ____.

Christ's Resurrection: The Foundation and Corner Stone of the Christian Faith

Christian News, April 8, 1996

The resurrection of Christ is the foundation and cornerstone of Christianity.[1] Apart from the resurrection of Christ as a real objective and historical event Christianity ceases to have meaning for human life, giving the latter purpose, meaning and hope for the future. Without Easter's unique event, Christmas (the incarnation), Good Friday (the death of Christ), would be purposeless.[2] If Christ was not physically and corporeally raised from the dead, the Christian religion would be a fraud. Then the Bible would be a book of misinformation and lies. It would also mean that all people who believed the claims of Christ about eternal life and died believing in Christ's resurrection would be lost, and all Christ's teachings would be vain and useless. If Christ's body decayed in the tomb of Joseph of Arimathea, believers in Christ's resurrection for the last nineteen hundred years believed in a myth.[3]

Throughout the centuries since A.D. 30 or 33 Christian theologians have been defending Christ's bodily and corporeal resurrection. Heberman claimed that "Jesus died and afterward rose again from the dead is both the central doctrine of the Christian faith and the major factor in a defense of its teachings. This was true in the earliest church and remains so today."[4] Vernon Grounds declared: "The Easter miracle is the heart of the Christian faith: the N.T., if anything, is even more resurrection-oriented than it is cross-centered. Indeed, it is the resurrection which interprets the cross and which therefore shapes the Church's theology and life."[5] Floyd Filson averred of the New Testament: "The interpreting clue and the organizing fact of Biblical theology is the resurrection of Jesus Christ."[6]

Rejection of Christ's Bodily Resurrection

Even though the New Testament is very clear and explicit relative to Christ's bodily resurrection, this has not prevented atheists, agnostics and enemies of Christianity from attacking the veracity of the crowning event of Christ's life, upon which Christianity is based. Christianity as a reliable religion, either it stand or falls with its actuality.

This century has witnessed attacks by prominent religionists, as Weatherhead, Bultmann, Brunner, Gloege, Blake, and now the Jesus Seminar, to which belong Roman Catholics, Lutherans, Episcopalians, Methodists and other denominational scholars.[6a]

The Biblical Sources for Christ's Resurrection

The primary sources are the Scriptures of the Old Testament, the Four Gospels, Book of Acts. The Pauline corpus of letters, the Catholic Epistles, Hebrews and The Apocalypse. There are hundreds of references to the Biblical doctrine of Christ's resurrection.[7]

Christ's Resurrection Foretold in the Old Testament

In his first Messianic promise God predicted the victory over Satan who would bring about Jesus' death (Gen. 3:15). David predicted in the Resurrection Psalm No. 16, that Christ's body would not see decay. The second part of Psalm 22 (22-36) spoke of Christ extending His life. The prophet Isaiah announced at least seven hundred years before Christ, that the latter would prolong His days (53:13). "Hosea predicted that the Messiah would be poison to death (13:14). Job, in speaking of the Messiah, stated that He would stand on the latter day upon the grave, thus presupposing a resurrection of a dead body.[8] According to the testimony of Jesus, the swallowing of the prophet Jonah by a great fish and his stay for three days and three nights in the great fishes stomach was a type of His own stay in the bosom of the earth and his subsequent being spewed out by the sea monster, was a type of His resurrection (Matthew 16:4; Mt. 12:38-41).

Related to those Old Testament passages that predicted Christ's resurrection are those passages that spoke of his elevation. Which presuppose a resurrection. Thus David said of the Messiah: "Sit thou on my right hand until I make your enemies your footstool (Psalm 110:1; Matthew 22:42). Or the verse that describes Jesus to be an eternal priest after the order of Melchizedek. Paul claimed that Christ rose again was predicted "according to the Scriptures," and at that time the apostle had the Old Testament in mind (I Cor. 15:1-2).

Christ's Resurrection as Taught in the New Testament

"The miracle of Easter is the heart of the Christian faith and message," so opined Killen.[9] During the days of His public ministry Jesus spoke to His disciples that He would be put to death and be resurrected. In John 16:16 Christ was referring to His departure from the world and that would be preceded by His resurrection. Thus he promised His disciples: "A little while, and you will see me no more; again a little while, and you will see Me." Christ predicted his resurrection to his enemies, when He said to them: "Destroy this temple, and in three days I will raise it up again (John 2:19)." As Christ and His disciples came down from the Mount of the transfiguration, Jesus instructed them, not to tell any man the vision until Christ had been raised from the dead (Matthew 17:9). After Peter's great confession at Caesarea Philippi, the Lord informs his disciples that He had to go up to Jerusalem to suffer all the things of the elders and chief priests and be killed and again to be raised on the third day (Matthew 16:21). To Martha he said: "I am the resurrection and the life (John 12:25)."[10]

Christ's Enemies Deny the Reality of His Death

By denying the reality of Christ's death, it would logically follow that no resurrection could take place. To do this, the enemies of the resurrection needed to ignore and reject Bible verses that assert that He did die and was resurrected. Christ Himself had predicted His death. Pontius Pilate gave the dead body of Jesus to Joseph of Arimathea, after he as governor, had ascertained from the soldiers that Christ was dead. To question, the reality of the Lord's resurrection means that the New Testament writers have perpetrated a hoax or deliberately are knowingly supporting a falsehood. Thus the apostolic writers are liars and deceivers.

Wilbur Smith called attention to the fact that on three occasions before His own resurrection, that on these occasions he raised people from the dead, and one may say this occurred in a progressive order.[11] The first was the son of the widow of Nain (Luke 7:11-18). All four Gospels record the raising of Jairus of Capernaum's daughter (Matthew 9:18-19, 23-26, Mark 5:22-24, 35-43; Luke 8:40-42,49-50). There is the extended account of the raising of Lazarus, who had died four days before Jesus came to Bethany. Calling Lazarus by name, Jesus said: "Lazarus, come forth (John 11:43)." One factor all three accounts have in common is that Christ spoke to the dead, they heard his command or voice and they responded.[12]

The Resurrection of Saints After Christ's Own Resurrection

After Jesus had died Matthew has this historical information: "Many bodies of the saints that had fallen asleep were raised; and coming forth after his resurrection, they entered into the holy city and appeared to many (27:52-53)." The careful reader will note that these resurrections occurred after Christ's own resurrection.

The Immediate Pre-Conditions for
Christ's Resurrection Appearances

Body and soul of Christ were reunited, which had been separated on Good Friday. Made alive by the Trinity, Jesus left the tomb, so that His resurrection occurred in calendar time (i.e. 30 or 33 A. D.).[13] It transpired on the first day of the week (Mark 16:2). Holy writ does not tell us the exact hour and minute of His resurrection but it took place sometime between Saturday 6 p.m. and before the rising of the sun on Sunday morning.

A demurer has been put forth by enemies of the Christian faith, claiming that the Easter morning events do not agree with Matthew 12:39-40, where Jesus stated that Jonah was three days and three nights in the stomach of the great fish which Christ claimed would be matched by His three days and three nights in the bosom of the earth. If three complete days and nights would be involved this would mean that Christ was placed in the grave on Thursday afternoon and His resurrection would have happened on Sunday afternoon. This contradicts the Biblical testimony that Christ was placed after 3 p.m. and 6 p.m. on Friday, and rose early on the first day. However, in Hebrew linguistic usage a day and night could be designated a day, although only a part of a day was involved. Thus Jesus was about three hours on the first day (Good Friday), twenty-four hours on the sabbath (Saturday) and a number of hours on the third day, (Sunday). Thus Christ was in the grave on three different days and two nights.

Christ rose from the dead according to His human nature.[15] Thereafter

Jesus appeared as the God-man. Christ effected His own resurrection and even in death used His divine majesty to cause his mortal body to become alive. To His disciple the Master said that He had the power to lay it down and had the power to take it again (John 10:18:2:19). Jesus reminded his opponents that the fact that he could restore His life was proof of His Sonship.

The Identity of the Crucified and Risen Christs

After His resurrection Christ had the same body that was nailed to the cross.[16] In John 20:27 the Savior showed the assembled disciples the wounds in His hands and in His feet and the wound on His side. Jesus, however, came forth with glorified body (Cf. Phil. 3:21). Christians in the hereafter will have glorified bodies like unto Christ's glorified One. That Jesus was corporeally raised from the dead, that He spoke with them and ate fish, as he also ate a meal with Cleophas and his friend, showed that Jesus had been bodily raised from the dead.

Although the post-Easter body was the same as Jesus had before His crucifixion, yet it was in some respects very different. In a mysterious manner He passed through clothes around His body leaving a rolled up encasement, passed through the walls of Joseph's tomb and on Easter eve through locked doors showed Himself alive to the disciples. The angel removed the huge stone, not to allow Jesus to escape but to show friend and foe that He was arisen from the dead.

That the uniqueness of Jesus of Christ's body was not limited by hard objects might be compared to Christ's transfiguration and Christ appeared with a glorified body. Schleiermacher, the father of liberal theology, put forth the inane idea that someone left the back door of the tomb open which Christ used to escape. The significant fact of Christ's resurrection was that His soul was free from all impediments and had entered into its complete Divine Majesty.[16a]

Christ's Resurrection: A work of the Triune God

The New Testament ascribes the resurrection of the Second Person of the Godhead to the cooperative effort of the three distinct persons of the Godhead.[17] Christ is said to have been raised by the Father (Col. 2:12; Acts 2:24) and by the glory of the Father (Rom. 6:4). Also the Holy Spirit participated in Christ's resurrection (Rom. 8:10). The Son is depicted as resurrecting Himself by His own power (John 10:18; 2:19). The explanation of the Trinity's action, lies, of course, in the fact that Father, Son and Holy Spirit are one God. According to I Peter 3:18 Christ was made alive by the Spirit. This passage declares the resurrection to be a proof of Christ's deity. John 5:19 testifies to the fact of His self-resurrection and revivification. What the Father does, the Son also does.

The New Testament Resurrection Accounts

The four gospels and I Corinthians 15 set forth the historical evidence for Christ's corporeal resurrection. Each of the five writers reported the events of Easter as he saw them and how they served his purpose and objectives.[18]

The following are appearances that occurred between early Easter morning and the day of the Ascension:

1. The appearance to Mary Magdalene, John 20:1-14.
2. The appearance to the women returning from the empty tomb, Matt. 29:9.
3. Appearance to Simon Peter, Luke 24:30; I Cor. 15:5.
4. The appearance to the Emmaus disciples, Luke 24:13-22; Mark 16:12.
5. The appearance to Ten Disciples, without Thomas (Mark 16:14; John 20:19-25).
6. The appearance to the eleven, including Thomas, John 20:26-31.
7. The appearance of Christ to seven disciples at a lake in Galilee, John 21:1-18.
8. Appearances of the eleven apostles, and over 500 disciples on a mountain in Galilee, Mark 16:15-18; I Cor. 15:7.
9. Appearance to James, the Lord's brother (1 Cor. 15:7).
10. The appearance for the last time on earth, Luke 24:44-53; Acts 1:3-12; Mark 16:19-20.

What a loss Christianity would have suffered if the historical events that transpired between the crucifixion and the Ascension were lost or never had not been recorded. Between the crucifixion and Christ's literal, ascent into heaven, there occurred Christ's death, burial, resurrection, ten different appearances, the Ascension, the outpouring of the Holy Spirit and the founding of the Christian Church on Pentecost day.

An Analysis of Christ's Appearances

It is clear from a number of sources that Jesus appeared at ten different times. Five were made on the day of the resurrection; one a week later, about three of them, the time of their appearances is not stated; one of them on the day of the Ascension. Observed Scroggie about the interesting facts divulged by the ten appearances: "These appearances were made to different people, singly and in groups and companies; and at different hours of the day. All these factors when correctly appreciated testify to the reality of these appearances."[19]

The Alleged Contradictions of the Resurrection Appearances

The rejecters of Christ's bodily resurrection realize the importance and the significance of these historical accounts and have examined them minutely, claiming that such investigation reveals contradictions. Arndt asserted a number of years ago: "Perhaps nothing in the Bible is pointed to with greater frequency in the attempt to prove that our sacred Book contains contradictions than the four accounts of the resurrection of our Savior. The respective passages are Matthew 28:1-10; Mark 16:1-11, Luke 24:1-12, John 20:1-18. We are told that in a number of points are at variance with each other. First of all, every person must admit without hesitation that not one of our four accounts of the resurrection is complete, reporting all the facts. None of them makes the claim of being exhaustive. Each one reports actual occurrences. Hence the reports may be fragmentary, incomplete, and yet true. If this simple principle is borne in mind, most of the difficulties contained in the resurrection story will vanish."[19a] A number of alleged differences have been used frequently by the deniers of Christ's resurrection from the dead,[20] which will now be examined.

Answer to Objection No. 1:
How Many Angels were in the Tomb?

Critics of the reliability of the Easter accounts claim that there are contradictions relative to the angels in the tomb. Matthew has one angel in the tomb who talks to the women (28:5), while Luke mentions two men in the tomb (24:4), one of whom testified concerning Christ's resurrection.

Matthew and Mark just report the fact that one of the two addressed the women. The first of the Gospel writers does not deny the presence of two angels, so there does not exist a contradiction on the issue of how many angels were in the tomb on Easter morning.

Answer to Objection No 2:
Which women visited the Tomb on Easter Morning

The four Gospels accounts are said to be contradictory relative to the issue of which women went to the tomb.[21] Matthew 28:1 states that Mary Magdalene, and the other Mary went to the tomb. Mark 16:1 reports that Mary Magdalene, and Mary the mother of James, and Salome went to the tomb. Luke 24:10 lists Mary Magdalene, and Joanna, and Mary the mother of James, and other women went to the tomb. Luke 24:1 refers to "they," whose antecedent is verse 53 of chapter 23 which informs: "The women who had come with him from Galilee, who on Good Friday afternoon saw where they deposited the dead Jesus." John 20 merely mentions Mary Magdalene, the center of John's resurrection account.

An analysis of these four accounts shows that all accounts mention Mary Magdalene; Matthew also refers to "the other Mary"; Mark also refers to Mary the mother of James and also Salome; Luke gives the visitors to the tomb as Magdalene, Joanna, and Mary the mother of James. Thus Mary appears in three of the gospels. It is she to whom Matthew alludes to in the term of "the other Mary" (27:56).[22] Thus this Mary occurs in three narratives of the gospels. There is therefore remarkable agreement between the accounts as far as the women who visited the tomb. It is true that Mark is the only one who mentions Salome as belonging to the group. Because Matthew and Mark do not mention her does not mean that they contradict the other gospels. Joanna is only mentioned once and that does not support the contention that the evangelists contradict each other. John in 20:1-10 mentions Mary Magdalene, but that does not mean that he is in disagreement with the other three gospels as to the number and persons who came to the tomb. In some respects the accounts corroborate each other or even supplement each other relative to who were the women who came to the tomb.[23]

Answer to Objection No. 3

Opponents of the reliability of the Easter morning events claim that the Gospel writers disagree with each other at what time the women came to the grave.[24] John and Mark are said to contradict each other. Mark asserts that the women came to the tomb when the sun had arisen. John claims that Mary Magdalene came to the grave while it still was dark.

Arndt averred that the difficulty can be resolved when one looks into the situation. In order to go to the tomb, the women had to walk quite a distance, whether they stayed in Jerusalem or Bethany. When they left their quarters it was still dark but by the time they reached the tomb, the sun was appearing on the horizon. John is thinking of the departure time, and Mark the time of arrival. (Arndt, *Bible Difficulties and Seeming Contradictions*, pp. 193-194).

Answer to Objection No. 4

One synoptic account says that the returning women told no person what they had seen or heard. Because of fear of the Romans and Jewish enemies, the women, according to Mark 16:8, hesitated to tell any person whom they met. Yet, Luke 24:9 informs that they told the disciples all that they had seen and heard.

This difference can easily be explained, namely, in one instance they refused to communicate with people they did not know, but did communicate with people who would not betray them. Another possible explanation would be that while at first out of fear for their lives they refused to tell any person of their experience, but by the time they reached Jerusalem, they had regained their composure, and were eager to tell of their experiences.

Answer to Objection No. 5

Mark 16:11 asserts that when Mary Magdalene told the disciples in Jerusalem that Christ was alive, they refused to believe her, thus rejecting the fact of Christ's resurrection, but on Easter afternoon Jesus, whom they did not expect to rise again, did rise. The women were given evidence by Jesus. They went to Jerusalem and reported to the disciples that Jesus was alive, but the disciples did not believe their report. However, the ten apostles themselves met the resurrected Christ in the evening.

The difference between the Markan and Lukan accounts is to the precise time that Jesus appeared; at one time rejecting the testimony of the women and of Cleophas and his friend. Later they have irrefutable evidence for Christ's resurrection. Different occasions and times account for the differences.

Why Did Jesus Not Appear to His Enemies?

Luke assured Theophilus, to whom he addressed the Acts volume, that Jesus appeared during a period of forty days after His resurrection, that much evidence was given during this time period (1:3).

But there is no Biblical evidence that He appeared to His enemies at any time. Obviously Jesus knew that they would deny the fact of His resurrection. The Sadducees on principle rejected the doctrine of the bodily resurrection. When then the leaders of the Jews heard that Christ had resurrected Lazarus, they took counsel to kill Jesus. To appear to such enemies would be like "casting pearls before the swine."

Denying the Factuality of the Resurrection

A number of different, but untenable theories, have been proposed to explain away Christ's resurrection from the dead.[25]

The Swoon Theory

The theory claims that Christ did not die, and was taken down from cross as a person who swooned, but while in this state survived; unraveled from himself the embalming cloths and then somehow pushed aside the great stone and then showed Himself as alive and as resurrected. How could a person weaken by the beating he endured by the soldiers and in a terribly weakened condition, remove the great stone and overcome the guard placed at the tomb?[26]

The Vision Theory

The proponents of this theory maintain that the disciples, brooding over Christ's death and His promised resurrection, became the victims of hallucinations, in which they imagined they saw Christ alive.[27] However, psychologists claim that a person might suffer hallucinations, but that it is not possible that the eleven, the Emmaus disciples and the women simultaneously would all at the same time suffer the same hallucinations. The Biblical resurrection accounts depict Christ speaking and eating as flesh and hone people usually do.

The Theft Theory

This theory holds that the disciples stole their dead Master from the tomb and hid Him somewhere and then boldly proclaimed that He had risen from the dead."[28] The Bible records that the disciples were in hiding in fear of their lives. These fear-ridden men supposedly overcame the Roman guard to remove Christ from Joseph's tomb. If the Jewish really believed that Christ's body had been stolen, the Roman government could have forced the tomb robbers to produce the dead body of Jesus[29] and thus end the deceptive and lying activity of Christ's followers.

The Transformed Spiritual Body

On the strength of I Corinthians 15:4 it has been claimed by the advocates of this view that the body of Jesus was not characterized by flesh and bones, but was a non-material body. This theory holds that consequently it is a misunderstanding of the New Testament data to believe in a bodily or corporeal reappearance of Christ.[30] If the body of Jesus was not the same one that He had before His death and burial, then John could not have depicted Christ showing His wounds in His hands, feet and His side. Jesus ate with the Emmaus disciples which a spirit does not do. At a lake in Galilee Jesus ate breakfast with the seven disciples. On a mountain in Galilee the 11 disciples heard Jesus give the great commission to evangelize the world. Luke, a doctor, who certainly knew the difference between a real body and a ghostly appearance told Theophilus that during a forty day period Jesus appeared to His disciples.

The Mistaken Tomb Theory
Former Harvard professor Kirsop Lake suggested that on Easter morning the women went to the wrong tomb, where they encountered a stranger from whom they fled. Lake ruled out the possibility of a corporeal resurrection of Christ but failed to explain what the soldiers did at the tomb or that the tomb was empty.[31]

Initial Doubts of the Disciples Concerning Christ's Bodily Resurrection
A study of the resurrection narratives shows that the disciples at first doubted Christ's bodily resurrection, even when they were told by others who saw and spoke with Christ. According to Mark 16:14 when Mary Magdalene told the disciples that she had seen Christ, they did not believe. When the two Emmaus disciples went to Jerusalem after meeting Christ on the Emmaus road and also eating with them the disciples did not believe their report (Mark 16:13). When Thomas was told that the ten had seen Christ Easter eve, he refused to believe (John 20:25). Matthew reports in 28:17 "And when they saw him they worshiped him; but some doubted." When on Easter morning the women went to Christ's tomb, they did not expect to find Christ alive, but they had come to embalm the dead body. Thus they were not looking for a resurrected Christ.

Post-Ascension Appearances of Christ
Luke reported in Acts that Jesus of Nazareth appeared to Saul[32], Stephen, the first Christian martyr was being lynched and stoned to death, he saw Christ in heaven (Acts 7:55). The apostle John, about sixty years later, received from Christ revelations about the future of the Church and about Christ's ultimate victory over all foes.[33]

Proofs for the Resurrection of Christ
Christ's resurrection is historically attested by (1) the fact of the sudden change in the lives of the apostles. At the crucifixion the 11 were cowards but 50 fifty days later at Pentecost prepared to die for Jesus, bodily proclaiming His resurrection[34] (2) the descent of the Holy Spirit on the day of Pentecost was a fulfillment of Jesus' promise (John 14:16; 15:26; cf. 7:37-39; Acts 2:32-33); (3) the changing of the day of worship from the seventh to the first, a testimony to the fact that Christ arose on the first day of the week; (4) the sudden and amazing growth of the Christian Church; (5) the existence of the New Testament, whose very message hinges upon the authenticity of Christ's resurrection; and (6) a Chris-

tian's conviction as a result of the Holy Spirit's conversion activity by which a Christian knows experientially that Christ is now alive and active in His church.[35]

The Results and Benefits of Christ's Resurrection

(1) By the resurrection Christ was declared to be the Son of God; (2) All Old Testament prophecies about the Messiah were fulfilled by His resurrection from the dead; (3)[36] The resurrection established the truthfulness of His claims and the veracity of the Savior's message; (4) It proves that God accepted Christ's atoning sacrifice for the reconciliation of the world; and (5) Christ's resurrection guarantees the believer's resurrection on the Last Day.[37]

Christ's Resurrection and the Christian Hope

As a result of Christ's resurrection, the believer can sing with Peter: "Blessed be the God and Father of our Lord Jesus Christ! By his great mercy we have been born anew to a living hope through the resurrection of Jesus Christ from the dead and to an inheritance which is imperishable, undefiled, and unfading, kept in heaven for you (1 Peter 1:3-4)."[38]

Footnotes

1. Walter Kueneth, *The Theology of the Resurrection* (St. Louis: Concordia Publishing House, 1965), pp. 23-24; R. Killen, "Resurrection of Christ, "in C.F. Pfeiffer, H. Y. Voss and J. Rea, *Wycliffe Bible Encyclopedia* (Chicago: Moody Press, 1975), II, p. 1458.

2. "Auferstehung Christi," in C. Eckhardt, *Homiletcisches Realexikon nebst Index Rerum* (St. Louis: Success Printing Company, 1907), A, p. 21.

3. Haberman, "Resurrection of Christ, "in Walter Elwell (Editor) *Evangelical Dictionary of Theology* (Grand Rapids: Baker Book House, 1984), p. 938b.

4. **Ibid.**, p. 938b.

5. Vernon C. Grounds, "The Resurrection of Christ, *Zondervan Pictorial Bible Dictionary*, Merrill C. Tenney, (Grand Rapids: Zondervan Publishing House, 1963), p. 714.

6. Floyd Filson, *Jesus Christ, The Risen Lord*, (New York: Abingdon Press, 1956), p. 29.

6a. Herman Otten, *Baal or God* (New Haven, MO: Leader Publishing Co., 1965), pp. 104-109.

7. *The Zondervan Expanded Concordance* (Grand Rapids: Zondervan Publishing House, 1968), p. 1144.

8. Joseph Stump, *The Christian Faith* (New York: The Macmillan Company, 1932), p. 174.

9. Killen, **op. cit.**, II, p. 1458.

10. Stump, **op. cit.**, pp. 174-175.

11. Wilbur M. Smith, "Resurrection," in Harrison, Bromiley and Henry Baker's *Dictionary of Theology* (Grand Rapids: Baker Book House, 1960), p. 450.

12. **Ibid.**, p. 450.

13. H. Wayne House, *Chronological and Background Charts of the New Testament* (Grand Rapids: Zondervan Publishing House, 1981) pp. 104, 129.

14. William F. Arndt, *Bible Difficulties and Seeming Contradictions* (St. Louis: Con-

cordia Publishing House, 1987), Revised Edition, pp. 93 and 176-177.

15. Stump, **op. cit.**, p. 176; E. Hove, *Christian Doctrine* (Minneapolis: Augsburg Publishing House, 1930), p. 202.

16. Stump, **op. cit.**, p. 176.

17. *Joh. Guiliemi Baeri Compendium Theologae* edited by Carol Fer. Guil Walther (St. Louis: Concordia Publishing House, 1889), III, p. 95.

18. Cf. Eckhardt, "Auferstehung Christi, "**op. cit.**, pp. 222-223.

19. W. Graham Scroggie, *A Guide to the Gospels* (London: Pickering & Inglis, 1948), pp. 578-579.

19a. **Op. cit.**, p. 191.

20. K. F. Scott, "Resurrection of Jesus," Virgilius Ferm, *An Encyclopedia of Religion* (New York: The Philosophical Library, 1945), pp. 658-659.

21. As cited by Eckhardt, **op. cit.**, p. 222.

22. **Ibid.**, p. 222.

23. Arndt, **op. cit.**, pp. 191-195.

24. Eckhardt, "Widersprueche," Homiletisches Reallexikon, **op. cit.**, V-Z, p. 316.

25. Scott, **op. cit.**, p. 659.

26. Stump, **op. cit.**, p. 177; Killen, **op. cit.**, pp. 1458-1459.

27. Killen, p. 1459; Stump, p. 177; Canon B. Streeter adopted the visionary theory of Keim.

28. Killen, p. 1459; The lie theory was already proposed on the first Easter day, by the Jewish leaders, Matthew 28:11-15.

29. M. Blaiklock, *The Archaeology of the New Testament* gives in scriptural evidence that it was a crime to steal a body, p. 170.

30. Killen, **op. cit.**, p. 1459.

31. Cf. Kirsop Lake, *The Historical Evidence for the Resurrection of Jesus Christ* (New York: Putnam and Sons, 1907), pp. 268-269.

32. Acts 9:5; (I Corinthians 15:8).

33. *The New International Version-Interlinear Greek-English New Testament, The Nestle Greek Text With a Literal Translation* by Reverend Alfred Marshall (Grand Rapids: Zondervan Publishing House, 1976), pp. 838,840-842.

34. Acts 2:iff.; Acts 3:4:10.

35. W.G. Scroggie, *A Guide to the Gospels* (London: Pickering & Inglis, 1948), pp. 605-606.

36. Lawrence O. Richards, *Expository Dictionary of the Bible* (Grand Rapids: Zondervan Publishing House, 1985), p. 528b.

37. Francis Pieper, *Christian Dogmatics* (St. Louis: Concordia Publishing House, 1951), II, p. 321; D. Garlington, "Resurrection of Christ, "in D. F. Wright and J. I. Packer, *New Dictionary of Theology* (Leicester, England: InterVarsity, 1988), p. 583.

38. *Revised Standard Version,* 1947 Edition.

Questions

1. Apart from the resurrection Christianity ____.
2. Who has attacked the resurrection ____?
3. Where was Christ's Resurrection foretold in the Old Testament? ____
4. Who did Jesus raise from the dead? ____
5. In Hebrew linguistic usage a day and a night could be designated as

____.

6. Where are the Resurrection accounts set forth? ____
7. How many times did the resurrected Jesus appear? ____
8. How many angels were in the tomb? ____
9. What time did the women come to the grave? ____
10. Why did Jesus not appear to his enemies? ____
11. What are some theories denying the Resurrection? ____
12. What are some proofs for Christ's Resurrection? ____
13. What are the results and benefits of Christ's Resurrection? ____

The Bible and the Quran

The Alleged Similarities of the Conservative Understanding of the Doctrines Revelation and Inspiration with the Same Teachings in the Quran: A Comparative Study

Christian News, May 20, 1996

The charge has frequently been made by liberal scholars that certain conservative Biblical doctrines have similarities with those in Mohammedanism. Those who hold the books of the Bible were inspired by the Holy Spirit and consequently are inerrant or errorless are said to have adopted Muslim concepts of inspiration and revelation. Thus Guillaume in his classic book, *Islam*, wrote as follows: "Apart from a minority of fundamentalists, Jewish and Christian scholars recognize that there is a human as well as a divine element in their sacred books. In Islam the doctrine of the infallible word of God is an article of faith, and the few who have questioned it, have for the most part expressed their doubts in enigmatic language, so as to leave themselves a way of retreat from a dangerous position."[1]

Raymond Abba, former professor at the University of Durham, England, in his *Nature and Authority of the Bible*, declared: "If we regard the Bible as the Mohammadans regard the Koran—as a faultless and inerrant oracle, an authority for intellectually honest men and women will of necessity be contingent upon the accuracy of its factual statements. Should it be proved wrong in any detail of history, physics or astronomy, its status, on such a presupposition, must become that of a fallen oracle. That is the dilemma in which Fundamentalists find themselves. By beginning with an unbiblical assumption—the inerrancy of the Scriptures—he is ultimately driven either to vindicate every factual statement of the Bible or to abandon all belief in biblical authority. It is not surprising that, in such a dilemma between obscurantism and skepticism, many sincere people who have been brought up on these views finds themselves impaled upon its latter horn."[2] C. H. Dodd pointed out in a work dealing with Biblical authority and warned: "The Bible does not make any claims to infallible authority for all its parts."[3] To hold otherwise is espousing teachings that create skeptics.

The Purpose of this Essay

It is not true that Christians down the centuries, that Roman Catholics,[4] Protestants,[5] Lutherans[6] who have held to the inerrancy of the Old and New Testaments, have borrowed its concepts regarding revelation, inspiration, also the view on error-less-ness of the Bible from Mohammedanism. The writer will portray and contrast the Islamic doctrines of revelation, inspiration with those revealed in the Judeo-Christian Scriptures.

The Doctrine of Revelation in the Quran

The most important doctrine of the Quran and of Islam is the doctrine of God,[7] whom Mohammad identified with Allah. The latter always remains inaccessible. Man can know nothing of Allah himself. But Allah has deigned to give men revelations. In Surah 42, verse 56, it is stated: "It is not for any mortal that Allah should speak to him except by revelation (Arabic: wahy) or from behind a veil, or by sending an apostle."[8] The Quran itself is an example of Allah's direction, ultimately communicated to men through the prophet Muhammad, thus Allah made himself known through revelations which constitute the contents of the Quran. Surah 13, verse 39, asserts: "To every age its book."[9] In the course of the ages Allah abrogated what he may have given before, as Allah desired to do in the seventh post-Christian century. Allah made known his decrees in writing. According to Muslim tradition, the next important source for Mohammedan theology next to the Quran, the method as to how Allah proceeded to give his revelations happened like this. Thus it is depicted as occurring as follows: First of all Allah created the Pen and with the Pen he caused all decrees to be written, thus disposing of all things until the Last Judgment.[10] Before the birth and coming of Mohammad, Allah had given revelations to both Jews and Christians. Allah is said to have sent apostles to the earth in the centuries prior to the coming of Mohammad. In Surah 3:181 the believer is told: "Allah was not to leave the faithful in their present plight, but only to separate from the good.

Nor was he to reveal to you what is hidden. But he chose those of His apostles whom he will. Therefore have faith in Allah and His apostle: for if you have faith and guard yourselves against evil, your reward will be rich."[11]

The Quran and the decrees of Allah are kept on deposit upon the Preserved Tablet in heaven. One hundred of these writings have been given to the prophets in the form of books (Arabic: kutub) and nine hundred in form of leaves (Arabic: suhuf). The four books which are referred to in the Quran are the **Taurat** (Arabic: Law), which was given to Moses by Allah (Surah 3:44); the **Zabur** (the Psalms) given to David (Surah 4.161) and the Gospel to Jesus (Surah 5, verse 50) and finally the Quran given to Mohammed and the Muslims. The Quran, it is asserted, is the final and most complete revelation of Allah. Of it, Surah 43:22 asserts: "Allah surrounds them all. Surely, this is a glorious Koran, inscribed on a preserved table."[12] Surah 43:2 begins: "Ha min. By the Glorious Book![13] We have revealed the Koran in the Arabic tongue so that you may grasp its meaning. It is a transcript of our eternal book, sublime and full of wisdom," Surah 12, the Joseph Surah begins: "Alif lam ra. These are the verses of the Glorious Book. We have revealed the Koran in the Arabic tongue so that you may understand it. In revealing this Koran we will recount to you the best of histories, though before we revealed it to you they were heedless of our signs."

The foregoing is the information given in the Quran and in Muslim Traditions about the origin of Islam's sacred book. The Quran is considered to be part of the eternal word of Allah. According to Muslim theology, Allah has seven attributes: Power, Knowledge, will, Hearing, Seeing and Speech. All these are eternal attributes and are a part of Allah's Speech which was communicated to the Prophets.[14] The speech of Allah was sent down by means of leaves (Arabic: shuf) and books (Arabic: kutub). Since Allah is inaccessible these revelations were thus given mankind through the angel Gabriel and through the prophets.

The Christian View of the Origin of Holy Writ

The books of the Old Testament were given by means of prophets, psalmists, wise men. Some of the Old Testament authors were priests, kings, scribes, shepherds, individuals from various walks of life.[15] The twenty-seven books of the New Testament were written by evangelists, apostles, and certain anonymous writers. The books, sixty-six in number of the Bible, were written between 1400 B.C. and A.D. 100, thus roughly covering sixteen hundred years. Different countries and lands constituted the places where these books were composed, written and published.[16]

In the Judeo-Christian Scriptures a distinction needs to be made between revelation,[17] inspiration[18] and illumination. Unger has correctly defined revelation as "the divine act of commuting to man what he could not know. Revelation may be oral or written. Most commonly God spoke His revelations audibly or communicated His message by supersensory impressions upon human agents (inspiration). In rare instances He Himself wrote His revelation, as He did upon tables of stone on Sinai in the case of the first draft of the Ten Commandments (Ex. 34:28, Deut. 13; 5:22; 10:4). Since, however, God's spoken message was usually written down, revelation is most generally understood to be the communication in written form."[19]

The Pentateuch of Moses contains important data for the study of "revelation." The Book of Genesis is particularly important for information about **oral** revelation. Jehovah (Hebrew: Yahweh) gave direct verbal revelations to Adam, Eve, Cain, Noah, Abram-Abraham, Sarai, Isaac, Rebekah Jacob-Joseph. In the Books Exodus to Deuteronomy, God gave Moses revelation; in subsequent books, one finds Joshua, various of the Judges, Samuel, Eli, Saul, David, Solomon, many of the prophets, such as Isaiah, Jeremiah, Ezekiel, Daniel and a number of the Minor prophets as the recipients of direct revelations. Many of the oral revelations were later written down and are available to this day.[20] At least thirty-six authors were involved in the composition of the entire Bible.

During Old Testament times Jehovah employed different ways and methods to give the people of the Old Covenant a knowledge of who He was and what He demanded of His people. God employed direct discourse, vision, dream, the Angel of the Lord, through Urim and Thumim (priestly oracle), through the Divine Spirit, revelation by ecstasy, divine miracles in history, and by God's glory reflected in nature.[21]

The Biblical Doctrine of Inspiration

Since the Bible in its entirety was inspired by the Holy Spirit, it is also correct to refer to the Bible as "God's written, revelation." The fact that

Holy writ was "God-breathed" (theoneustos) is the equivalent of asserting that the Holy Bible was inspired by God. That raises the question: what does "inspiration" mean? P. E. Kretzmann defined inspiration in this way: "Inspiration refers to that activity of the Holy Spirit whereby the Third Person of the Godhead caused selected individuals to record infallibly for all time those truths, historical facts, salvational events and instructions necessary that the man of God may be perfect thoroughly furnished unto all good works. The truth which God has supernaturally revealed to men has been recorded by inspired men in Holy Scriptures. These Holy Scriptures consist of the Canonical Books of the Old and New Testaments."[22]

Stump observed about the Bible: "They possess primarily the character of a record of the divine revelation. The revelation, which was given historically in connection with God's dealings with the human race, existed before the written revelation of His dealings with Abraham was written by Moses. So also the Gospel of the crucified and risen Savior was known and widely preached before New Testament books were written. But if the revelation which God gave was to be preserved in its purity, a divinely inspired record was indispensable. The inspired record is given to us in sixty-six books of the Bible."[22a]

Benjamin Warfield defined "inspiration" as "the supernatural influence exerted on the sacred writers by the Spirit of God, by virtue of which writings are given Divine trust-worthiness."[23] Steinman has pointed out that three factors must he kept in mind in defining the Biblical doctrine of inspiration. First, there is the primary efficient **Cause**, the Holy Spirit, who acts upon man. Secondly, there is the subject of inspiration, man, the **agent** upon whom the Holy Spirit acts directly.[24] Finally there is the result of inspiration, a written revelation once for all given and thoroughly accredited, attested by miracle and fulfilled prophecy.[25]

In opposition to Mohammedanism, the Bible demands divine illumination on the part of the believer so that he will accept God's word and further illumination is also a prerequisite for the Bible's proper interpretation.

The Biblical Doctrine of Illumination
Unger has defined illumination "as that influence or ministry of the Holy Spirit which enables all who are in right relation with God to understand the objective written revelation."[25]

Contrasts Between Biblical Revelation, Inspiration and Illumination
Unger has pointed the contrasts of these three Biblical terms: "Revelation comprehends God given **truth**. Inspiration embraces man, under divine guidance accurately **receiving** the truth thus given. Revelation on God's part involves the origin of truth; inspiration on man's part deals with inerrant reduction of that truth into writing under the influence of the Holy Spirit. In brief, inspiration is help from God to keep the report of divine revelation free from error.[26] Illumination is to be distinguished

from revelation and inspiration, for the later applies only to certain books and their writers, while illumination promised to all believers, as Paul stated it in I Corinthians 2:14. "Revelation, as it concerns the Holy Scriptures, had a specific time period and involves the inspiration of certain sovereignly chosen individuals as the recipients of the revelation. Both of these divine operations have ceased, whereas illumination is continuously operative in behalf of all who qualify for this ministry of the Holy Spirit. In summary, it may be said, revelation involves **origin**, inspiration relates to **reception and recording** and illumination concerns **understanding or comprehension** of the written objective revelation."[27]

Prophecy and Apostleship in the Islamic Doctrine of Revelation

Relative to prophets and apostles the Quran has taken positions different from those found in the Bible.[28] Mohammad made a distinction between apostleship and prophet-hood. In the Judeo Christian Scriptures both are depicted as recipients of God's revelations and as individuals who put God's revelations down in an objective written form. According to the Quran a prophet (Arabic: nabi) is a person who receives a message from Allah but is not commanded to proclaim it, while an apostle not only receives revelations but he is commanded to proclaim them.[29] Only a freeman could serve as a prophet. He cannot be a male jinn but must be a male of the children of Adam. By contrast, to the Bible a number of women are denominated "prophetesses," such as Miriam (Ex. 15:20-21); Huldah (II Kings 22:14-20), Anna, the aged woman who held Christ in her arms in the Temple (Luke 2:36-38). In contrast to the New Testament, Mary the mother of Jesus, was however, given the status of a prophetess in the Quran. Mary was considered a prophetess because she heard the message of the angel, but she was not commanded to proclaim the message to her age, and was not entrusted with any dispensation. Most orthodox Muslims hold that the sign of apostleship and prophet-hood is such that these should perform miracles. Since Muslims also believe that a magician may perform wonders, they have added that Allah would not permit a magician to substantiate a false claim to apostle-hood or prophet-hood in this way! Spencer has pointed out that Shibli Numani, the modern (unorthodox) Muslim author, declares however in his book **al Kalam**, that this condition of miracle working cannot be supported by the Quran itself.[30] Muhammad brought no signs but only came to warn, declares Shibli, and he quotes Suras 13, vv. 827, and 29 v. 49 in support of this statement.

Mohammad, Allah's Prophet and Apostle

While there had appeared many prophets before Muhammad's was sent, the latter was the last and final of all prophets.[31] The Quran recognized individuals like Adam, Abraham, Moses, David and Christ and others as belonging to the order of prophethood.[32] It is Muslim contention that Christ predicted the coming of Muhammad. Some Muslim sects have given the number of others as 144,000 prophets sent to the nations of the world to guide them. Each prophet was sent to the nations of the world to guide them for a long period of time. In the Quran twenty-eight different prophets are mentioned by name. Miller has claimed: "Each was given a Book of God containing laws of civil and religious for the regulating of life of men. The laws given by one prophet were in effect till they were abrogated by the laws contained in the Book brought from God by the succeeding prophet. Each of the Great Prophets foretold the coming of the Prophet who would succeed him. The last and greatest of the Prophets is Muhammad, who is Sura 33:400, and is called the "Seal of the Prophets ."[33] No other prophet will come between Muhammad's time and the Day of Resurrection.

The Quran also teaches that Muhammad was called to his apostleship and that before this call, the Quran was brought down to the lowest heaven and stored in the House of Honor (Bait ul Izza). The Muslim Bible also informs its readers that before the Angel Gabriel appeared to Muhammad in a vision, he heard the words of the Quran. The Muslim Creed reads: "There is no God but Allah and Muhammad is His Prophet." The latter was the promoter of a religion which, he believed, had originated with Allah.[34]

Mohammad's Contribution in the Formation of the Quran

There are basically two different views as to the authorship of the Quran, that of Muslims and that of non-Muslims. According to Muslim theology Muhammad was the receiver and publisher of the Quran. The Prophet of Mecca was in the habit of meditating upon Mount Hira, in Arabia, where Mohammad spent some time in meditation. Then one day Gabriel appeared to the prophet in a vision calling upon Muhammad to warn the idolatrous Arabian tribes to give up their idolatrous practices. Gabriel held before the prophet Muhammad a scroll, on which was written the 96th Sura, which was read to the Arabian prophet and thereby Muhammad was given his mission. After receiving his mission there followed three years (Arabic: fatra) during which the prophet received no revelations.[35] The fact that nothing happened after the Gabriel experience, greatly disturbed Mohammad and he began to believe that jinn had taken possession of him; he felt himself to be mad, but his wife Khadijah "comforted him and assured him that he was the apostle of Allah. After a long wait Muhammad saw Gabriel standing between heaven and earth. The prophet ran to his wife Khadija and asked her to wrap him up in a mantle. Then the words of Sura 74 (which has the title of "The Cloaked One") were communicated to him with the revelation of the Surah the **fatra** was broken and Mohammad was on his way as Allah's prophet.[36]

The Quran has one hundred and fourteen Suras or chapters.[37] The present arrangement does not give the Suras in chronological order. The complete Quran contains 77,639 words, which are mechanically arranged by length rather than logically or chronologically.[38] The Quran is about the length of the New Testament. In printed texts the length varies from two lines and the longest has some 710 (Sura 2). Each Sura is made up of short verses (ayahs, "signs," "wonders") the exact delineation has varied slightly in different editions. Western language translations follow one of two different systems of versification, which makes it hard to find a given numerical reference; the variance can be up to seven (ayahs). The shortest Surah 108, has three brief **ayahs**, the longest, is Sura 2, has 268 longer ones.[39] The Surahs were written either in Medina or Mecca.[40] According to the superscriptions of the 114 Surahs, twenty were revealed at Medina, the others at Mecca. While it is important to maintain this distinction, scholars have by a comparative study been able to establish a less summary chronological distinction. A study of styles, mode of composition and the subject matter, has permitted specialists to establish two categories within the Mecca or pre-Hijran period.[41]

The oldest Surahs are animated, very lyrical and also abrupt. Interjections, exclamations, striking images and unfinished sentences very often characterize the early chapters. Another fact of the first Surahs to be spoken by the Arabian prophet was the piling up of oaths. These oaths grow less as Mohammad acquires authority in the Islamic community.[42]

The Meccan Surahs reflect the vicissitudes of Mohammad's preaching. In the least ancient Surahs, one finds the stories concerning the Biblical prophets. It is also at this time that the name of Rahman was introduced with which to designate Allah. The verses which at first were short in the earliest Surahs, became longer in the later ones. In fact, the latest Meccan Surahs foreshadow the Medinese period. The oratorical "O men!" is supplanted by "O believers!" Mohammad only attacks the heathen during the first ten years of his career, refraining from attacking Jews and Christians, because he believed himself in agreement with them.[43]

The Medinese Surahs

The style of the Medinese Surahs is more prosaic, having many legal stipulations. Very often the sentences become periods with the tone in them completely different from the Meccan Surahs. The reason for this the new position acquired by Mohammad, for he is now leader and lawgiver. Imperatives abound. The Jews take the place of the heathen in the polemics of Mohammad, and they are called "hypocrites" and "infirm of heart." Some Surahs contain military addresses. There are also found allusions to contemporary events, as for instance, the war between the Romans and Persians.

Reference to Mohammad's Family Affairs

References to family affairs are found in the Medinese Surahs, as for example: Mohammad's marriage with the divorced wife of Zaid, his adopted son (Surah 33:37), the accusation brought against his favorite wife Ayesha (24,10,11), the statute imposed on his wives after his death, that no other man was to marry his wives. Islam is called the Faith of Abraham "and the latter is made founder of the Ka'ba (22:76fr.) Abraham was neither Jew nor Christian" (3.60).

The Sources of the Contents of the Quran

A number of Jewish and Christians scholars have written to show that the Biblical material, relating to both the Old and New Testaments, was borrowed from Jewish and Christian sources, books written by scholars embracing a different view on Christian theology, have written books, that show that Judaism and Christianity were used by Mohammad when setting forth his theological beliefs, found in the Quran and in "the Sunnah" or the Traditions of Islam. We refer to St. Clair Tisdall, *The Original Sources of the Quran*[44] and Henry Preserved Smith's *The Bible and Islam*.[45]

Cf. also Bell, *The Origin of Islam in Its Christian Environment*."[45a] In addition to Judaism and Christianity Mohammad also borrowed from Arabian heathenism.

Smith asserted: "The dependence of the Koran upon the Bible is apparent at a glance. There is not a page whose language does not remind us of the Old Testament or New. This is party accounted for by similarity of the Arabic language to the Hebrew, and also by the resemblance of the civilizations represented in the two books."[46]

Since there did not exist an Arabic translation of the Bible at Mohammad's time, his sources were Jews and Christians. The practice of Mohammad was to subordinate the stories from the Bible to his main purpose.[47]

The Pentateuch furnished the Arabian prophet with most stories he employed as those of Adam, Noah, Abraham, Lot, Joseph and Moses. The destruction of Sodom is given eight times, as is the account of the flood. Other stories are those of David, Saul, Solomon! He has allusions to Elijah, Job, Gog and Magog.

Among the Biblical characters who interested the Prophet of Meccah was Abraham, whom he made the builder of the Kaaba and the founder of Mecca. Mohammad sees in Abraham a predecessor and a model. He calls Abraham a Hanif, a man turned from Idols and became a monotheist. The idea that Abraham was the Father of believers, he most likely obtained from St. Paul in the New Testament. The name of Moses occurs thirty-four times in the Quran. Moses made the magicians of Pharaoh repent and confess the true God. This was probably obtained from the Christian apocryphon,[48] the *Liber Poenitentiae, Jamnae et Mambrae*.[49] Dr. Smith listed a number of episodes in the history of Moses in the Quran which are due to his imagination, and witness to his ignorance of the Biblical story.

The Quran's Use of the New Testament

Turning to the New Testament one observes that only two of its histories are known to Mohammad. These are the history of John the Baptist and the life of Jesus. Two facts particularly impressed Smith; one is that Jesus is always mentioned in connection with Mary. "It almost seems that Mohammad were more impressed with Mary than with Jesus."[50]

Concerning Mary, Mohammad tells us that she was dedicated by her parents to the service of *God,* and thus came into the care of Zecharias, to whom she was assigned by the sacred lot. She resides in the temple where she is fed by angels. These and other facts are not found in the

1579

New Testament, Mohammad obtained from Syriac Christian sources.

W. St. Clair Tisdall, in his *The Original Sources of the Quran*, has given eleven different Koranic items which are based upon the apocryphal Gospels.[51]

Christ in the Quran

Muhammad denied the deity or divinity of Christ, while affirming His apostleship. The belief of Christians that Jesus was the Second Person of the Trinity was rejected. Mohammad denied that Christ died. He appears to have secured his views about the non-death of Jesus from the Docetists. Christ was not a person, although God has spoken in a final form as Hebrew 1:1-2 teaches. There is no place for the vicarious substitution. In Surah 5:78 Muhammad asserted: "The Messiah, the son of Mary is only a prophet." Those who call Jesus the son of God are liars (Surah 9:30).[52]

The Purpose of Islam

The object of religion according to the Quran is to believe what the Quran has said about Allah and keep his commands. A Muslim must respect Muhammad as Allah's final messenger to mankind. Five times a day the Muslim must say his prayers directed toward Mecca. He must observe the month of Ramadan yearly. He must give alms to the poor and during his lifetime make a pilgrimage to Mecca. He must accept the essential beliefs found in the Quran.[53]

Non-Muslim writers are agreed that Mohammad was the author of the Surahs of the Quran, and therefore, is an Arabic volume (written in classical Arabic) that originated in the seventh century A.D. Since Allah is the creation of Mohammad's religious imaginations and feelings, the Quran claims to be eternal and uncreated—it is a creation of Mohammad. Like the Bible, the Quran is a product of space and time. Its contents were formed out of pre-Islamic religion and beliefs and concepts borrowed from the Old and New Testaments. The Quran, contradicts the New Testaments on serious issues about Jesus Christ; the Quran depicting Jesus as a prophet and not God. In fact, Mohammad has been elevated over Jesus (Arabic: Isa). Archer, in his *Survey of Old Testament Introduction* has as Appendix 2, a chapter entitled "Anachronisms and Historical Inaccuracies in the Koran," shows that the Quran is not errorless nor inerrant.[54]

The Bible the Sources for Inspiration and Revelation

The Bible is the word of God, written by inspiration of the Holy Ghost, every word is infallibly true.[55] That makes the Bible the source for Christian doctrine and the supreme and only authority. The LCMS and other conservative denominations teach that the Bible does not merely contain the word of God, but that every word is true, because of verbal inspiration. The Nicene Creed asserts that the Holy Ghost spake by the prophets.[56] The Lutheran Confessions identify the Holy Scripture with God's Word.[57] There are many Biblical passages that teach the truth, that even though God employed human authors, they wrote what God wanted them to write. Paul said that God gave the Jews the oracles of God, Romans 3:2. Of the Old Testament Paul, declared that it was "God-breathed" (2 Tim. 3:16). The same apostle averred "which things we also we speak, not in words which man's wisdom teacheth, but which the Holy Ghost teacheth" (1 Cor. 2:13). Christ promised His disciples the gift of the Holy Spirit who would guide them into all truth (John 16:13). Jesus declared: "Thy word is the truth."

Since Islam appeared on the world scene centuries after the completion of the New Testament, the Islamic concepts of revelation and inspiration could have been borrowed from the Bible and not vice versa.

Footnotes

1. Alfred Guillaume, *Islam* (Middlesex England: Penguin Books, 1954), p. 55.
2. Raymond Abba, *The Nature and Authority of the Bible* (Philadelphia: Muhlenberg Press, 1958), pp. 304-305.
3. Marcus Dods, *The Bible: Its Origin and Nature.*
4. John E. Steinmueller, *A Companion to Scripture Studies* (New York: Joseph Wagner, 1941), I, pp. 6-12; 13-20.
5. Rene Pasche, *The Inspiration and Authority of Scripture.* Translated by Helen I. Needhem (Chicago: Moody Press, 1960), pp. 233-247.
6. Robert Preus, *The Inspiration of Scripture* (Mankato, Minn.: Lutheran Synod Book Company, 1955), pp. i-xvi.
7. H. Spencer, *Islam and the Gospel of God* (London: I.S.P.C.K., 1976), p. 1.
8. N. J. Dawood, The Koran (Middlesex: Penguin Books, 1956.) In this edition of the Koran, it is 42:51.
9. **Ibid.**, p. 146.
10. Spenser, **op. cit.**, p. 17.
11. Dawood, **op. cit.**, p. 423.
12. Dawood, p. 150.
13. **Ibid.** Muslim and Christian scholars do not know the meaning of "ha Min" and of "Alif lam ra" at the beginning of a Surah.
14. Spence, **op. cit.**, p. 17.
15. Cf. John H. Walton, *Chronological and Background Charts* (Grand Rapids: Zondervan Publishing House, 1978), p. 13 for listing of books of Old Testament. Their authors and time of composition.
16. Cf., H. Wayne House, *Chronological and Background Charts of the New Testament* (Grand Rapids: Zondervan Publishing House, 1981), pp. 16-17.
17. On the Biblical doctrine of "Revelation," cf. Adolf Hoenecke, *Ev. Lutherische Dogmatik* (Milwaukee: Northwestern Publishing House, 1909), II, pp. 1-15.; C. Eckhardt, "Offembarung,"in
Homiletisches Reallexicon nebst Index Re run (St. Louis: Success Publishing Company, 1912), M-R, pp. 225-229.; Olaf Norlie, *The Open Bible* (Minneapolis: Augsburg Publishing House, 1918), pp. 25-26.
18. Joseph Stump, *The Christian Faith* (New York: The Macmillan Company, 1932), pp. 319-332; Norman L. Geisler, and William Nix, *A General Introduction to the Bible* (Chicago: Moody
Press, 1968), pp. 27-447; 29-31; Norlie, **op. cit.**, pp. 29-37.

19. Merrill F. Unger, *Introductory Guide to the Old Testament* (Grand Rapids: Zondervan Publishing House, 1951), pp. 22-23.

20. Leon Morris, "Revelation," in J.D. Douglas, Organizing Editor, *The New Bible Dictionary* (Grand Rapids: Wm. B. Eerdmans Publishing Company, 1963), pp. 1092-1096.

21. E. Barton Payne, *The Theology of the Older Testament* (Grand Rapids: Zondervan Publishing House, 1962), pp. 48-62.

22. P. E. Kretzmann, *The Foundations Must Stand* (St. Louis: Concordia Publishing House, 1936), pp. 63-64.

22a. Stump, **op. cit.**, p. 313.

23. B.B. Warfield, "Inspiration," James Orr, *The International Standard Bible Encyclopedia* (Grand Rapids: W. B. Eerdmans Publishing Co., 1939) III, p. 1473.

24. John E. Steinmueller, *Companion to Scripture Studies* (New York: Joseph Wagner, 1941), I, pp. 5, 14.

25. Unger, **op. cit.**, p. 24.

26. **Ibid.**, pp. 24-25.

27. **Ibid.**, p. 25.

28. Spence, **op. cit.**, p. 17-18.

29. William M. Miller, *A Christian Response to Islam* (Philadelphia: Presbyterian and Reformed Publishing Company, 1976), p. 47.

30. Shibli Numani, al Kalam, 4th Edition, Azamgaath, pp. 71ff as cited by Spence, **op. cit.**, p. 18.

31. Cf. J. Christy Wilson, *Introducing Islam* (New York: Friendship Press, no date), p. 38.

32. Jack Budd; Red Sea Mission Team, *Studies on Islam. A Simple Outline of the Islamic Faith* (London: Red Sea Mission, 1967), pp. 30-31.

33. Miller, **op. cit.**, p. 47.

34. Cf. D. S. Margoliouth, *Mohammad and the Rise of Islam* (New York: Putnam Sens, 1905), pp. 1-481.

35. Robert S. Ellwood, "Quran," in William H. Gentz, *The Dictionary of Bible Religion* (Nashville: Abingdon Press, 1986), p. 864; W. A. Graham, "Quran," in Keith Crim; General Editor, *Abingdon Dictionary of Living Religions* (Nashville: Abingdon Press, 1981), p. 593.

36. Paul E. Johnson, "Koran or Quran," Vergilius Ferm, *An Encyclopedia of Religion* (New York: The Philosophical Library, 1945), p. 422.

37. **Ibid.**, p. 422.

38. **Ibid.**, p. 422b; George Sale, *The Koran* (Truebner's Oriental Series, London, 1882-1886).

39. Cf. M. F. Shakir, *The Qu'ran* (Elmhurst, New York; Tahrike Tarsike Quran, 1991), p. 422 and pp. 1-30 (Surah 2).

40. Sale, *The Koran*, **op. cit.**, Cf. Section 111, "Details of the Koran," H. Lammens, *Islam, Beliefs and Institutions* (New York: E . P. Dutton, 1226), Chapter III.

41. Sale, **op. cit.**

42. **Ibid.**

43. **Ibid.**

44. W. St. Clair Tisdall, *The Original Sources of the Quran* (London: SPCK, 1905), 287 pp.

45. Henry Preserved Smith, *The Bible and Islam* (New York: Charles Scribners, 1897),

319 pp.

45a. R. Bell, *The Origin of Islam in Its Christian Environment* (London, Macmillan, 1926), 224 pp.

46. **Ibid.** p. 60.

47. Smith, **op. cit.**, Chapter III "The Koran Narratives."

48. Smith, **op. cit.**, cf. "Quran Narratives."

49. **Ibid.**

50. **Ibid.**, p. 87.

51. Tisdall, **op. cit.**, pp. 1-287.

52. S. M. Zwemer, *The Moslem Christ* (New York: American Tract Society, 1912), cf. chapter 1 and 2.

53. These duties are called **ibadat** and are known as the fire pillars of Islam; Lammens, **op. cit.**, chapter III.

54. Gleason L. Archer, *A Survey of Old Testament Introduction* (Chicago: Moody Press, 1974), pp. 498-500.

55. Dr. John Warwick Montgomery, *God's Inerrant word—An International Symposium on the Trustworthiness of Scripture* (Minneapolis; Bethany Fellowship 1974), cf. especially pp. 15-42; 43-62; 63-94; 178-200; 242-281. James Montgomery Boice (Editor), *The Foundation of Biblical Authority* (Grand Rapids: Zondervan Publishing House, 1978), cf. especially Gleason Archer, (*The Witness of the Bible to Its own Inerrancy*), pp. 85-102. Vilhelm Koren, *The Inspiration of Holy Scriptures* (Minneapolis: Augsburg Publishing House, no. date), pp. 3-16.

56. "The Nicene Creed" as found in Theodore G. Tappert, *The Book of Concord* (Philadelphia: Fortress Press, 1959), p. 19.

57. **Ibid.**, pp. 464-465; pp. 444; 57; pp. 483.13

Questions

1. Some liberal scholars claim certain conservative Biblical doctrines have similarities with ____.
2. C. H. Dodd warned ____.
3. At least ____ authors were involved in the composition of the Bible.
4. Mary is given the status of a ____ in the Quran.
5. The Muslim Creed reads ____.
6. The Quran has one hundred and fourteen ____ or ____.
7. The Quran is about the length of the ____.
8. Muhammed denied the ____ of Christ.
9. What is the difference between the Quran and the Bible? ____

Did Zoroastrianism Influence the Religion of the Old and New Testaments?

Christian News, August 5, 1996

Zoroastrianism is often referred to as one of the four ancient monotheistic religions of Asia. The other three supposedly being Christianity, Judaism and Islam. Christianity is a continuation of the religion of the Old Testament; Judaism is also based on the Old Testament but after A. D. 70 its path was especially determined by the views and beliefs as expressed in the Mishnah and Gemara of the Palestinian and Babylonian Talmuds. When Mohammad wrote his Quran, he embodied a number of items from Judaism (from the Old Testament)[1] and from certain aspects of Christianity as found in the New Testament as reflected by Nestorianism.[2]

Critical scholars also believe that that both Judaism and Christianity show the influence of Zoroastrianism in their theological beliefs as reflected in the Old Testament, the literature of the intertestamental period and upon the writings of the New Testament.[3] Roman Catholic,[4] Protestant,[5] and Lutheran and Jewish scholars have expressed the views that some religious concepts found in both Testaments owe their origin to Zoroastrianism. Lutheran scholars, influenced by historical-critical views about the Old Testament, have postulated an influence from Zoroaster's religion on the Jews during the exilic and postexilic periods of Old Testament history. Thus Hanz Schwartz in a number of his books has emphasized the fact that the later books of the Old Testament show Iranian influence.[6] The Lutheran scholar Gerhard Rosenkranz, in an article on "Zoroastrianism," stated: "It is well-known that this rigidly doctrinaire dualism exerted a strong influence on later Judaism (e.g. the Book of Daniel, Enoch) and that traces of it may be also found in the New Testament."[7] The Roman Catholic scholar J. Duchesne-Guillemin claims that there are striking parallels between Zoroastrianism and Judaism and that they took shape during the Exile.[8]

The views put forward by Roman Catholic and Lutheran scholars are not new. A.V.W. Jackson, now deceased, but an authority on Zoroastrianism in the early part of this century, a professor at Columbia University, wrote a series of articles for the former *Biblical World* in which he endeavored to point out possible parallels between Zoroastrianism and Christianity. Thus the Indo-Iranian Columbia professor wrote: "Avesta, the Bible of Zoroaster," *Biblical World*, June 1893, 421-422; vol. VIII, 140-163; and "Zoroastrianism and the Resemblances Between It and Christianity," *Biblical World*, XXVII, May, 1906, 355-343.

Jackson pointed to Isaiah, Kings, Chronicles, Ezra, Esther and the Apocrypha and the New Testament as showing likenesses to Zoroastrianism. Since 722 B. C., 605,597 and 587 B.C. people from the Northern and Southern Kingdom respectively were taken to Babylonia, Assyria, Elam and Persia. From Ezra, Esther and Nehemiah it is evident that Hebrew people came into contact with members of the Zoroastrianism. Some scholars hold that Cyrus the Great, Cambysis, Darius I were adherents of Zoroastrianism.[8a] Those Old Testament scholars and students of Comparative Religions who do not believe the Biblical doctrine of revelation and who also advocate late dates for a number of Biblical books logically assume that any resemblances between Zoroastrianism and Christianity must be due to borrowing from Zoroastrianism.

The Early History of Zoroastrianism[8b]

Zoroaster or Zarathustra was from the first half of the seventh century before Christ in Media, where, a century previously, "the king of Assyria had placed captive Israelites in certain cities of the Medes," and his death occurred about the time that the Jews were carried into captivity at Babylon. The dates assigned by direct tradition for Zoroaster's life are 660-583 B.C., as given in the Pahlavi books. However, there are some modern scholars and modern Zoroastrianians of Bombay who believe the prophet was born much earlier.[9] Zoroaster's birth has been placed as early as about 1000 B.C. His teachings appear to have taken root in the hearts of some Persian people before the rise of Cyrus, whom, in the words of Isaiah, the LORD called "His anointed" and "His shepherd," and whom Yahweh had raised up in righteousness, although the Persian conqueror of Babylon in 539 B.C. did not know the "the God of Israel." Jackson held that the Achaemenian rulers were devotees of Zarathustra. In 331 B.C. the conquests of Alexander the Great terminated the history of Persia and at the same time shook the foundations of the Persian religion.[10] However, in the third century A.D. the Parthian rulers revived Zoroastrianism and sustained the fabric of the ancient faith for a half millennium until A.D. 650, when Mohammed overthrew the Zoroastrian faith. The faith of Allah and the Quran took over in Persia, with the result that eventually only 10,000 Gabars remained in Persia and about 90,000 Parsis in Bombay, India and its vicinity.[11]

The Main Tenants of Zoroastrianism

To appreciate the claim that Judaism and Christianity are allegedly indebted to Zoroastrianism, it is important to outline the main tenets of the faith promulgated by Zoroaster. One of the most prominent features of Zoroaster's faith is found in the metrical sermons, or psalms, called Gathas[12] (i.e. Songs, Canticles), and in other parts of the Avesta, is that of dualism. However, it must be noted that the idea of two opposing principles, one of good and the other of evil is not alone a characteristic of Zoroaster's faith.[13] Jackson asserted about Zoroastrianism's dualism that "this attempt, earlier than Plato's, to solve the problem of the existence of evil in the universe laid stress on the independent origin of evil, personified as a great principle, by the side of Good, and conceived as a vital factor against which man must contend in daily life." (Yasha 3:3; 45:2 (Avesta)[15] Zoroaster says that these two "primeval spirits" possess each a special sphere of activity, and are endowed with a marked individuality, and contend constantly with each other in a strife which pervades the world. Although the struggle between the Good Principle and the Evil Principle is intense, eventually the Good Spirit will triumph over the Evil One. The victory will be effected by Man; the establishment of the Wished-for Good Kingdom will then be completed, a Restoration of the world will take place, the Resurrection of the dead will occur, and the Righteous will reign supreme. According to Jackson, these are the main points of Zoroaster's theology and Zoroaster taught his followers to make the right choice in all questions of religious doubt (cf. Yasna 31:2). This Yasna was incorporated in Jackson's **Hymn of Zoroaster**.[16]

The Bible of Zoroastrianism

The Bible of Zoroastrianism is known as the **Avesta**. Its oldest part, the Gathas, dates back to Zoroaster himself its most recent portions are said to be as late as Sassanian times (3rd century to 7th century A.D.) The Avesta is composed of two different ancient dialects, which are believed to be as old as the Sanskrit of the Indian Veda. However, by Sassanian times this dialect was obsolete and had to be replaced by the language used by the Sassanians, namely, Pahlavi.[17] These commentaries in Pahlavi on the Avesta were called **zand** and the Avesta has erroneously been called **Zand-Avesta** at times. There is a summary of the Avesta that has survived Sassanian times which shows that the Avesta was four times as long once than it is now. The Avesta contained at least 21 books, of which only one has survived intact, the **Videvdat** or "Code against Demons." The rest of the Avesta is composed of fragments of the other 20 books, arranged for liturgical purposes.[18]

The Avesta is comprised of three parts. The first has the Yasna or Sacrifice, a text which is chanted during the performance of the chief ceremony of the Zoroastrian ritual, which Duchesne-Guillemin suggested is reminiscent of the sacrifice of the Roman Catholic mass. In the Avestan ritual sacred liquor and water are offered before an ever burning fire. The Gathas are found in the Yasna and are metrical discourses and said to be revelations given to Zoroaster. The **Visparat** or "All the Psalms," is a collection of additions to the Yasna, recited under certain circumstances.

The second part of the Yasna contains a series of 21 Yashts or Hymns addressed to many different divinities, including ancient gods whom Zoroaster had ignored or combated, but it represents a cult which crept into Zoroastrianism. Mithra, the goddess Anamita, the star-god Tishtriya, are the main gods among these deities. The reader of the Avesta will also notice that their hymns are devoted to the spirits of the deceased, to the Fravashis, to Xvarnah or Royal Fortune.

The third part of Avesta is the **Videvdat** or "code against Demons." In additions there are also found several minor sections. While the Avesta is valuable for students of comparative religions, only the Gathas are said by Duchesne-Guillemin to have any literary value. Certain Yahsts do however, have literary value.[19]

Basic Theological Concepts in Zoroastrianism

Zoroaster envisioned a combat between the powers of light and darkness, Ahura Mazda, "the Lord Wisdom" (later Persi "Hormazd"), is helped by six ministering angels, **Amesha Spentas**, "Immortal Holy Ones," who are created by Ahura Mazda and who stand before the latter's throne to fulfill his divine commands. Their names are personifications of abstract ideas: "Good Thought" (Vohu Manu), "Righteousness" (Asha), "Holy Harmony" Spenta Armaiti), and the like. Iranian scholars have compared them with the Christian idea of archangels, and in the campaign against the reality of evil, they are assisted by minor divinities, **Yazatas**, "Worshipful Ones," a celestial order only slightfully lower than the **Amesha Spentas**, and are portrayed as performing a role similar to that of the legions of angels in the Bible.

The Devil, **Angra Mainyu**, "Enemy Spirit" (later Persian Ahriman) operates in the kingdom of darkness. The essence of the devil is **Druj**, "deceit, falsehood" and he wages a fight against Righteousness with the help of a great group of evil spirits and arch-fiends. Angra Mainyu is attended by a number of members comprised of a group of six arch-demons, who offset the sixfold—band gathered around Hormazd. The most devilish of them is called Asha, "Demon (daeva) of Wrath," whose name supposedly appears in the Book of Tobit as **Asmodeus**. The diabolical foes of the good angels are a crew of **Daevas** and **Drujes**, "demons and fiends," and they join the archfiends in their attempts to pervert and mislead mankind and destroy the world (cf. Avesta,[20] (Yasna 30:6; Vendidad 19:1; Pahlavi Bd. 2:11; 3:1-27) The Daevas play a kindred part to that which religionists have assigned to unclean spirits and minor devils.

The Alleged Influence of Zoroastrianism on Judaism

Professor Jackson wrote over ninety years ago: "Without entering into a detailed discussion of the subject I may state that Judaism, and through it Christianity, is generally believed to have been influenced in the doctrine of demonology and also of angelology and it is the tendency of modern scholarship to trace this influence back to the period of contact, between the religions during the Babylonian captivity and the following centuries."[21]

Higher-critical scholars have proposed the view that the subject of both of good and bad angels was taken over from Zoroastrianism.[22] However, already in the Book of Genesis angels are depicting as acting as messengers of God in the days of the patriarchs, 2100-1875 B.C. According to the interpretation of the New Testament the Devil, the leader of the world of demons, was the actual tempter of Eve and Adam. St. Paul, St. John have made the identification of the Serpent and the Devil as active in Paradise. According to the inspired Apostle Jude: The angels which kept not their first estate and left their domain, are held by him in blank darkness, in everlasting chains for judgment of the Great Day (verse 6). The fall of the Devil and the evil angels occurred before the temptation of man. The Old Testament writers whenever they refer to Satan and evil angels are referring to historical realities and to actions and deeds performed by these against God's creatures. The Bible did not borrow these spirit beings from Zoroaster's religion!

One of the results of the espousal of negative Biblical criticism has been the denial of the existence of the devil or Satan in the Old Testament before the Persian period (559-331 B.C.).[22] But the Lutheran Confessions associate Satan with the very beginning of human history as a powerful spirit, who as a personality had existence. Article XIX of the Augsburg Confession states: "Nevertheless, the cause of sin is the will of the devil and of men turning away from God, as Christ said about the devil" (John 8:44).[23] "When he lies he speaks according to his own nature." *Formula of Concord*, solid Declaration asserts: "Hence, they say, we must preserve the distinction between the nature and the essence of fallen man (that is, between his body and soul, which are God's handiwork and creatures in us even after the Fall) and original sin (which is a work of the devil by which man's nature has been corrupted).[24] Again in the Solid Declaration of the *Formula of Concord* it is declared:

> The source of cause of evil is not God's foreknowledge (since God neither creates nor works evil, nor does he help it along and promotes it), but rather the wicked, and perverse will the devil and of man, as it is written, "Israel, thou hast plungeth thyself into misfortune, but in me alone is thy salvation (Hos. 13:9). Likewise, "Thou are not a God who delights in wickedness" (Ps. 5:4).[24a]

In a number of places the Lutheran Confessions quote the New Testament statement to the effect that "the devil was a liar and a murderer from the beginning."

Genesis 3 is the first place in Holy Writ where the personality of Satan manifests itself by the use of the serpent through whom he spoke and tempted Eve and Adam. Luther believed in the literal accuracy and the truthfulness of the Genesis account and was convinced that in the garden of Eden the real Devil spoke to Eve. He did not hold that Genesis 3 was to be understood allegorically, poetically, symbolically or parabolically. For Luther there was no poetic personification or primitive myth or folklore to be found in Genesis 3. His understanding of this crucial chapter was stated by the Reformer in a sermon of 1923, treating Genesis 3:1-6. Here is the Reformer's interpretation as given by Plass, in *What Luther Says*:

> The first thing... Moses here describes is how the serpent spoke with the woman. We cannot escape that fact. We must, as I constantly say, allow Scripture to retain the simple, plain meaning supplied by the words and must provide no comments. For it does not behoove us to explain God's Word as we please. We must not direct it but must allow ourselves to be directed by it and accord it the honor of being put down better than we can express it. We must, therefore, let it stand that what the woman saw with her eyes was a real, natural serpent. . . . But the devil dwelt in it, because Moses says it talked with her. For speaking is not an endowment of any animal, but only of a man. Thus Moses makes it clear enough for our understanding that the devil in the serpent spoke through its tongue. And this should not

surprise anybody, for the devil is a powerful spirit. God has not forbidden him to deal with physical things. We still see that he is the lord and prince of the world and speaks not only through animals but also through human beings nowadays for the greater part through the latter. (Weimar edition, 24,42f.)[25]

Luther, however, found in Genesis 3:15 the Prot-evangelium, an announcement of God's declaration to Eve, that from her offspring there would come a Descendant Who would crush the Devil's head (Weimar, 42, 144).[26] Historical Lutheran theology has taught the fall of the Devil and of the evil angels occurred between the end of the seventh day of creation and the time of the temptation of Eve and Adam in Genesis 3.

The Lutheran Cyclopedia, in both its 1954 and 1975 editions states:

Satan himself, for whose subjugation Christ came, is the originator of all wickedness, Eph. 2:2, an opponent of the kingdom of God. He is the tempter of the faithful, 1 Peter 5:8ff., who led Eve into sin and so became the originator and king of death, Heb. 2:14. Originally created good, evil spirits, through their own fault, fell, 2 Pet. 2:4, and are destined to a fearful sentence (p. 209 of 1954 edition; p. 233 of 1975 edition).[27]

Guillemin, one time professor at the Oriental Institute, University of Liege, Belgium, claimed that "although a definite borrowing is still impossible to prove, the resemblances between Zoroastrianism and Judaism are numerous and important and probably took shape during the Exile."[28] Resemblances of course do not necessarily demand borrowings. The possible borrowings of Judaism from Zoroastrianism are all dependent on a late dating of various Biblical writings, all of which according to the testimony of the text and of Scripture are earlier, if there is any borrowing one could just as logically argue that Zoroastrianism borrowed or was influenced by Judaism.

The Doctrine of God in Zoroastrianism Compared With That in Judaism

A. V. William Jackson in his monograph on "Ormazd" in *The Monist*, Vol. IX (1899), 161-178 claimed that Zoroastrianism has a lofty conception of the Supreme Being. He wrote about the god of Zoroastrianism: In another "we have before us a divinity exalted far above human passions, an all-wise being, an omniscient ruler, a spirit divine and unchanging, a giver of rewards and punishments, merciful and just, righteous and holy, the father and creator of all good things, who was, and is, and ever shall be; his throne is in the heavens, where the company of ministering angels stand ready to perform his decrees."[29]

Among the Aryans Zoroaster's concept of the spirituality and individuality of deity has not been exceeded, "although some reminiscences of the primitive sky God can be traced in it."[30] Jackson, however, admitted The Holy One of Israel alone transcends it, and extended comparison might be drawn if space would allow. An important point of difference however, between Ormazd and Jehovah must be noticed; it is a fundamental one, and grows out of the opposite conception of the relation of the power of evil to the good, and is traceable ultimately to the doctrine of dualism. In Israel, the Holy One is omnipotent, the devil is subordinate to him, and is suffered to exist; in Persia, on the other hand, Ahura Mazda is not endowed with absolute omnipotence. It is only at the time of the millennium when mankind, through constantly choosing what is right, shall have become perfected, that Ormazd will be supreme, the power of Angra Mainyu be over thrown, wickedness (Druj) delivered into the hands of Righteousness (Asha), and evil eliminated from the world.[31]

Because of the ultimate triumph of Good over Evil, Zoroastrian scholars have labelled Zoroastrianism a monotheistic religion, even though during most of the world's historic existence Ahura Mazda does not have absolute control of Angra Mainyu and it will be because of man's efforts that Ahuran Mazda will be victorious. Any person acquainted with the Old or New Testaments knows that Yahweh and the Triune God of the New Testament are all powerful and that the catastrophic events that will conclude the age will be brought about by an omnipotent God.

A Comparison of the Evil Spirit of Zoroastrianism and Christianity

Professor Jackson claimed that the Zoroastrian conception of the Evil Spirit does not look upon Angra Mainyu as a fallen angel, but as an independent power. In the Gathas, the oldest part of the Avesta, Angra Mainyu is presented as opposed to Ormuzd's Holy Spirit, personified as an emanation of the godhead, while in the later Avesta the opposition to Andra Mainyu to Ahura Mazda is direct. (cf. Jackson's discussion in Geiger and Kuhn, *Grundriss der Iranischen Philologie*, Vol. II,pp.621-31;647-49, where full references to the various treaties on the subject are given.)[32]

In Gathic theology, the development of the universe is portrayed. Angra Mainyu is coeval with God, and is practically coequal but not coeternal (Yasna 30:31; Yt. 19:12 (mairyo rotush); Bundahisn 1:3). Cf. Casartelli, *Religion under den Sauanaiden* (p. 53).[33] Because Angra Mainyu is not coeternal his influence and existence are limited. He is depicted as evil by choice from the very beginning, and all evil in the world emanates from him, since Ormazd at no time ever made anything that was not good. Angra Mainyu is a malicious being, a cowardly tempter and betrayer, the very soul of death. Like Satan, his dwelling place is in the Abyss of darkness, and he appears to be omnipresent and always marring the good, although he is ignorant and lacks prescience.

Scholars have found analogies between Ahriman and Satan in the Bible, and have drawn the erroneous conclusion that the Satan or Devil of the Bible ultimately owes his origin to a borrowing from Zoroastrian religion.[34]

In Zoroastrian theology the soul of man is placed between the warring elements of good and evil.[35] On man depends the final outcome of this conflict. Zoroaster's religious tenets are designed to help man to be victorious. The devotee of Zoroasteris strongly encouraged to follow a doctrine of moral responsibility and he was told that he possesses freedom of the will (Cf. Yasna 31:11; 30:2; Denkart 9:30, and others). Every deed increased the evil, while each good deed increases the good. (Yasna 31:15; 35:5).

Ormazd's creatures are told to choose the right (Yasha 31:2; 30:2). In the Gathas Zoroaster depicted himself as guiding man in his choices, and by right choices to advance the "Good Kingdom," "the Wished-for Kingdom," the "Kingdom of Desire." Every action of man is recorded in a lifebook, and a man will be judged hereafter and be rewarded or punished accordance with this reckoning (Cf. Yasna 31:13; Ys. 32:6 and others).[36]

Zoroastrian Eschatology and
Christian Eschatology Compared

It is especially in the area of eschatology that Judaism and Christianity are alleged to have borrowed prominently from Zoroastrianism.[37] The later as a faith that is supposed to have taught an optimistic hope of a regeneration of the world and a general resurrection of the dead in a marked way. In the Gathas there is set forth the expectation and promise of a new order of things and this it is said that it echoes through all the ancient Iranian Scriptures. The Zoroastrian religion teaches that a great crisis is impending, and therefore it behooves each person to seek after the ideal and the eternal; in the end mankind will be made perfect and the world will be regenerated (frashem aham, frashotema; frashokereti). The new dispensation (vidaiti, ie. "division") will be accompanied by the advent of a savior (saoshyant) and the establishment of the sovereignty of the kingdom of righteousness.

Judaism is supposed to have borrowed the idea of the Messiah from the Iranian Scriptures. Thus Duchesne-Guillemin, as gives the Messiah concept possible borrowing of Judaism from and Zoroastrianism and wrote: "Secondly, the figure of the Messiah, originally a future king of Israel who would save his people from oppression, evolved, in Deutero-Isaiah, for instance into a universal Savior very similar to the Iranian Saosbyant."[38]

The critical theory about the development of the Old Testament Messiah does not agree with the teaching of the Scriptures itself. St. Paul states in Galatians 3 that through the seed of Abraham, through One Seed, the nations were to be blessed and this seed was Jesus Christ. The coming of the Messiah was not first a development of the time of 540 B.C. (the critical date for the Deutero-Isaiah), but was intimately connected with the eternal decree of redemption, determined in eternity, before the world ever was created. The many Messianic prophecies which constitute the heart of the Old Testament, were not the result of human development, but were a matter of divine revelation. A number of very important prophecies about the Messiah had been given many years before the birth of Zoroaster in Persia in the seventh century A.D. (750-680 B.C.), who lived a century before Zoroaster, has many outstanding Messianic prophecies. There is no need for Judaism to have borrowed its ideas of a Messiah from the **saoshant** teaching of Zoroastrianism.

The Jews of the Old Testament allegedly borrowed their ideas of concept of a millennium from Zoroastrianism.[39] But the Old Testament does not speak of a millennium. The Book of Revelation is the only place in the Bible, where the concept of a millennium occurs, where the Greek text speaks of a thousand years (chilia ete) six times. In Revelation 20 the millennium begins with the binding and confinement of Satan (20:1-3) in the bottomless pit, followed by the resurrection and reward of the martyred dead of the period immediately preceding (20:4). The millennium, the period between the birth of Christ and the Second Coming, ends with Satan's being loosed in rebellion against Christ which is crushed in a fiery destruction from heaven (20:9). The language of Revelation is symbolical and every detail is not to be understood literally. Premillennialists have worked out a very complicated eschatology, in which Christ is depicted as ruling in peace on earth from Jerusalem, with the converted Jewish people His followers. This complicated and involved eschatology is read back into the Old Testament, claiming that many Old Testament passages picture a kingdom of righteousness and peace on earth ruled over by the Son of David. This is incorrect because Christ's kingdom is spiritual in nature, for as Christ told Pontius Pilate: "My kingdom is not of this world." Christ, as object of faith, rules over a spiritual kingdom, the kingdom of grace and will end in the kingdom of glory.

There is absolutely no borrowing from Zoroastrianism, just as the whole Biblical doctrine of eschatology, of the last things, is unrelated to that of Zoroastrianism. Both Joel and Isaiah have in their prophetic writings passages which predict eschatological events. Both of these prophets, one in the ninth and the other in the eighth century, B.C., made eschatological predictions, about the future Christian Church (Joel 2-3) and in chapters 24 and 34 of Isaiah the great events of the general judgment. Both of these men prophesied before Zoroaster's time.

The Names of the Elect Written in the Book of Life

The idea of the names of the elect written in the book of life is found already in the Mosaic period (Exodus 32:32) and in the Davidic 10th-century 69th Psalm (V. 28) and also in (Isaiah, 4:3) all instances before pre-Zoroastrian times. The division of mankind into two classes the Judgment is a logical deduction from both the Old and New Testaments, where it is taught that the evil will be punished and the believers rewarded. The last two verses of Isaiah speak of the eternal punishment of the wicked (66:23- 24). While there seems to exist interesting parallels between Zoroastrianism and Judaism and also between the New Testament religion, one must reject the statement of the author of "Zoroastrianism" in The Encyclopedia Britannica: "After the exile, the traditional hope in a messiah—king of the House of David who would reestablish Israel as an independent nation and make it triumph over all enemies gave way gradually to a concept at once more universal and moral. The salvation of Israel was still essential, but it had to come about in a framework of a general renewal; the appearance of a savior would mean the end of this world and the birth of a new creation; his judgment of Israel would become a general judgment, dividing mankind into good and evil. This new concept, at once universal and ethical, recalls Iran so strongly that many scholars attribute it to the influence of that country."[40]

These alleged relationships are particularly conditioned by the system of dating adopted for Old Testament writings, namely, the dating of the Pentateuch, the multiple authorship of Isaiah (Proto-, Deutero- and Trito Isaiahs), the late dating of various prophetic books, the placement of anything apocalyptical into the third and second centuries B.C. makes possible to date pre-Zoroastrian Biblical data in the postexilic period and thus assume the influence on later Judean books by Zoroastrian writings. It might be appropriate at this juncture to remind the reader that some modern scholars concerning with the Iranian religion differ widely about the time of Zoroaster's birth, his character and theological views. On the one hand Zoroaster has been seen as a kind of primitive sorcerer or shaman, intoxicating himself with hemp fumes, and on the other hand, he has been represented as a lofty moralist and social reformer.[41]

It is necessary to state the judgment of Manasce relative to Zoroastrianism's influence: "to approve the influence of Zoroastranism on Judaism one would need detailed knowledge of Zoroastrianism as it was when the two religions were in contact during the exile. Its apparent polytheism would have repelled the Jews, but the overwhelming personality of Ahura Mazda may have reminded them of Yahweh. Zoroastrian eschatology seems to have stimulated the development of their own, but it is very unlikely that KASHTRA gave rise to the idea of the Kingdom of God."[42]

Alleged Zoroastrian Influence on the Intertestamental Jewish Literature

Instances of Zoroastrian influence during the intertestamental period are allegedly "the myth of the two spirits in The Rule of the Community and certain ideas in the Book of Tobit.[43] The Qumran book of The War of the Sons of Light With the Sons of Darkness is supposed to reflect the influences of Iranian dualism.[44] However, the dualism of the sect of Qumran can be justified on the basis of the same dualism which is found in the Old Testament. It can just as logically be explained as an outgrowth of the dualistic ideas found in the Old Testament writings.

Alleged Zoroastrianism's Influence on the New Testament

John Hinnelli in an article, entitled "Zoroastrian Savior Imagery and Its Influence on the New Testament,"[45] claimed that the Jesus's defeat of demons, his gathering of man for the judgment scene, his raising of the dead, and his administration of judgment is due to Zoroastrian influence. Unfortunately, Hinnelli comes to the New Testament with anti-supernaturalistic bias, he does not recognize the uniqueness of the person of Christ, who was not only a man, but God. As God he did not need to borrow ideas from other faiths, but as the Creator, Preserver and Judge of the world he knew the future and He revealed to His disciplines and to the readers of the Gospels what was the purpose of life, how to be redeemed and how to live a Godpleasing life. Jesus, the Giver of life, also had the power over death and life. The raising of the daughter of Jairus, of the young man of Nain, of Martha's and Mary's brother Lazarus were demonstrations of his deity. The events that will accompany the end of this age, the general judgment, the doctrine of a general resurrection and the judgment are announced by Jesus in the Mount Olivet Discourse. Various New Testament writers, such as Paul and Peter, apostles of the Lord, as inspired spokesman for Jesus, also have been the recipients of revelations dealing with eschatological themes.

Carnoy in his articles in *The Encyclopeclia of Religion and Ethics*, in writing about the influences of Zoroastrianism on other faiths, admitted that it was difficult to show direct borrowings of other religions taken from the Iranian religion. Carnoy believes that the teachings of Zoroaster could not have reached the Greeks and the Jews in any pure form. Thus he wrote:

> In the question of the relation of Judaism to Mazdaeism one must distinguish between Gathism, late Zoroastrianism, the doctrines which are transmitted only through Pahlavi books of Sasanian times, and the doctrines which are common to Magism and to the various sects of Iran.[46]

Zoroastrian Ethics Compared With Biblical Ethics

While Zoroaster emphasized having "good thoughts, good words, good deeds," being good according to Zoroastrianism meant chiefly to abstain from demon worship and to worship Ahura Mazda and to follow the latter's precepts. Body and soul must be kept pure.[47] In addition, it is man's duty to foster agriculture, cattle raising and irrigation, and to protest the especially the cow and dog, to abstain from lying and robbery. The elements of the earth, fire and water must be kept from defilement.[48] The Hebrew ethical teachings are found in the Ten Words (Exodus 20:1-17); Deuteronomy 5:6-21), ethical directives in Leviticus and many ethical directives in the Psalms, Proverbs and the sixteen prophetic books. Actually, there exists a gulf between Zoroastrianism's ethics and those of the Old Testament and surely those of the New Testament. That Judaism and Christianity borrowed religious conceptions from Zoroastianism is sheer speculation and completely unfounded.

Eschatological Beliefs Rejected by Judaism

The conflict between the two forces, of the good and evil, Zoroaster's faith taught that these two forces would continue until the end of the world cycle, which consists of 12,000 years, when Ahura Mazda will finally triumph and Ahriman overthrown. The last cycle of 3,000 years began with Zoroaster's life. The Bible knows nothing of such a period of 12,000 years. When a Zoroastrian devotee dies he could not be buried because something dead would defile the earth. So burial and cremation were forbidden, and it was a Zoroastrian custom to place the dead on "towers of lence,"[49] where vultures would consume the flesh. The Hebrew and Christians buried their dead in graves or sepulchers.

Zoroastrianism had no place for the forgiveness of sins, for which provisions were made for penitent Jews through the use of the sacrificial system and by personal confession of sins and trusting God's forgiving promises. In connection with the individual judgment Zoroastrianism taught that three days after death the souls cross the Cinvat bridge to be judged,[50] the righteous going to heaven, and the wicked to the tortures of hell. If the good and evil deeds balance exactly, the soul passes to an intermediate place, called Hamestakan. In the general resurrection men are subjected to another ordeal. They must go through molten metal, which causes the good joy, but gives extreme pain to the wicked. Finally, all souls now purified will be taken to heaven and a new world will be established, one that will last to eternity. Judaism has a different Eschatology when compared with Zoroastrianism. That is especially true when compared with New Testament Eschatology.

Footnotes

1. Henry Preserved Smith, *The Bible and Islam* (New York: Scribners, 1897), chs. 3-10.

2. Richard Bell, *The Origin of Islam in Its Christian Environment* (London: The MacMillan Company, 1928), chapter 7.

3. R. C. Zaehner, *Living Faith of Zoroaster*, 1959), p. 209; Zaehner, *The Dawn and Twilight of Zoroastrianism* (Oxford: At the Clarendon Press, 1958; A.V.W. Jackson, Zoroastrian Studies (New York: Columbia University Press, 1928). John A. *Religions of the World* (New York: Doubleday Company, 1968), I, p. 204.

4. J. D. R. Menasce, "Persian Religion: Ancient," *New Catholic Encyclopedia*, XI, pp. 165.

5. W. G. Oxtoby, "Zoroastrianism," in Keith Crim, *Abingdon Dictionary of Living Religions* (Nashville: Abingdon Press, 1981), p. 829b.

6. Hans Schwartz, *The Secrets of God*, pp. 124-17; Hans Schwartz, "Eschatelogy," in Carl Braaten, and Robert W. Jensen, Editors, *Christian Dogmatics* (Philadelphia: Fortress Press, 1984), II, pp. 484-485.

7. "Zoroastrianism," in Julius Bodensieck, Editor, *The Encyclopedia of the Lutheran Church* (Minneapolis: Augsburg Publishing House, 1965), III, p. 2573.

8. J. Duchesne-Guillemin, "Zoroastrianism," *The Encyclopedia America*, 1990), 29, p. 85 cf also his article in *The New Catholic Encyclopedia*, I, p. 223b.

8a. E.M. Yamauchi, "Ancient Religions," in *The International Standard Bible Encyclopedia*, Merrill C. Tenney, General Editor, (Grand Rapids: Wm. B. Eerdmans Publishing Company, 1988), IV. p. 126a.

8b. Cf. M.N. Dhalia, *History of Zoroastrianism* (New York: oxford University Press, 293 pp. 9. Cf. James Hope Moulten, Early Zoroastrianism (London: Constable, 1913).

10. Keith Crim and Martin Crim, "Zoroastrianism," in William Genz, Editor, *The Dictionary of the Bible and Religion* (Nashville: Abingdon Press, 1986), pp. 1141-1142.

11. Harden **op. cit.**, pp. 205-206.

12. Cf. "Avesta," in *Abingdon Dictionary of Religion*, **op. cit.**, p. 83; H. Humbach, Die Gathas des Zarathustra (Heidelberg, 1960).

13. This yasna is incorporated in A.V.W. Jackson, *Hymn of Zoroaster* (Stuttgart and Boston).

14. "Avesta," *The Dictionary of Religion*, **op. cit.**, p. 83.

15. Yasna 3:3; 45:2 in *Hymn of Zoroaster*, **op. cit.**, p. 83.

16. Jackson, *The Hymn of Zoroaster*, **op. cit.**, Yasna 31:2.

17. "Pehhlevi," in Mario Pei and Frank Gaynor, *Dictionary of Linguistics* (Totowa: Littlefield Adams and Company, 1969), p. 163; cf. West, Pahlavi Texts I-V, (Sacred Books of the East, vols. 5,18, 24,37, 47 (Oxford, 1880-1889).

18. "Avesta," Vergilius Ferm, Editor, *An Encyclopedia of Religion* (New York: The Philosophical Library, 1961), p. 163.

19. J. Darmsteter and L. H. Miller, Avesta, *The Sacred Book of the East*, vols. 4, 23,31 (Oxford, At the Clarendon Press, 1880 ff.)

20. Crim, "Zoroastrianism," Dictionary of the Bible and religion, **op. cit.**, pp. 1141-1142.

21. A. V. W. Jackson, "Zoroastrianism and the Resemblances Between it and Christianity," *Biblical World*, **op. cit.**, p. 338.

22. "Devil," *The Dictionary of the Bible and Religion*, **op. cit.**, pp. 267-268.

23. Theodore G. Tappert, Editor, *The Book of Concord* (Philadelphia: Fortress Press, 1959), pp. 40-41.

24. **Ibid.**, p. 509.

24a. **Ibid.**, p. 617.

25. Ewald Plass, *What Luther Says* (St. Louis Concordia Publishing House, 1959), I, no. 115 1153, p. 393.

26. *Martin Luther's Werke. Kritischeausgabe* (Wreimar, 1883), 42, p. 144.

27. Erwin Lueker, *The Lutheran Cyclopedia* (St. Louis: Concordia Publishing House, 1954), p. 209 and in the 1975 edition 233.

28. Guillemin, "Zoroastrianism," *The Encyclopedia Americana*, 24, pp. 815a.

29. Jackson, *The Biblical World*, 27, May, 1906, p. 338.

30. **Ibid.**

31. **Ibid.**, p. 331.

32. Cf. Jackson in Geiger and Kuhn's *Grundrias der Iranischen Philologie*, II, pp. 621-631.

33. Casartelli, *Religion der Sasaniden*, p. 53.

34. Crim, "Zoroastianism," *The Dictionary of the Bible and Religion*, **op. cit.**, p. 1142.

35. W. L. La Sor, "Zoroastrianism," Walter Elwell, Editor, *Evangelical Dictionary of Theology* (Grand Rapids: Baker Book House, 1984), p. 1202 a.

36. Iraschi, J.S. Tara porewala, "Zorastiran Thought," in Vergilius Ferm, *A History of Philosophical Systems* (New York: The Philosophical Library, 1950), pp. 26-27.

37. Julian Morgenstern, "Eschatology," Ferm, *Encyclopedia of Religion*, **op. cit.**, p. 255; W. O. E. Osterley, *A History of Israel* (Oxford: At the Clarendon Press, 1945), pp. 184-185; cf. also Oesterley, *Hebrew Religion*, chapter 15.

38. Guillemin, "Zoroastrianism," *The Encyclopedia Americana*, 29, p. 815.

39. **Ibid.**

40. "Zoroastrianism, and Parsiism," *The Encyclopedia Britannica*, Macropedia, 19, p. 1176.

41. Guillemin, "Zoraster," *New Catholic Encyclopedia*, 14, p. 1134; W. Hennig, *Zoroaster: Politician or Witch Doctor?* (Oxford: At the Clarendon Press, 1951).

42. J. P. De Manasce, "Persian Religion, Ancient," *New Catholic Encyclopedia*, XI, p. 165.

43. F. F. Bruce, *Second thoughts About the Dead Sea Scrolls* (Grand Rapids: Eerdmans Publishing Company, 1856), p. 104.

44. C.F. Pfeiffer, *The Dead Sea Scrolls and the Bible* (Grand Rapids: Baker Book House, 1979), pp. 65-66.

45. John Hinneli, "Zoroastiran Savior Imagery and Its Influence on the New Testament," Numen, 16:161-185 (1969).

46. Carnoy, "Zoroastrianism," last volume of *The Encyclopedia of Religion and Ethics*, p. 866.

47. "Zoroastrianism," *Lutheran Cyclopedia*, 1954, p. 1159.

48. **Ibid.**

49. S. Lang, *A Modern Zoroastrian* (London: Chapman and Hall, 1890), ch. 13.

50. Jack Finegan, *The Archaeology of World Religions* (Princeton: Princeton University Press, 1952), p. 90.

Questions

1. Zoroastrianism is often referred to as ____.
2. Zoroaster's teachings seem to have taken root in the hearts of some ____.
3. ____ overthrew the Zoroastrian faith.
4. The Bible of Zoroastianism is known as ____.
5. Zoroaster envisioned a conflict between ____ and ____.
6. Higher critics have proposed that the subject of good and bad angels was taken from ____.
7. Did the Bible borrow these spirit beings from Zoroaster's religion? ____
8. The Lutheran Confessions associate Satan with the very ____.
9. What did Luther believe about the Genesis account? ____
10. In Persia Ahura Mazda is not endowed with ____.
11. The many Messianic prophecies were not the result of ____ but were a matter of ____.
12. Does the Old Testament speak of a Millennium? ____
13. The language of Revelation is ____.
14. Christ's kingdom is ____ in nature.
15. Is there any borrowing in Christianity from Zoroastrianism? ____
16. That Judaism and Christianity borrowed religious conception from Zorostrianism is ____.

Christ's Resurrection and the Christian's

The Reality of Christ's Resurrection and Its Significance for the Christian's Future Eternal Life

Christian News, March 31, 1997

First Corinthians is a wonderful epistle and in it are found some outstanding passages which have always impressed Christian readers. Of these must be mentioned those that speak of the foolishness of the Gospel and the Church (1-2), the reference to the Lord's Supper (11) which is the fullest on this sacrament in the New Testament; the Church, the body of Christ, manifesting unity in diversity (12), and the incomparable passage about Love (13) and the Resurrection of Christ (15), a chapter which is the crowning glory of I Corinthians, a demonstration of a future resurrection body for Christians.[1]

Christianity owes a great deal to individuals, who, influenced by Greek philosophy doubted the resurrection of the body [2] and by implication the corporal resurrection of Christ.[3] I Corinthians is one of the greatest apologetic documents in the literature of the New Testament. The fundamental doctrines of the Christian faith are contained in the fifty-eight verses of chapter 15, the longest chapter in all of Paul's epistles.[4] Our knowledge of the Last Things would be greatly impoverished had not Paul refuted the doubts of certain Corinthian members. In proving the reality of Christ's corporal resurrection Paul rests the whole argumentation upon the indisputable fact of Christ's resurrection from the dead. Paul informed the Corinthian congregation that he was writing them to remind them of this truth of the Christian faith and specifically that of the resurrection of the body, denied by Plato and other Greek philosophers and writers.[5]

The Greek doubt that Paul in I Corinthians combats is one that strikes at the root of Christianity which concerns the fundamental fact of the Gospel's truth. In a lofty and sustained eloquence pastor and teacher Paul again sets forth instructions about the heart of the very essence of the Christian faith, without which Christianity would be a riddle. In chapter 15 Paul only discusses the physical resurrection of believers, and not that of unbelievers. The doctrine of the bodily resurrection is peculiar and distinctive of the Christian faith and must be distinguished from the teaching of the transmigration of souls, taught in some religions or the Greek notion of a vague and dim existence in the afterlife.[6]

Jesus had taught that all in the graves will hear the voice of the Son of Man and will come forth, they that have done good to everlasting life, and they that have done evil to everlasting damnation (John 5:25). Paul like Jesus also believed in the resurrection of the wicked and that they will be punished (Acts 24:15) and speaks of a "resurrection of the just and unjust." In Chapter 15 of I Corinthians Paul is describing an event that will occur at the Second Coming of Christ. Before the resurrection of the body on the Last Day, Christians will exist in a disembodied state as spirits, who are with Christ. At Christ's Return Christians will be clothed in a final state, in a perfect state. The body will be a glorious body like unto Christ's resurrection body, one with which He ascended into heaven.[7] This teaching, therefore differs from the belief that the soul sleeps from time of death unto the day of the resurrection, a view held by Seventh Day Adventists.[8] It also corrects the conception that at death the soul exchanges a mortal state for an immortal body. When a Christian dies, he enters a state of blessedness and is in conscious fellowship with Christ but the believer's future existence will reach its culmination at the Second Coming of the Lord. In this chapter Paul says nothing about the intermediate state, the time between death and the resurrection of the body.[9]

The Organization of Chapter XV of I Corinthians

First Corinthians may be divided into two parts. Part I dealing with the fact of Christ's Resurrection and the Christian's Resurrection (w 1-34) and Part II treats of the mode and nature of the resurrection body (w 35-58).[10]

Part I The Fact of the Resurrection Ch. 15:1-34
The Resurrection of Christ Ch. 15:1-11

Paul argues for the undisputed fact of Christ's corporeal resurrection from the dead. Since Christ was raised from the dead, it follows that Christians likewise will be raised. In I Corinthians 15 Paul proclaims to the Corinthians that Christ's resurrection is also intimately connected with other doctrines.

He places the resurrection of Christ among the ABC truths of Christianity. Christ's death, burial and resurrection belong to the foundational doctrines of Christ's religion. Erdman correctly asserted: "These truths he declared to be the essential elements of Gospel, he preached, which they had received, which they still hold, and upon which, if it is a valid Gospel, their salvation depends."[11]

Among the fundamental facts of Christ's life, received by Paul, either by direct revelation, or by tradition from other apostles, was the truth that Christ died for mankind's sins and also the fact that the Old Testament (I Cor. 15:4) had foretold Christ's burial and His resurrection. The apostle clearly asserted that Christ died for "our sins," which means on behalf of our iniquities in order to atone for them, so that sinners might be forgiven and delivered from their punishment.

The fact that Paul mentions Christ's burial testifies to the fact of the reality of Christ's bodily resurrection. The readers are reminded of the empty tomb. Christ had really not swooned nor had his dead body not been deposited in the tomb of Joseph of Arimethea. If Christ had not been raised from the dead, then Christianity is a fraud and delusion. If Christ had not been raised by the Father (Col. 2:12; Acts 2:24) and by the Holy Spirit (Rom. 8:10) and Christ by His own power (John 10:18; 2:19), then the enemies could have found Christ's body in the tomb, but they did not find Him.[12]

Verses 5-7 of 1 Corinthians lists a number of people who saw Christ on Easter Sunday and on other occasions during the forty days between

Easter and the Ascension.[13] In writing to the Roman congregation: Paul said, "Christ died for our sins and was raised up for our justification" (4:25). The resurrection was necessary to make the atonement effective (V.17). Both the death and resurrection of Christ were pre-told in the Old Testament Scriptures, Cf. Psalm 2 and Isaiah 53 for Scriptural prophecies of His death and resurrection, predicted also in Psalm 16:10-11.

Paul argues in his great resurrection chapter that the corporeal resurrection of Christ was an historical event, is shown by the fact that real living people saw Him after His burial. The Apostle mentions four instances: First to Peter or Cephas (both names meaning the Rock man). Paul may have heard Peter tell about this appearance of the Lord while Paul was in Jerusalem for 15 days (Gal. 1:18). Then Christ appeared to the Twelve (minus Judas Iscariot). Then He appeared to "more than five hundred brethren." This may have occurred on "mountain in Galilee or on the day of the Ascension on the Mount of Olives (Luke 24:50-51), more likely earlier than the Ascension Day. Then Jesus was seen by all the apostles and also by James, the half brother of Jesus. Paul also claims that most of the five hundred were still alive and could be consulted for verification purposes."[14]

Various theories designed to deny or explain away Christ's resurrection are shown to be false. The enemies of Christ put forth the lie that Christ's disciples had stolen the Lord's body but that is totally repudiated by the ten different appearances of the LIVE Christ. The swoon theory does not hold up for the Scriptures of both Testaments assert that Christ died. A man resuscitated from a swoon and who would naturally be weak was a pitiful sight to look at. The vision theory also is not plausible. The swoon theory maintains that the disciples, by brooding over Christ's death and His promised resurrection became victims of hallucination. However, the conditions for hallucination were not present. One person might be a victim, of hallucination but not five hundred people at the same time. The resurrection appearances occurred on different days, at different times of the day, at difference places and to both men and women, all pointing to the reality of Christ's bodily resurrection.

The New Testament also reports that Christ appeared to at least three different individuals between A.D. 33 and A.D. 100. Post-resurrection appearances were granted Stephen, the first adult Christian martyr (Acts 7:56), to Paul (I Cor. 15:8) and to St. John (Rev. 1:18). While Paul was on the way to Damascus to persecute Christians, Jesus appeared to Saul of Tarsus (I Cor. 15:8) and Jesus converted him and told him that he was to bring the Gospel especially to the Gentiles.

This appearance to him, Paul called "like a child untimely born." The Tarsian compared himself to the other apostles who more gradually and naturally matured from disciples, by years of nurture, into fully developed messengers of Christ.[15]

The mention of the appearance to Paul, in A.D. 33 was not anticlimactic, because to begin with Paul was a persecutor of the church of God. How can one account for Paul's personality reversal that suddenly he preached Christ whom he had hated till near Damascus[16] and for whose sake he suffered decapitation in A.D. 68? Further how is it to be explained that Paul outshone all other apostles in devotion and in missionary accomplishments?

b. The Refutation of the Denial of
Christ's Resurrection (w 12-19)

There appear to have been certain converts in Corinth who rejected the idea of a resurrection of the body. If Christ was not raised from the dead, then there is no resurrection for Christ when He was bodily raised up. A number of implications would flow from such a denial. First of all, the preaching of the Gospel would be pointless and useless. The rejection would empty the Christian message. If Christ was not raised on the third day, the apostles would be false witnesses, in other words, they were liars. It would also follow that if Christ had not been raised, the resurrection account would be a lie and a fraud, because it would mean that the Christian faith would be futile and ineffectual for Christ's death did not achieve the justification of the sinner. Further, it would mean that those who had died believing Christ's atoning death would still be in their sins. A denial of the corporeal resurrection of the Lord would be tantamount to asserting that the dead saints had perished.

Finally, living Christians, who suffered persecution and privation, will have engaged in futile activities and because of their self-deception were engaging in dreaming and were to be pitied.[17]

c. The Assurance of the Resurrection of Believers (w. 19-28)

Now Paul proceeded to the positive aspects of the resurrection. By His resurrection Christ has become "the first fruits of a great harvest." The same principle that brought death brings life from the dead. Just as all men descended from Adam must die, so all who share Christ and have His nature will live (w 20-22). In the order of resurrection Christ is the first and when He returns all Christians will have life. At the end time Christ will defeat all enemies and then deliver the kingdoms of this world to God the Father. All death will be abolished (w 27-28).[18]

d. The Hope of the Resurrection
in the Christian's Life (w 29-34)

At this juncture in his presentation Paul changes the direction of his thought and advances a further reason for belief in the Resurrection.[19] The apostle reminds the Corinthians of their most sacred hopes. Here Paul states what would be lost if Christ had not been raised. How highly certain men held the doctrine of the resurrection is seen by their participation in a "baptizing for the dead." This practice is one of the puzzling assertions in the Pauline corpus of letters. Many different explanations have been given for this Corinthian practice. It does not refer to proxy baptisms as practiced by the Mormons. Erdman claims that "it can hardly mean a vicarious baptism where one confesses the name Christ with the view of saving a friend who had died in unbelief nor can it refer to accepting baptism for a believer who had died without receiving baptism. There is no evidence that such practices existed in New Testament times."[20]

Probably one good explanation would be that those who turned to Christ with the hope of reunion with the dead through the resurrection of the dead were baptized at their graves. This would also be the motive for those who stand in jeopardy every hour and suffer for Christ as Paul had done, when he was forced to fight beasts in Ephesus (vv. 30-32). Paul warned his readers to avoid worldly pleasures and instead to live for spiritual things.[21]

Part II The Nature of the Resurrection (15:35-58)
a. The Resurrection Body

Some members at Corinth had raised the question of the nature of the body which will be resurrected (v. 35). Paul points out by a number of analogies, as illustration the bodily changes and manifestations (Vv. 36-39).[22] While there will be a continuity of the personality in the resurrection, between the earthly and heavenly, there will be a difference in the resurrection body. The inspired writer Paul shows that there will be a difference in the spiritual body. The natural body is governed by the animal life and it belongs to an earthly order and is subject to corruption. The spiritual body is a part of the heavenly order, and is adapted for and under control of the Spirit and is not subject to the problems of this present world. As Christians have the image of the earthly in their natural bodies, so Christians bear the image of the heavenly in their spiritual bodies.

b. The Change of Dead and Living (w. 50-58)

What Paul wrote about the believer's resurrection will not occur at the Christian's own death, but at the Second Coming. Those who depart believing in Jesus Christ are like sleepers awaiting the reveille of the trumpet which will be heard at Christ's Second Coming.[23] The resurrection of believers will consist in their changing of corruptible and perishable bodies, so as to become fit organs of the Holy Spirit in the Kingdom of God. In a split second those that are alive of Christ's appearing will be aroused and raised and clothed with incorruption and immortality. This Paul announced as a Divine revelation (vv. 51-53). Then at last the prophetic saying "death is swallowed in victory" will be fulfilled (Is. 25:8). All that this means, is that for the Christian death has lost its sting which is sin and which Christ the sin bearer has removed. Death's strength is the law whereby sin is judged and punished. The praise for the victory over death is owed to Jesus Christ (vv. 54-55). This being the case, the Christian's logic ("therefore") is to continue unmoved and immovable in the work of the Lord, no matter what sufferings be involved, knowing that Christ is leading to an everlasting, heavenly harvest.

Paul's Resurrection Doctrine in Other Epistles

In Paul's preaching the death of Christ and His resurrection from the dead are the central truths of the Church's early kerygma or message.[24] Paul teaches the centrality of Christ's resurrection in other epistles. In the beginning of his great theological epistle Romans he stated: "Concerning His Son Jesus Christ our Lord, who was born of the seed of David according to the flesh, and declared to be the Son of God with power, according to the Spirit of holiness, by the resurrection from the dead (vv. 1:4-5)." At the end of the chapter setting forth justification by faith Paul wrote it: "It shall be imputed to us who believe in Him who raised up Jesus our Lord from the dead, who was delivered for our offenses, and was raised for our justification (4:24-25)." In the practical part of Romans Paul asserted: "For to this end Christ died, that He might be Lord of both the living and dead" (14:9). In chapter 10, setting forth the truth that Israel needs the Gospel, Paul made this significant statement: "That if you confess with your mouth the Lord Jesus and believe in your heart that God raised Him from the dead, you will be saved, for with the heart one believes to righteousness, and with the mouth confession is made to salvation" (10:9). In 2 Corinthians, written after 1 Corinthians about A.D. 55, Paul assured the Corinthian Church in chapter 4:14: "Knowing that He who raised up the Lord Jesus will also raise us up with Jesus and will present us with you." In his discussion of Christ's Second Coming, the apostle told the Thessalonians: "For if we believe that Jesus died and rose again, even so God will bring with Him those who sleep in Jesus" (I Thess. 4:14).

Haberman correctly declared: "That Jesus Christ died and afterward rose from the dead is both the central doctrine of Christian theology and the major factor in a defense of its teachings. This was true in the earliest church and remains so today."[25]

Footnotes

1. W. Graham Scroggie, *Know Your Bible*. Volume II, *Analytical - The New Testament* (Pickering and Inglis, no date), p. 131.

2. Charles R. Erdman, *The First Epistle of Paul to the Corinthians* (Philadelphia: The Westminster Press, 1956), p. 151.

3. Charles B. Williams, *A Commentary on the Pauline Epistles* (Chicago: Moody Press, 1953, p. 160.

4. Cf. The discussion of 1 Corinthians in C.M. Zorn, *Die Korinthebriefe* (Zwickau: Verlag des Schriftvereins, no date), pp. 129-159; P. E . Kretzmann, *Popular Commentary* (SL Louis: Concordia

Publishing House, 1924), New Testament II, pp. 160-168; G. G. Findlay, "I Corinthians," *The Expositor's Greek New Testament* (Grand Rapids: Wm. B. Eerdmans Publishing Company, 1979), II, pp. 917-943.

5. John R. Hinnells, *Dictionary of Religions* (Baltimore: Penguin Books, 1984), p. 26.

6. "Transmigration," in Vergilius Perm, *An Encyclopedia of Religion* (New York: The Philosophical Library 1945), p. 792.

7. St. Paul in *The Letter to the Philippians*, 3:20,21 (R.V.) Cf. also E. Hove, *Christian Doctrine* (Minneapolis: Augsburg Publishing House, 1930), p. 453.

8. Th. Engelder, W. Arndt, Th. Graebner and F. E. Mayer, *Popular Symbolics* (St. Louis: Concordia Publishing House, 1934), Par. 352.

10. Erdman, **op. cit.**, p. 153. Cf. also John McNichol, *Thinking Through the Bible* (Grand Rapids: Kregel Publications, 1976), pp. 259-260.

11. Erdman, **op. cit.**

12. Unbelief has proposed a number of theories denying Christ's corporeal resurrection, cf. Joseph Stump, *The Christian Faith* (New York: The Macmillan Company,

1932), p. 177.

13. Cf. George Stoeckhardt, *Exegetical Lectures on the First Epistle to the Corinthians* presented by H. W. Degner (Fairmount, Minnesota, 1969), pp. 93-96.

14. For a complete listing of all of Christ's resurrection appearance, cf. W. Graham Scroggie, *A Guide to the Gospels* (London: Pickering and Inglis, 1948), pp. 608-619.

15. Erdman, **op. cit.**, p. 156.

16. Luke in Acts gives two accounts of Paul's conversion; Acts 9:1-9; 26:9-18. Worthy of reading is William Arndt, *Bible Difficulties and Seeming Contradictions* (St. Louis: Concordia Publishing House, 1987), Revised by Robert G. Hoerber, and Walter Roehrs, 1987), pp. 191-196.

17. For discussion of this paragraph, cf. McNichol, **op cit.**, p. 260a. Findlay, **op cit.**, 922-925; A. T. Robertson, *Word Pictures of the New Testament* (Nashville: Broadman Press, 1931), pp. 188-190.

18. For a discussion of this paragraph cf. F. Godet, *Commentary on Paul's First Epistle to the Corinthians* (Edinburg: T. & T. Clark, 1893), II, pp. 348-381.

19. For this paragraph, cf. Clarence Tucker Craig, "I Corinthians," in *The Interpreter's Bible* (Nashville: Abingdon Press, 1925), pp. 232-240.

20. Erdman, **op cit.**, p. 162.

21. Philip Edgcumbe Hughes, I Corinthians in C.F. Henry, *The Biblical Expositor*, (Philadelphia: A. J. Holman Company, 1960), III p. 276.

22. Williams, **op cit.**, pp. 165-166; McNichol, **op cit.**, p. 260.

23. Hughes, **op cit.**, p. 277.

24. H. Wayne House, *Chronological and Background Charts of the New Testament* (Grand Rapids: Zondervan Publishing House, 1981), p. 120.

25. G. R. Habermas, "Resurrection of Christ," Walter A. Elwell, Editor, *Evangelical Dictionary of Theology* (Grand Rapids: Baker Book House, 1984), p. 938b.

Questions

1. What chapter is the crowning glory of First Corinthians? ____
2. Who denied the resurrection of the body? ____
3. At Christ's return Christians will be clothed with ____.
4. Does the soul sleep from the time of death until the day of resurrection? ____
5. When a Christian dies he enters into ____
6. Paul argues for the undisputed fact of ____.
7. What are the foundational doctrines of Christ's religion? ____
8. The enemies of Christ did not find ____.
9. Where was the resurrection of Christ predicted in the Old Testament? ____
10. Who saw Christ after His burial? ____
11. What is the swoon theory? ____
12. Would 500 people at one time be victims of hallucination?____
13. "Baptizing for the dead" does not refer to ____ held by the Mormons.
14. One good explanation would be that those who turned to Christ with the hope of reunion with the dead were ____.
15. The spiritual body is part of ____.
16. Where else does Paul preach about the resurrection? ____

The Holy Trinity, Islam and Its Denial of the Biblical Doctrine of the Trinity

Christian News, May 19, 1997

Trinity Sunday is the dividing line between the first half of the Christian church year and the non-festival second half of the Church Calendar.[1] Trinity Sunday or the Festival of the Holy Trinity was the last festival to be adopted by the Christian Church. Reed claimed "The early centuries witnessed local diocesan celebrations, particularly throughout the West, in honor of the Holy Trinity. Popes Alexander II in the 11th Century and Alexander III in the 12th Century regarded this as unnecessary, and discouraged its observance, contending that the Holy Trinity was acclaimed in every day's worship. Pope XXII ordered the festival observed by the Universal Church, and on this Sunday."[2] The teaching of the Trinity distinguishes Christianity from all other world religions[3] and Christianity's concept of God differs from the God of Talmudic Judaism, Zoroastrianism Buddhism, Jainism, Confucianism, Islam, the religions of Japan, Hinduism, Gnosticism, Bahaism, Mazdeism, classical Greek and Roman religions, as well as the numerous cults such as: Christian Science, Jehovah Witnesses, Buchmanism, Theosophy, The Way International, Mormonism, Sun Myung Moon's Unification Church, Herbert W. Armstrong's Worldwide Church of God. The Unity School of Christianity, Seventh-day Adventism and Rosicrucianism, also with the philosophical schools of monism; pantheism, deism, agnosticism, idealism and theological dualism.[4]

The Trinity of the Godhead is clearly taught in Holy Writ. Peter told the assembled multitude on Pentecost Day, A.D. 30, in Jerusalem: "Whoever shall call upon the name of the Lord shall be saved (Acts 2:21)." It is a matter of life or death to know the God of mankind's salvation.

A true Biblical follower of God believes in God the Father, Almighty Maker of heaven and earth. He is the Lord. He is the God of eternal power and love, who so loved the world that He gave His only begotten Son. True devotees of the God of the Bible believe in Jesus Christ, the Father's only-begotten Son, who is the God of power and love, and so loved the world that he gave His life to redeem all the descendants of Adam and Eve. Further, a true Bible believer confesses: "I believe in the Holy Spirit. He is the Lord. He is the God of eternal power and love, who so loved the world that he labors to convince, enlighten, convert, regenerate and sanctify and keep the elect in the saving faith."[5]

There are not three distinct persons in the Godhead. The Father is not the Son, nor is the Holy Spirit part of the Father or another name for the Father's influence. God the Father is always depicted and different and separate from the Son (John 3:16; Gal. 4:6). We do not know how the Father, a spirit, can beget a Son, nor how the Eternal Father begets the Eternal Son (Psalm 2:7). We do not know what is involved that breath is proceeding from the mouth of a spirit (John 15:26; 2 Thess. 2:8) nor how the Holy Spirit can have a distinct personality. But this is clear from Holy Writ, namely, that there are three divine persons, who are mighty and eternal.[6]

Still Christianity does not worship three Gods (That would be polytheism, as Muhammed contends in his Koran).[7] Clearly the Jew, Paul, once a Pharisee, stated to the Corinthian congregation:

"There is none other God but one." He that filleth heaven and earth so fills them that there is no other God beside Him. If there were three Gods, in which one would a true believer trust? 1 John 5:7, in the Byzantine text, reads: "There are three that bear record in heaven, the Father, the Word, and the Holy Spirit and the three are one." Many textual critics consider this verse spurious and not a part of the original text of 1 John, even if this rejection were true, the Comma Johaneum is Bible-speaking correct and would be in agreement with doctrinal statements found in the Constantinopolitan and Athanasian Creeds.[8]

The Lord is the Triune God; there are not three gods, but three persons. There is not one personality, but one God. Jesus is a different person from the Father, but the Eternal Godhead which is the Father's the same, unchanged and undivided, is the Son's. It is the One Divine Essence that is concerned about our salvation, for "the Son can do nothing of himself, but what He seeth the Father do (John 5:19)." Yet there are distinctly three who are concerned about us, for Jesus said: "I will pray the Father, and he shall give you a comforter (14: 16)."[9]

The Mystery of the Trinity

The doctrine of the Trinity is a mystery of mysteries,[10] but thus God has revealed Himself in both the Old and New Testaments. The Scripture teaches the Trinity in Unity, because there is none other God. Mankind may be certain that if the doctrine of the Trinity had been invented by human beings, they would not have proposed a doctrine with which intellectuals could attack as unreasonable.[11] However, since God has revealed it, it behooves God's creatures to accept what the Word of God has said about it. The scientist does not know what electricity is, yet he studies it, and derives benefits from its use. God is the Source of life and the Trinity is the source of spiritual life and blessed is that individual who calls upon the name of the Lord.[12]

The Necessity for the Acceptance of the Holy Trinity

"Whosoever denies the Son, the same hath not the Father (1 John 2:23)." "All men should honor the Son as they honor the Father, he that honoreth not the Son honoreth not the Father (John which sent him 5:23)." "No Man can call Jesus Lord (i.e. God) except by the Holy Spirit." (1 Cor. 12:3) Jesus clearly taught the Doctrine of the Trinity. Said Jesus of Nazareth: "I will (Second Person) pray the Father (First Person of the Trinity), and He will send you another Comforter to be with you forever, the Spirit of Truth (Third Person of the Godhead) (John 14:16-17). It is God the Father, the God of our salvation, who out of love sent His Son

1606

into the world to suffer and die and atone for mankind's sins, and Jesus sends the Holy Spirit to apply to sinners the benefits of the Son's atoning sacrifice."[13]

The church composed of infants, children, adolescents, young adults, middle-aged people, and seniors cannot be happy except through the grace of the Lord Jesus Christ, the love of the Father, and the communion of the Holy Spirit (2 Cor. 13:14).

Islam's Rejection and Attack on the Trinity

Islam, which has borrowed materials from both the Old Testament and the New Testament,[14] claims to be the true monotheistic religion of the World, consequently rejects the Biblical doctrine of the Holy Trinity.[15] In a number of Surahs (i.e. comparable to chapters in the New Testament) the doctrine is attacked by Mohammad. Thus in Surah 5:76-79 the reader will find this denunciation: Infidels now are they who say: "God is the Messiah, son of Mary," for the Messiah said: "O children of Israel, worship God, my Lord and your Lord." Whoever shall join other gods with God, God shall forbid him the garden and his abode shall be fire and the wicked shall have no helpers. They surely are the infidels who say: "God is "the third of three" for there is no God but one God, and if they refrain not from what they say, a grievous chastisement shall light on the infidels . . . the Messiah the son of Mary is but an apostle, other apostles have flourished before him, and his mother was just a person. They both ate food."[16]

Again in Surah 4:169, the Quran warns: "O ye people of the book, overstep not bounds in your religion, and of God speak only the truth. The Messiah, Jesus, son of Mary, is only an apostle of God, and his word which he conveyed to Mary and a Spirit proceeding from himself. Believe therefore in God and his apostles, and say not 'Three' (Three is a Trinity), forbear, it will be better for you."[17]

Dr. John Elder, who served in Iran from 1922 to 1964 as an evangelistic missionary commented about these Quranic assertions: "There are two misconceptions here: First, that Christianity teaches that there are three gods. Second, that the Virgin Mary is considered one of the three gods. When the Quran says that Mary was a just person and ate food, the meaning is that she would not claim to be a goddess and that the fact that she ate proves that she was not a goddess. The passage also indicates that the greatest of sins is the sin of polytheism, i.e., of worshipping many gods."[18] The great fault of the Christians, according to Muhammad, was teaching an erroneous doctrine about God and thereby rejecting and insulting Allah. At first Muhammad was favorably inclined toward Christians, but when they refused to believe there was only one God, Allah, and further accept Muhammad as God's last, and greatest of all prophets and the receivers of Allah's book, the Quran the relationship between Muhammad and Christians ended in the latter to be punished for in Surah 9:29ff the prophet of Mecca and Medina commanded Muslims to make war on both Jews and Christians.[19]

Where Did Muhammad Obtain His Views About Christ?

On his numerous travels Muhammad probably came into contact with different Christian Churches who had different views about the Trinity, especially about the person and work of Christ.[20] A number of scholars believe that Muhammad's knowledge of Christ and Christianity was not based on a New Testament he saw and read, but upon the aberrant views of Nestorians and Monophysites. He became acquainted with the apocryphal gospels, which contained many puerile miracles,[21] allegedly performed by Christ. Christ is called: Isa, the son of Mary, who is confused with Miriam, the sister of Moses (Surah 3:31; 19:29). The Quran defends Christ against the calumnies of the Jews relative to the Virgin Birth (Surah 4: 155). From the cradle Jesus performed miracles (Surahs 13:8,27; 17:95; 25:8; 29:44).

The Arabian prophet rejected the New Testament teaching of the crucifixion and death of Jesus.[22] Muslims deny that Jesus was crucified, they say that "Allah raised him to himself" (Surah 4:156) and that he was with his mother "on a hill peaceful and well-watered" (Surah 23:52). For Muhammad and his followers the atonement was unnecessary, because according to the Quran Allah decrees recognize no atonement. According to Muslim theology God determined the fate of all man from the beginning and people are helpless to it . . . This belief is set forth in permanent passages in the Quran. Thus Surah 6:39 declares: "Allah will mislead who he pleaseth and whom he pleaseth he will place upon the strait path." In yet another Surah, Allah has recorded: "And whoso willeth taketh the way to his Lord. But will it shall not unless Allah will it (Surah 76:29-30). Furthermore, the atonement was unnecessary for God does not love man, in fact Allah cannot be affected by man's actions. Elder cites Al Ghazali, outstanding Muslim theologian, to the effect, that Allah should love because as Al Ghazah wrote: "When there is love there must be in the lover a sense of incompleteness, a realization that the beloved is needed for complete realization of self."[23] This is impossible according the reasoning of the Muslim theologian, since Allah is perfectly complete. There is no reaching out by God. He only affects man so that man turns and goes out to him. The Christian concept that God's love goes out to man, is totally foreign to the Muslim mind. Consequently, the ideas of John 3:16 are unacceptable to Allah.

Surah 4:158 has this view about Jesus's death. The Quran teaches: "And for their saying 'verily we have slain the Messiah Jesus Son of Mary, an apostle of God' yet they slew him not and crucified him not, but they had only his likeness. And they who differed about him: no sure knowledge had they about him but followed only an opinion and they did not really slay him. But God took him up to himself."[20] It is possible that Muhammad borrowed his views from the Gnostics, who denied the death of Christ on the cross on Calvary. Because Muhammad denied the crucifixion and death, he also rejected the teaching on the resurrection, with which Christianity stand or falls.[24]

Muhammad knew that the Christian had the Gospel, called in Arabic jil. The Jews were berated for changing the Taurat (the Law) and the

Christians were castigated for having changed the Gospel. Had the Arabian prophet seen a correct version of the four Gospels and been able to read the other twenty-five books he would have found clear evidences for the Trinity. Here are some major examples of verses clearly setting forth the Trinity: Matthew 3:16-17 Father, Son and Holy Spirit present at the same time: "And Jesus when He was baptized, went up straightway out of the water; and lo, the heavens were opened unto him, and he saw the Spirit of God and descending like a dove, and lighting upon Him. And lo, a voice from heaven, saying, 'This is my beloved Son, in whom I am well pleased'."

Matthew 17:5 - Father and Son at one time on the Mount of Transfiguration—

"While he yet spoke, a bright cloud overshadowed them; and behold, a voice out of the cloud which said, This is my beloved Son, in whom I am well pleased, hear ye him."

Romans 8:26-27- Father and Spirit working together in Prayer—

Likewise, the Spirit also helpeth our infirmity; for we know what we should pray for; but the Spirit himself maketh intercession for us with groanings, which cannot be uttered. And he that searcheth the hearts knoweth what is in the mind of the Spirit, because he maketh intercession for the saints according to the will of God. John 10:30-Father and Son are one—

"I and the Father are one."

Matthew 28: 19 - One name for God yet three persons active in fulfillment of the Great Commission—"

Go ye, therefore, and teach all nations, baptizing them in the name of the Father, and of the Son, and of the Holy Spirit."

John 14:16 - The Spirit distinct from the Father and the Son—

"And I will pray the Father, and he shall give you another Comforter, that he may abide with you forever."

II Corinthians 13:14 – Benediction of the divine God in threefold form—

"The grace of the Lord Jesus Christ and the love of God, and the communion of the Holy Spirit be with you all."

"Amen."

I Peter 1:2 - All three persons active on redemption—
"Elect according to the foreknowledge of God, the Father, through sanctification of the Spirit, unto obedience and sprinkling of the blood of Jesus Christ: Grace unto you, and peace be multiplied."

John 17:5,24. Relationship between the Father and Son existed before the foundation of the world—

"And now, O Father, glorify thou me with thine own self with the glory which I had with thee before the world was."

"Father, I will that they also, whom thou has given me, be with me where I am, that they behold my glory, which thou hast given me for thou lovest me before the foundation of the world."[25]

The Doctrine of the Trinity A Practical Teaching

Very correctly, Stump has written, calling attention to the practicality of the doctrine of the Holy Trinity, when he wrote: "The doctrine of the Trinity is not merely a theoretical or speculative one. On the contrary, it is intensely practical. Our salvation is bound up with it. It was revealed to us historically in most intimate and vital connection with the redemption of our race, and not as an abstract theological or metaphysical conception. God the Father sent His Son in to the world to save us; and God the Holy Ghost applies the redemption of Christ to our souls. If there is no Trinity, then we are not saved; for then there is no Father to send His Son, no Son to make atonement for our sins, and no Holy Ghost to bring us to living faith in Christ our Savior. It is, of course, from the practical side and not from the metaphysical, that the doctrine is to be presented in the pulpit."[26]

Footnotes

1. James L. Brauer, "The Church Year," in Fred Precht, Editor, *Lutheran Worship* (St. Louis: CPH, 1993), p. 149.

2. Luther D. Reed, *The Lutheran Liturgy* (Philadelphia: Muhlenberg Press, 1947), p. 472.

3. Joseph Stump, *The Christian Faith* (New York: The Macmillan Company, 19321, p. 45.; P.E. Kretzmann, *The God of the Bible and Other "Gods"* (St. Louis: Concordia Publishing House, 1943), pp. 183-195.

4. Irvine Robertson, *What the Cults Believe* (Chicago: Moody Press, 1979), pp. 17-148; Kenneth Boa, *Cults, World Religions and You* (Fullerton, Ca., 1978: Victor Books, 1978), pp. 12-62, 63-1108; 156-203.

5. Theodore Engelder, L. Fuerbringer, P.E. Kretzmann, *The Concordia Cyclopedia* (St. Louis: Concordia Publishing House, 1927), pp. 291-298.; Th. Engelder, W. Arndt, Th. Graebner, F.E. Mayer, *Popular Symbolics* (St. Louis: Concordia Publishing House, 1934), pp. 33-36.

6. The Athanasian Creed," in Theodore Tappert, Editor), The *Book of Concord* (Philadelphia: Fortress Press, 1959), pp. 19-21.

7. J. Elder, *Biblical Approach to the Muslim* (Houston: International Headquarters, 1353 DeLuxe, Houston 77047), p. 78.

8. Bruce Metzger, *A Textual Commentary on the Greek New Testament* (London and New York: The United Bible Societies, 1971), pp. 716-718.

9. "The Nicene Creed," "The Athanasian Creed." Tappert. **op. cit.**, pp. 18-21.

10. A W. Tozer, *The Knowledge of the Holy cited by Reaching Muslim & Today, A Short Handbook*, published by North African Mission, 239 Fairfield Avenue, Upper Darby Pa., 19082), p. 28.

11. Theodore Engelder, "The Holy Trinity," Doctrinal Essays (St. Louis: Concordia Publishing House, no date), pp. 9-10.

12. **Ibid.**

13. Richard C. Caemmerer, "God." *The Lutheran Cyclopedia*, **op. cit.**, pp. 418b-419a Richard C. Caemmerer, "The Nature and Attributes of God," in Th. Laetsch *The Abiding Word* (St. Louis: Concordia Publishing House, 1947). II, pp. 59-72.

14. Richard Bell, *The Origin of Islam in its Christian Environment* (London: Macmillan, 1926 cf. especially Lecture IV.

15. D. A. Dawood, *The Koran* (New York: Penguin Books. 1977), p. 383.
16. **Ibid.**
17. **Ibid.**, pp. 383-384.
18. Elder, **op. cit.**, pp. 76-77.
19. Richard Bell, *The Origins or Islam in Its Christian Environment*, on cit.; Lecture 5.
20. H. Spenser, *Islam and the Gospel of God*, (Delhi, India: ISPCK, 1976), p. 10.
21. William H. Miller, *A Christian Response to Islam* (PNutley, N.J.: Christian and Reformed Publishing Company, 1976), p. 16.
22. Elder, **op. cit.**, p. 71; Reaching Muslims Today, North Africa Mission Darby, pa; p. 21.
23. Elder, **op. cit.**, p. 73; W. W. Cash, *Christendom and Islam* (New York: Harper and Brothers, 1937), of Chapter 3.
24. Cf. Henery Preserved Smith, *The Bible and Islam* (New York: Charles Scribner and Sons), end of chapter 3.
25. Taken from *Reaching Muslims Today*, **op. cit.**, pp. 37-38.
26. Stump, **op. cit.**, p. 47.

Questions

1. What is the dividing line between the first half of the church year and the second half? ____
2. What distinguishes Christianity from all other world religions? ____
3. Christianity does not worship three ____.
4. Islam has borrowed material from ____.
5. What is a Surah? ____
6. Where did Muhammad obtain his views about Christ? ____
7. Muslims deny that Jesus was ____.
8. What is foreign to the Muslim mind? ____

Morphew on Buddhism and Christianity

Christian News, September 29, 1997

Clark Morphew, liberal and humanistic (Evangelical Lutheran Church in America) clergyman, who is known for his dislike of traditional and conservative theology, in his syndicated column for August 22, 1997, has tried to link together Buddhism and Christianity because of a number of teachings that appear to set forth the same religious ideas. The St. Paul Pioneer Press writer was inspired to write his: "Jesus, Buddha: Sayings Show Parallel Thoughts," after reading Marcus Borg's "Jesus and Buddha: The Parallel Sayings."[1] Professor Borg is a member of the Jesus Seminar, engaged in rewriting the facts of the New Testament and demoting Jesus to human status alone. In the introduction of his attack on Christianity he pointed out the similarities that Buddha and Christ share: The primary importance of compassion, the love of enemies and the admonition to find a new way to live—to embrace a "world-subverting wisdom that goes beyond human appetites and desires."[2]

The fact that there are a number of sayings found in Christ's and Buddha's writings, averred Morphew, "seems almost spooky. We look for a reasonable explanation, but there is none." From Borg's book he cites parallels on "love," "diets," "sin," "salvation," and "spiritual practice . . ." Morphew admits that what he has borrowed from Borg does not constitute a "modern day melding of Buddhism and Christianity." However, "the parallel sayings do energize one's imagination, particularly when one ponders what may become of these two religions in the future." The **St. Paul Pioneer Press** columnist closes his column: "Entire religions are being swallowed by hungry pilgrims. Young people are searching for moral wisdom in places we never considered. Buddha and Jesus must be smiling."[3]

While there may exist a similarity of sayings in Buddhism and Christianity, no person knowledgeable with the tenets of Buddhism would suggest that these two world religions parallel each other or even suggest that the theology of Buddhism would be worth consulting or even embracing.

The Life of Buddha

Buddhism like Jainism began as a revolt and reform movement in Hinduism. Siddhartha Gautama, the founder of Buddhism, was born in 560 B.C.[4] in the Kshatriya (ruther class) of Hinduism. His father Suddhodana was a feudal lord and his mother was Maya. Gautama's father tried to protect him from seeing any form of evil and suffering. At 19 years of age Gautama married Yasodharma and later had a son named Rahula. Buddha lived a life of luxury till 29, when he became dissatisfied.[5]

At the age of 29 he left his family in search of enlightenment. He placed himself under Hindu masters and devoted himself to asceticism for six years, practicing self-mortification and finally came close to death. Convinced of the futility of asceticism, Buddha developed the principle of the Middle Path. He avoided the two extremes of asceticism and indulgence and became involved in meditation. During one of these meditations under a fig tree, he reached a state of enlightenment. Gautama believed that he had entered nirvana while still alive. In 522 B.C. he became the Buddha, the Enlightened One and the tree under which he had experienced nirvana came to be known as Bo or Bodhi (Wisdom) Tree.[6]

Gautama then proceeded to gather a core of disciples in northern India. His disciple called him Sakayumi (the sage of Sakyas), Tathagata (the one who comes thus), the Truth-Winner and Bhara (Lord). In 480 B.C. Buddha died of food poisoning.[7]

Buddhism was considerably different from Hinduism out of which it had come. In describing the difference between the two faiths of India, Boa wrote: "Hinduism had degenerated on the one hand to empty philosophical speculations and disputes, and on the other hand to a crass polytheism, rituals, magic and superstition. Authority for truth was a special possession of the highest caste. Gautama attacked the caste system and rejected all forms of speculation, ritual and occultism."[8]

The Distinctive Features of Buddhism

Buddha advocated the Four Noble Truths and the Eightfold Path. A knowledge of the Fourfold Truths and the Eightfold Path are basic and foundational for a comprehension of Buddhism.[9] The First Noble Truth is the recognition of suffering. Birth is painful as is death, disease and old age. Not having what we desire is painful and having what we desire is painful. The second Noble Truth is caused by **tanha,** the desire or thirst for pleasure, existence and prosperity. The third Noble Truth is knowing that suffering can be overcome by the elimination of these cravings. The Fourth Noble Truth is accomplished by following the Eightfold Path.[9]

The Eightfold Path

The Eightfold Path is a system designed to develop habits which will release people from the restrictions caused by ignorance and craving. The Eightfold Path is constituted as a therapeutic for aiding individuals to achieve the goals of Buddhism. Each devotee must join the order (the Sangha). The first step of the Eightfold Path is accepting the Four Noble Truths and practice the Eightfold Path. The second step is Right Resolve. The Buddhist disciple must renounce the pleasures of the senses; the devotee must harbor no ill will toward anyone and harm no living creature. The third step is Right Speech, overcoming falsehood, and promoting truth. The third step is Right Behavior. Do not destroy any living creature; take only what is given you; do not commit any lawful sexual act. The fifth step is Right Occupation, certain occupations were forbidden to a follower of Gautama, such as slave trader, tax collector, or butcher. The sixth step is Right Effort. The true Buddhist must strive heroically to prevent evil qualities from arising and must encourage the good qualities possessed. The seventh step is Right Contemplation achieved by self-analysis. The eighth step is Right Meditation. After all pleasures have been surrendered, the Buddhist must enter the four degrees of meditation.[10] During the holy season and in certain monasteries five additional precepts are followed: Do not eat at forbidden times, do not dance to music, do not indulge in personal adornment (the use of cosmetics and jewelry), do not use a high or broad bed.[11]

Requirement of Buddhist Monks

The monastic community is called Sancha and its members were required to have a shaven head, wear a yellow robe, practice meditation and affirm the Three Refuges: (1) I take refuge in Buddha; (2) the Dharma (doctrine),[12] and (3) the Sangha (the order).[13]

The Goal of Buddhism

The goal of Buddhism is to attain the state of nirvana. About this word "nirvana" Boa remarked: "This word means 'to extinguish' or 'to blow out.' It does not simply mean release from suffering, desire and finite self."[14] In his original teachings Buddha stated that "nirvana" was not God or heaven, for his theological system had no place for God. The absolute is described as completely impersonal, and salvation is solely attained by self-effort.

Borrowing From Hinduism

From Hinduism Buddha adopted two Hindu teachings, namely, the transmigration of souls (samsara)[15] and karma.[16] However, different from Hinduism the transmigration of souls was Buddha's denial of man having a soul (anatta). For Gautama there is no enduring self-continuing after death, which he depicted as not experiencing rebirth, but only as undergoing a set of feelings, impressions and present moments. The Buddhist saint (arhat) aims at becoming independent of the casual realm of nature, to have total awareness and total being. This is considered enlightenment. When a Buddhist devotee has overcome five mental hazards and ten fetters, he finally achieves nirvana and is freed from rebirth and from the law of karma.[17]

The History of Buddhism After Gautama's Death

In northern India Buddhism developed slowly but in 265 B.C. Asoka, emperor of India, accepted it and was responsible for the acceleration of the spread of Buddhism not only in India but its spreading in other adjacent countries.[18]

In the course of time Buddhism developed different beliefs and practices. There came into existence two major forms: A conservative and a liberal school of thought. The conservative school is known as Theravada (The Way of the Elders)[19] or Hinayana (The Lesser Vehicle). The latter is also denominated the Southern Vehicle, because it was strongest in Sri Lanka, Thailand, Laos, Burma (formerly) and Cambodia. Theravada Buddhism is based upon and guided by the Tripitika or Three Baskets, written in the Pali language.[20] The Tripitika consists of Vinaya (the law and the rules of monastic Buddhism, to Sutra (the sermons and teachings of Buddha) and the Addiharma (philosophical interpretations of Buddhist teachings). The Tripitika was first transmitted orally but written down only four centuries later.

Liberal Buddhism

The liberal school is known as Mahayana (The Greater Vehicle) or Northern Buddhism. It became strong in Nepal, China, Korea and Japan, Tibet and Indonesia. Boa claimed "'There are so many clear contrasts between Theravada and Mahayana Buddhism that they are like two separate religions."[21]

Essential Differences Between
Theravada and Mahayana Buddhism

The two major forms of Buddhism are different in many respects. Theravada is founded upon the Pali canon and Gautama's early teachings.[22] Theravada emphasizes insight and wisdom, while Mahayana stresses feelings and compassion. Theravada claims that by self-effort a person must achieve his own salvation, by contrast Mahayana advocated the idea that salvation is dependent upon other people's grace.[23]

Theravada is the way of the few because it centers on renunciation and monastic system. For this reason the Theravada is referred to as the Lesser Vehicle. By contrast, the Mahayana is the path of the many (The Greater Vehicle became the religion followed by the majority).[25]

In the Theravada Vehicle the ideal of achievement is to become a saint (arhat) and Gautama is the ideal to be followed. Buddhahood (Buddha seen as savior) is the ideal in Mahayana Vehicle or the Bodhisattva, described as one who comes to the verge of nirvana but renounces it in order to become a helper of a savior of mankind.

Salvation in Mahayana is accomplished by faith in Gautama and the many Bodhisattvas. Thus in Mahayana Buddhism Gautama has been raised to the status of being a god. One of the significant developments of Mahayana Buddhism has been the adoption of a whole line of Buddhas, all of whom are traced back to Dharmakaya the eternal Buddha. The next Buddha to appear on this earth is Maitreya.[26]

The Theravada School does not engage in cosmological speculations, while the Mahayana gives graphic portrayals of heaven and hell. In the conception of God the two schools differ radically, with the Theravistic School being essentially atheistic, the Mahayanistic favoring polytheism.

Relative to the source of authority the Theravada bases its tenets on the Tripitika or Three Baskets (the original Scriptures), while the Mahayana has added many books. The situation in Buddhism is confusing for as Boa has noted: "Unfortunately, most of these foregoing descriptions of Mahayana can be only general, for Mahayanism has many forms, sects, and subjects. The difference between some of the sects is sometimes as great as the difference between Theravada and Mahayana Buddhism."[27]

The Most Important Sects of Mahayana Buddhism
1. Pure Land Buddhism.
This school of Buddhism reached China shortly after Christ's birth,[28] and did not grow till the fourth century A.D., when Buddhism spread to Korea. From Korea it was introduced into Japan. In fact, as Buddhism spread in other countries, it died out in India, the land of its origin. By the 13th century Buddhism was absorbed by Hinduism, because Mahayanism lost many of its distinctive teaching promulgated by Gautama.

Pure Land Buddhism became popular in China and Japan, and in those lands it focuses on Buddha called Amitabha in India, Omito Fu in China and Amida in Japan.[29] In this form of Buddhism salvation consists of placing faith in Buddha and chanting his name. Doing these things a Buddhist can achieve the pure land, a paradise created in the West by Amida. This Pure Land Buddhism is said to appeal to people because of its approach to salvation and its vivid descriptions of hell and heaven.

2. Zen Buddhism
It resembles Theravada Buddhism and is the opposite of Pure Land Buddhism. Salvation is only possible for the Buddhist who must do it himself. Zen Buddhists go beyond the limitations of language and reason, and hope to have a supernatural experience known as **satori.**

The word "zen" is the Japanese equivalent to the Chinese word **ch'an,** which is a translation of the Sanskrit word **dhyana,** which meditation leads to insight.[30] Boa described this special feature of Zen Buddhism like this: "Zen Buddhists practice seated meditations using the lotus posture. They also use a number of irrational problems (koans) to baffle the mind of the mediator. These eventually exhaust reason and open the way for the intuitive flash (satori). Though Zen is very individualistic, the serious student must follow a Zen master, with whom he consults concerning his meditation (this is called sanzen)."[31] Zen Buddhism has exerted a strong influence on Japanese culture, including landscape painting, gardening and flower arrangement.

3. Nichiren Buddhism
Nichirin (1222-1282) based his brand of Buddhism on the Sutra of the Lotus of Truth. He attacked all other forms of Buddhism and claimed that salvation could only be found in the Lotus Sutra. It has been characterized by a nationalistic, militant and emotional religion.[32]

4. Tibetan Buddhism (Lamaism)
Buddhism reached Tibet about the seventh century, where it combined with the religion of Tibet to form a new kind of Buddhism.[33] It came to be known as Lamaism, with its priests called lamas. At the head of the religious hierarchy was the Dali Lama, a man who was worshipped as an incarnation of a Bodhisattva. Lamaism was characterized with many Buddhas, Bodhisattvas and demons. The Buddhist of the Tibet employ prayer wheels, and secret formulas.[33]

Buddhism, a World Religion
Buddhism is the fourth largest religion in the world with 311,836,170 people.[34] The Buddhist Church of America has a membership of 100,000

members in 100 congregations. The Nichiren Shoshu branch of Buddhism has 20,000,000 members with 30,000 in the United States.[35]

Buddhism and Christianity Compared

No form of Buddhism has a place for the distinctive tenets of Christianity. Its concept as to who God is as radically different as can be imagined. The atheistic or pantheistic views about the nature of God never could be harmonized with the doctrine of the Trinity.

The Biblical teachings of the doctrine of man, sin, salvation and the resurrection could have no place in any form of Buddhism. It may be true that some forms of Buddhism, namely, Amida Buddhism speaks about salvation by faith, networks. But this Buddhistic religion is a counterfeit of Christianity and even uses the phrase "new birth" and "changed lives." It is a counterfeit because it does not confront the sins of mankind nor suggest a true way of overcoming sin.

Central to Christianity is Jesus Christ, of whom Peter asserted: "There is no salvation in no other, for there is no other name under heaven given among men, in which we must be saved (Acts 4:12). Apart from Jesus Christ there is no salvation. Jesus claimed: "I am the Way, Truth and the Life no man comes to the Father, except by me" (John 14:6).

Blum-Ernst in the revised edition of his *Handbook* appropriately wrote: A comparison of the two founders of religions, Buddha and Mohammad, with Jesus Christ is possible only when we divest the Christian Redeemer of His divine essence and consider only the purely human side of his essence and of His doctrine. And even so neither Buddha nor Mohammad are able to stand a comparison with singularity and majesty of the person of Jesus. It is the Christian faith which in this instance a priori must decline every comparison, because it holds the firm conviction that Christianity according to its innermost essence is of a supra-historical nature, a deed of God and not of men, that its origin is-from above and not from below, and that in it, through Jesus Christ, the Son of God, the reality of God is revealed. In this way the Founder of the Christian religion is removed from every comparison with the other two founders of religion and the demand of Christianity for absolute authority is postulated."[36]

When Morphew in closing his column depicts both Buddha and Jesus smiling upon the efforts of future generations, he fails to remember that Buddha is dead, while Christ was raised from the dead and is alive.

Conclusion of the Comparison of Buddhism and Christianity

In concluding his discussion of Buddhism, Kretzmann summarized the result of his study with these words: "No matter, then, from what angle we approach Buddhism, we find its teachings hopelessly inadequate even from the standpoint of ordinary human intelligence and common sense. At its best, in the inarticulate groping for the truly divine, it does not approach the idea of monotheism as revealed in the Bible, it does supply an adequate picture of a deity, one that would satisfy a soul searching for the truth. And even the most earnest of such souls could not be satisfied with the vagaries of Buddhism, but requires revelation of the truth as found in the Bible."[37]

Hardon's Statement About Christianity's Influence on Modem Buddhism

Father Hardon concluded his chapter of "Buddhism" by asserting: ". . . Christianity is making a lasting impression quite apart from its own evangelization. The religions of the East, but especially Buddhism, are reevaluating their positions against the background of the Christian faith whose foundation in history and theological structure are a challenge to the inquiring mind."[38]

A Hypothesis for the Parallels Between Buddhism and Christianity

The parallels found in the New Testament, similar to certain sayings in Buddhism, can be duplicated from the Old Testament, most of which were written and known by Jews who lived in the Persian empire by 400 BC. Since Buddha was born in 560, there were Jewish people living in Mesopotamia and Persia. India was a part of the Medo-Persian Empire. Wrote Finegan: "The homeland of Persia was the western and larger part of the Iranian plateau, which stretches from the India on the east to the Tigris and Euphrates on the west."[39] Cyrus the Great defeated the Babylonians and controlled from India to Egypt, as Esther 1:1 indicates that Xerxes I (484-464 B.C.) ruled "from India to Ethiopia, over 120 provinces." The Persian kings established an excellent road system throughout their empires so that royal couriers could travel speedily to the various provinces to foster the king's business. Those same roads could be used by travelers and missionaries. It is not impossible that Buddha came into touch with people who had their Hebrew Old Testaments and may have learned from them some religious concepts promulgated. Since it took at least 400 years for the contents of The Tripitika to be recorded, this could mean that not everything Buddha did and said has been recorded in The Three Baskets, the earliest form of the Buddhist canon. A part of Buddha's life was spent in searching for the truth and it may have happened because of travel which existed in the Persian Empire that Buddha came into contact with the Jewish Scriptures that had similar ideas later found in the New Testament, now set forth as parallels to certain sayings of Buddhism by students of comparative religions, as Borg and Morphew.

Footnotes

1. Clark Morphew, "Jesus, Buddha sayings show parallel thought," *The News-Sentinel*, August 22,1997, 2 F.
2. **Ibid.** 2 F.
3. **Ibid.**
4. Jack Finegan, *The Archaeology of World Religions* (Princeton: Princeton University Press, 1952), pp. 234-235; C. George Fry, James R. King, Eugene R. Swanger, and Herbert C. Wolf, Great Asian Religions (Grand Rapids: Baker Book House, 1984),

pp. 66-68. John A. Hardon, *Religions of the World* (Garden City, New York: Image Books, 1968), pp. 98-102.

5. Huston Smith, *The Religions of Man* (New York: Harper & Row, 1965), pp. 90-96.

6. "Buddha," John R. Hillells, *The Penguin Dictionary of Religions*, From Abraham to Zoroaster (Baltimore: Penguin Books, 1984).

7. **Ibid.**

8. Kenneth Boa, *Cults, World Religions and You* (Wheaton, Illinois, Victor Books, 1978), p. 26.

9. Josh McDowell and Don Stewart. *Handbook of Today's Religions* (San Bernardino, California: Campus Crusade for Christ, 1989), p. 309.

10. McDowell and Stewart, **op. cit.**, p. 307.

11. Boa, **op. cit.**, p. 27.

12. John B. Noss, *Man's Religions.* (London: The Macmillan Company, 1969), The Fourth Edition, pp. 129f, 131, 136, 140-141, 152, 164, 165, 167-168.

13. Boa, **op. cit.**, p. 27; *The Penguin Dictionary of Religions*, **op. cit.**, pp. 32-33.

14. Boa, **op. cit.**, p. 27.

15. Noss, *Man's Religion*, **op. cit.**, pp. 134-137.

16. *The Penguin Dictionary of Religions*, **op. cit.**, p. 180.

17. Boa, **op. cit.**, p. 27.

18. Finegan, **op. cit.**, pp. 234-235.

19. K.K.S. Chen, "Buddhism," in Keith Crim, *Abingdon Dictionary of Living Religions* (Nashville: Abingdon Press, 1981), pp. 124-136.

20. *The Penguin Dictionary of Religions*, **op. cit.**, p. 332. Smart, *The World's Religions*, **op. cit.**, p. 79.

21. Boa, op. cit,, p. 28.

22. Ninian, **op. cit.**, pp. 78-79.

23. **Ibid.**, pp. 80-82.

24. W. Gentz, *The Dictionary of Bible and Religion* (Nashville: Abingdon Press, 1986). pp. 161-163, Finegan, **op. cit.**, p. 279.

25. For a comparison of these two schools of Buddhism cf. Boa, **op. cit.**, pp. 28-29, McDowell and Stewart, **op. cit.**, pp. 308-309.

26. *A Penguin Dictionary of Religions*, **op. cit.**, p. 198.

27. Boa, **op. cit.**, p. 29.

28. *A Penguin Dictionary of Religions*, **op. cit.**, cf. "Chinese Buddhism," pp. 84-85. Ninian Smart, **op. cit.**, p. 120.

29. Boa, **op. cit.**

30. On Zen Buddhism cf. Noss, **op. cit.**, pp. 173-177. *The Penguin Dictionary of Religions*, **op. cit.**, pp. 359-360.

31. Boa, op cit., p. 30.

32. McDowell and Stewart, **op. cit.**, p. 315.

33. Noss, **op. cit.**, p. 161; *A Penguin Dictionary of Religions*, **op. cit.**, pp. 357-358.

34. Charles R. Manske, *World Religions* (Irvine, California, Institute for World Religions, 1991), 4.00.

35. **Ibid.**, 4.01 "Buddhist Churches of America."

36. Wurm-Blum, *Handbuch der Religionsgeschichte* (Calvwerverlag), p. 536fr.

37. P. E. Kretzmann, *The God of the Bible and Other "Gods"* (St. Louis: Concordia Publishing House, 1943), p. 86.

38. John Hardon, *Religions of the World* (Garden City: Doubleday Doran Company,

1968), p. 146.

39. Jack Finegan, *Light From the Ancient East* (Princeton; Princeton University Press, 1946), p. 192.

Questions

1. Clark Morphew was a ____.
2. He tried to link together ____ and ____.
3. Marcus Borg was a member of ____.
4. Buddhism began as ____.
5. Who is the founder of Buddhism? ____
6. When did he become the Buddha? ____
7. Buddha died in ____ of ____.
8. What are some distinctive features of Buddhism? ____
9. By the thirteen century Buddhism was absorbed by ____.
10. Is there any form of Buddhism that has a place for the distinctive doctrines of Christianity? ____
11. Buddha is ____ while Jesus Christ is ____.

The Lord's Supper

A Comparison of the Lutheran and
Reformed Doctrine of the Lord's Supper
(A Study for Holy Thursday)
Christian News, April 6, 1998

The Thursday of Holy Week, known as Maundy Thursday, is yearly observed in liturgically oriented churches as the anniversary of the institution of the Lord's Supper or Eucharist by Jesus. Recently the largest Lutheran Church in the United States (ELCA) has established altar and pulpit fellowship with a number of churches that represent Reformed tradition as reflected in the Confessions of denominations following either Zwingli's or Calvin's views on the Last Supper. It might be appropriate to inquire on what grounds churches holding divergent positions can now recognize each other and give the impression that now these doctrinal differences are inconsequential for Lutherans to reject the Lutheran stance on Holy Communion as not important and recognize positions Luther rejected and are not consonant with the Lutheran Confessions of 1580 is puzzling and baffling. It also involves espousing a system of hermeneutics that would permit to be held diametrically opposed views in the Lord's Supper.

The Beginning of the Differences
Between Lutherans and Reformed

The Lord's Supper controversy in Lutheranism became acute at the Colloquy at Marburg, held October 2-4, 1529, at Marburg-Hesse-Nassau. The Zwinglian and Calvinistic views were the subject of discussion and debate. At the Marburg Colloqium the Lutherans were represented by Luther, Melanchthon, Jonas, Cruciger, Veit, Dietrich, and Roerer from Wittenberg, Myconius and Eberhard von der Than from Eisenach, Osiander, Brenz and Steven Agricola from South Germany, while the Reformed part was represented by Zwingli, and Ulrich Funk from Zurich, Oecolampadius and Rudolph Fry from Basel, and Bucer, Hedio and Jacob Storm from Strasburg.[1] The chief point of discussion was the Lord's Supper which had agitated Protestants for three years.[2] Luther and his collaborators held to the plain meaning and understanding of the word "is." Zwingli and his adherents insisted on the metaphorical understanding of the words of institution. Ultimately the discussion at Marburg resulted in a debate about the real presence of Christ and of the ubiquity of Christ also in the Holy Supper. In order to strengthen himself, Luther wrote the Greek word "esti" on the table and argued that he was not able to depart from them because the words were clear. Even though Zwingli argued that Luther's understanding was contrary to reason, the Wittenberg Reformer refused to yield to Zwingli's metaphorical interpretation. At the Marburg Colloquy the participants, Lutheran and Reformed, agreed on the doctrines of the Trinity, the person of Christ, faith and justification, the Word of God, Baptism, good works, confession, secular authority, human order, infant baptism, but on the Lord's Supper no agreement could be reached. Luther said to Zwingli: "Yours is a different spirit from ours." The Marburg Colloquy of 1529 marked a permanent split between the Zwinglians and the Calvinists on the Holy Communion.[3]

The Teaching of the Augsburg on the Lord's Supper

Article X of the Augsburg Confession reads: "Of the Lord's Supper they teach that the Body and Blood of Christ are truly present and are distributed to those who eat in the Supper of the Lord and they disapprove of those who otherwise teach."[4]

John Calvin believed that Christ was spiritually present and not bodily. In the Smalcald Articles Luther wrote: "Of the Sacrament of the Altar we hold that bread and wine in the Supper are the true body and blood of Christ (Part II, Article VI)."[5] The *Formula of Concord* of 1577 asserted: "That on account of the sacramental union the bread and wine are truly the body and blood of Christ." Already in 1529 the Wittenberg Reformer wrote in his Small *Catechism*: "It is the true body and blood of our Lord Jesus Christ, under the bread and wine, given us Christians to eat and drink, as it was instituted by Christ Himself." In the *Apology of the Augsburg Confession*, Melanchthon said: "In the Lord's Supper the body and blood of Christ are truly and substantially present."[6] Luther rejected the Roman Catholic doctrine of transubstantiation, namely, that the officiant changed the bread into the very body of Christ, and likewise turned the wine into the blood of Christ. This view Luther already expressed early in his career, in the Babylonian Captivity of the Church (1520). Yet Luther held that the body and blood of Christ were present when the Lord's Supper is celebrated (Augsburg Confession, Article X). However, Luther never was interested in the mode of the Real Presence.

The relationship of the earthly things: Bread and wine to the body and blood has been a question for Lutherans and those who do agree with the Real Presence in the Lord's Supper. The Real Presence was rejected by the Reformed because Calvin and others who believed that Christ was at the right hand of God and therefore could not be present in the Lord's Supper. Thus Christ could only be spiritually present.[7]

The Sacramental Union,
Reaction of Calvinists and Enthusiasts

While Luther rejected the Roman theory of transubstantiation, he and the Lutherans insisted on the fact "that the body and blood of Christ are truly present and are distributed,"[8] then the question arose: "How are the elements related to each other, namely, the earthly bread and (Christ's body and the wine to the blood?) The *Formula of Concord* rejected impanation namely that body is enclosed in bread as the mode of the presence. Consubstantiation means the creation of a third substance (tertium quid), that out of two a third substance is formed, this consubstantiation view attributed by the Reformed as the Lutheran stance has been rejected by the historic Christian Church. Wrote Neve relative to

Lutheranism's rejection of impanation: "This is implied in the whole manner in which the *Formula of Concord* treats of the sacramental union by which our Confessions simply understand that in the Holy Supper the earthly elements (bread and wine) remain what they are, retaining not only their color, taste, odor but their very substance and heavenly elements (body and blood) of Christ also remain what they are; but that during the celebration of the Communion after the commandment of Christ, there exists between the two elements such a relation that where the bread there is also the body and where the wine there is also the blood of Christ, and as our article says 'Both are distributed to those who eat in the Supper of the Lord.'"

The Oral Eating in the Sacrament and the Sacramental Union

The doctrine of the Lord's Supper, as understood by CONFESSIONAL Lutheranism, is that Sixteenth Century Lutherans believed in the oral eating of the body and blood of Christ and that this view is consistent with the sacramental union. The *Formula of Concord* states: "We believe, teach and confess that the body and blood of Christ are received with the bread and wine, not only spiritually by faith but also orally yet not in a Capemaitic, but in a supernatural union." The Augsburg Confession declared: "The body and blood of Christ are truly present and distributed to those who eat in the supper of our Lord."[9]

The Difference Between The Reformed and Lutherans as to the Reception of Christ's Body

There is a difference between the Lutheran and Reformed interest as to how the communicant comes into contact with the body and blood of Christ as a pledge and seal for the forgiveness of sins. Calvin held that the predestined person needs to lift himself up by a strong faith to the right hand of God in order there to participate in Christ's body and blood which was located in a certain place in heaven.[10] Calvin held that the elect person needed to draw Jesus down from heaven through the individual's faith. Luther, however, argued that inasmuch as Christ's humanity was glorified, Christ was omnipresent with His divinity, and therefore by the Word of God, in the words of institution, Christ's body and blood were "truly" present. In the Lord's Supper it is through eating and drinking that the communicant receives the body and blood of Christ.

The Difference in the Eating and Drinking

The eating of bread and drinking of the wine occur in a natural manner, while the eating of Christ's body and drinking His blood occurs in a supernatural manner, in a manner human sense and man's reason cannot comprehend. It is not a Capernaitic eating and drinking, as though the flesh were rent with our teeth and digested like other food. Luther contended that the Sacramentarians endeavored to force this view upon Christians against the testimony of conscience, despite the fact that Lutherans frequently protested this misconception, endeavored to force it on Lutherans and make the latter's doctrine odious to their hearers.

1623

The Lord's Supper as a Sacrament

The doctrine of the Lord's Supper must be considered in connection with the doctrine of the Sacraments. The word "sacrament" is not a Biblical term but an ecclesiastical one. But what do Lutherans mean by a sacrament? An LCMS handbook on Christian doctrine offered this answer: "A sacrament is a sacred act instituted by God Himself, in which there are certain visible means connected with God's Word, by which God offers, gives, and seals unto us the forgiveness of sins which Christ has earned for us."[11]

The Augsburg Confession, Article XIII "The Use of the Sacraments" states this: "It is taught among us that the sacraments were instituted not only to be signs by which people might be identified outwardly as Christians, but that they are signs and testimonies of God's will toward us for the awakening and strengthening of our faith."[12] Article V "The Office of the Ministry" states this: "To obtain such faith God instituted the office of the ministry, that is, provided the Gospel and the sacraments. Through these means, he gives the Holy Spirit, who works faith, when and where he pleases, in those who hear the Gospel. And the Gospel teaches that we have a gracious God, not by our own merits but by the merit of Christ."[13]

The Means of Grace in Reformed Theology

Relative to the Means of Grace, the Word of God in Reformed theology is not a means of grace. Zwingli severed the influence of the Holy Spirit from the instrument of the Word and held that the Holy Spirit worked directly and immediately upon the heart."[14] Zwingli distinguished between the external word and the internal word; only the latter, he said, is effective. Calvin even went further than Zwingli—in the case of the elect—he held to a real divine energy connected with the word. He averred: "that our ignorance, slothfulness, vanity of our minds requires eternal aids, in order to the production of faith in our hearts, and its increase (Institute IV, 1.1)."[15] However, for the stance it must be concluded that this interpretation "prescribes a perpetual rule for God, precluding his employment of any other method; which he certainly has used in the calling of many, to whom He has given a true knowledge of Himself in an internal manner, by preaching (ib. 16,19).[16] In the same way "infants are regenerated by the power of God which is as easy to him as it is wonderful and mysterious to us"[17] (ib. 18). Since God can produce faith in the elect and save them without external means, why could he not save also heathen who never heard the Gospel? Calvin did not draw this inference, but Zwingli believed that among the elect were unbaptized infants and even heathen and therefore would be saved.

The Sacraments in Reformed Theology

Since according to Zwingli the Holy Spirit worked directly upon people's hearts, he did not attach too much significance to the sacraments. The sacraments are mere symbols and of no sacramental significance, but merely of obligatory sacrificial obligation in and in a symbolic fashion remind the users of the blessing of salvation.[18] The Presbyterian West-minster Larger *Catechism* added prayer "to the outward and ordinary means whereby Christ communicated to His church the benefits of His mediation." The Reformed Confessions do not teach that the Word of God, Baptism and the Lord's Supper are means by which God bestows faith and forgiveness of sins.[19] The Helvetic Confession, states in XXI: "They are visible, holy signs and seal appointed of God for the end and that by the use thereof He may the more fully declare and seal to us the promise of His Gospel."[20] These words can be properly understood as correct, but what is known from historic Reformed theology, the words "seal" and "sign" were employed by them to signify the absence of power. Because of this interpretation the Reformed theologians hold that faith belongs to the essence of the supper, and as a result of this position unbelievers do not receive the body and blood of Christ. Thus the Reformed reject the **oral manducatio** (oral reception of the body and blood) and thus would not be taking the Lord's body to their damnation. The Reformed theologian Hodge wrote: "These symbols of the Reformed churches on the continent of Europe agree with those of our own church not only in representing the Sacraments as Means of Grace, but also in denying that their efficacy is due to their inherent virtue or to him who administers them, and in affirming that it is due to the attending operation of the Holy Spirit and is conditioned on the presence of faith in the recipient."[21]

Significant Differences Between Lutheran and The Reformed on the Lord's Supper

Zwingli and Calvin, the Reformed churches of Europe and Scotland, could not agree with the Wittenberg Reformer on many aspects of the Lord's Supper teachings. The Reformed could not accept the Lutheran teaching that God would give the penitent sinner His body and blood for the forgiveness of sins, because for the Reformed it would mean that God would be deceiving people whom he had predestined to damnation. Calvin did not believe that the grace of God was available for all people, denying the efficacy of the atonement would then be misleading those not elected from eternity because Calvin sponsored the teaching of a limited atonement, Christ only saved the elect but not the non-elect. No wonder that Zwingli withdrew his signature from a document which had accepted Luther's doctrine of universal grace. If according to Calvin human beings were chosen for eternal life, it did not matter if they received the Lord's Supper as a pledge of God's promise of the forgiveness of sins.

One of the arguments advanced by Zwingli against Luther's doctrine of the Lord's Supper was: Even if the interpretation of Luther was possible, it must be rejected for Christ cannot really be present for the communicant in a special way. Zwingli's position also involved a wrong view about the Person of Christ, because according to the Reformed understanding of the person of Christ, the two natures of Christ are not really united in a unity, but connected. The human nature since the Ascension is supposed to be confined to heaven according to the Reformed understanding.[22]

Against the presence of Christ in the Lord's Supper according to both natures, the Reformed as well as the Crypto-Calvinists of Luther's time denied the Lutheran doctrine of the omnipresence of the God-man. They were very bitter in their attacks upon the doctrine of the ubiquity and claimed Luther's doctrine was based upon it. The same charge was later made against the *Formula of Concord*, which does not even mention the word ubiquity in the positive part.[23]

The Lutheran doctrine of the Person of the Christ was attacked by the Crypto-Calvinists in a number of writings, especially in the Exegesis perspicuum. In the *Formula of Concord* the Lutheran doctrine of the person of Christ and the communication idiomatum were set forth in Article VIII and it also contains the refutation of the charge that the Lutheran doctrine of the Lord's Supper contradicts the fact of the Ascension. The correct doctrine of the union of the two natures in Christ and the present state of glorification establish the fact that Christ can do as he has promised to do in the Lord's Supper.[24]

After Luther's death, Melanchthon and other Lutherans wished to keep all Protestants together, were willing to adjust to Calvinism. For Zwingli and the Lord's Supper were a mere confessional act, by which Christians show faith. Calvin went further and taught that the Sacrament is an assurance of divine grace to the elect. Thus Calvin taught that Christ was also in the Lord's Supper as a spiritual food for the soul of the believer. Calvin adapted himself to Luther as far as possible and even employed such phrases as "the body" and "blood" are given to us, but by it Calvin meant through faith we receive the blessings and benefits acquired by Christ in His death. According to the Geneva Reformer everything depended upon the faith of the communicant. If the latter had faith, then Christ is present for him; if he does not have it, then he receives merely bread and wine. This doctrine of the Holy Supper was developed in the Consensus Trigurinus (1550).[25]

The Interpretation of the Greek "est!" = "is" in the Words of Institution

The Lutheran and the Reformed (Zwinglian and Calvinistic) have vital differences between their understandings of the key word "esti" = "is." The Reformed claim that the words: "Take eat this is my body" and "Take drink this is my blood" are to be taken metamorphically and not in their proper sense. Krauth in his *Conservative Protestant Reformation* claimed that in his day there were more than twenty different interpretations in Reformed writings dealing with the Lord's Supper.[26] All Reformed theologians were persuaded that "esti" does not mean that the recipient in, with and under does not receive the body of Christ which is orally received. Since Christ according to his humanity is in heaven, it is impossible for Him to be in the many celebrations of the Lord's Supper occurring at the same time throughout Christendom. The *Heidelberg Catechism* says: "What does it mean to eat the crucified body of Christ and to drink His blood that he shed?" Answer: "It means not only to receive with a believing heart the whole PASSION and death of Christ and thereby lay hold on the forgiveness of sins and life eternal, but besides through the Holy Spirit, who lives at the same time in Christ and in us, to be united more and more with His blessed body in such a way that although He is in heaven and we on earth, we are nevertheless flesh of His flesh and bone of His bones and live and are governed throughout the soul."[27] "The Reformed denominations, whether of Zwinglian or of Calvinistic origin contend that the Greek 'esti' does not indicate that the body and blood are given in the Lord's Supper, but that the communicant when eating the bread and drinking the wine spiritually receives Christ's body and blood. Thus the *Westminster Confession* declares of the body and blood of Christ that they are not 'corporally' or 'carnally,' in, with and under the bread and wine, yet as really, but spiritually present to the faith of the believers in that the ordinance as the elements themselves to their outward senses."[28]

Different Reformed Interpretation of "Esti"

Zwingli took the words of institution as meaning "signify" or "represents my body." Calvin, Oecolampadius sought the figure of speech in the word "my body," explaining them as follows: "That which I give you is the sign of my body (signum corporis)."[29] Carl Stadt asserted that the word "this" does not refer to the bread, but to the body of the present Christ, who at the Last Supper pointed to His own body, while pronouncing the words of institution.[30]

Krauth in his book, *The Conservative Reformation and Its Theology*, in answering the Reformed interpretations of "esti" asserted that language would commit suicide if it should tolerate the idea that the substance verb is should not express the substance, but symbol.[31] Meyer, a Reformed Protestant exegete, in his *Corinthian Commentary* claimed that **esti** never means anything less than **est**, never does it mean "significant"; it is a copula, always expressing that which exists (die Kopula des Seins).[32]

The Lutherans rejected the Corpus Christi processions, which presupposed the theory of transubstantiation, on the grounds that the Christ is only present while the Holy Supper is being celebrated.[33] The Reformed were also opposed to the Roman Catholic doctrine of transubstantiation and praying to the elevated host which occurred in the celebration of the Mass.

LCMS's *Popular Symbolic* of 1934 might be cited as a summary of the differences between the Reformed and Lutheranism's beliefs relative to the Holy Eucharist: "On no subject has there been a persistent controversy between the Reformed and Lutherans as on the Lord's Supper. It is here that the rationalism of the former, which chiefly divides them from the latter, becomes more apparent. Not only have the Reformed teachers overemphasized certain external matters, as the one of unleavened bread, rejecting wafers (as a protest against Roman Catholicism), declaring furthermore, the breaking of bread essential, saying that it symbolizes the breaking of Christ's body on the cross, insisting besides, that the elements must be given into the hands of the communicant and not be conveyed directly to the mouth; but they have also inculcated a low view of the blessed Sacrament of the Eucharist, denying the real presence of the Lord's body. Concerning the position of the Reformed on the minor, external features mentioned a few quotations may be submitted." On the breaking of the bread the *Heidelberg Catechism* says Qu. 75: "How is it signified and sealed unto thee in the Holy Supper that thou dost partake of the one sacrifice of Christ on the cross and its benefits?

Answer: Thus, that Christ has commanded me and all believers to eat this broken bread and to drink of this cup and has joined there with these promises: First, that His body was offered and broken on the cross for me and His blood shed formed as certainly as I see with my eyes the bread of the Lord broken for me and the cup communicated to me; and further, that with His crucified body and shed blood He himself feeds and nourishes my soul to everlasting life as certainly as I receive from the hands of the minister, and taste with my mouth, the bread and cup of the Lord which are given to me as certain tokens of the body and blood of Christ."[34] With regard to the use of **unleavened** bread and the rejection of the wafer those Reformed churches which have this peculiarity declare that they wish to oppose papistical superstition, forgetting that they are legislating where God has not made any laws.[5]

The Benefits of the Lord's Supper
The main benefit of the Lord's Supper is that the communicant receives the forgiveness of sin. Every time the Eucharist is celebrated the Christian recalls the death of Christ for the salvation of the world. It is truly a great blessing also to receive body and blood of Christ in a supernatural manner. Among the Reformed Churches there were two views promoted in the days of the Reformation by Zwingli and Calvin.[36] Zwingli emphasized the memorial purpose as the main reason for the celebration of the Last Supper, reminding the communicant what Christ had achieved for him through His suffering and death; for Calvin[37] the main benefit was that the communicant spiritually received Christ's body and blood. In the *Confession of the Ministers of the Church at Zurich* one reads: "We teach that the memory of the body offered for us and of the blood shed for the remission of our sins is the chief thing, the beginning and end, toward which the entire ceremony of the Eucharist is directed."[38] Calvin's view is reflected in the words of the Geneva *Catechism*: "Why is the body of Christ represented by bread and His blood by wine? This teaches us that the same power which bread possesses, to nourish our bodies for this present life, is exerted by the body of our Lord to nourish our souls in a spiritual way; and again, that, just as the wine delights the heart of man and renews his strength and fills the whole man with vigor, so a like benefit will come from the blood of the Lord for our souls."[39] They deny that the unworthy communicant receives the body and blood of Christ. Thus Article XXIX, chapter 8 of the Westminster Confession declares: "Although ignorant and wicked men receive the outward elements, yet they do not receive the thing signified thereby."

A valuation of the Reformed Luther "A Formula of Agreement" made between The Evangelical Lutheran Church in America and The Presbyterian Church (USA), the Reformed Church in America, and the United Church of Christ indicates that these four Protestant Churches were willing to accept positions that historically the Reformed and Lutherans disagreed on over 400 years ago.

The Lutheran ELCA and the three Reformed Churches are now willing to accept as legitimate doctrinal views, ones which were once considered unscriptural. What brought about such a complete reversal? How is it possible to give recognition to teachings which are in disagreement with each other? If the Lutheran believes Reformed theology is in error on a number of teachings, how can they recognize the Reformed central principle, namely the sovereignty of God with its doctrine of double predestination? How can the Reformed accept as possibly true the Lutheran teachings of the Means of Grace, baptism and the Lord's Supper and many other teachings rejected by Reformed theologians of the past? Both ought to take seriously the warning of Paul. To avoid those teaching contrary to the doctrine they believe is correct (Romans 16:17).

These four churches give the impression, by the recognition of each other's contrary stances, that Holy Writ is not perspicuous and clear on fundamental doctrines. To recognize each other's doctrines what are considered erroneous, means that they were willing to compromise their own theological positions.

It is possible that the theologians recommending pulpit and altar fellowship were motivated by the assertion of Christ in His high priestly payer "that they all may be one" (John 17:11). However, how is this unity of Christians to be achieved? "Jesus said to the Father that He has given His body, the Church, His Word, which alone is able to sanctify them, because God's Word is the truth. Unity in the Body of Christ can only be effected by faithfulness to the Word of God. That certainly cannot be the case when four churches recognize contrary positions on the plan of salvation, Baptism and the Lord's Supper and also practice unethical views. The United Church of Christ tolerates ordination of active homosexual persons and supports a pro-abortion position."[40]

The Recognition of Altar and Pulpit Fellowship is Unionism

The merger of the Reformed Churches with ELCA represents unionism, which as church union lacks unity of doctrine. What happened in 1997 was adding to the syncretistic and union movements which have characterized the twentieth century, and has resulted in the mainline churches losing many members and become religiously ineffective, because of unionism's sin in violating God's command to cherish and uphold pure doctrine and to hate and shun false doctrine (Cf. Matt. 7:15; Rom. 16:17, I Tim. 6:3-5; Titus 3:10; John 8:31,32; 2 Thess. 2:15 and many other passages).

Footnotes
1. Luther D. Reed, *The Lutheran Liturgy* (Philadelphia: Muhlenberg Press, 1947), p.

400.

2. "Marburg Colloquy," L. Fuerbringer, Th. Engelder, P.E. Kretzmann, *The Concordia Cyclopedia* (St. Louis: Concordia Publishing House, 1927), pp. 438-439.

3. **Ibid.**, p. 439.

4. Theodore Tappert, *The Book of Concord* (Philadelphia: Fortress Press, 1959), p. 34.

5. E.H. Klotsche, *Christian Symbolics* (Burlington, Iowa; The Lutheran Literary Board, 1929), p. 238; F.E. Mayer, *Religious Bodies in America* (St. Louis: Concordia Publishing House, 1956), p. 215.

6. Tappert, **op. cit.**, p. 34.

7. John Calvin, *Institutes of the Christian Religion*, translated from the Latin by John Allen (Philadelphia: Presbyterian Board of Christian Education, no date, Book IV, ch. 17:10; 18:30,36. Vol. II, p. 651, cf. also Consensus Trigurinus, 21-25.

8. Tappert, **op. cit.**, *Formula of Concord*, VII "Lord's Supper," par. 63, p. 581.

9. **Ibid.**, p. 179.

10. Calvin's Institutes, **op. cit.**, Book IV. Chapters 17,10,11,12.

11. A Short Explanation of Dr. Martin Luther's Small *Catechism* (St. Louis: Concordia Publishing House, 1943)k p. 169.

12. Tappert, **op. cit.**, p. 35.

13. **Ibid.**, p. 31.

14. Mayer, **op. cit.**, p. 213a; *Calvin's Institutes of the Christian Religion*, **op. cit.**, IV, 1,1. Hugh Thompson Kerr, *A Compend of Calvin's Institutes* (Philadelphia; Presbyterian Board of Education, 1939), p. 93.

15. Calvin's Institutes, **op. cit.**, IV 1,1.

16. **Ibid.**, IV, 16,19.

17. Institutes, **Ibid.**, IV, 18.

18. Jim Neve, *Christian Symbolics*, **op. cit.**, p. 235.

19. As cited by *Popular Symbolics*, p. 215.

20. Philip Schaff, *The Creeds of Christendom* (New York: Harper and Brothers, 1919), III "The Creeds of the Evangelical Churches," pp. 223-224.

21. Charles Hodge, *Systematic Theology* (Grand Rapids: Wm. B. Eerdmans Publishing Company, 1940), III, p. 501.

22. J. Neve, *Introduction to the Symbolical Books of the Lutheran Church* (Columbus).

23. *Lutheran Book Concern* (1926), p. 36.

24. **Ibid.**, 436-437.

25. Schaff, *Creeds of Christendom*, **op. cit.**, III, pp. 225-226.

26. C.P. Krauth, *The Conservative Reformation and Its Theology* (Minneapolis: Augsburg Publishing House, 1879, reprint 1963), p. 607; *Concordia Cyclopedia*, op. cit., 219.

27. As quoted in *Popular Symbolics*, op. cit., p. 215.

28. Chapter XXIX, 7 of *Westminster Confession*.

29. J.T. Mueller, *Christian Dogmatics* (St. Louis: Concordia Publishing House, 1975), p. 514.

30. **Ibid.**, p. 514.

31. Krauth, *Conservative Reformation*, op. cit., p. 619.

32. As cited by Mueller, op. cit., p. 514.

33. **Ibid.**, pp. 509, 511, 520.

34. *Popular Symbolics*, op. cit., p. 218, gives the *Heidelberg Catechism* quotation.

35. **Ibid.**, p. 218.

36. **Ibid.**, p. 220.
37. **Ibid.**, p. 220.
38. As cited by *Popular Symbolics*, p. 220.
39. H. A. Niemeyer, *Collectionum in Ecclesiis Reformatis Pulicatarum* (Leipzig, 1840), p. 164.
40. Cf. "President A. L. Barry Responds to ELCA," *Affirm*, September 1997, p. 7.

Questions
Part I
1. Thursday of Holy Week is known as ____.
2. Luther and his compatriots held to the plain meaning of ____.
3. Luther told Zwingli ____.
4. Calvin believed that Christ was ____ present but not ____.
5. What is transubstantiation? ____
6. What is impanation? ____
7. Consubstantiation means ____.
8. The eating of the bread and drinking of the wine occurs in a ____ manner.
9. Does the Bible teach Capernaitic eating and drinking? ____
10. What is a sacrament? ____
11. Zwingli believed that among the elect are ____.
12. Calvin did not believe that the grace of God ____.
13. The Reformed maintain that since the Ascension the human nature of Christ is ____.
14. According to Calvin everything depended upon the ____ of the communicant

Part II
1. The Reformed claimed that the words "Take eat this is my body" are to be taken ____.
2. Zwingli took the words of institution as meaning ____.
3. Does the Bible teach praying to the elevated host? ____
4. What is the main benefit of the Lord's Supper? ____
5. Calvinists deny that the unworthy communicant receives ____.
6. ELCA and three Reformed Churches are now willing to accept ____.
7. The Reformed central principle is ____.
8. Unity in the Body of Christ can only be effected by ____.
9. The United Church of Christ tolerates ____.

An Evaluation of Dietrich Bonhoeffer's Life and Theology After Half of a Century

Christian News, May 15, 2000

On April 9, 1995 certain Christians observed the 50th anniversary of the execution by hanging of Dietrich Bonhoeffer, who did not even reach the 40th year of his life.[1] It is averred that he was only active as a theologian for about less than 15 years and despite this brevity he has exercised a great influence on Protestants, Roman Catholics and other religionists ever since his premature death. Webster of Toronto College claimed that Bonhoeffer remains one of the most provocative voices in contemporary Christianity despite the fragmentary and occasional character of much of his writings.[2] Martin E. Marty and Dean Peerman in their *Handbook of Christian Theologians* discussed him with such recent theologians as Anders Nygren, Gustav Aulen, C.H. Dood, Oscar Cullmann, Reinhold Niebuhr, Karl Barth, Emil Brunner, Friedrich Gogarten, Rudolf Bultmann, Paul Tillich.[3] Franklin Sherman, Lutheran Tutor at Oxford University, wrote: "Since his death in 1945 and largely due to the impact of his posthumous published *Letters from Prison, and Other Papers* and fragmentary *Ethics*, his influence has spread throughout the Christian world."[4]

How great a Lutheran theologian was Bonhoeffer? Did he have a correct understanding of Christianity and the purpose of human existence? Since on a number of occasions, members of the Lutheran Church-Missouri Synod have honored Bonhoeffer and commemorated him and his theology, it might be profitable to examine his life and theology.[5]

A Survey of Life of Bonhoeffer

Dietrich Bonhoeffer was born in Breslau on February 4, 1906, the son of a university professor and leading authority on psychiatry and neurology.[6] He could boast that in his previous lineage there were theologians, professors, a great church historian (Carlvon Hase), artists and some even had aristocratic blood in their veins. Dietrich had three brothers, a twin sister, and three other sisters. Leibholz claimed that from "his father Dietrich Bonhoeffer inherited goodness, fairness, self-control, and ability; from his mother his great human understanding and sympathy, his devotion to the cause of the oppressed, his unshakeable steadfastness." In Breslau and Berlin Dietrich's parents raised their offspring "in the Christian humanitarianism and liberal tradition which was to Bonhoeffer as native as the air he breathed. It was the spirit which determined Dietrich Bonhoeffer's life from the beginning."[7]

Already at 14 he planned on becoming a theologian. He attended lectures at Tuebingen in 1923-1924 but came back to Berlin where he completed his education. Here he completed his education. Here he encountered the liberal Troeltsch and also had opportunity to listen to Lietzmann's New Testament lectures.[8] Gradually he felt himself drawn to Karl Barth's theology. Bonhoeffer visited America, and attended

Union Theological Seminary, New York, where he imbibed some of the religious and political ideas of Reinhold Niebuhr. After a year internship in a German-speaking congregation in Barcelona, he was appointed as lecturer (Privatdozent) in theology at Berlin (1930).[9] After a year at Union Seminary in New York, he served in Berlin both as lecturer at the university and as pastor to the students of the School of Technology. The year 1933 found Bonhoeffer as pastor of two German-speaking congregations in London. While in London there began a life-long friendship with Bishop of Chichester, G.K.A. Bell. When Hitler came to power, he at once opposed National Socialism and its attempt to use the Church for political purposes. In 1935 Bonhoeffer was back in Germany and joined the Confessing Church Movement, representing one third of Protestantism. With Martin Niemoeller, he became a leader in the Confessing Church Movement, which protested the inroads of Nazism on German Christianity.

Karl Barth was the principal author of the Barmen Declaration, a document to which Bonhoeffer contributed.[10] He became director of a short-lived seminary, founded to prepare pastors for the Confessing Church. His license to teach was revoked in 1936 and Himmler closed the seminary in 1937. Bonhoeffer traveled a lot to apprise other Christians of the dangers of Nazism. He did secret work for the churches and traveled to Sweden and Switzerland and had dealings with church officials outside of Germany.[11]

He joined the Resistance Movement and once was involved in a plot to prevail upon the British to accept a conditional German surrender, but the British demanded unconditional surrender. His brother-in-law Dohnany brought him into the anti-Nazi resistance movement and served as a double agent in the German military office. For aiding 14 Jews to escape he was imprisoned, as well as for his anti-government activities and speeches. Bonhoeffer spent a little over two years as a Nazi prisoner (1943-1945). He was confined for 18 months in the Tegel Military Prison, near Berlin. When it was discovered that he had participated in the plot which unsuccessfully attempted to kill Hitler on July 20, 1944, he was removed to the Maximum prison and executed by hanging on April 9, 1945,[12] just a few days before the liberation of the Flossenberg prison. On April 5, 1945 his sister and Dohnany were executed by the Gestapo.

Although Bonhoeffer did not reach 40 when he died, he is supposed to have exercised a great influence on his contemporaries in the 50 years since death. His writings, both before and after his imprisonment, are alleged to have exercised a great influence on certain individuals. He has also been gratefully remembered for opposing Nazism on its Jewish extermination policy and for the fact that he aided the smuggling out of Germany of 14 Jewish people, for which he was imprisoned. Outside of Germany Bonhoeffer was hailed as a martyr.

Bonhoeffer's political and religious ideas were determined by the fact that he saw everything through the glasses of the political developments in Germany. His books also did not fail to reflect his parental training, his university experiences and the thinking of the great theologians of the day. His writings contain divergent views. In his Sanctorum Communion, his doctoral dissertation, he endeavored to bridge the theology of revelation as held by Barth and the philosophy of sociology,[14] a work which reflected Troeltsch, in that Boonhoeffer tried to center the church concept in its corporate life. This thesis is said to have set forth some of his future ideas appearing in his *Ethics, Christology,* and *Letters from Prison.*

While Bonhoeffer lectured at the University of Berlin, he issued *Creation and the Fall*, a writing dealing with Genesis 1-3. In this book the German opponent of Nazism claimed that Genesis presents the "ancient world picture in all its scientific naiveté." Genesis he averred "contains myths." Bonhoeffer was an exponent of higher criticism. Bonhoeffer's Christology was different from the traditional Christology as set forth by Roman Catholicism, Lutheranism and for the most part of Calvinism. In his *Christology* he showed the influence of Karl Barth. His Christology is totally removed from Luther and the Lutheran Confessions. During his directorship of the seminary for the Confessing Church at Finkenwalde, he published *Life Together* and *The Cost of Discipleship.*

Bonhoeffer and the Ecumenical Movement

Leibholz claimed that Dietrich Bonhoeffer after October 1933, after six months of the Church struggle, realized that the situation in which the world and the Churches found themselves in the 1930s that nothing was to be gained any longer for the churches by citing their old Creeds as statements.[15] He believed that the ecumenical movement was the only way to unite the various churches of the body of Christ. Bonhoeffer believed that the Christian Church needed to listen to a new message of the Bible and place themselves in the context of the whole Church. Even before this he had become a member of the Youth Commission of the World Council of Churches and of the World Alliance for International Friendship through the Churches. He was elected (with Praeses Koch) to be a member of the Ecumenical Council for Life and Word at Froena, Denmark in 1943. Bonhoeffer tried to use the ecumenical movement to oppose Hitler's Nazification of the Christian Church. Together with Niemoeller (1890-1894) he organized the First Synod of Barmen in the Buhr. The Barmen Declaration was directed at the "German Christians" movement, which fostered extreme nationalism and anti-Semitism, to which was also added a liberal theological stance. Represented at Barmen were both Lutherans and Reformed. Douglas claimed: "The declaration did not purport to be a comprehensive statement against common deviations; it stressed the headship and finality of Christ, and the preeminence of Scripture belief and practical actions for Christians. There was a pointed repudiation of the German Christians subordination of Christ's Church to the State."[17]

An Evaluation of Bonhoeffer's Theology

Although he claimed to be a Lutheran and is so-called in the literature of our day, Bonhoeffer was not a Lutheran, for he rejected and abandoned the theology and teachings of the *Book of Concord*. Luther was opposed to unionism, to wrong ecumenical altar and pulpit fellowship which Bonhoeffer advocated.[18] When one reads the religious writings of Bonhoeffer one finds that there is not much they have in common with Luther's Small *Catechism*, "the layman's Bible." The fact that the Lutheran, Reformed and Enthusiasts has diametrical differences, did not disturb Bonhoeffer. Although he knew the concept of "purity of doctrine," which himself used, he was totally unconcerned with it. He agreed with the "melancholy Dane of Denmark, Soren Kierkegaard, who in the early 1900's averred that if Luther had lived later, he would have completely abandoned his views." With this erroneous judgment Bonhoeffer agreed. One might retort: How does any person hundreds of years after, know what Luther would have done? They are ascribing to Luther apostasies of which they are guilty.

In *The Cost of Discipleship* Bonhoeffer constantly refers to many Bible verses, hundreds of Bible passages are employed and he seems to insist upon the importance of following Christ. A portion of this writing was an exposition of the Sermon on the Mount (Matthew 5-7).[19] While much of it sounds orthodox and Biblical, Bonhoeffer looked at religion through his spectacles of opposing Hitlerism and the people's worship of the German Chancellor as God. The followers of Christ must take a stand at all costs against National Socialism and if necessary be willing to die in opposing it.[20]

The Cost of Discipleship denounces the doctrine of the churches as a hindrance to attracting the unchurched to Christ.[21] He accused the Protestant Churches of offering "cheap grace." Under the cover of Luther's doctrine of justification by faith alone, he charges, believers have been relieved of the obligations of discipleship.[22] Even though he quotes from Romans, Bonhoeffer never sets forth the true doctrine of justification by faith and its implications for the Christian life. Here is Dietrich's famous statement about "cheep grace": "Cheap grace" means grace as a doctrine, a principle, a system, it means forgiveness of sins proclaimed as a general truth, the love of God taught as the Christian "conception" of God. . . .The Church which holds the correct doctrine of grace has, I suppose, **ipso facto** a part in that grace. In such a Church the world finds a cheap covering for its sins; no contrition is required, still less any real desire to be delivered from sin.[23] Bonhoeffer claimed that "costly grace" accompanies a life of discipleship. Here he fails to distinguish properly between justification and sanctification! His misunderstanding of faith and works is further shown by his claim that the Church has wrongly insisted that obedience follows faith. Bonhoeffer challenged the following statement: "Only he who believes is obedient (i.e. only works done in faith are truly good)." Dietrich, by contrast, puts it this way: "Only he who is obedient believes (only a faith expressed in good works is truly faith.)[24]

Bonhoeffer's Concept of The Church

Already in his doctoral dissertation, already alluded to, he tried to emphasize the fact of the social character of the Church. Sherman pointed out that Bonhoeffer held that if man's very nature as a created being is social, "the fall" is equally communal. The doctrine of original sin implies the solidarity of the guilt of the whole human race. Hence man's redemption must be equally corporate in character; and so it is, since it consists precisely in the creation of a community of the redeemed. In Jesus Christ as "a collective person" "as deputy" or "representative for all mankind," the humanity of Adam is transformed into the humanity of Christ.[25] Through his life, death and resurrection, the communion of saints is realized. Bonhoeffer identified the Church with Christ. Thus he wrote: "The Church is the presence of Christ, as Christ is the presence of God." Adapting the phrase from Hegel Bonhoeffer claimed that the church was "Christ as existing community."[26]

Bonhoeffer misunderstood and misinterpreted Paul's comparison of the first and second Adam as set forth by Paul in Romans 5:12-17. The German martyr had a wrong understanding of what Christ meant and did for the Church and what membership in it involved and how one becomes a member of the spiritual body of the Church. He ignored the Lutheran distinction between the invisible and visible Church.

Bonhoeffer's Christology

He wrote a series of lectures on Christology which were lost, but his friend Bethge has reconstructed Bonhoeffer's views from student notebooks and are now found in Volume 3 of his *Gesammelte Schriften*. His Christology is considerably different from that found in Luther and in the Lutheran Confessions. He divided his Christology into the following topics: The contemporary Christ, the historical Christ and the eternal Christ.[27] Instead of beginning with the eternal, followed by the historical he began his Christology with the contemporary, namely, with Christ as existing as the Church. Relative to the historical Christ he averred that Christ did not exist for himself but as a Christ "for me." "This is the deputyship that he assumed not only for me but for the whole of nature and of history." Bonhoeffer's Christology is not what the New Testament sets forth, but was the product of his philosophical thinking. For one thing, he made a strange distinction between Christ's humiliation and his incarnation. Thus the German theologian claimed "that humiliation pertains to the fallen, creation, the incarnation to primal creation. The humiliation was temporary, but the incarnation is permanent. With the return of Christ to the Father, humanity has been assumed into the eternal life of God himself."[28] The doctrine of the communication of attributes was ignored by him. Only the Son now has His human nature, but the Father and Holy Spirit are spirit, and do not have a body.

In *The Cost of Discipleship* Bonhoeffer made this assertion: "As they contemplated the miracle of the Incarnation, the early Fathers passionately were wrong to say that God took human nature upon him, it was fallacious to say that God chose a perfect individual man and united himself to him. God was made man. This means that he took upon him our entire human nature with all of its infirmity, sinfulness, and corruption, the whole of apostate humanity."[29] This contradicts who the real human Christ was. According to Bonhoeffer redemption involves not merely the plurality of human individuals duals, but rather humanity it its entirety, which meant that redemption was universal in scope. Redemption was the restoration of the "form" of man as he was originally created in God's image. In Jesus Christ the Incarnate, Crucified, and Risen One, this true image has again taken form in human history. Bonhoeffer's Christological views are pure speculation and are anti-Biblical.[30]

Van Til's Critique of Bonhoeffer's Christology

Van Til in his book, *The Great Debate Today* contended that the Christ of Bonhoeffer is not the Christ of Scripture and can hardly be distinguished from the Christ of Karl Barth (pp. 42-76 of Van Til's Book). Asserted the former Westminster professor of Christian Apologetics: "The ecumenical significance of all this is far reaching. Bonhoeffer was quite consistent with his own theology and with that of post-Kantian, neo-Barthian tradition when he did his best to further the cause of modern ecumenicism. Why should not Lutherans, Calvinists and Armenians unite under a banner of new primacy of a new Christ projection? Surely all of us want to make the message of the saving grace of Christ relevant to the needs of modern man. So we must all demythologize and then re-mythologize or allegorize not only the Genesis narrative but the New Testament as well. After that we must give an existential interpretation of its message in terms of the ideal man calling him Christ and interpreting all reality of him as its center."[31]

Bonhoeffer and Biblical Hermeneutics

Bonhoeffer abandoned the hermeneutics of Luther, Melanchthon, and Chemnitz, the authors of the Lutheran Confessions and the theologians of the era of Lutheran Orthodoxy.[32] He utilized and reinterpreted Scriptures to fit in with his false religious and philosophical views. In Genesis 1-3 he opposed the literal meaning of the opening chapters as may be seen from *Creation and Fall: A Theological Interpretation of Genesis 1:3*. Later he adopted the Bultmann view on "mythology" of the New Testament. Concerning the mythological elements of Christianity, he wrote: "I am of the view that the full content, including mythological concepts, must be maintained. The New Testament is not a mythological garbing of universal truth, this mythology (resurrection and so on) is the thing itself – but the concept must be reinterpreted in such a way as to made religion a pre-condition of faith (cf. circumcision of Paul)."[33]

In Bonhoeffer's opinion Bultmann in his mythologization of the Bible did not go far enough. Thus the German asserted: "It is not only mythological conceptions such as miracles, the ascension and the like (which in principle are not in principle separate from the conception of God, faith and so on) that are problematic, but the religious conceptions themselves."[34]

His View of the Resurrection

In one of his prison letters Bonhoeffer declared: "Belief in the resurrection is not the problem of death."[35] However, Bonhoeffer ignored all those scripture verses that speak of the blessedness of a life after death, and the statement of Christ, that on the Last Day all men will rise, either to condemnation or to eternal life. (John 5:28-29) Bonhoeffer's great error was that he concentrated on this life instead on the life to come. The subject of eschatology is absent from the purview of his theology.

Old Testament Contains No Religion of Salvation

Wrote Bonhoeffer: "To resume our reflections on the Old Testament. Unlike other oriental religions the faith of the Old Testament is not a religion of salvation. Christianity, it is true, has always been regarded as a religion of salvation. But isn't a cardinal error, which divorces Christ from the Old Testament? And interprets him in the light of the myths of salvation. Of course it could be argued that under Egyptian and later Babylonian influence, the idea of salvation became just as prominent in the Old Testament e.g. Deutero-Isaiah. The answer is, the Old Testament speaks of Historical redemption on this side of death whereas the myths of salvation are concerned to offer men deliverance from death. Israel was redeemed out of Egypt in order to live before God on earth. The salvation myths deny history in the interest of an eternity after death. Sheol and Hades are not metaphysical theories, but images which imply that the past, while it still exists has only a shadowy existence in the present. It is said that the distinctive feature of Christianity is its proclamation of the resurrection hope, and that this means of a genuine religion of salvation, in the sense of release from the world. The emphasis falls upon the far side of the boundary drawn by death. But this seems to me to be just the mistake and danger. Salvation means salvation from cares and from fears and longing, from sin and death into a better world beyond the grave. But is this the distinctive feature of Christianity as proclaimed in the Gospels and St. Paul? I am sure it is not? The difference between the Christian hope and resurrection and a mythological hope is that the Christian hope sends a man back to life on earth in a wholly new way which is even more sharply defined than it is in the Old Testament."[36]

Bonhoeffer and the Existence of God

While in prison Bonhoeffer came to the conclusion that mankind no longer needs God. What amounts to theoretical atheism was Bonhoeffer's assertion that God is teaching us that we must live as men who can get along very well without him. "The God who is with us is the God that forsakes us (Mark 15:34)."[37]

While in prison Bonhoeffer asked: What do we mean by God? His answer was: not in the first place in an abstract belief in the omnipotence, etc. That is not a genuine experience of God is not a religious relationship to a Supreme Being, absolute in power and goodness, which is a spurious conception of transcendence, but a new life for others, through the participation in the Being of God. This is the starting point for the interpretation of biblical terminology (creation, fall, atonement, repentance and faith, the new life, the last things)."[38]

Bonhoeffer's Perverse Views about God's Presence in the World

Again in prison he declared: "God is being increasingly pushed out of a world that has come of age, out of the sphere of our knowledge and life, since Kant has been relegated to a realm beyond the world of experience. Theology has on the one hand resisted this development with apologetic and has taken up arms in vain against Darwinsim."[39] Traditional Darwinism, it should be noted was atheistic and denied God's part in creation and preservation.

No Need for God, Bonhoeffer's Claim

"God as a working hypothesis in morals, politics or science, has been surmounted and abolished and the same thing happened in philosophy and religion. For the sake of intellectual honesty, that working hypothesis should be dropped, or as far as is possible eliminated."[40]

Man No Longer Needs God

Man has learned, so the German opponent of National Socialism claimed, "to deal with himself in all questions of importance without recourse to the 'working hypothesis' called God. In questions of science, art, and ethics this has become an understood thing at which one now hardly dares to tilt. But for the last hundred years or so it has also become increasingly true of religious questions; it is becoming evident that everything gets along without 'God' – and, in past just as well as before."[41]

"If in fact the frontiers of knowledge are being pushed further and further back and that is bound to be the case, then God is pushed with them and is therefore (constantly) in retreat. The God of Jesus has nothing to do with what God, as we imagined, could do or ought to do."[42]

Mankind No Longer Needs Religion

Bonhoeffer argued that mankind has become of age and no longer needed religion, which is only a deceptive garment of true faith. He sought to acknowledge Christ not as an object of religion, but as in truth the Lord of the world. He spoke of Christian worldliness, which he claims was quite different from religion, which retreats from the world into inward life and speculation. The Christian is identified not by his beliefs but by his actions, by his participation in the suffering of God in the life of the world."[43]

Bonhoeffer's Plan for The Reconstruction after the War

One writer has called Bonhoeffer "a Christian humanist."[44] A humanist was a person who emphasized the importance and centrality of man and a Christian humanist one which utilized certain Christian ideas, but was basically anthropocentric.[45] While in prison in the Tegel Military

Prison he was worried as to what would happen to the culture of Western civilization, whose demise Hitler could bring about. During his imprisonment he looked forward to the reconstruction of Christian thought after his death. He suggested the need for "a religionless Christianity."[46] While the world was going to decay, he claimed that "God had come of age." God as a concept had been used by people as a stop gap for their embarrassment, and God was superfluous, in that in Europe the world had become of age. A Christian was a man who had concern for others and that the church must be for others.[47]

Bonhoeffer's Negative Influence
Bonhoeffer's final writings, especially as reflected in views found in Letters from Prison and Other Papers, exercised a tantalizing power over thinkers since his death and have given impulse to Marxist theologians sponsoring 'liberation theology' and on those who contended that "God was dead" in the world and those wishing to promote a this worldly social Gospel.[48] Both Zerner[49] and Webster[50] endeavor to save Bonhoeffer's respectability as a Christian theologian, but Bonhoeffer's statements are too clear and explicit.

Roger Shinn of Union Theological Seminary claims that the Tegel Prisoner proposed views radically different from his earlier writings.[51] *The Westminister Dictionary of Church History* asserts about Bonhoeffer's *Ethics and Letters* and *Papers in Prison*: "In these he reinterprets Biblical concepts for a world that has become of age, in which neither the usual metaphysical categories are adequate. He teaches a revolutionary understanding a Christian belief in which there is no separation of the religious realm. The Christian identifies with and suffers for the world as Christ did. Bonhoeffer's worldly commitment and execution tend to illustrate and emphasize his written ideas."[52] H.D. McDonalds made this judgment of Bonhoeffer's writings: "So varied and opposing are the theories deriving from his that a meaningful sketch of his ideas presents difficulty. Among his most fruitful insights were his total rejection of natural theology, and of a 'religious aprior' in man; the reality of God absolute self-disclosure in Christ; as God revealed incognito; Christ interpreted in terms of 'the-man-for-others'; and particularly, his much discussed and misunderstood concepts of 'religionlesss' and 'worldly Christianity,' and 'man come of age.'[53]

Bonhoeffer Memorials and Eulogies
When the execution of Bonhoeffer became known in Europe and in America, there were held memorial services and eulogistic pronouncements were made for a man who had died for his opposition to National Socialism and for the exercise of humanitarianism. Leibholz made this hyperbolic statement that Bonhoeffer's death could not be measured by human standards. They felt that God himself had intervened in the most terrible struggle the world had witnessed so far by sacrificing one of his most faithful and courageous sons to expiate the crimes of a diabolical regime and to revive the spirit in which the civilization of Europe had to

be rebuilt."[54] Leibholz claimed that Bonhoeffer's life and death belong to the annals of Christian martyrdom, or as Neibuhr said "to the modern Acts of the Apostles."[55] On July 27, 1945 a memorial service was held at Holy Trinity in London, sponsored by the Bishop of Chichester.

Bonhoeffer Versus Jesus and Paul on Duty to Human Government

In *The Cost of Discipleship* Bonhoeffer discussed Christ's statement on Matthew 22:21[56] and Paul's instruction on Romans 13:1-8.[57] Both Jesus and St. Paul lived in the first century A.D. Roman Empire. At their time there were 60,000,000 slaves in the Roman Empire, who were treated like chattel, with no rights.[58] Neither Christ nor Paul denounced slavery and encouraged slaves to rebel. In fact, Paul made a runaway slave, Onesimus, return to his master Philemon. Crucifixion was used as a death penalty, involving Jews and Gentiles. Neither Jesus nor Paul condemned this practice. The Jews of Palestine suffered indignities at the hands of Roman procurators. Yet neither Christ nor Paul called upon the Jewish people to revolt. Emperor worship (the deification of a man) began with Augustus and later caused the death of many Christians who refused to offer up incense to Caesar as divine. One might say that there were great similarities between the first century Roman world and 20th century Germany. Claudius banished Jews from Rome in A.D. 49.

Christ told his enemies who wanted to entrap him: "Render unto Caesar the things that are Caesar's and to God the things that are God's." Writing at the time, when Christian-murdering Nero was emperor, to the Roman congregation penned the God-inspired directives: "Let every person be subject to the governing authorities. Or there is no authority except from God and those that exist have been instituted by God. Therefore he who resists the authorities resists what God has appointed, and those who resist will incur judgment."[59]

Bonhoeffer joined German resistance movement and traveled on its behalf. In May 1942, Bonhoeffer met with the Bishop of Chichester at Sigunta, Sweden, and conveyed plans for over-throwing the Nazi regime, together with proposals for the subsequent establishment of peace. Anthony Eden was given these proposals and later rejected them, demanding unconditional surrender.[60] Bonhoeffer was also involved in a plot to assassinate Hitler, which was unsuccessful. He did the opposite of what St. Paul said Christians were to do in dealing with the government. As a churchman Bonhoeffer was using the sword to save the church, concerning which Jesus had told Peter: "Put thy sword into the sheath for all they that take the sword shall perish by the sword." (Acts 5:29) It is true that man must obey God rather than men, when the latter demands the doing that which violated God's laws. However, when he takes a stand against the government, he must take the consequences, which Niemoeller did when he protested Jewish persecution and the anti-religious views of Hitler and his minions.[61] Niemoeller together with Bonhoeffer had been one of the writers and promoters of the Barmen Declaration. Niemoeller was a prisoner from 1937-1945 at Sachsenhausen and then at Dachau. But after his release he became active as a church leader in German Protestantism. The difference in the end between these two religious leaders was: one engaged in hostile activity against his constituted government, the other did not.

Ed. *The Doubled Life of Dietrich Bonhoeffer – Women, Sexuality, and Nazi Germany* by Diane Reynolds (Cascade Books, 2017) and *Strange Glory – A Life of Dietrich Bonhoeffer* by Charles Marsh (Alfred A. Knoff, a Division of Random House, 2014) show that Bonhoeffer was a homosexual (*Christian News*, March 27 ,2017).

Footnotes

1. "Turret of the Times," *Christian News*, July 10, 1995, p. 3.
2. J.B. Webster, "Dietrich Bonhoeffer," Sinclair B. Ferguson, David Wright, *New Dictionary of Theology* (Downers Grove, Illinois: Intervarsity Press, 1081) p. 107.
3. Franklin Sherman, "Dietrich Bonhoeffer," in Martin Marty and Dean G. Peerman (Cleveland and New York: The World Publishing Company, 1967), pp. 464-484.
4. **Ibid.**, p. 464.
5. "Defender of God is Dead Theologian to Speak at Concordia Seminary, St. Louis," *Christian News*, March 20, 1992, found also in *Christian News Encyclopedia*, Vol. 5, p. 3466. At a Reformation Festival in Detroit Bonhoeffer's theology was honored, CF. *Christian News Encyclopedia* V, p. 3466. Bolton Davidheiser criticized The Council of Inerrancy for Utilizing Bonhoeffer, cf. *Christian News Encyclopedia*, III, p. 1900.
6. Cf. G. Leibholz, "Memoirs," Introduction to Dietrich Bonhoeffer, *The Cost of Discipleship* (New York: Macmillan Publishers, 1963), pp. 11ff.
7. **Ibid.**, p. 11
8. Sherman **op. cit.**, pp. 464-465.
9. R. Zerner, "Dietrich Bonhoeffer," Walter A. Elwell, *Evangelical Dictionary of Theology* (Grand Rapids, Baker Book House, 1988), p. 168b.
10. J.D. Douglas, "Barmen Declaration, in Elwell, **op. cit.**, 76b. also Sherman **op. cit.**, p. 465.
11. Roger L. Shinn, "Bonhoeffer, Dietrich," *Encyclopedia Americana* (Danbury: Grollier Incorporated, 1994), p. 208.
12. Sherman **op. cit.**, p. 466.
13. Reinhold Niebuhr, "Death of a Martyr," *Christianity and Christ*, June 25, 1945.
14. The complete title of this dissertation was: Sanctorum Communio: eine Dogmatische Untersuchung Zur Sociologie der Kirche, 1930.
15. Bonhoeffer, *The Cost of Discipleship*. Revised and Unabridged Edition (Macmillan Publishing House, 1963), pp. 38-339.
16. Leibholz, "Memoir," **op. cit.**, pp. 14-15.
17. Douglas, **op. cit.**, p. 465.
18. Cf. F.E. Mayer, *The Religious Bodies in America* (St. Louis: CPH, 1956), p. 134; 212-216.
19. Bonhoeffer, "Cum grano Salis," pp. 13-32.
20. Bonhoeffer, *The Cost of Discipleship*, **op. cit.**, pp. 37-352. The book is characterized by the use of many Biblical passages, cf. listing of verses cited on pp. 348-342.
21. *The Cost of Discipleship*, **op. cit.**, pp. 38-39.

22. **Ibid.**, pp. 45-46.

23. **Ibid.**, p. 46.

24. Thus Sherman, **op. cit.**, p. 467.

25. **Ibid.**, pp. 468-469.

26. **Ibid.**, p. 469.

27. **Ibid.**, p. 474.

28. **Ibid.**, pp. 474-475.

29. **Ibid.**, p. 475.

30. For a correct presentation of Christology, cf. John Schaller, *Biblical Christology. A Study in Lutheran Dogmatics* (Milwaukee: Northwestern Publishing House, 1981), 287 pp.

31. Cornelius van Til, *The Great Debate Today* (Nutley: Presbyterian and Reformed Publishing Company, 1970), pp. 42-47.

32. Cf. Ralph Bohlmann, *The Hermeneutics of the Lutheran Confession* (St. Louis: CPH, 1983), 163 pp. Revised Edition: Raymond F. Surburg, "The Significance of Luther's Hermeneutics for the Protestant Reformation." *Concordia Theological Monthly*, 24, 243-261, April, 1953.

33. Dietrich Bonhoeffer, *Letters and Papers from Prison* (London: SCM press, Fontana Books, 1953), P. 110.

34. **Ibid.**, p. 94.

35. **Ibid.**, p. 93.

36. **Ibid.**, p. 112.

37. **Ibid.**, p. 122.

38. **Ibid.**, p. 163-165.

39. **Ibid.**, p. 341.

40. **Ibid.**, p. 360.

41. **Ibid.**, p. 325.

42. **Ibid.**, p. 311.

43. Sherman, **op. cit.**, p. 483; *Prison Letters and Other Papers*, pp. 163ff.

44. Leibholz, **op. cit.**, p. 35.

45. Jerald C. Brauer, Editor, *The Westminster Dictionary of the Bible* (Philadelphia: The Westminster Press, 1971), Cf. article on "Humanism," pp. 415-416. Shinn, "Bonhoeffer, Dietrich,": *Encyclopedia Americana*, **op. cit.**, p. 208.

46. Sherman, **op. cit.**, p. 483; *Bonhoeffer, Prison Letters and Other Papers*, pp. 163-165.

47. Bonhoeffer, *Prison Letters and Other Papers*, pp. 163-165.

48. Shinn, **op. cit.**, p. 208.

49. Zerner, **op. cit.**, p. 169.

50. Webster, **op. cit.**, p. 109a.

51. Shinn, **op. cit.**, p. 208.

52. *Westminster Dictionary of Church History*, **op. cit.**, p. 124.

53. McDonald, "Bonhoeffer, Dietrich," Douglas, Editor, *The New International Dictionary of the Christian Church*, (Grand Rapids: Zondervan Publishing House, 1974), p. 142b.

54. Leibholz, "Memoir," **op. cit.**, p. 32.

55. **Ibid.**, p. 33.

56. *The Cost of Discipleship*, **op. cit.**, p. 296.

57. **Ibid.**, pp. 292-296.

58. Frank E. Gaebelein, *Philemon, the Gospel of Emancipation* (New York: Our Hope Publications, 1939), p. 17.
59. *Good News for a New Life* (New York: American Bible Society, 1964-1965), p. 263.
60. Sherman, **op. cit.**, pp. 465-566.
61. John P. Dever, "Niemoeller, Martin," in Douglas, *The New International Dictionary of the Christian Church*, **op. cit.**, p. 712a.

Questions

1. When was Dietrich Bonhoeffer executed? ____ .
2. Bonhoeffer attended Union Seminary in ____ .
3. Bonhoeffer felt drawn to ____ theology.
4. The Confessing Church Movement protested ____.
5. Who was the principal author of the Barmen Declaration? ____
6. Who brought Bonhoeffer into the anti-Nazi resistance movement? ____
7. Bonhoeffer was confined for 18 months in the ____.
8. Bonhoeffer was executed in ____ on ____.
9. According to Bonhoeffer, Genesis contains ____.
10. Bonhoeffer was an exponent of ____.
11. Bonhoeffer's Christology is totally removed from ____.
12. What did Bonhoeffer believe about the ecumenical movement? ____
13. Was Bonhoeffer a Lutheran? ____
14. What did Soren Kierkegaard aver about Luther? ____
15. Luther's Small *Catechism* is the ____.
16. Bonhoeffer accuses the Protestant Churches of offering ____.
17. Bonhoeffer fails to distinguish properly between ____ and ____.
18. Bonhoeffer ignored the Lutheran distinction between ____.
19. Was Bonhoeffer's Christology that of the New Testament? ____
20. Van Til contended that the Christ of Bonhoeffer is not the Christ of ____.
21. Bonhoeffer adopted Bultmann's view on ____.
22. This mythology includes the ____.
23. Bonhoeffer declared that "Belief in the ____ is not the problem of ____."
24. According to Bonhoeffer the Old Testament contains no ____.
25. While in prison, Bonhoeffer came to the conclusion that ____.
26. According to Bonhoeffer there was no need for ____.
27. Bonhoeffer argued that because mankind has come of age it no longer needed ____.
28. Neither Paul nor Christ denounced ____.
29. As a churchman, Bonhoeffer was using the ____ to save the ____.
30. What was the difference between Bonhoeffer and Niemoeller? ____

www.ingramcontent.com/pod-product-compliance
Lightning Source LLC
Chambersburg PA
CBHW030633150426
42811CB00048B/89